Leibniz on Time, Space, and Relativity

Leibniz on Time, Space, and Relativity

RICHARD T. W. ARTHUR

OXFORD
UNIVERSITY PRESS

OXFORD
UNIVERSITY PRESS

Great Clarendon Street, Oxford, OX2 6DP,
United Kingdom

Oxford University Press is a department of the University of Oxford.
It furthers the University's objective of excellence in research, scholarship,
and education by publishing worldwide. Oxford is a registered trade mark of
Oxford University Press in the UK and in certain other countries

First Edition published in 2021

Published in the United States of America by Oxford University Press
198 Madison Avenue, New York, NY 10016, United States of America

British Library Cataloguing in Publication Data
Data available

Library of Congress Control Number: 2021943456

ISBN 978–0–19–284907–6

DOI: 10.1093/oso/9780192849076.001.0001

For Thomas and Alexander

Preface

I first discovered how interesting a philosopher Leibniz was while working on my PhD dissertation on time and the foundations of physics at Western in London, Ontario in the late 1970s. Reading his discussion of space and time in his controversy with Samuel Clarke, I found I was finally getting some insight into how to interpret the '*t*' that physicists manipulated in their equations. On my return to Western after a year in Calabar, Nigeria, my teaching appointments in the Department of Applied Mathematics allowed me leeway to follow up this interest in earnest. I audited Robert Butts' graduate seminar on Leibniz, and gave a talk on Leibniz's theory of time to the Philosophy Department.

Initially I had assumed, perhaps naively, that when Leibniz wrote about "the phenomena" in physics he simply meant "what are observed," and that his account of phenomenal bodies as resulting from more basic entities, his monads or simple substances, was analogous to the situation in modern physics. The idea that the extendedness of bodies is not fundamental, but derives from more primitive entities having the nature of force, did not seem a far cry from the situation as described by quantum physics. Similarly, his notion of the states of substances following one another in a continuous series as a result of their "appetition" (or tendency towards subsequent states) seemed to have an analogue in quantum theory, where the Hamiltonian operator acts on a given quantum state of an isolated system to generate later states of the same system.

The more I learned of Leibniz's metaphysics, however, the more perplexing I found it. As is well known, Leibniz insisted that substances do not strictly speaking interact with one another. He equated their states with perceptions, where perception is taken in the broad sense of a monad's representation of the universe (more or less confusedly) from its particular point of view, making their appetition more like a generalization of desire, and rendering monads decidedly mind-like. Yet, Leibniz held, physical bodies are infinite aggregates of monads, and any change occurring in such composites presupposes change in the qualities of the simple. How could this be? Bodies could not be composed from minds, nor physical changes from psychic ones (as Leibniz himself stressed).[1]

[1] For the impossibility of such compositions, we have the authority of Leibniz himself. As he wrote to Johann Bernoulli (September 30, 1698): "You were afraid that matter would be composed of non-quanta. I respond that it is no more composed of souls than of points"; and to Michelangelo Fardella in 1690: "it should not be said that indivisible substance enters into the composition of a body as a part, but rather as an essential internal requisite" (A VI 4, 1669/AG 103), and "a soul is not a part of matter, but a body in which there is a soul is such a part" (AG 105).

Puzzlement about the way bodies and their changes result from monads is, of course, par for the course. But at least as perplexing for me was the general consensus that Leibniz excluded relations from his fundamental ontology. Following Russell, it is widely believed that, appearances to the contrary, Leibniz denied relations at the deepest level of his metaphysics. Since it is incontestable that he regarded space and time as relational, this would (it is thought) account for his regarding them as ideal. Monads, on this interpretation, could have no location in space and time, and would exist timelessly, like Kant's noumena, only in the intelligible realm. But such an interpretation, it seemed to me, was directly contradicted by Leibniz himself in many places. In 1703 he assured his correspondent De Volder, for example, that there is a place for all changes of both spiritual and material things both "in the order of coexistents, that is, in space," and "in the order of successives, that is, in time" (LDV 266/267). Even though they are not themselves extended, simple substances cannot exist without a body, "and to that extent they do not lack situation or order with respect to other coexisting things in the universe" (LDV 266–269). It was on this foundation—namely, on the mutual situations of coexisting substances through their extended bodies—that Leibniz built his theory of space, as he explained (all too briefly) to Clarke. Here, it is true, one may argue that since monads are only situated through their bodies, and bodies are phenomena, then these relations are only among the phenomena and not among monads themselves. It is different with time, however, since there (as I have long argued) Leibniz bases temporal relations directly on relations among monadic states. This calls into question the idea that the ideality of relations precludes the existence of monads in time, or that temporal succession applies only to the states of phenomena. But if monadic states are ordered in time, and each state expresses the situations of the bodies of coexisting monads, providing the basis for their spatial ordering, this suggests that space and time are not mere mental constructions, but also have some basis in reality. How this could be so, and in what sense, has motivated the line of research I have pursued that has culminated in this book.

It began as three chapters of a projected volume on Leibniz's Labyrinth of the Continuum, which I had originally titled *Ariadnean Threads*. The idea was to have each chapter corresponding to one of the topics Leibniz himself had included under the rubric of a book project he had conceived in 1676 '*de Compositione continui, tempore, loco, motu, atomis, indivisibili et infinito*' (A VI 3, 77/DSR 90)—that is, on the composition of the continuum, time, place, motion, atoms, the indivisible, and the infinite. That project, however, became too big and unwieldy, so I separated off what was pertinent to the theory of substance as a solution to the labyrinth of the continuum, and published that in 2018 as *Monads, Composition, and Force*, postponing the treatment of time, space, and the more mathematical topics for another volume. Now that remainder has undergone a further fission, as I recognize that a treatment of time, space, and motion—all of them relational—would form a coherent monograph all by itself, saving treatments of the mathematics of the infinite and the infinitely small for further projects. How the arguments of the present work relate to and depend on those of the previous one I describe in detail in the introduction below.

In addition to that introduction, this book is comprised by three substantial chapters, each of seven sections, with its own introduction and conclusion, and supplemented by four appendices and a glossary of the technical terms Leibniz used, particularly in relation to the infinite. I have chosen to write a conclusion specific to each chapter rather than writing a general conclusion, and I include in each some observations on how Leibniz's views relate to modern thinking on the same subject.

Chapter 1 is built around my first publication on Leibniz ('Leibniz's Theory of time', 1985), which I had extensively reworked for intended inclusion in *Ariadnean Threads* in 2008–9. It is supplemented by material from my treatment of the causal theory of temporal precedence (2016) in response to criticisms, as well as from a forthcoming paper on vague states and discontinuous change to appear in a forthcoming *Festschrift* for Massimo Mugnai (thanks are due here to Peter Momtchiloff for granting me permission to use much of the material in §1.5 for that paper). This chapter also includes substantial new material on time and contingency, and on reduction and the nature of Leibniz's nominalism about time. I present a formal exposition of the theory in Appendix 1: the relational core in two versions, compossibility, temporal counterparts, and Leibniz's complex and innovative views on change and the continuity of time. On this last topic in particular, I believe I have broken new ground here.

Chapter 2 builds upon my 'Leibniz's Theory of Space' (2013b)—which itself drew on ideas from my (1987) and (1994b)—although it mainly consists in new material. I expand upon the genesis of Leibniz's views on space, present a succinct account of the main features of *analysis situs* as a mathematical treatment of space, and two sections on how this relates to his metaphysics of space. Leibniz's *analysis situs* remained an unfinished project, and our understanding of it will almost certainly undergo changes and improvements as the collection and editing of his manuscripts on it proceeds. But if I have succeeded in giving some semblance of an account of it compatible with my reading of Leibniz's metaphysics, illuminated by the contrast with De Risi's phenomenalistic interpretation, I will be well satisfied.

Chapter 3 is a substantial reworking of a paper I finished in the summer of 2019, 'Causes and the Relativity of Motion in Leibniz'. That paper drew on my (1994a), and incorporated elements of other papers I published while working on it, (2013c), (2015a), and (2015b), but it turned out to be too long for publication in a journal. It is now about twice as long as it was in 2019, since it incorporates a new section on Copernicanism and instrumentalism, and a substantial treatment of the whole question of whether Leibniz's space could accommodate motion through space and time, and what kind of 'spacetime' is implicit in this. I doubt if this is the last word on Leibniz's views on the relativity of motion, but I believe I have at least made it seem far more coherent than it is generally portrayed to be.

Since the status of relations in Leibniz's thought is both crucial to the interpretation I give in the chapters, and yet too involved for inclusion in the main text, I present an essay treating this question in the second appendix. In the third, I give translations of extracts from Leibniz's writing on *analysis situs* over the years, since there is very little available in English translation. In the fourth, I give translations of three drafts Leibniz wrote in Rome in 1689 on the question of the relativity of motion, Copernicanism and the Censure. Finally, in the glossary I explain some of the technical terms Leibniz used, particularly in connection with the infinite.

It is a pleasure to acknowledge the generous feedback I have received from colleagues on drafts of this work. Preeminent among these has been Vincenzo De Risi, with whom I have been discussing and corresponding about Leibniz's metaphysics of space (and learning from him) ever since I was an examiner for his PhD thesis at the *Scuola Normale Superiore* in Pisa in 2005. In response to material I had asked him to look over (the penultimate versions of chapter 2 and section 3.5), he sent me an exquisite 13-page essay, which was hugely helpful for me in clarifying my own views as well as his; he also provided emendations for the glossary. Osvaldo Ottaviani also read through the whole manuscript and provided me with extremely valuable responses, sources, links, and corrections. Many thanks, too, to the OUP readers, for their feedback on the draft manuscript I submitted in September 2020, and suggestions for its improvement. That helped me to clarify my thought and my expositions of several points, and also prompted me to provide the introductory chapter and glossary of technical terms. I am also very grateful to David Rabouin, Lucia Oliveri, Laurynas Adomaitis, Jeffrey Elawani, and Angela Axworthy for their substantial critical responses to samples I sent them; to Filippo Costantini for welcome advice and commentary, especially on the mereology in Appendix 1; to Paul Lodge, Jeffrey McDonough, and Mattia Brancato for their suggestions and comments on some of the material; and to Massimo Mugnai, Samuel Levey, Ed Slowik, Nico Bertoloni Meli, Pauline Phemister, Daniel Garber, Tzuchien Tho, Jan Cover, Stefano Di Bella, Ohad Nachtomy, Ursula Goldenbaum, Don Rutherford, Doug Jesseph, Jean-Pascal Anfray, Enrico Pasini, Stephen Puryear, Martha Bolton, Mic Detlefsen, Marco Panza, Laurence Bouquiaux, Arnaud Pelletier, and Gianfranco Mormino for fruitful exchanges of views over the years on various aspects of what is discussed here. Thanks, also, to David Rabouin for drawing my attention to texts on *analysis situs* recently prepared from manuscript sources by his team in the ANR MATHESIS project in collaboration with the Leibniz Research Centre in Hanover (Leibniz-Archiv), and to him, Siegmund Probst, Vincenzo De Risi, and Michael Kempe for permission to publish translations of three of them here (they are to appear under the Creative Commons licence CC-by-NC 4.0.). I am particularly indebted to Siegmund Probst for his prompt and invaluable expert help in dating these manuscripts.

Thanks, finally, to Peter Momtchiloff of Oxford University Press for his unstinting help in seeing this project through to completion, and to family and friends for their forbearance and support in the creative process.

Contents

Abbreviations and Conventions

A	G. W. Leibniz. *Sämtliche Schriften und Briefe*, ed. Akademie der Wissenschaften (Leibniz 1923–); cited by series, volume and page, e.g. (A VI 2, 229).
AG	Ariew and Garber, eds. Leibniz: *Philosophical Essays* (Leibniz 1989).
AT	*Oeuvres de Descartes*, 12 vols., Nouvelle présentation, ed. Charles Adam & Paul Tannery. Paris: J. Vrin, 1964–76; cited by volume and page, e.g. (AT VIIIA 71).
BA	*The Complete Works of Aristotle*, 2 vols., (Aristotle 1984).
C	Couturat. *Opuscules et fragments inédits de Leibniz* (Leibniz 1903).
CG	G. W. Leibniz. *La caractéristique géométrique*, ed. Echeverría and Parmentier (Leibniz 1995a); cited by fragment number and page number, e.g. (CG ix 148).
DSR	G. W. Leibniz. *De Summa Rerum: Metaphysical Papers 1675–1676*. Translated with an introduction by G. H. R. Parkinson. New Haven: Yale University Press, 1992.
E	G. W. Leibniz. *Latina Gallica Germanica Omnia*. Ed. Joannes Eduardus Erdmann. Berlin: Olms, 1840.
GM	Gerhardt. *Leibnizens Mathematische Schriften* (Leibniz 1849–63); cited by volume and page, e.g. (GM II 157).
GO	Gassendi, Pierre. *Opera Omnia*, 6 vols. Lyon. 1658. (repr. Stuttgart-Bad Canstatt: Friedrich Frommann, 1964); cited by volume and page, e.g. (GO I 280).
GP	Gerhardt, *Die Philosophische Schriften von Gottfried Wilhelm Leibniz* (Leibniz 1875–90); cited by volume and page, e.g. (GP II 268).
H	G. W. Leibniz. *Theodicy: Essays on the Goodness of God, the Freedom of Man, and the Origin of Evil*, trans. E. M. Huggard. La Salle, Ill: Open Court (Leibniz 1985).
L	Loemker. *Leibniz: Philosophical Papers and Letters* (Leibniz 1976).
LAV	G. W. Leibniz. *The Leibniz-Arnauld Correspondence*, trans. Voss (Leibniz 2016a).
LBr	*Der Briefswechsel des Gottfried Wilhelm Leibniz in der Königlichen Öffentlichen Bibliothek zu Hannover*, ed. Eduard Bodemann (Hannover, Hann'sche Buchhandlung, 1889). (A catalogue of handwritten manuscripts from Leibniz's correspondence.)
LC	G. W. Leibniz and S. Clarke. *The Leibniz-Clarke Correspondence*, ed. Alexander (Leibniz 1956).
LLC	G. W. Leibniz. *The Labyrinth of the Continuum*, trans. Arthur (Leibniz 2001).
LDB	G. W. Leibniz. *The Leibniz-Des Bosses Correspondence*, trans. Look and Rutherford (Leibniz 2007).
LDV	G. W. Leibniz. *The Leibniz-De Volder Correspondence*, trans. Lodge (Leibniz 2013).
LH	*Die Leibniz-Handschriften der königlichen öffentlichen Bibliothek zu Hannover*, ed. Eduard Bodemann (Hannover and Leipzig, Hann'sche Buchhandlung, 1895). (A catalogue of Leibniz's handwritten manuscripts, other than from correspondence.)
LSC	G. W. Leibniz. *The Leibniz-Stahl Controversy*, trans. Duchesneau and Smith (Leibniz 2016b).

MP	*Leibniz: Philosophical Writings*, trans. Morris and Parkinson (Leibniz 1995).
MT	+ text number: a text in the *Mathesis Texts* volume (Leibniz 2021).
RB	*New Essays Concerning Human Understanding*, trans. Remnant and Bennett (Leibniz 1981) of *Nouveaux essais sur L'entendement humaine*, which has page numbers keyed to A VI 6.
WFT	Woolhouse and Francks, G. W. Leibniz: *Philosophical Texts* (Leibniz 1998).

All the translations from the Latin, French, and German are my own.

I translate the Latin *seu* or *sive* by 'ôr' when this denotes an 'or of equivalence', in order to discriminate it visually from 'or' denoting an alternative.

I cite Leibnizian texts by the original language source first, followed by a readily available English translation after a backslash, thus (GP VII 400/LC 70). If the same sources are repeated consecutively, I abbreviate thus: (401/70).

Calendars: the dates indicated for these writings are keyed to the calendar in use at the source. The Catholic countries in this period had already adopted the Gregorian calendar, or New Style (NS) of dating, which was only adopted in the Protestant states in Germany and in the provinces of the Dutch Republic in 1700, and was not adopted in Great Britain and its Dominions until 1752. Until those times they still used the Julian calendar (Old Style, OS), whose dates are 10 days behind NS until March 1700, and 11 days behind thereafter.

Introduction

Since the arguments in this book are premised on the general line of interpretation of Leibniz's metaphysics that I proposed in *Monads, Composition, and Force*, I should briefly describe that at the outset. First, though, I need to provide a sketch of the main elements of Leibniz's metaphysics of substance and force, and some of the problems of interpretation they present to commentators.

As is well known, Leibniz claimed in his mature philosophy that all that exist in the created universe are substances, the true unities that (from the mid-1690s onwards) he called *monads*, together with everything that results from them.[1] These unities are simple, that is, partless, although they have internal qualities and actions, namely *perceptions* (defined as representations of the composite or external in the simple) and *appetitions* (principles of change by the action of which one perception passes continually into another). As substances, they are essentially active; this activity consists in a primitive active force, Leibniz's reinterpretation of Aristotle's first entelechy or the Scholastics' substantial form,[2] which needs to be completed by a primitive passive force, his reinterpretation of the Aristotelian primary matter. These *primitive forces* are manifested in bodies as the *derivative forces* treated by Leibniz in his new science of dynamics: the active ones being, for instance, *vis viva*, and quantity of progress (momentum), the passive ones being forces of resistance to penetration and to new motion (inertia). *Composites*, such as bodies, are strictly speaking not substances, but aggregates of simple substances, the monads; they are *many* not *one*, and are therefore phenomena. The monads, being partless, cannot be material; they are not parts of bodies, but are presupposed by them.[3] In fact, monads are presupposed in every actual part of a body, rendering a body an infinite aggregate of monads. Each monad, moreover, has an organic body belonging to it, and of which it forms the first entelechy or substantial form; and the monad together with its organic body make up a corporeal substance, that is, a *living thing* or animal. Consequently,

[1] The exposition in this paragraph largely follows that given in Leibniz's two essays of 1714, the so-called *Monadology*, and the *Principles of Nature and Grace*, supplemented by passages in the *Theodicy* of 1710. As he explains in the second of those essays, "*Monas* is a Greek word which means unity, or that which is one." (GP vi 598/WFT 259).

[2] "The active substantial principle is usually called substantial form in the Schools, and primary Entelechy by Aristotle." Draft of a letter to Rudolf Wagner, June 4, 1710; LBr. 973, Bl. 326; transcription sent to me by Osvaldo Ottaviani.

[3] In this respect, Leibniz is contesting the Cartesian view of corporeal substance as consisting in an extended body whose parts are themselves extended substances.

Leibniz on Time, Space, and Relativity. Richard T. W. Arthur, Oxford University Press. © Richard T. W. Arthur 2021.
DOI: 10.1093/oso/9780192849076.003.0001

"there is a world of creatures—of living things and animals, entelechies and souls—in the smallest part of matter" (*Monadology*, §66).

That is the system in very broad outline, with all kinds of important detail left out. Nonetheless, some problems of interpretation are apparent even from this sketch. How can a material body be an infinite aggregate of monads if the latter are immaterial? How can it be a phenomenon if its constituents—the living things contained in any of its parts—are real? If a corporeal substance is a monad together with its organic body and the latter is an infinite aggregate of monads, then how does it differ from a body, which is not strictly speaking a substance? Such problems, together with Leibniz's characterization of monads as consisting in perceptions and appetitions alone, have led commentators to ascribe to Leibniz a variant of phenomenalism, at least in the last years of his life, rendering his philosophy a thoroughgoing idealism. On such an interpretation, immaterial monads alone are real, and their primitive active force is simply appetition, the internal principle bringing about transition to a substance's future perceptions, while their passive force is at best a limitation of appetition; bodies are phenomenal because they exist only in the perceptions of monads; and corporeal substances, in the final analysis, do not differ from phenomenal bodies; motion, too, is a mere phenomenon, since it consists in the change of place of phenomenal bodies, while place is a relation, and relations exist solely in the mind.[4]

Although one can find quotations in Leibniz's writings that seem to invite such an interpretation, there is much in his numerous descriptions of his views that does not easily cohere with this reading. Concerning the substantial forms decried by most of his contemporaries, for instance, part of Leibniz's motivation for rehabilitating them was precisely to distinguish the organic body of a substance from a mere aggregate of bodies, like a pile of wood. If he had abandoned this distinction, one might have expected him to declare this, as opposed to continuing to write of the created monad as the entelechy of its organic body, as he does in the *Monadology* (§62).[5] Many leading interpreters have found the idealist reading sufficiently compelling, however, to dismiss that distinction (as well as his continuing positive references to corporeal substances) as "heteronomous" to his metaphysics. All that exist, they maintain, are the immaterial monads, and everything else, even the living creatures so dear to Leibniz, are, like all bodies, merely phenomena, existing only in the monads' perceptions.[6] Even those who,

[4] The essentials of this reading can be traced all the way back to Baumgarten's 1739 *Metaphysica* (Baumgarten 2013, §§198-199). But the most recent (and very erudite) interpretation along these lines is that of Robert M. Adams (1994).

[5] This reading also makes problematic Leibniz's claims about the harmony between two realms—that of souls governed by the laws of final causes, and that of bodies by the laws of efficient causes. Donald Rutherford (1995) attempts an explanation, ceding that "to talk of 'bodies' at all in this scheme must be regarded as a type of shorthand" (217).

[6] The claim that corporeal substances are "heteronomous" to Leibniz's own metaphysics is made by Robert Adams, who classes them as "an accommodation to traditionalist concerns of others, especially

like Daniel Garber (2009), have presented Leibniz as an Aristotelian realist in his middle period, have portrayed him as gradually, although not always consistently, adopting such an idealistic metaphysics in his final years.

In opposition to these readings I have tried to show in my recent books, (2014) and (2018), how Leibniz's metaphysics appears in a different light when it is viewed genetically, rather than as anticipating aspects of Kantian philosophy. His commitment to the mechanical philosophy was early and lasting, but it was overlaid on certain principles, both Platonic and Scholastic, that privileged minds as sources of activity in the world. On the one hand, Leibniz was as convinced a mechanist as Pierre Gassendi and Robert Boyle, holding that all natural phenomena are explicable in principle, without appeal to substantial forms, by efficient causal explanations involving the motions of bodies; on the other, he subscribed to a view emanating from the Scholastic doctrine of the plurality of forms, whereby the seeds of all living things were created at the beginning of the world, each seed consisting in an immaterial form or active principle dominating the organic body containing it, with the body containing within itself other bodies activated by their own subordinate forms. This view was popular among Lutheran philosophers, who held that God's providential plan for his creation would naturally unfold from within by the activity of these forms. Seen in this light Leibniz's famous doctrine of pre-established harmony was intended as a solution not just to the mind–body problem bequeathed by Descartes, but to the deeper problem of how the teleological activity of forms, leading to the increased perfection of the world, could be reconciled with the impossibility of the action of immaterial forms on matter.

His commitment to active principles within matter led Leibniz to find fault with the foundations of the mechanical philosophy. At the forefront of his thought were problems about the composition of matter and motion, which his contemporaries had taken as continua requiring a foundation in elements from which they are composed. The key thing to understand in this connection, I maintain, is Leibniz's sharp division between the continuous and the discrete. The continuum is not an existing thing, but rather an order according to which existing things are arranged; it is divisible, but has no actual parts. Matter, on the other hand, is actually divided to infinity by its internal motions, leaving it an aggregate of actual parts in contrast to the merely possible parts into which a continuum can be divided. As something perpetually divided, a body cannot constitute a unified whole, a true unity, but is only perceived as one. If it is not to be a mere aggregate

Roman Catholics" (Adams 1994, 307). A subtler form of idealism is attributed to Leibniz by Donald Rutherford (1995, 218), who recognizes that for Leibniz "matter is essentially a multitude of monads" (221) external to the perceiver, which "happen to give the appearance of being an extended object when apprehended by other finite monads" (218). With that much I agree; but see chapter 2 of my (2018) for an analysis and criticism of Rutherford's further claim that monads are not actually in bodies, but are only essential requisites of the concept of body.

of parts within parts, however, it must be an aggregate of true unities. Therefore, he maintains, each actual part of a body presupposes something that is a true unity, and an enduring one, and the reality of body is constituted by the reality of these presupposed unities or entelechies; each such constituent unity, moreover, must be nonmaterial (or else it would be divided). This is one way in which a body is phenomenal, rather than substantial: it derives its unity and continuity from being perceived as one continuous thing, although it is in fact an aggregate of discrete unities. A second way in which body is phenomenal derives from the fact that it is a different aggregate of parts from one moment to the next. So it does not continue to exist as the same aggregate of parts from one moment to another. True substances, on the other hand, remain the same thing throughout their existence.

Secondly, Leibniz insists that substances are things that act, and must have a principle of activity in them responsible for their changes. He therefore interprets the entelechies as "things that by acting do not change," sources of action which remain self-identical through time. A corporeal substance, such as an animal, must therefore have an entelechy adequate to it, an immaterial principle of unity that is responsible for its continuance through time as the same thing. This differs from the Thomist interpretation of Aristotle, favoured for instance by the Jesuits, according to which a corporeal substance is rendered an actual, continuous whole by its possession of a substantial form, which confers on it a synchronous unity for as long as it continues to exist, its matter being made actual by this form. For Leibniz, by contrast, a body without a dominant form—secondary matter—is still actual, since its constituents are made actual by their own forms.[7] This applies also to the body of an animal, which is still an aggregate of discrete parts containing substances: it is a 'many" and not a "one," a *substantiatum* (substantiated thing), not a *substantia* (substance). What makes an animal (or any substance) a living thing is not the sum of these parts, but the principle responsible for its remaining the same thing despite its body's being constituted by a different aggregate of substances at each assignable time.

From these considerations, the picture we have of Leibniz's theory of substance is as follows. Substance is something which by acting does not change. This requires a permanent basis for its changing accidents, as well as an active principle, one by which the substance's changes are brought about by its action. Leibniz finds an exemplar of this in the human self.[8] We experience ourself as remaining the same self through our various perceptions, and we experience those

[7] Also, in agreement with the criticisms of such as Robert Boyle, Leibniz found the idea of forms being created or annihilated unintelligible; so for him all forms are coeval with the created universe. See chapter 4 of my (2018).

[8] See my (2018, ch. 7). This analysis derives support from what Leibniz wrote in a preliminary study for his letter to Rudolph C. Wagner of June 4, 1710: "And in every living thing the substance is conceived to be like that which I understand in myself when I say: *I*; for, even if the mass of my body is in a continuous flux (so that in my old age I will probably not retain anything of the mass I received

perceptions as passing from one to another, even while our bodies undergo changes and do not remain the same. On this model, Leibniz takes a substance to be a primary entelechy like the soul, with perception and appetition as its defining attributes.[9] This characterizes a substance in essence: an entelechy always perceives and always has appetition (albeit, not necessarily consciously). In order to exist, however, a substance must constitute a subject; that is, it must include a principle of individuation, a basis for distinguishing itself from other substances. This will depend on the particular content of its perceptions and appetitions, which will depend on the situation of its own body at different times. Thus although a monad must have perception and appetition as permanent attributes, it will only constitute a principle of individuation by virtue of the concrete perceptions and appetitions which give it its point of view in the world, situating it spatially in relation to other substances through its body at each time. Thus having a body is another essential characteristic of any concrete created substance, along with perception and appetition.[10]

Furthermore, it is only through its body that a substance is capable of being acted upon. Primary matter is the *principium passionis*, the principle of being acted upon; "it is related to the whole mass of the organic body."[11] That is, the primitive passive force of a monad stands in an essential relation to the organic body of that monad. This force is only manifested in its body, however, as a derivative passive force: it is the power of resisting the (derivative) active forces of other substances external to the body, responsible for its resistance to being penetrated or taking on new motion. In this way the extension of the body is a result of the diffusion of this passive force, and this in turn requires a body consisting in a multiplicity of monads as sources of the active and passive forces in it—what Leibniz calls secondary matter. There is, then, no created monad that does not have associated with it an organic body of which it is the entelechy, and

when I was born), nonetheless *I* do remain the same, and the same holds in the case of all living, sensing and reasoning beings, that they persist even though their mass flows" (LBr 973, Bl. 327r; again, thanks to Osvaldo Ottaviani for the transcription).

[9] Indeed, one can argue (as was done in a paper I recently refereed) that the experience of self establishes the possibility of such a definition of substance, in keeping with Leibniz's requirement for a real definition, as opposed to a merely nominal one (as something that acts, for example, or something that can be distinctly conceived).

[10] Anne-Lise Rey makes much the same point in her introduction, "L'ambivalence de l'action," to her edition of the Leibniz-De Volder correspondence: "*Si la Machine est bien le situs de la monade qui exprime la relation d'ordre, il faut indiquer que le situs est la manière dont le point métaphysique donne son point de vue à la substance et lui permet de s'exprimer par l'entremise des corps. Le situs fonctionne comme un principe d'individuation de la substance simple dans les corps, qui atteste, par la, de la présence des substances simples dans les corps*" (Leibniz and De Volder 2016, 76).

[11] This phrase occurs in the immediate preamble to Leibniz's famous five-part schema of the composition of corporeal substance in his letter to Burchard De Volder of June 20, 1703 (GP II 252/ LDV 265). See Pauline Phemister's illuminating discussion of this schema (Phemister 2005, 50), and of the whole question of primary matter in chapter 2 of that book.

through which it exists in relation to the other monads in the world. And the monad together with this organic body is a corporeal substance.[12]

Such an analysis provides the basis for my answers to the problems of interpretation sketched above (namely: how can a material body be constituted by an aggregate of monads if the latter are immaterial? How can it be a phenomenon if its constituents are real? And how does a mere aggregate differ from an organic body?) Even though monads are essentially constituted by immaterial principles, a created monad is never purely immaterial: its material aspect, the primitive passive force, can only be manifested in reality through the derivative passive and active forces of the other monads making up its organic body. All secondary matter consists in aggregates of monads with their organic bodies. A body that is a mere aggregate, such as a woodpile, is "semi-mental," in that its unity is provided through being perceived as one thing. An organic body, by contrast, is the body associated with a principle of unity; it is called *the same* organic body despite its constituents constantly changing, by virtue of its contributing to the actions and purposes of its principle of unity, the dominant monad.

What this means, I have suggested, is that there is no neat separation into separate monadic and phenomenal levels, as is supposed by most modern interpreters.[13] It is certainly true that substances are more fundamental than phenomena. But among the phenomena are the derivative forces in bodies that give rise to their extension and motion. These forces are phenomenal in a sense that would be accepted by all Leibniz's contemporaries: they are perceptible, accessible to the senses. But they are not mere appearances in the mind, and neither are bodies: centrifugal force, for example, is produced in a rotating body independently of anyone perceiving it, even if for Leibniz it is not independent of the possibility of being perceived. At the same time, however, these derivative forces are phenomenal according to Leibniz because they are transitory modifications of the powers of the substances from which they arise, rather than being enduring existents, like the primitive forces that are enduring attributes of the substances from which the bodies are aggregated.[14] Thus, as Leibniz pointed out to

[12] As Leibniz writes in one drafted passage of his letter to Rudolph Wagner, "Corporeal substance is a being in itself, for instance, a living being, a man, an animal. For it consists of the primary Entelechy and the organic body" (LBr. 973, Bl. 326). This is hard to square with some of the things Leibniz says to Des Bosses in their correspondence concerning substantial bonds; for an attempt, see my (2018), ch. 6, §3.

[13] A chief proponent of this levels view is Glenn Hartz, who describes it as follows: "after 1695 Leibniz endorsed a *fundamental* level where the monads and their states reside; just above that he has bodies, derivative force, motion, extension, and duration at the *phenomenal* level; and finally at the top he has the items that are furthest from being taken seriously ontologically. These include space, time, and "mathematical bodies," which are consigned to the *ideal* level" (Hartz 1992, 518).

[14] "Therefore, in secondary matter there arise derivative powers, through the modification of primitive ones; and from this it happens that matter acts in different ways and resists in different ways.... Primitive powers are something substantial, whereas derivative powers are only qualities. Hence, primitive power is perpetual, and it cannot be naturally destroyed; but derivative power can naturally begin and cease, and usually does. Substance persists, quality changes" (from the draft of the letter to R. C. Wagner of 1710 referenced in fn. 2 above). He says much the same thing in the *Theodicy*, §87 (H 170).

De Volder, there could be no derivative forces in a body if there were no primitive ones there of which they are the accidental modifications.[15] But the very fact that they are modifications of substance means that they cannot be on a different ontic level from it, any more than shape or figure, which is an accidental modification of matter, must exist on a different ontic level from matter itself. Leibniz makes this point himself in what appears to be a study for his letter to De Volder of October 7, 1702:

> The most distinguished De Volder acknowledges derivative force ôr an impetus added to substance, although not primitive force. But it must be recognized that everything accidental is a modification of the substantial... It should be said, then, that all accidents are nothing but modifications of substance, whereas modifications add nothing real and positive to substance, but only limitations, just as in the variation of figures nothing is varied but limitation.
>
> (A VI 5, N. 2567, 00)

To be sure, substances are more fundamental than the phenomena resulting from them, such as bodies, their derivative forces, and their motions; but created substances can no more exist without their organic bodies than can an extended figure without a shape. So there is no ontic level on which there are just created monads, distinct from the phenomenal bodies and motions that are their immediate results, nor does Leibniz ever write of such distinct levels. The ontic level framework, I contend, is a facet of a Kantian interpretation of Leibniz, where monads are conceived as denizens of a "noumenal world," in contrast to the "phenomenal world" in which bodies are pure appearances in the experiences of individual subjects.

These considerations are of direct relevance for the correct interpretation of Leibniz's views on time, space and motion. For example, as we shall see in chapter 3, most interpreters of Leibniz's views on the relativity of motion take for granted that primitive force must exist "at the metaphysical or monadic level," while motions occur "at the phenomenal level." Motions, on such a view, are mere appearances of bodies moving in the perceptions of monads; indeed, nothing really moves, since the monads are not in space (which is ideal). Instead, it is held, all we have is the appearances of motions, these being the changing relations of bodies to one another, with these relations constituting phenomenal space (whose relation to mathematical space, incidentally, is thereby rendered problematic, as we shall see in chapter 2). Such a construal leaves out of account the derivative forces of Leibniz's physics, so these have been held to occur at an intermediate level of reality (at least during the 1680s and 90s when Leibniz was actively

[15] "And indeed derivative forces are nothing but modifications and results of primitive forces.... Every modification presupposes something lasting" (to De Volder, June 20, 1703; LDV 262–3).

formulating and defending his dynamics). So on the latter reading, defended for instance by Garber (2009) and Anja Jauernig (2008), derivative forces are qualities of corporeal substances existing on this intermediate level of reality, so that if we suppose that Leibniz came to reject the existence of corporeal substances in his maturity, then he would have had to abandon his dynamics too. Yet Leibniz continues to uphold his dynamics in his maturity, linking it with his metaphysics of monads consisting in primitive active and passive forces, manifesting themselves in the created world as the derivative forces of his physics, as explained above. It is through the derivative forces that bodies are constituted as extended, and derive their cohesion and inertia, and it is through the laws governing the derivative forces (such as the laws of conservation of active force and quantity of progress) that motions through space are estimated. Just as there could be no transient modifications without the permanent attributes of the substances they modify, so there could be no primitive forces in created substances if they could not manifest themselves externally to any subject as derivative forces acting on bodies and producing their motions.[16]

The transient and momentaneous status of derivative forces has important implications for Leibniz's theory of time, which I shall be exploring in detail in chapter 1 below. Monadic states have been thought to exist on a separate level from phenomenal states, so that with time pertaining only to the phenomenal level, relations among the states of monads would have to be atemporal and ideal. But there is passage within a monad: Leibniz defines the state of a simple substance or monad as "the transitory state which incorporates and represents a multitude within a unity" (*Monadology*, §14). That is his definition of a *perception*, while *appetition* is the principle that "brings about change, or the passage of one perception to another" (*Monadology*, §15). Moreover, notably, there is no distinction to be found here between monadic and perceptual states. A monad, rather, "cannot continue to exist without being in some state, and that state is nothing other than its perception," each such perception containing "a great multiplicity of smaller perceptions" (*Monadology*, §21). So the picture we have is of a monad passing through a sequence of transitory states or perceptions, each containing smaller ones, and tending by appetition towards future ones in the same series. The appetition or tendency to pass from one perception to another is identified by Leibniz with the primitive force of a substance,[17] whereas "derivative

[16] Appetitions do, of course, still manifest themselves *internally* as desires and other affects, but even these must correlate (perfectly) with external phenomena, the pain with the thorn in the flesh, and so forth: "anyone who has some perception in his soul can be certain he has received some effect of that in his body" ("Metaphysical Consequences of the Principle of Reason" §4, C 12/MW 173).

[17] "It believe it is evident that primitive forces can be nothing other than the internal tendencies of simple substances, by which they pass from perception to perception by an internal law, and that they agree [*conspirant*] with one another at the same time, relating the same phenomena of the universe in a different arrangement, something which necessarily originates from a common cause" (to De Volder, January 1705 (?); LDV 319).

force is the present state itself insofar as it tends towards a following one" (to De Volder, January 21, 1704; LDV 287). That is, the individual tendencies or appetitions toward future perceptions that accompany each perception are the basis not only for the internal desires and aversions of our inner experience, but also for the derivative forces of physics, and the motions consequent on this. For the changes in the situations of external phenomena themselves must correspond perfectly with how they are represented internally to each monad in its perceptions.

It is important to realize, however, that even though substances are inherently active, they do not possess the attribute of primitive force through themselves, any more than they could continue to exist by themselves, without divine concurrence. As Leibniz insists in texts from 1676 till the end of his life, every created substance has power (*potentia*), perception (or knowledge, *scientia*), and an appetite or tendency towards perfection, only as *limitations* of the corresponding divine attributes of omnipotence, omniscience, and omnibenevolence. In this respect Leibniz explicitly aligns himself with the Platonists, where "things exist by the participation of Being itself," or through the participation of the divine attributes.[18] Thus even though derivative forces are momentary accidents of the enduring primitive forces from which they arise in succession, the primitive forces themselves are in another sense derivative, since they are limitations of divine omnipotence. They are enduring attributes, but they derive their reality independent of (created) perceivers by the participation of the absolute: they are finite limitations of divine attributes that are infinite. Thus although they are absolute with respect to the derivative forces that are their modifications, they are derivative with respect to God, the infinite substance:

> The infinite substance is GOD, in whom there is no passive power, no antitypy...; rather, in God there is just the Entelechy alone... Therefore, only God is pure act. And since all things flow from him, he can be distinguished from the other things not in terms of space or figure (or other modifications), but in terms of a kind of primitive nature itself. For, even though all forces or powers are primitive with respect to the subject in which they inhere, nevertheless, absolutely speaking, they are all derivative.[19]

[18] Leibniz states this Platonic foundation particularly clearly in an important manuscript from around 1698, insisting "that things exist by the participation of Being itself, that is, through the participation of the First Being, and that unities, good things, and beautiful things exist by the participation of the one itself, of the good itself and of beauty itself, that is, by benefit of absolute reality or goodness, which is in the prime substance." From "Towards a Science of the Infinite" (LH 35, 7, 10, Bl. 5r–8v; 5v), transcribed and translated by Osvaldo Ottaviani and myself for a proposed forthcoming volume on Leibniz on the infinite.

[19] From the preliminary study for the letter to R. C. Wagner (LBr 973, Bl. 327v).

This, then, is a deeper Neoplatonism at the heart of Leibniz's metaphysics.[20] Not only are substances distinguished from the fleeting phenomena of bare matter by their enduring attributes of (primitive) force, perception, and appetite for perfection, but they possess these attributes only as finite limitations of the corresponding infinite divine attributes of omnipotence, omniscience, and omnibenevolence. Insofar as a substance is created, its activity must be limited: unlimited activity would be *actus purus*, pure act, and would therefore be God, the Absolute; likewise a substance's unity must also be limited or negated, otherwise it would be the absolute One.[21] From this Neoplatonic perspective, matter connotes not only a passive principle, the limitation or negation of activity, but also a negation of unity, and therefore a multiplicity: thus Leibniz calls matter not only the *principium passionis* or passive principle, but also the "principle of multiplicity."[22] This entails that a created substance is a principle of unity and of activity that is necessarily limited by the creation of other unities with their own principles of activity, and these necessarily constrain its own tendencies to action.

As I have argued, the fact that matter is constituted by a multiplicity of other substances is of great relevance to Leibniz's mature metaphysics of contingency, which I shall be exploring further in these pages. That is, the fact that the actions of any particular substance are contingent on those of all the other substances in the same world, constitutes a crucial aspect of contingency. Moreover, although Leibniz had initially conceived the actions of animate things as being determined by the laws of mechanics, and thus as geometrically necessary, once he had developed his possible worlds ontology in the early 1680s he came to see that the physical laws pertaining to one world might differ from those governing another, since they would depend on the relationships among the individuals that were possible in each such world. Thus actions in the created world, construed as tendencies towards greater perfection, would depend not only on all the

[20] For a detailed examination of the Platonist currents at the heart of Leibniz's metaphysics, see Christia Mercer's (2001). This is valuable in particular for showing the degree to which Neoplatonic themes were common property in Leibniz's milieu. But whereas I follow Leibniz's own description of his philosophy as taking the best of Plato on the one hand and Democritus on the other (A VI 6, 71-73/RB 71; Mercer 2001, 465), Mercer sees him as developing his own system in conscious opposition to them (471), and I think hugely underestimates his commitment to and contributions to the mechanical philosophy.

[21] See Maria Rosa Antognazza's analysis in a forthcoming paper "Leibniz seems to turn to a metaphysical model inspired by the Neoplatonic 'One'. Only what is *beyond* all determinations (or, as Leibniz puts it, what is *hyper-categorematic*), while containing *eminently* all determinations, can be the ontological grounding of all things (*omnia*) without being tainted by the negation which comes with any determination."

[22] We see these descriptions as early as 1678–79 in the important text "Metaphysical Definitions and Reflections": "Substantial form ôr Soul is the principle of unity and of duration, whereas matter is the principle of multiplicity and change"; "Matter is the principle of passion [*principium passionis*], Form the principle of action" (A VI 4, 1399/LLC 245, 247). He elaborates: "Since it is necessary that a principle of passion effectually contains within itself a multiplicity [*multitudinem in se potestate continere*], it follows that matter is a continuum containing several things at the same time, ôr an *extensum*" (A VI 4, 1400/LLC 247).

other substances in the actual world, but also on that world having been chosen by God as the best possible, the one maximal in perfection. Not only would the existence of the individuals created be determined by a divine comparison with those that might have existed instead of them, the laws of the actual world would themselves be contingent on God's deciding that they would provide an optimal fit. (This, Leibniz believed, made it possible to derive the physical laws holding in the actual world through principles of optimization.) All this is of crucial relevance for understanding how Leibniz conceived determinism as not entailing geometric necessity. This in turn is critical for seeing how the determinism underlying Leibniz's theory of time is not incompatible with his teachings on contingency and free will, as I explain in chapter 1 below.

The above considerations also cast Leibniz's theory of space in a novel light. I have explained above how it is the concrete perceptions and appetitions of a substance which give it its point of view in the world, situating it spatially in relation to other substances through its own organic body. In fact, Leibniz insists, if a simple substance did not have an organic body corresponding to it, it could "not have acquired any kind of order to the other things in the universe, nor could it act or be acted upon in an orderly way."[23] Proponents of an idealistic interpretation of Leibniz see no problem with the idea that a substance must have an organic body, reading this just as a set of representations internal to a monad of things existing externally to it, but represented as standing in closer relation than those outside the body itself. They interpret Leibniz's insistence that "there is no nearness among monads, no spatial or absolute distance" (LDB 251) as supporting their view. As a matter of fact, Leibniz makes that remark in an effort to rule out Des Bosses's suggestion that monads exist directly in absolute space, like the continuous "mathematical body" which Des Bosses believed must be "actualized" in order for bodies not to be reduced to mere appearances.[24] Leibniz is not denying that monads have situations through their organic bodies: they may be said to exist within certain spatial boundaries relative to other bodies (as our minds do not exist directly in space, but their mental contents still depend on our

[23] "Metaphysical Consequences of the Principle of Reason," *c.*1712 (C 14/MP 175). "And therefore since every organic body from the whole universe is affected by determinate relations to each part of the universe, it is no wonder that the soul itself, which represents the rest to itself according to the relations of its own body, is a kind of mirror of the universe representing the rest according to its own, so to speak, point of view" (C 15/MP 176).

[24] Thus in his letter of May 20, 1712, Des Bosses urges Leibniz to concede that over and above monads and their phenomena there needs to be "superadded to monads a certain unifying reality which is something absolute (and thus substantial), even if it adds a flux to the things to be unified (the monads)" (LDB 236/237). In his letter of June 12 of the same year, reading Leibniz as advancing a pure phenomenalism, he asks "Do you think that the thoughts that we now have would be true if the monads of the whole world were compressed into one point, as it were, or separately carried away from each other in a vacuum?" (GP II 448/LDB 250/251). It is to this remark that Leibniz directs his comment about monads having no absolute distance, adding, "To say that they are massed together in a point or strewn about in a vacuum is to employ certain fictions of our mind" (to Des Bosses, June 16, 1712/LDB 257/258). See my discussion of this correspondence in (2018, ch. 6, §3).

bodies' situations).[25] "Mass and its diffusion result from monads," he tells Des Bosses in an earlier letter (31 July 1709), "but not space.... For space is something continuous, but ideal; mass is discrete, namely an actual multiplicity ôr being by aggregation, but from infinitely many unities" (LDB 140/141). It is from this infinite aggregation of unities that extension results, by the repetition of the principles of action and passion existing in as small a part of body as one wishes. But, I contend, that depends on the real existence of the organic bodies containing these principles as subjects, having real effects through their (albeit transient) derivative forces, not simply on their appearing in monads' perceptions.

There is, nevertheless, a close connection between space and perception through the notion of coexistence. Those things coexist which can be perceived together (Latin: *simul*, "together," or "at the same time"); moreover, they coexist in a certain order, and this order of their possible coexistence is what for Leibniz constitutes space. As we shall see in chapter 2 below, Leibniz defines a situation as a mode of coexisting, so that space, as the order of possibly coexisting things, is thereby the order of all possible situations. This connects Leibniz's theory of space with his mathematical theory of *analysis situs*, the analysis of situation. Vincenzo De Risi has given a masterly treatment of this novel approach to geometry, and has erected on this basis a powerful defence of a phenomenalist interpretation of Leibniz's metaphysics (at least for the period 1712-1716). According to De Risi's interpretation (which I discuss at length in chapter 2), space is "transcendentally determined" (De Risi 2007, 427); it is a diffusion of situation, where situation is not "an objective material property," "but only a formal property of the possibility of experience" (2007, 415). The importance of Leibniz's geometry of situation for understanding his views on space has been very much underappreciated by commentators on his metaphysics; and given my debt to De Risi's skilful and cogent reconstruction of the main features of the *analysis situs*, my reading of Leibniz's theory of space would be deficient if it did not address the phenomenalist interpretation he gives of it in connection with Leibniz's late metaphysics. Consequently I found myself obliged to pay more attention to the work of a rival interpreter than is typical in a study of this kind.

On the reading I have proposed in my (2018), phenomenal bodies are not simply the contents of perceptions, but are substantiated things (*substantiata*). Their phenomenality consists not in their being mere appearances, but has dual sources in traditions that were typically opposed, one nominalist-conceptualist,

[25] "Even though the places of monads are designated through the modifications ôr boundaries of the parts of space, monads themselves are not modifications of a continuous thing" (to Des Bosses, July 31, 1709; LDB 140/141). More succinctly, as Leibniz says in response to Gabriel Wagner: "I would admit position of incorporeals, by reason of their organic bodies; their extension I would not admit" (On §2), and "The mind is not in the brain or some other determinate place, but belongs to the whole machine" (On §10) (*Ad Schedam Hamaxariam*; LH 4, 3, 5c, Bl. 1r (1703?); transcription and translation by Osvaldo Ottaviani and myself to appear in the *Leibniz Review*).

the other Platonist. According to the nominalist strain, pluralities only exist as the individuals they are aggregated from, so that the unity of an aggregate is something added in the act of perception. And according to the Platonist strain, bodies are phenomena in the sense that they do not remain precisely the same thing from one moment to another. This motivates Leibniz's insistence that there need to be permanent substances out of which bodies are constituted, possessing enduring primitive forces of which the derivative forces in bodies are transient modifications.

There is, I contend, a similar creative tension between these two strains in Leibniz's thinking about space. Once he begins to conceive space in terms of relations of situation in the late 1670s, he firmly rejects it as an existing thing. It is an order by which things are really situated to one another, but the order itself is something ideal. This is similar to what he says about relations generally, under the influence of the kind of nominalism-conceptualism espoused by William of Ockham. From this point of view, relations are analogous to aggregates: they derive what reality they have from the things related, but with the relation itself contributed by the perceiving mind. This has led commentators to ascribe a phenomenalist reductionism to Leibniz, where spatial facts are said to consist only in spatial relations being part of the representational content of monads' perceptions.[26] But space is not simply an ideal order imposed by the human mind on things (as it is in Kant, "the form of outer sense"). For although relations are not modifications of substances which can be produced and destroyed in their own right, they "result from the creation of other things," and "have reality without regard to our understanding, for they are truly there when no one is thinking them. They receive it from the divine understanding, without which nothing would be true."[27] As Massimo Mugnai has observed, here again we find Leibniz's thought determined by his commitment to the two typically opposed traditions: "to nominalism-conceptualism, on the one hand, and to the Neoplatonic claim that every individual reproduces or reflects in itself the entire universe, on the other" (Mugnai 2012, 204).

Being instantiations of the divine understanding would be enough, one supposes, to guarantee the reality of relational truths, which would arise as consequences of the modifications of substances. But this would leave substances only conceptually related spatially through their bodies, whereas Leibniz seems to

[26] Thus Robert Adams writes: "There are no spatial facts at the ground floor level of Leibniz's metaphysics, except insofar as facts about monads' perceptions having spatial relations as part of their representational content may belong to that level" (1994, 255). This reading presupposes the levels schematism that I find inappropriate. Monadic states, for instance, succeed one another in a serial order, which should be impossible if the actual (the states) and the ideal (their order) are on two separate ontic levels.

[27] "On Temmik"; quoted from the passage cited in Mugnai (1992, 155). The latter work, as well as Mugnai's more recent (2012) or (2018) should be consulted for a thorough treatment of Leibniz's byzantine philosophy of relations.

intend a more robust meaning for the reality of relations of connection in order to underwrite his deep-seated belief in "the universal connection among all things."[28] For, as he explains to Des Bosses (July 21, 1709), "even though a simple substance does not have extension in itself, nonetheless it has position, which is the foundation of situation, since extension is the continuous repetition of position, just as we say a line comes to be from a point, since its different positions are conjoined in the trace of this point" (LDB 98/99). Here we see Leibniz pointing up the profound role of situated substances for producing the extension of bodies through the repetition of their forces (see my 2018, ch. 6, §2). In the present work I show how Leibniz's Neo-Hobbesian interpretation of the situations of bodies as representations of geometrical figures allows him to see them as limitations of the extended—indeed, as limitations of the divine attribute of *immensity*, or divine omnipresence. This connects in turn with his interpretation of God's omnipresence as being realized (insofar as it is natural, and not miraculous) by the continuous creation of substances everywhere in matter. What is real in space is divine immensity, so conceived. The resulting extensions of bodies are not parts of divine immensity, however, but limitations of it.[29] Nor are they parts of space, which is indifferent to the ways in which it might be partitioned. The introduction of boundaries into space marks out the possible figures, and thus the situations. But there is no actual situation before the introduction of boundaries, which result from the introduction of the appetitions of substances into what is otherwise an ideal order. When active substances are introduced, extension is a product of their resistance to being penetrated, i.e. their antitypy. Thus without bodies, space would be wholly ideal;[30] but the spatial relations among any existing bodies are instantiations of the ideal order of all possible situations existing in the divine mind. Space in itself, abstract space, is a diffusion of possible situations; but these are the situations of the bodies whose extension would be constituted by the continuous repetition of the derivative forces of the substances they contain. All this is discussed in chapter 2 below.

[28] For an examination of Leibniz's debt to Johann Heinrich Bisterfeld on the universal connection of all things, see Mugnai (1973), and Rutherford (1995, 36-40). Leibniz himself cites the Stoics in this connection, whose views about tranquillity have been "rescued from scorn" by the moderns in the form of "the optimum connection among things" (Leibniz 1695, 146; GM VI 235/WFT 155). Elsewhere he often invokes the Hippocratic notion of sympathy. In any case, his metaphysics should not be regarded as wholly Neoplatonic.

[29] As Leibniz explains in "Towards a Science of the Infinite," "the absolute should not be thought to be like a whole which comprises limited things of its own kind (as certain people think the immense substance is the universe of things itself), for what is constituted by parts has a nature posterior to its parts, whereas the absolute is the origin of limited things" (LH 35, 7, 10, Bl. 5v).

[30] In a letter to Louis Bourguet of July 2, 1716 (GP III 595/Adams 1994, 254) Leibniz explains that to Clarke's objection about space being "indifferent to where God places bodies," he had responded that "the same thing proves that space is not an absolute being, but an order, or something relative, and which would be merely ideal if bodies did not exist in it."

In a similar way, time too receives its reality from the participation of a divine attribute, in this case that of eternity. "Just as we conceive space as a thing of maximal extension, even though nothing in that concept is real but the immensity of God, [so] in infinite time there is nothing but the eternity of God."[31] The maximum in duration is infinite in its own kind, but without any limits or determinations in itself.[32] On this Neoplatonic conception, succession and duration are conceived as limitations or negations of eternity: they are limited by the introduction of changes, which are the boundaries of durations. So durations are not parts of time, which is indifferent to the ways in which it might be partitioned, but particular limitations of that divine attribute. The particular durations are specific to the individual things that are created. As Leibniz explains to Clarke, "if there were no created things, there would be neither time nor place" (to Clarke, GP VII 415/L 714). In respect of the spatial and temporal orders, the attributes of immensity and eternity "signify only that God would be present and coexistent with all the things that would exist" (415/714). In the case of time, God's presence is manifested by continuous creation, but successively, not synchronously. This grounds the reality of the succession of states within each substance, so that the temporal relations among them are instantiations of the ideal order of all possible successions existing in the divine mind.[33] Time in itself, abstract time, is the ideal order according to which all such successions of states can occur.

So again we have the creative fusion of strains referred to earlier. Taken in abstraction from the things in it, time, like space, is merely ideal, in keeping with the nominalist-conceptualist strain. Yet they are both "real relations": "whatever is real in space and time consists in God comprising everything" (A VI 4, 629/LLC 275). As we have seen, this is given a Neoplatonic reading. But these relations of connection are underwritten not just by God thinking them, but by his being able to create substances in those relations. This is how space can be at the same time a condition for bodies to be conceived or perceived together in mutual situations, yet also how the extension of mathematical bodies in space can be a limitation of divine immensity; and how time can be at once the order of all possible states or perceptions, and also the foundation for the real succession of things.

That should be sufficient to explain the general framework for this book. Now let me proceed to the details of Leibniz's views on these topics.

[31] From a cancelled draft in "Towards a Science of the Infinite" (LH 35, 7, 10, Bl. 7r), which is repeated in almost the same words in the main text.

[32] In 1676 Leibniz writes that in a temporal sense, eternity as an attribute "will be duration through an unbounded time"; but that in itself it "is the very necessity of existing, which does not in itself indicate succession, even if it should happen that what is eternal should coexist with all things" (A VI 3, 484/DSR 41).

[33] As Leibniz wrote in the mid-1680s, "the root of time is in the first cause, potentially containing in itself the succession of things, which makes everything either simultaneous, earlier or later" (A VI 4, 629/LLC 275).

1

De tempore: Leibniz's Theory of Time

"Since the nature of every simple substance, soul, or true monad, is such that its following state is a consequence of the preceding one; it is there that the whole cause of the harmony is to be found."

(to Clarke, V §91; GP VII 412/LC 85)

Introduction

Concerning the main features of Leibniz's philosophy of time, the following three theses are beyond dispute:

(i) *Time is relational.* That is, time is not an independently existing entity, but is rather a *relation* or *ordering* of successives.
(ii) *Time is ideal.* Time has no existence apart from the things it relates in order of succession; it is therefore an *ideal entity*. This is clearly consonant with Leibniz's beliefs that *continuity* is a concept that applies to things considered as ideal, and that
(iii) *Time is a continuous quantity.*

But there are radical divergences of interpretation about how to interpret these claims, especially in relation to Leibniz's theory of substance and philosophy of relations.[1]

It is widely maintained that the *simple substances* or *monads* that Leibniz took to be the fundamental constituents of matter *do not exist in time*. Some have held, following Bertrand Russell's highly influential interpretation of Leibniz's metaphysics (Russell 1900), that this follows directly from Leibniz's supposed denial of any reality to relations; others, that it even follows independently of the philosophy of relations. For if time is ideal, it does not apply to actual things, which are therefore timeless.[2] Likewise, if it is continuous, it cannot apply to actuals, which

[1] So I began my first paper on Leibniz (Arthur 1985). I believe it still describes the situation accurately.

[2] Thus J. E. McGuire writes "If time is an ideal notion, it cannot apply to the actual" (1976, 312). Granted, he also recognizes that for Leibniz "States being successive are by nature inextricably temporal" (309–10), but finds in this an unresolvable contradiction at the heart of Leibniz's

Leibniz on Time, Space, and Relativity. Richard T. W. Arthur, Oxford University Press. © Richard T. W. Arthur 2021.
DOI: 10.1093/oso/9780192849076.003.0002

are discrete.[3] Either way, this is held to imply that Leibniz's theory of time, like Kant's, must be a theory of the *temporal order of phenomena*, not of the order of monadic states.

According to Russell, Leibniz held that all relations must ultimately be reduced to unary or monadic predicates of the related things, since it is only predicates of this type that can be said to be completely contained in the subject. Consequently, relations, insofar as they are anything apart from such unary predicates, are merely ideal entities that must be superadded by the perceiving mind, so that no two monadic states can stand in a temporal relation to each other in actuality. On this reading, time is only an ideal relation holding among perceived phenomena.

The latter conclusion was also endorsed by Hermann Weyl in his authoritative (1949), without reference to Russell's interpretation of relations. Weyl suggested that the idea of "a *monad* existing beyond space and time" would be "in line with Leibniz's ideas" (Weyl 1949, 175), and suggested that Leibniz was driven to "epistemological idealism" by his struggles with "the 'labyrinth of the continuum'..., which first suggested to him the conception of space and time as orders of the phenomena" (1949, 41). In a similar vein Glenn Hartz and Jan Cover have since argued in an influential article (Hartz and Cover 1988) that Leibniz's advocacy of the ideality and continuity of time in his mature work precludes its applicability to monads and their states, which are, on the contrary, truly actual and discrete.[4]

Other authors, following Reichenbach (1958, 14–15, 25), have read Leibniz's theory of time as a causal theory of time like Kant's. Leibniz claims that one state is before another if the former 'contains the ground for' the latter. Interpreting this as a causal relation among phenomenal states, Reichenbach—and, after him, Grünbaum and Van Fraassen[5]—assumed that temporal relations for Leibniz, as for Kant, would therefore have to be relations among phenomena. A more explicit aligning of Leibniz's theory with Kant's was given by Jacques Jalabert ([1947] 1985). Arguing that Leibniz's use of the term "phenomenon" should perhaps be interpreted in "an almost Kantian sense," Jalabert held that "on the plane of veritable reality" of monads and their states, "the substance would deploy, independently of time, the entire series of its virtualities; between the terms of the series there would subsist the priority of nature,...but there would not be any

metaphysics. Similarly, Heinrich Schepers, while acknowledging that "substances have a tendency toward internal change" (2018, 420), can still claim that "Things do not act in space and time but constitute the orders that we can recognize as space and time" (422).

[3] Again, McGuire: "Moreover, the actual cannot be continuous, as it is simple and indivisible 'and not formed by the addition of parts' [GP V 144]" (1976, 311).

[4] Cover in his (1997) interprets Leibniz's doctrine that space and time are abstract or ideal entities as constituting an eliminative reduction, rather than an identificatory one. He argues that the further reduction of temporal relations to causal ones make Leibniz's time "non-basic," so that temporal relations do not apply "on the monadic level," but are part of an "ideal world accessible via abstraction or by thought alone" (1997, 303).

[5] For discussion see Grünbaum (1963, ch.7) and Van Fraassen ([1970] 1985, ch. IV).

chronological priority."[6] John Whipple (2011, 14) has recently endorsed Jalabert's interpretation.[7] J. E. McGuire sounds a similar theme in his article "*Labyrinthus continui*." While noting many passages where Leibniz appears to attribute temporal change to monads, he maintains that Leibniz "denied that monads are divisible in the dimension of time as they are truly simple substances" (McGuire 1976, 314).[8] "If they are simple," he asks, "how can they be divided in the dimension of time? For strict simplicity involves a denial of successive differentiation" (315).[9] Vincenzo De Risi, finally, interprets Leibniz as "embrac[ing] a radical phenomenalism relative to time. Thus . . . he seems to think that monads are outside any temporal order, while only their phenomenal manifestations occur in time" (De Risi 2007, 271).

Against such interpretations, in my (1985) I quoted several passages where Leibniz describes monadic states as chronologically ordered. For instance, in the course of clarifying the nature of substance to Burchard de Volder in January 1704, he explained:

> The succeeding substance is held to be the same when the same law of the series, or of continuous simple transition, persists; which is what produces our belief that the subject of change, or monad, is the same. That there should be such a persistent law, which involves the future states of that which we conceive to be the same, is exactly what I say constitutes it as the same substance.
>
> (To De Volder, 1704/1/21: GP II 264; Arthur 1985, 273–4)

No doubt it is possible to read this passage as involving only a priority of nature. But it seemed to me then, and still does now, that in referring to "continuous transition" and "future states" Leibniz clearly meant a temporal succession. The same seems true of his talk of "following" and "preceding states" in his fifth and last letter to Clarke, where he wrote "the nature of every simple substance, soul, or true monad, is such that its following state is a consequence of the preceding one" (Fifth Paper for Clarke §91; GP VII 412; Arthur 1985, 274).

[6] «. . . on est autorisée, semble-t-il, à prendre ici le mot phénomène en une acception voisine du sens kantien. . . . Dans cette interprétation seul l'Acte indivisible et intemporel appartiendrait au plan de la réalité véritable. . . . La substance déploierait, indépendamment du temps, la série entière de ses virtualités ; entre les termes de la série subsisterait la priorité de nature, que Leibniz déclare ordinaire en philosophie, mais il n'y aurait aucune priorité chronologique.» (1985, 208).

[7] It was John Whipple's endorsement of Jalabert's interpretation that led me to it. But I am not persuaded by Whipple's defence, and agree with Michael Futch's criticisms of this interpretation (Futch 2008, 163–5).

[8] McGuire (1976, 314) cites in evidence Leibniz's letter to De Volder of November 19 [he cites it as "10 November"], 1703 (GP II 258). I can find no such denial there.

[9] On the contrary, Leibniz writes in the "Monadology" that "as every natural change takes place by degrees, something changes and something remains the same; and consequently it is necessary that in a simple substance there is a plurality of affections and relations, even though there are no parts in it" (GP VI 608). As we shall see below, Leibniz argues that as unities, monads are not further resolvable, not that they are indivisible: "Unity is divisible but not resolvable," as he explains to Louis Bourguet, August 5, 1714, (GP III 583/L 664).

Some commentators have tried to dismiss Leibniz's papers for Clarke as not representing his "deep and considered metaphysics."[10] The idea (actually contradicted by the preceding quotation) is that in arguing with Clarke he avoided talk of monads, restricting his discussion to the order of succession of phenomenal changes. But in the "Monadology" (usually taken as giving his deep metaphysics) Leibniz was adamant that monadic change is presupposed by change in composite things: "I also take it for granted that every created thing is subject to change, and therefore the created monad as well; and indeed that such change is continual in every one" (§10; GP VI 608/WFT 269). Accordingly, his justification for applying the same time concept to changes in composite things as to monadic changes of state is that the former are results of, and are grounded in, the latter. All things change, and composite things (phenomena) change because of the changes in simple things from which they result. Leibniz was perfectly explicit on this point in a previous letter to De Volder (June 1703):

> You doubt, distinguished sir, whether a single simple thing would be subject to changes. But since only simple things are true things, the rest being only beings by aggregation and thus phenomena, and existing, as Democritus put it, νὸμω not φυσει, it is obvious that unless there is a change in the simple things, there will be no change in things at all. (GP II 252)

So Leibniz explains phenomenal changes as resulting from monadic changes, which form a continuous series of succession in every simple substance. It is very difficult to see how that could be so if time were only an ideal relation holding among perceived phenomena.[11] Thus it appears that Leibniz's theory of time is primarily a theory of the ordering of changes in monadic states, applicable also to the states of the things aggregated from them.[12]

[10] Thus A. T. Winterbourne claims that the discussion in the correspondence 'is not intended as a general metaphysics of space and time which we may take as Leibniz's definitive and complete viewpoint' (1982, 202) Cover concurs: "the Clarke Correspondence is not written as an expression of Leibniz's deep and considered metaphysics" (Cover 1997, 289, n.15). I see no cogent basis for this allegation. Clarke himself is able to illuminate the relevant Leibnizian metaphysical theses occurring in the correspondence with liberal quotations from Leibniz's published writings.

[11] Russell, for his part, cedes that Leibniz supposes monadic series to be temporally ordered, and recognizes that this is incompatible with time's applying only to phenomena. But he flatly concludes that the inconsistency is Leibniz's: "time is necessarily presupposed in Leibniz's treatment of substance. That it is denied in the conclusion is not a triumph, but a contradiction" (Russell 1900, 53).

[12] Thus, contrary to Michael Futch's charge that in my exposition I "implicitly assign to monadic states places in time" and that my analysis "is concerned exclusively with the relations among monadic states" (Futch 2008, 119, fn.11), I would say that on the contrary I *explicitly* assign times to monadic states, and by extension *also* to phenomenal ones. He makes the same charges about of Cover's analysis. But among the various reductionist interpretations of Leibniz's causal theory that Cover considers is one that construes it as a supervenience theory, where "expressions of temporal relation figure in truths on the phenomenal level because they supervene on *prima-facie* non-temporal

Accordingly, in my (1985) paper I set about giving a construction of Leibniz's theory of time on this basis. I argued, moreover, that accepting that monadic states are temporally related, in defiance of Russell's interpretation, allows one to appreciate the intimate connection between his theory of time and the pre-established harmony. For the latter is the epitome of a relational hypothesis: each state of the world (or of any possible world) is a distinct aggregate of monadic states, each of which, as mandated by their harmony, *involves* (or *is compatible with*) all the others in the aggregate, and *involves the reason for* (or *contains the ground for*) all those in its world which succeed it in time. It is on these two relations of compatibility and reason-inclusion of monadic states, I argued, that Leibniz's mature theory of time is based. I present a revised version of the core of that construction in section 1.1 below, with formal details relegated to Appendix 1.

The main merits of this construction, as I see it, are these: (1) in allowing relations among states of different monadic series, it obviates the need for supposing a distinct *monadic time* specific to each series, as has been proposed by some authors; (2) it validates Leibniz's idea that to be *at the same time as* a given state is to be a member of an equivalence classes of states compatible with it, the class of states contemporary with it, without presupposing a pre-existing time; (3) it supports that aspect of Ishiguro's and Hintikka's reading of the *ideality of time* according to which the ideality of relations pertains only to relations considered as abstract objects, abstracted from all particular relata.[13] In the following sections I will build on this core, amplifying my previous arguments and responding to criticisms. In so doing I shall try to show that the attempt to reconstruct Leibniz's philosophy of time is not "an impossible task."[14]

Of course, this same relation of "involving the reason for" or "containing the ground for" is what many commentators have taken to be a causal relation among phenomenal states. I have given reasons above for doubting that Leibniz was presenting a theory of time that was restricted to the ordering of phenomena. But in my (1985) I was also sceptical of attributing a causal theory of time to Leibniz in which the temporal priority of one state to others in the same monadic series is explained by that state *causing* the other states. For Leibniz construed causes in

relations one level down" (Cover 1997, 315). On the other hand, John Whipple appears to conflate my account with Cover's when he says that both of us "have argued that Leibniz is not committed to real intra-monadic temporality" (Whipple 2010, 384).

[13] The leading articles by both authors, Hintikka's "Leibniz on Plenitude, Relations and the 'Reign of Law'," and Ishiguro's "Leibniz's Theory of the Ideality of Relations," are to be found in Harry Frankfurt's still valuable collection (Frankfurt, ed., 1972), pp. 155–90 and 191–214, resp.

[14] This was the opinion of Vincenzo De Risi in his analysis of Leibniz on time in *Geometry and Monadology* (2007, 270–7), according to whom "[Leibniz's] metaphysics of time has always remained incomplete, uncertain, and definitely obscure. At this early point in our study, we will not even hypothetically attempt to reconstruct Leibniz's philosophy of time (presumably, an impossible task), or to deduce the necessity of time in Leibniz's phenomenalism." As is evident here, De Risi interprets Leibniz as committed to a form of phenomenalism, although this is a phenomenalism more nearly indebted to Cassirer's interpretation of Leibniz than to Kant.

terms of reasons, rather than the other way round. This very fact, however, is a symptom of the close connection of reason with cause in his thought, and indeed Leibniz himself promotes a type of causal theory of time in the 1680s based on the idea of requisites, conditions that must be in place for something to occur. This necessitates a careful reconsideration of the question of the causal theory, which I give in §1.2. There I conclude that the relation I have taken as basic in my reconstruction, "involving the reason for," should be understood in terms of requisites as "is the mediate requisite of," whereas two states are simultaneous if each is the other's immediate requisite. Mediate requisites are efficient causes; they are productive of their effects, and require intervening changes. A full cause, meanwhile, is the sum of all the requisites, including the states of other coexistents.

I also argue that although the temporal order maps onto the causal order, so conceived, this is not to eliminate time by reducing it to a merely rational order among perceptions. For the production of an effect by a cause presupposes the activity of the substances: Leibniz posits appetition, in addition to perception, as a defining characteristic of monads. As a result, Leibniz's philosophy of time is dynamic, not static (like most modern "B-theories"), and presupposes the reality of becoming, that states come to be out of their predecessors.

Nevertheless, the idea of each state being produced by all those constituting its full cause raises the question of necessitarianism: if something happens when all the requisites are in place, does it not happen necessarily? Russell is not the only author to have claimed that Leibniz's premises commit him to a necessitarian position, notwithstanding his attempts to avoid this consequence. Leibniz tries to make room for contingency by appeal to other possible worlds where things might happen differently. But if within the actual world everything that is to happen is determined by the laws of the series of each substance in it, then in what sense can anything happen contingently? These are the issues I tackle in § 1.3, where I defend the consistency of Leibniz's account of contingency, and give a construction to show how one can make sense of temporal counterfactuals in the framework of his theory of time.

According to that construction events in each different possible world are ordered according to distinct laws of general order peculiar to that world. Time in general nevertheless pertains to all possible such orderings. This therefore presupposes a distinction between two senses of time: there is the particular temporal order of each possible world, a 'concrete' time determined by the relations among particular states of that world; and there is time as the order of successives in general, in abstraction from the relations of reason-inclusion specific to any particular world. To this it has been objected that the latter, time as the structure of all possibles, is something merely ideal, a being of reason, and as such inapplicable to actual existents. This necessitates a systematic treatment of the issue of the ideality of time, which I undertake in §1.4. Holding a view similar to Ockham's, Leibniz denies that time has to be posited in addition to things occurring in a temporal order. Being "at the same time as" does not presuppose existing times, but only that there is an equivalence relation of simultaneity. This involves a reduction to equivalence relations that is typical of Leibniz's "provisional nominalism." But it is not an "eliminative reduction" in the sense of precluding temporal relations among actual things, since it presupposes enduring things whose accidents or states occur successively, and succession is, of course, a temporal relation. It is only by means of such *concreta* that instants can be individuated. This is how abstract time applies to concrete processes, something which is inexplicable on an eliminativist reading.

The difficulties connected with Leibniz's theory of temporal continuity are profound.[15] According to him, matter and the changes occurring in it are in actuality discrete. How, then, can a continuous time apply to them? And doesn't this contradict the Law of Continuity, which he claimed, holds everywhere in nature? These issues are the subject of the last section of this chapter, §1.5, where I examine the development of Leibniz's views on change from its beginnings in 1676. According to this analysis (of which I give a formal rendition in Appendix 1.5), all enduring states are vague: they are necessarily represented as extended at every level of analysis, yet every state contains changes that in fact divide it into further substates. The result is a conception of durations as actually infinitely divided, analogously to matter. In this way, the arbitrarily small substates can be treated as infinitesimals, in complete accord with Leibniz's understanding of the differential calculus. Changes are therefore discrete, and dense within any time; but time itself is continuous.

But before confronting the intricacies of Leibniz's views on these matters, I want first to consider the prevalent but (I believe) wholly mistaken claim that his account of time in any case presupposes a monadic time, or intra-monadic time series for each monad. This will afford us an opportunity to disclose the full power and coherency of Leibniz's relational theory.

1.1 The Relational Core

Bertrand Russell was among the first in the twentieth century to recognize the merits of the relational theory of time, showing how one can formulate it mathematically using the modern theory of relations in a way that makes the separate postulation of instants redundant.[16] I shall argue that a similar analysis is

[15] Their resolution was the original focus of the last section of my (1985), which eventually became too extensive to be included in the published article. See Arthur (1986), (1989), my introduction in LLC, and Arthur (2008b).

[16] See Russell (1914), (1915), and (1936).

implicit in Leibniz's work. This makes for an acute irony in Russell's relation to Leibniz. For at the beginning of his career, and prior to his seminal work in logic and the foundations of mathematics, Russell proposed an influential interpretation of Leibniz (*Critical Exposition of the Philosophy of Leibniz*, 1900) in which he accused Leibniz of failing to appreciate the importance of relations and of trying to eliminate them.[17]

To compound the irony, it seems that when he wrote these criticisms in 1900 Russell had a curious neo-Hegelian understanding of the relational theory, one that owes much more to the views of Hermann Lotze than to Leibniz. He does not yet appreciate there, for instance, that points and instants can be defined relationally, but rather appears to have believed that the fact that they are not themselves relations somehow constituted Leibniz's ground for rejecting the composition of space and time out of them (Russell 1900, 112–14). Also, in keeping with the Lotzean conception of relations he attributes to Leibniz, Russell believes that temporal relations must be analysed into monadic temporal predicates, taking states to be such predicates. But, he objects,

> The definition of *one* state of a substance seems impossible without time. A state is not simple, on the contrary it is infinitely complex. It contains traces of all past states, and is big with all future states. It is further a reflection of all simultaneous states of other substances. Thus no way remains of defining one state except as the state at one time. (Russell 1900, 52)

Russell assumes here what he is supposed to be proving: that simultaneity cannot be defined except by presupposing an absolute time. But his reason for endorsing the absolute theories of time and space in his (1900) is that he thought he had demonstrated the absurdity of the relational view: intermonadic relations cannot exist if they are to be reduced to predicates of the related substances. Thus it is no surprise to find him concluding against Leibniz that "time is necessarily presupposed in Leibniz's treatment of substance. That it is denied in the conclusion, is not a triumph, but a contradiction" (53).

Such has been the influence of Russell's interpretation, however, that similar claims can be found in the more recent expositions of Leibniz's theory of time by distinguished commentators who are well aware of the modern theory of relations and of Leibniz's anticipation of its application in the theory of time. Nicholas Rescher, for example, follows Russell in believing that Leibniz intended to reduce

[17] It is interesting to note that Russell retracts none of his criticisms of Leibniz's relationalism in the preface to the second edition of his *Critical Exposition* in 1937, and never acknowledges indebtedness to him on this score in his papers on temporal order (1915, 1936), nor in his treatment of the relational theory in *Our Knowledge of the External World* (1914). I have conducted a detailed examination of this circumstance in two papers which are intended to be reworked as chapters of a co-authored book with my colleague Nicholas Griffin, *Russell on Leibniz*.

relational properties to non-relational ones, although he makes an exception for what he calls "intra-monadic relations," relations that are internal to any one given monad. This leads him to claim that

> Time...has a dual nature for Leibniz. There is the essentially private, intra-monadic time of each individual substance continuing, by appetition, through its transitions from state to state. There is also the public time obtaining throughout the system of monads in general, made possible by the inter-monadic correlations established by the pre-established harmony. Leibniz's standard definition of time as *the order of non-contemporaneous things* would be vitiated by an obvious circularity if it did not embody a distinction between intra- and inter-monadic time, carrying the latter back to (i.e. well-founding it within) the former.
> (Rescher 1967, 92; 1979, 88)

Similarly, J. E. McGuire objects:

> If time is an ideal notion, it does not apply to the actual. But actual substances have expressed states, are expressing states, and will express states. Moreover, as they are states of one and the same individual substance, that substance is programmed to unfold a unique history. But such action implies not only some notion of continuity but some conception of monadic time.
> (McGuire 1976, 312)

But, disregarding for now the question of time's ideality, *does* the activity of monads in time imply that they have their own *private* time? Certainly, it must be granted that monadic states precede and succeed each other in time, since Leibniz explicitly claims this, as we have seen. The implication of the above criticisms, however, is that unless the time in which they succeed each other is *different* from the time "obtaining throughout the system of monads in general" (Rescher), or from phenomenal time (McGuire), there will be a vicious circularity. This echoes Russell's criticism that in Leibniz's treatment of substance an individual state of a monad can only be defined as one occurring *at a given instant*, so that time cannot be defined in terms of relations among monadic states on pain of circularity.

As I have already suggested, I think these charges betray a serious misunderstanding of the relational theory of time, not just in its post-Russellian manifestation, but in Leibniz's own version. To demonstrate this, in my (1985) I formalized Leibniz's theory using the theory of relations of modern logic and set theory, along the lines of John Winnie's set-theoretic rendition of the causal theory of time,[18]

[18] John A. Winnie (1977). This excellent article prompted much of my initial thinking on Leibniz's theory of time. I should add that Winnie presents his account not as an interpretation of Leibniz, but as a modern exposition of the causal theory with an eye to Leibniz's contribution.

although my interpretation is rather different, as I shall explain. In fact, I have modified my exposition a little, noting that it is inappropriate to saddle Leibniz with infinite sets, given his eschewal of infinite collections. I believe the resulting formal construal of the relational core of his theory follows the letter of Leibniz's texts quite closely, and successfully captures his intent with a bare minimum of anachronism or distortion. It shows how temporal relations among diverse monads arise without presupposing that each monad has its own monadic time, and without presupposing the existence of instants. I have relegated the mathematical details to an appendix so as not to interrupt the flow of the argument. Here I shall attempt to convey the cogency of the resulting theory with an informal sketch.

Leibniz gives the most complete account of his mature theory of time in the *Initia rerum mathematicarum metaphysica* (that is, "Metaphysical Foundations of Mathematics," hereafter abbreviated as the *Initia rerum*), which he penned in April 1715, a year and a half before he died. There he explicitly defines simultaneous states of things as those which do not "involve opposite states"; temporal precedence, on the other hand, is defined in terms of one state "involving the reason for" or "containing the ground for" another.[19] (Leibniz does not specify what these "things (*res*)" are; I hold that these must comprise simple substances, as well as composites, as I argued above.)

> *If several states of things are supposed to exist, none of which involves the other, they are said to exist* ***at the same time***. Thus we deny that those things which happened last year and those happening presently exist at the same time, since they involve opposite states of the same thing.
>
> *If one of two states that are not simultaneous involves the reason for the other*, the *former* is held to be the *earlier*, the *latter* to be the *later*. My earlier state involves the reason for the existence of my later state. (*Initia Rerum*, GM VII 18)

It is tempting to try to capture the sense of these remarks by simply defining one state as temporally preceding another if and only if it involves the reason for it according to the law of the series of that monad. But this approach would only give us an intrinsic temporal order for each monad, and simultaneity of states of different monads would then have to be introduced in terms of these orders, thus falling afoul of Russell's criticism—of course, some such reasoning may well have been a factor in leading Russell, Rescher, and McGuire to accuse Leibniz of having presupposed monadic times.

[19] The term "involve," as Massimo Mugnai has noted, is a technical term of Leibniz's logical ontology. Whereas one thing is said to be explicable and to *imply* another when the second can be inferred from the first in a finite series of substitutions, "it is inexplicable when the series of substitutions is infinite, and the inference [from the first to the second] is said to *involve* it" (A VI 4, 862; Leibniz 2008, 187).

But Leibniz introduces simultaneity first, defining it in terms of the *non-opposition* of the states of things: *one state is simultaneous with another if and only if it is not incompatible with it.*[20] The beauty of this strategy (like the later strategies of Whitehead and Russell, in which an instant is defined in terms of events through extensive abstraction) is that we can now determine which states occur at the same time without any recourse to instants—provided, of course, that we can give a consistent rationale for the compatibility of states which does not presuppose their temporal relations.

This can be achieved as follows. First we consider the relation G of "involving the reason for" or "containing the ground for" applied to an individual series of states. It is natural to assume from Leibniz's talk of the present being "pregnant with the future"[21] that the relation is asymmetric (Axiom G1), and transitive (Axiom G2) too.[22] It also seems natural to lay down that for any two different states of the same monadic series, one state must contain the ground for the other, or vice versa: this is the property of simple connectivity (Axiom G3).

But what of states from different series? As defined, G is an intra-monadic relation: it only relates states of each individual series as ordered by the law of that series. As we shall see, however, Leibniz extends the notion of ground containment so that it may also relate states x and y in different monadic series in the same world. Now two states of substances are in the same possible world if and only if the substances are *compossible* with one another. (We will define compossibility in due course; for now we take it as a primitive.) A world W_m will then be the aggregate of all monadic series of states S_n that are compossible with S_m. Let us then begin again, with a new relation R of "involving the reason for" defined on W_m, the aggregate of all monadic series compossible with series S_m, and lay down axioms for this relation. Here we may note that if it is only pairs of states in the same series that are related by R, then the first two axioms will automatically apply to all the states in a given world. For in that case if xRy, then x and y will be in the same series, and if xRy and yRz, then x, y, and z will be in the same series. But it still makes sense to assume that the first two axioms will hold even when R is extended to apply to all states in a given world (Axioms 1 and 2): for, even if the reason for succeeding states is not given only by the law of each individual series, there must (as we shall see) be an analogous foundation for the order of reasons in

[20] In his (1970), van Fraassen gives a nice analysis of Leibniz's idea of incompatibility, relating it to its Aristotelian precedents.

[21] Cf. "Monadology" §22: "And since every state of a simple substance is a natural consequence of its preceding state, the present is pregnant with the future" (GP VI 610).

[22] Although this is indeed natural, given Leibniz's many pronouncements to this effect, it is not wholly unproblematic. For he is capable of saying also that "it is essential to substance that its present state involves its future states *and vice versa*" (to de Volder, January 19, 1706; LDV 333). As we shall see in the next section, Leibniz does tackle this issue: for one state's to involve another, taken simply, is a mutual relation; but the kind of involvement underlying causation also involves not only priority by nature but the existence of intervening changes, which renders the relation of one state's "involving the reason for" another both asymmetric and transitive.

the whole world of compossibles. Simple connectivity, of course, will no longer hold: two monadic states neither of which involves the reason for the other may be in different series. In fact, it seems clear from what Leibniz says in the *Initia Rerum* that if neither state involves the reason for the other, they do not involve each other's opposite, and are therefore simultaneous. This motivates the definition of one state's involving the opposite of another iff either of them involves the reason for the other (Def. 1). Simultaneity is then defined à la Leibniz as non-opposition or compatibility (Def. 2): two states are simultaneous iff neither involves the other's opposite, i.e. if neither involves the reason for the other (Theorem 4).

But in order for this strategy to give us unique and well-defined classes of simultaneous states, we need some axiom to connect compatible states of different monads. For what we cannot rule out with the axioms we have so far is the situation where one state is compatible with (and thus simultaneous with) a second, and the second with a third, even though the third is incompatible with the first. In other words, it is possible for simultaneity to be non-transitive. Now (as John Winnie first pointed out) it is precisely this possibility that Leibniz precludes by invoking the "connection of all things" in the continuation of the passage from the *Initia Rerum* quoted above:

> My earlier state involves the reason for the existence of my later state. And since, because of the connection of all things, my earlier state involves the earlier state of the other things as well, it also involves the reason for the later state of these other things so that my earlier state is in fact earlier than their later state as well. And therefore *whatever exists is either simultaneous with, earlier than, or later than some other given existent.* (*Initia Rerum*, GM VII 18)

One state "involves" a second one if the second can be inferred from the first (albeit by an infinite series of reasons). It follows that they are not incompatible, i.e. are simultaneous. So the gist of this passage in captured by Axiom 3: If x does not contain y's opposite, and y involves the reason for z, then x also involves the reason for z. What this axiom guarantees is that (so long as there exists a plurality of monads and their states), any of these states is a member of a unique class of states all of which are compatible with it. Now the quality which all these states have in common is that they are simultaneous, i.e. occur at the same moment of time. To paraphrase what Leibniz said about how Euclid dealt with ratios, instead of defining what a moment is, Leibniz has been content to say what it is *to occur at the same moment*. A moment is therefore effectively an equivalence class, the class of all states simultaneous with a given state. (I reserve the term "moment" for such a simultaneity class, not "instant," since we have not yet shown its punctual nature. In the last part of Appendix 1 (A 1.5), I show how Leibniz's strictures about change and continuity entail the punctual nature of moments, and thus their equivalence to instants, properly speaking.)

This, of course, is very similar to how Russell proceeds in his *Our Knowledge of the External World* of 1914. There he defines an instant as a group (i.e. class) of possibly overlapping extended events, which are "such that no event outside the group is simultaneous with all of them, but all the events inside the group are simultaneous with each other" (Russell 1914, 126). He then lays down axioms on the relation of "wholly precedes" sufficient to guarantee that they form a series, and this enables him to "say that an event is 'at' an instant when it is a member of the group by which the instant is constituted" (127).[23] The correspondence is obvious: both Leibniz and Russell abstract from the entire class of simultaneous states what they have in common, namely their membership in this class: to occur at a given time is to be a member of such a simultaneity class.

There are, however, some salient differences which we will come to later. Russell holds (quite defensibly, in my view) that what happens are extended events;[24] for Leibniz, what happens are changes; a state is an interval between changes, with other changes occurring in it. Also, as we will discuss in detail in §1.4 below, whereas instants for Russell are classes of actual events, out of which time is constituted, for Leibniz instants are the times at which any possible changes could occur. Changes divide the continuum, and actual instants designate points in the continuum at which change actually occurs; but instants considered in themselves are entities abstracted from the changes occurring in them. As he tells Clarke, "instants considered without the things [that happen at them] are nothing at all, and . . . consist only in the successive order of those things" (to Clarke III, §6: GP VII 364/LC 27). Thus time is an abstract continuum, in contrast to the duration of each thing constituted by the series of its states, divided by a particular sequence of changes. Moreover, the axiom of connection ensures that the orderings of states of all the individual monads in any one world correspond with one another, so that there is just one order of reasons governing each possible world. Correspondingly, there is just one temporal ordering in any world, and, as Leibniz claims, this is a *total ordering*: any given state is either simultaneous with, earlier than, or later than, any other state in the world.

Returning to our initial problem, I think the above analysis vindicates the consistency of Leibniz's theory against the charges of vicious circularity. Certainly, there is no circularity in defining temporal succession in terms of the relation of reason-inclusion, nor simultaneity in terms of relations defined in

[23] Russell gives two ways of demonstrating the punctual nature of the instants he has defined—see his (1914, 125–9). The first is in terms of the notion of the "initial contemporaries of a given event—all those events which are simultaneous with that event, and do not begin later" (127). Hellman and Shapiro (2018) adopt a similar strategy for defining points or instants in a "gunky continuum." The second is in terms of Whitehead's notion of an "enclosure relation." My argument for the punctual nature of Leibniz's moments in A 1.5 follows a similar strategy.

[24] In my book on the philosophy of time (Arthur 2019c) I argue (along with Whitehead and the Russell of 1914) that all events are indeed extended, and that point-instants are not ontologically basic, but rather useful constructions.

terms of this. For once Leibniz's axioms are accepted, it follows that if one state involves the reason for another then it precedes it in time, and that any two states of compossible monadic series which do not involve the reason for each other are thereby simultaneous. Thus Russell is wrong in claiming that time is presupposed in Leibniz's treatment of substance through the reference to future states, and that one state can only be defined through reference to an intra-monadic time. Similarly, since instants of time are only defined in terms of the compatibility of the states of *different* monadic series, there are no intra-monadic instants, and therefore, *contra* Rescher, there is no "private intra-monadic time," despite the fact that each monadic state does indeed occur at some given time (is a member of some equivalence class of compatible monadic states).

Moreover, the preceding account allows us to appreciate the intimate connection between Leibniz's theory of time and the pre-established harmony. Hegel had criticized Leibniz for declaring the absolute independence of monads from one another, and then artificially pasting on harmony afterwards as externally imposed by God.[25] Similarly, Russell referred to "the paradoxes of the pre-established harmony" (Russell 1900, 15) in which Leibniz embroiled himself by insisting that "every relation must be analysable into adjectives of the related terms" (Russell 1900, 46), making his assertion of the plurality of substances self-contradictory. On the above interpretation, by contrast, Leibniz does not deny the relatedness of the states of different substances. Rather, this relatedness is the essence of pre-established harmony, since each monadic state is a representation of the universe from its own particular point of view. This is clear in the passage whose beginning was quoted above:

> Since the nature of every simple substance, soul, or true monad, is such that its succeeding state is a consequence of the preceding one; it is there that the whole cause of the *harmony* is to be found. For God has only to make a simple substance be once and from the beginning a *representation of the universe*, according to its point of view; since from this alone it follows that it will be so perpetually, and that all simple substances will always have a harmony among themselves, because they always represent the same universe.
>
> (To Clarke, V §91; GP VII 412/LC 85)

In other words, the fact that a state is a representation of the same universe from its own point of view guarantees that there exists a unique class of

[25] As Hegel wrote, "There is therefore a contradiction present, which remains unsolved in itself—that is, between the one substantial monad and the many monads for which independence is claimed—because their essence consists in their standing in no relation to one another" (Hegel 1896, 342). He then derides Leibniz's claim "that it is God who determines the harmony in the changes of individuals" as an evasion producing only "an artificial system" (Hegel 1896, 348). See Arthur (2018b) for discussion.

possible monadic states that correspond with it, or are in harmony with it, and are thereby simultaneous with it. And since one state involves the reason for all states subsequent to it in the same monadic series, it thereby involves the reason for all future monadic states of the same universe. This is an unambiguous statement of the axiom of connection for all the monadic states of any given universe.[26]

Now, the fact that states representing the same universe from different points of view are harmonious suggests that there may be a more economical way of defining simultaneity than the one given above in terms of non-opposition. If states which are representations of the same universe from their own points of view are necessarily simultaneous, one can define simultaneity directly in terms of monads' perceiving or representing the same state of affairs. Jan Cover has outlined such an alternative approach to Leibniz's theory of time in his (1997), which he calls the "Second Version."[27] I give my own exposition of this version in Appendix 1b.

We assume that there exist objective phenomenal *states of affairs* represented or expressed by each monadic state. In fact, what each monadic state expresses, perceives or reflects (Cover prefers another synonym, "*contains*") is "the same state of the universe." Let us call this, following Cover, a "world state" (Cover 1977, 312); the idea is that the monadic states express these world-states partially and confusedly from their own particular point of view, and that this is what accounts for the difference among the diverse but harmonious monadic states. The aggregate of harmonious states is *maximal* in the sense that there is no representation of the same world-state from a different point of view not included in the aggregate.[28] The gist of the construction is that "two states are simultaneous iff the world-state contained by one is identical with the world-state contained by the other" (312). Since identity is symmetric, transitive and reflexive, it follows that simultaneity will be too, establishing at a stroke that simultaneity is an equivalence relation. Now *G* can be interpreted as the relation between a representation from a single point of view and all future representations from that point of view: given one such representation or state and the law of that series, all future states in that series will be entailed. And given that each state represents the same

[26] Cf. also Leibniz's letter to de Volder of 1704/1/21: "But all individual things are successive ôr subject to succession . . . Nor for me is there anything permanent in them other than that very law which involves a continued succession, the law in each one corresponding to that which exists in the whole universe" (GP II 263/LDV 289).

[27] I have given my own construal here, as I do not follow the details of Cover's. According to his (1997, 312), we "help ourselves to states of affairs, and define a *world-state* as a maximal state of affairs." It is not clear to me, if a "state of affairs" is phenomenal, what it is that is maximized here. Moreover, monadic states are said to "contain" (= express or represent) world-states; whereas world-states are said to "include" states of affairs.

[28] Cover claims that it is the world-states themselves that are maximal: "Since they are maximal, world states *x* and *y* are identical iff neither includes a state of affairs the other does not" (1977, 312).

world-state ("reflects the same world state"[29]), as do all simultaneous states, the pre-established harmony is guaranteed.

Cover's view is that this second way of construing the Leibnizian theory reflects the way Leibniz explicates "common judgments about pseudo-causal change" in terms of "two (let us call them) metaphysical principles—real ('immanent') causality of the intra-substantival sort, by which new states of a substance arise from previous ones in accordance with an inner law of the series, and inter-substantival expression or reflection" (307). On this basis he criticizes the original construal I gave of Leibniz's theory of time in Arthur (1985) as "awkwardly uneconomical" (Cover 1977, 312) in that, in extending the relation *G* from applying only within diverse monadic series to applying to all states, it then becomes a disjunction of the original "real causal" relation *G* and a relation of "reflection of real cause": a disjunction of an intra-monadic relation with an inter-monadic one. The Second Version is thus to be preferred as a "less disruptive way of defining simultaneity" (312), since it "accomplishes the definition of simultaneity by more economical means than the previous version, appealing to the single relation *contains the same world state* rather than the two relations of *real cause* and *reflects a real cause*" (312–13).

Deferring the question of *G* as a "real cause" for the moment, I want first to examine the claim that, given that it is properly only an intra-substantial relation, it is therefore a misconstrual to extend it to apply to states from different monadic series. To these criticisms I would reply on two counts.

First, Cover's "Second Version" does indeed give a more direct account of simultaneity, but the overall theory of time can hardly be described as more economical: it requires the postulation of a series of world-states, each of which is reflected by all the states in each simultaneity class. I have no objection to imputing to Leibniz a belief that there exist objective states of affairs or "world states" that each state of a monad reflects from its own point of view, and that these states differ at different times. The question is whether this existence assumption needs to be built into the theory of time. It surely detracts from the ontological economy of the theory to have to make this assumption, especially when the theory can be made to work without it. And it requires the positing of two basic relations, "real causation" and "reflection," where I have assumed only one.[30]

[29] Cover acknowledges Franklin Mason for suggesting the construal in terms of "reflects the same world state as" (311, n.40).

[30] As we will see in section 1.4, there is a motive behind this: Cover, like Rescher and Mates before him, believes that Leibniz's doctrine of the ideality of relations means that time, as a relation, supervenes on denominations intrinsic to the individual monads. If one believes with Rescher that intra-monadic relations are exempt from this reduction, Mates' model for Leibniz's mirror thesis in terms of intrinsic properties possessed by individual monads suggests that the relation of "expression" need not be taken as basic, but supervenes on these properties.

Second, this criticism fails to appreciate that it is not I but Leibniz who introduces the idea of states from differing monadic series involving the reason for each other, in what he refers to as the "interconnectedness of all things," which I have formalized as the Axiom of Connection. As I pointed out in my (1985),[31] and have made more explicit in the construal in Appendix 1, the introduction of this axiom can only occur if the relation 'involves the reason for' is reinterpreted not as restricted to states within individual monadic series (as is *G* in Appendix 1), but as holding between pairs of states across all series (as is *R*).[32] Cover presumably calls *R* "pseudo-causation" when it relates states of different monads because there is no causal action ("real causation") between states of different substances. But clearly Leibniz did not think that this precluded such states as being *related* by *R*. A substance acts through its changes of state, and according to the Axiom of Connection its earlier state contains the reason for its future states and those of other monadic series alike.[33] But according to the "Second Version" outlined by Cover, this axiom will not hold, since "involves the reason for" is interpreted by him as a "real causal" relation that is purely intramonadic. This failure to account for Leibniz's own statement of "the connection of things" must therefore count against this "second version" of Leibniz's theory.

Turning now to Cover's description of the relation *G* as a "real causal" relation internal to each monadic series, I do not think that this usage is in keeping with Leibniz's notion of causation. Active force is not a causal action of states upon states. One state of a substance may be said to be a "real cause" of another of its states in the sense that it involves the reason for it; but it does so by virtue of the law of the series of that individual substance, and the substance's appetition towards its future states, not by virtue of the first state's acting on the second.[34]

[31] "Assuming now that we can somehow extend this conception of incompatibility to the set of states of all the monads in a given universe or world *W*..." (Arthur 1985, 302); "What we need, in fact, is some way of extending the concept of incompatibility across states of *different* monads" (303).

[32] One can express Cover's criticism about "pseudo-causality" failing for individual monadic series as follows: according to Axiom P3, defined on the states of a monadic series S_m, if neither of two states involves the reason for the other, $\neg(xGy \vee yGx)$, then the states are identical: $x = y$; but if the relation of involving the reason for is redefined as *R* on W_m, then $\neg(xRy \vee yRx)$ instead implies non-opposition, $\neg xOy$, by Definition 1, and therefore simultaneity, xSy, by Definition 2; on the assumption of Axiom 3. So *R* is not the same as *G*; but, I would insist, there is nothing 'pseudo' about it.

[33] In the same vein, Anfray (2007) says of the relation "containing the reason for": "Cette relation semble à première vu assez différente de la relation causale et pourrait suggérer une certaine précaution quant à l'attribution d'une théorie causale du temps à Leibniz..." (2007, 102).

[34] There has been some controversy among commentators whether states or perceptions bring about succeeding states by themselves (the "efficacious perception view"), or whether substances, endowed with a primitive force of action, are what are causally efficacious (the "monadic agency view"). (See Whipple 2011, 388–89, for a discussion and references.) I concur with the reading of Don Rutherford, as reported by Whipple, according to which one state or perception produces another by virtue of its derivative force (which is a determinate expression of its primitive force); so that a given state is "only causally efficacious to the extent that the substance itself is" (Rutherford 2005, 165; Whipple 2010, 389); although I would add that such causal efficacy does not involve an action of one state on another.

A substance acts through its changes of state, which it produces "from its own store"; but its earlier state contains the reason not only for its future states but also those of other monadic series in the same world.[35] Moreover, if a state of a monad is the result not only of an earlier state with some tendency in the same monad, but of states with possibly opposing tendencies in all the monads in that world, then perhaps "real causality" is not simply "intrasubstantial" as Cover contends (1997, 311), but in fact intersubstantial.

But these matters concerning the attribution to Leibniz of a causal theory of time deserve a full discussion in their own section.

1.2 The Causal Theory of Temporal Order

Leibniz explicitly sketches a causal theory of time in manuscripts probably composed between summer 1683 and early 1685. In one of these he writes:

> Change occurs if two contradictory propositions are true of the same thing, and then the two propositions are said to differ *in time*. . . .
>
> If one thing is the cause of another, and they are not able to exist at the same time, the cause is *earlier*, the effect is *later*. Also earlier is whatever is simultaneous with the earlier. (A VI 4, 568)[36]

and in a passage from another piece written in mid-1685:

> Those things are *simultaneous* one of which is the condition of the other absolutely.[37] Whereas, if the first is the condition of the second by an intervening change, then the first is *earlier*, the second *later*. Now the earlier is understood to be that which is simultaneous with the cause, the later that which is simultaneous with the effect. (A VI 4, 628/LLC 273–75)[38]

[35] I might add that this idea of earlier things being conditions or requisites for all later ones in the same possible world, not just states of the same substance, is not unique to the *Initia Rerum*. As we shall see in §1.3, it is made explicit by Leibniz in a draft of his theory of time in 1685, and it also underlies Leibniz's conception of "laws of general order." These laws determine the notion of the whole possible world, and are thereby the foundation for one state's involving the reason for all succeeding ones in the same world, and thus for the Axiom of Connection.

[36] This is from a manuscript the editors have titled *Genera Terminorum. Substantiae*, A VI 4.

[37] As Heinrich Schepers points out (Schepers 2018, 413), this definition is anticipated in Leibniz's reflections in *De Magnitudine* of Spring 1676 (A VI 3, 484): "It must be seen whether we shouldn't say that those things are simultaneous of which, if one exists, the other also exists."

[38] This and the following passage are from my translation volume. The first is excerpted from the *Divisio terminorum ac enumeratio attributorum* (A VI 4, 559–65/LLC 265–71), dated summer 1683 to early 1685, the second from the *Definitiones notionum metaphysicarum atque logicarum* (A VI 4, 627–9/LLC 271–5), dated mid-1685.

In a third manuscript from the same period he describes the earlier state as being 'prior by nature' to the later state, rather than its cause:

> Next, from two contradictory states of the same thing, that is *earlier in time* which is prior by nature, i.e. which involves the reason for the other, or what amounts to the same thing, which is more easily understood. For example in a clock, in order to understand completely the present state of its hands, it is required that we understand its reason, which is contained in the preceding state; and so on. (A VI 4, 563/LLC 269–71)

Taking such passages together, it seems clear that Leibniz is equating one thing's *being a cause of another* with its being a more easily understood condition of it, or *involving a simpler reason than is involved in the other*. This is confirmed by other manuscripts of this period, where he defines a cause as a requisite or "a simpler condition, or, as they commonly say, prior by nature" (A VI 4, 627/LLC 271).

This appeal to priority by nature accords with what may be Leibniz's earliest statement of his causal theory of time, from *Elementa juris naturalis* of the second half of 1671 (?):

> A *cause* is a producer prior by nature to what is produced (*inferens natura prius illato*). There are producing things which are posterior to the things produced, for an effect often produces the cause. When I say: if *A* exists, then *B* also exists, then *A* is a *producer*, B a *produced. Naturally prior*, although not temporally prior, is whatever can be clearly thought before the other, whereas the other cannot be thought before it. In the same way, temporally prior is whatever can be sensed before the other, whereas the other cannot be sensed before it.
>
> (A VI 1, 483)[39]

We will return later to the distinction Leibniz makes here between priority in time and priority by nature. First, let us concentrate on *priority by nature*. As Stefano Di Bella (2005) and Michael Futch (2008, 109) have both noted, this notion harks back to Aristotle, who in his *Metaphysics* 1019a3 defined as prior by nature "those things which can be without other things, while the others cannot be without them." This should be compared with Leibniz's definition, in a paper dating from 1671–72, of a *requisite* as "that which, not being given, the thing does not exist." A requisite is thus a necessary condition for a thing's existence.[40]

[39] I am indebted to Osvaldo Ottaviani for bringing this passage to my attention (email of November 21, 2020).

[40] In his early writings, Leibniz follows Spinoza in conceiving everything as following, by logical necessity, from the first cause as the requisite for all existents. After meeting with Spinoza at the end of 1676, he realizes that he must articulate the objective contingency of existential priority (requisites for existing), as opposed to the logical implication involved in essential priority (conceptual requisites). On this, more below; but see also Ottaviani (2016), and my account in chapter 4 of Arthur (2014).

Despite the Aristotelian provenance of priority by nature, however, Leibniz's appreciation of the importance of the notion of *requisite* might well have been due to his assiduous studies of Hobbes in the early 1670s. For the latter definition of it occurs in the course of a demonstration of the Principle of Sufficient Reason, a demonstration that he repeats in an appendix to the *Theodicy* with explicit acknowledgment to Hobbes.[41] In the 1671–72 paper, having given that definition of requisite, Leibniz argues that whatever exists has all its requisites, since if it lacked any of them it would not exist. Now, if a "sufficient reason" is defined as "that which, being given, the thing exists," all the requisites constitute a sufficient reason for the thing's existence. It follows that "whatever exists has a sufficient reason."[42] Leibniz gives a similar account in a wide-ranging table of definitions edited by Couturat, and dated by him as from 1702–4 (C 437–509). In this manuscript, as Futch observes in his lucid discussion (2008, 107–15), Leibniz defines a necessary condition (*suspendens*) as "that which, unless it is posited, the other thing is not posited," and a requisite as a "*suspendens natura prius*," i.e. a necessary condition that is prior by nature to that for which it is a condition (C 471; Futch 112).

Thus there is more to priority by nature than simply being a necessary condition. Leibniz examines how it should be explicated in a paper of 1679, *Quid sit natura prius*. There he notes that if it is understood to mean simple "involvement" (i.e. entailment), then two states could be prior by nature to each other, thus undermining the asymmetry of the relation.[43] He writes:

> There is some difficulty in explaining what 'prior by nature' might be. For, just as the later state of any substance involves the earlier, so in turn the earlier state involves the later, since each can be known from the other. Whence it seems to follow that to be prior is not to be simpler than the later, but both involve the same things, and there is a kind of equivalence between them. (A VI 4, 180)

The solution Leibniz offers is to distinguish simple involvement[44] of *A* in *B* from ontological priority, which derives from *A* being "more easily understood" than *B*.

[41] In the essay, "Reflections on the work that M. Hobbes has published in English on Freedom Necessity and Chance," Leibniz writes: "He shows very clearly that . . . for each effect there must be a concurrence of all the sufficient conditions anterior to the event, not one of which, evidently, can be lacking when the event is to follow, because they are conditions; and that the event no more fails to follow when these conditions all occur together, because they are sufficient conditions" (GP VI 389/ H 394–5).

[42] (A VI 2, 483); for discussion, see Parkinson (DSR xxiii) and Arthur (2014, 91ff.).

[43] This difficulty, and Leibniz's solution to it, have been noted by both Di Bella (2005, 246–7) and Futch (2008, 111) in their discussions of Leibniz's causal theory of time.

[44] Leibniz is not always consistent in his terminology. In these papers from the mid-80s and elsewhere, he writes of simultaneous states being such that one involves the other, whereas in the *Initia Rerum he writes that "If several states of things are supposed to exist, none of which involves the other, they are said to exist* ***at the same time***" (GM VII 18). In the latter statement I have interpreted him to mean "none of which *involves the reason for* the other," and, as I shall argue below, "involves the reason for" needs to be interpreted as meaning "is the *mediate requisite* of."

"There are often many properties of the same subject," he writes, "one of which is more easily discovered and demonstrated, and nonetheless they are all reciprocal, and thereby involve the same things" (A VI 4, 180; Futch 2008, 111). Thus the idea of priority by nature, like the notion of cause itself, carries an aetiological meaning: for *A* to be prior to *B*, it must be more easily understood or more easily demonstrable than *B*:[45]

> And so prior by nature is that whose possibility is more easily demonstrated; ôr, that which is more easily understood. Of two states one of which contradicts the other, that is prior in time which is prior by nature. Two incompatible ôr contradictory existences differ in time, and that is earlier or later in time which is prior or posterior by nature. (A VI 4, 181)[46]

This relates to an objection that has been raised by Michael Futch which also turns on an alleged symmetry of the notion of priority by nature. Equating "being a cause of" with "being prior by nature to," he alleges that for Leibniz *a cause may be simultaneous with its effect*: "Leibniz denies that things prior or posterior by nature or causally related are *ipso facto* incompatible and non-simultaneous" (119). So temporal priority can be defined in terms of causal priority only for those states that have already been established as simultaneous. Futch claims that this undermines accounts of Leibniz's theory that define simultaneity in terms of causation—such as the ones Cover (in his "First Version") and I gave in our papers:[47]

> According to Leibniz, for any given instant of time containing an infinite plurality of monadic states, it is in principle possible that some or even all of those states stand in relations of natural priority or posteriority to one another (at the same time). As a result, a straightforward transposition of Winnie's analysis onto Leibniz inevitably leads to a fundamental error. (Futch 2008, 119)

45 These remarks can be read as comments on Aristotle's account in the *Posterior Analytics* Bk 1, ch. 2 (I am indebted to Lucia Oliveri for this reference). There Aristotle writes of the premises of an argument as 'causes of the conclusion' (71b30), and cautions that "'prior' and 'better known' are ambiguous terms, for that which is prior by nature is not the same as that which is prior in relation to us, and that which is (naturally) better known is not the same as that which is better known to us" (71b33–72a1).

46 Cf. the following from a fragment on temporal terms, probably from some time after 1700: "In a change, that of two contradictory states is *earlier* in which there is a reason for the other (the *later*)" (LH IV 7C, Bl. 92; I am indebted to Osvaldo Ottaviani for sending me his transcription of this fragment).

47 See Arthur (1985, 269–70, and Appendix), where I define simultaneity in terms of mutual lack of ground. In his "First Version" (1997, 310–12), Cover interprets this as amounting to "pseudo-causality," "a kind of universal vicarious causal connection, riding piggy-back on real causality plus reflection (expression)," and offers his "Second Version" (312–13) as a preferable alternative interpretation.

If this is correct, moreover, it also constitutes a serious difficulty for Leibniz's own theory. For if simultaneity has to be defined independently of cause, this seems to take us outside the remit of a strictly causal theory of time. Futch cedes the point:

> Employing the qualitative temporal relation of simultaneity between A and B seems to undermine Leibniz's causal theory of time insofar as the grounding base—the causal structure of the world—must be supplemented by this specifically temporal relation in order to provide a comprehensive analysis of things that are temporally (but not causally) related. (124)

It must be admitted, I think, that not everything Leibniz wrote about the causal theory is mutually consistent.[48] He did not publish a definitive theory, and the disparate attempts he made on separate occasions do not always agree in all specifics. Nevertheless, I think it is possible to construct a coherent account that is not susceptible to Futch's criticisms. The basic idea is consistent with what Leibniz had already written back in 1671: what is prior by nature concerns essences, not existents; essences are not temporally ordered, whereas existents are. Priority among existents concerns *requisites for existing*, not conceptual requisites. If one existent can be perceived before another, then the former must be prior, and separated from the latter by intervening changes. As Leibniz wrote in 1671,

> *Naturally prior*, although not temporally prior, is whatever can be clearly thought before the other, whereas the other cannot be thought before it. In the same way, temporally prior is whatever can be sensed before the other, whereas the other cannot be sensed before it. What is prior by nature is prior essentially, what is temporally prior is prior existentially. We estimate essence through thought, existence by sense. Thus the efficient cause is prior to the effect in time, but action is only prior to passion by nature. (A VI 1, 483)

I am not convinced, therefore, by the evidence Futch adduces for his claim that Leibniz held that a cause could be simultaneous with its effect. I have criticized the details of his arguments elsewhere.[49] The main point I think he misses is the

[48] For instance, as Osvaldo Ottaviani has objected (private communication), there seems to be some circularity in Leibniz's definitions of natural priority, in that "natural priority enters into the characterization of requisites, but, at the same time, when he has to explain what is naturally prior, Leibniz ultimately resorts to what has fewer requisites." As we have seen, in *Quid sit natura prius*. Leibniz says that "prior by nature is that whose possibility is more easily demonstrated; ôr, that which is more easily [*facilius*] understood." But Leibniz also characterizes '*facilius*' as what has fewer requisites (A VI 4, 29, 303).

[49] See Arthur (2016). Futch presents three main pieces of evidence. The first is from Leibniz's appeal in the *TMA* to the Scholastic doctrine of signs, one of which may be prior to the other in the same instant. Second, Leibniz then appears to uphold this doctrine in the *Theodicy* (§388–90) when, having

relevance of a distinction to which he himself draws our attention in his own exposition of Leibniz's analysis of conditions. This is the distinction Leibniz makes between *mediate* and *immediate requisites*: of these, only mediate requisites are causes. As Leibniz states it in the manuscript *Definitiones notionum metaphysicarum atque logicarum* of mid-1685, "Some requisites of things are mediate, and must be investigated by reasoning, such as causes; others are immediate, such as parts, extrema, and generally things which *are in* a thing" (A VI 4, 627; Futch 2008, 112). The distinction goes as follows.

When one thing is a condition of another by an intervening change, the first is a *mediate requisite* for the second (A VI 4, 628), and thus a cause of it. "A cause," he specifies further, "is a requisite according to that means [*modum*] by which the thing is produced. I would prefer to call it an efficient cause" (A VI 4, 629; Futch 2008, 123). This aligns with his earlier definitions, where a *cause* is defined as "a requisite for that means of producing a thing by which the thing is supposed produced," and a *full cause* as an "*inferens natura prius illato*, ôr what involves all sufficient requisites (that is, those from which the remaining requisites follow)," where it must be noted that "whatever involves all the requisites for a means of producing a thing also involves all the requisites for producing the thing itself" (A VI 4, 563–4; Futch 2008, 112).

A requisite or condition is *immediate*, on the other hand, if it involves no intervening changes. As we saw, the examples Leibniz gives are "parts, extrema, and generally things which are in [*insunt*] a thing" (A VI 4, 627; Futch 2008, 112). The relation of *inesse* ("being in") is of course of very general application in Leibniz's conception. A point may be in a line, and a monad may be in an organic body, in the sense that once the latter is given, the former is immediately understood to be given too; and a concept *A* may be in a concept *B* in the sense that *B* involves or entails *A*. An immediate requisite is a requisite, and so is a condition that is prior by nature; but such priority is one of *essence*, and involves no intermediate changes. Leibniz calls such conditions *absolute conditions*. This is of critical relevance to the theory of time, since simultaneous things are understood by Leibniz to be such that one is a condition of the other in precisely this absolute sense:

granted that "the instant excludes all priority of time, being indivisible," he nevertheless asserts that "the action by which God produces is prior by nature to the existence of the creature that is produced" (GP VI 345–6/T 357–8). But, I contend, Leibniz abandons his earlier theory of priority within an instant, and the priority by nature he is discussing in the *Theodicy* concerns creatures considered in their essence, not in their existence or the order of their accidents. There he is trying to defend God's consideration as involving "a natural order, but no order of time or interval." Futch's third item of evidence is the reciprocity of action and passion that Leibniz discusses earlier in the *Theodicy* (§66). But I do not accept that this implies their simultaneity; that would contradict his account of one thing's acting on another through intervening changes, as I argue below.

> Those things are *simultaneous* one of which is the condition of the other absolutely. Whereas, if the first is the condition of the second by an intervening change, then the first is *earlier*, the second *later*. Now the earlier is understood to be that which is simultaneous with the cause, the later that which is simultaneous with the effect. Or the earlier is understood to be that which is simpler than or what is the requisite of the second. A requisite I have defined as a condition simpler by nature than that whose condition it is. (mid-1685; A VI 4, 628)

Or, as he expresses it more succinctly in *Genera Terminorum* from 1683–85:

> If *B* follows from *A* absolutely, *B* is *simultaneous* with *A*. Likewise those things which follow absolutely from the same thing are simultaneous. On the other hand, those things that oppose one another are not simultaneous.
> If one thing is the cause of another, and they are not able to exist at the same time, the cause is *earlier*, the effect is *later*. Also earlier is whatever is simultaneous with the earlier. (A VI 4, 568).

Two states reflecting the same universe of co-existents, therefore, are conditions of one another absolutely, and therefore simultaneous.[50] If two states *A* and *B* are not simultaneous, and *A* is the requisite of *B*, then it will be its mediate requisite, which means that there will be intervening changes between *A* and *B*. And, according to Leibniz's distinctions, it is precisely such mediate requisites that count as efficient causes, or "producers." Since there is in these cases a series of intervening changes, cause and effect are precisely not simultaneous; for "if the first is the condition of the second by an intervening change, then the first is *earlier*, the second *later*" (A VI 4, 628).

Still, this analysis appears to support Futch's subsidiary point, that in defining two states as simultaneous if one is an absolute condition of the other, and then defining temporal precedence in terms of causation among non-simultaneous states, Leibniz is not giving a purely causal theory of time. The same would apply to Jan Cover's "Second Version" of Leibniz's theory of time, described in section 1.1 above, in which Cover explicitly begins by defining simultaneity first, before defining temporal precedence in terms of causal precedence. In fact, this Second Version fits very well with the above definitions given by Leibniz. According to Cover, two states are simultaneous iff they express the same world-state (Cover 1997, 312). If we interpret "expressing or containing the same world-state" as meaning that the two states "follow absolutely from the

[50] One difficulty that the account I have given does not resolve is that of the asymmetry of being a requisite. It is one thing for two simultaneous states to be requisites of each other, and thus necessary and sufficient conditions for one another, without either being temporally prior. But it is still difficult to see how each could be conceptually prior to or simpler than the other. (I am indebted to one of the readers of my manuscript for noting this.)

same thing" (or that once the world-state is given each state immediately follows), then this would make those states simultaneous according to Leibniz's definitions in the *Genera Terminorum* passage above.

Is Leibniz's theory then not a pure causal theory?[51] As just described, it appears not, as the relation of being an immediate requisite of, or of reflecting the same state of the universe as, has to be taken as basic, in addition to the (causal) relation of being a mediate requisite of states that are not simultaneous. But it is also possible to argue that each state in a given world is an absolute condition of every state reflecting it, since if a state reflects or represents the rest of the world, it must represent or reflect each state in it. The states would then "harmonize exactly among themselves" (GP II 57), and be simultaneous with one another. It also seems clear that Leibniz regards immediate and mediate requisites as mutually exclusive: a requisite for existing must be either immediate (involving no intervening changes) or mediate (involving intervening changes). On this picture, all the states in a given world are requisites of one another, either immediate (reflecting one another) or mediate (each involving the other's opposite). Any state that neither reflects another in the same world, nor involves or is involved by its opposite, must lie outside this world. But this then vindicates the construction given in section 1.1 above, provided "involves the reason for" is interpreted as meaning "is a mediate requisite of." For if two states in the same world involve one another absolutely (are each other's immediate requisites), they are precisely not causally related, since an immediate requisite cannot involve intervening changes; and conversely, if neither state is the mediate requisite of the other, then one must be the immediate requisite of the other, so that they are simultaneous. The very fact that mediate and immediate requisites are contradictories means that the latter can be defined in terms of the former.

According to Leibniz's definitions, a mediate requisite is "a requisite according to which the thing is actually produced" (A VI 4, 563), that is, it is a "producer" (*inferens*), an efficient cause (A VI 4, 629). It is a necessary condition for the thing and prior by nature to it, in the sense that it is "simpler by nature than that whose condition it is" (A VI 4, 628), i.e. that its "possibility is more easily demonstrated" (A VI 4, 181). That makes it a [*sufficient* or] *partial cause*. A *full cause*, on the other hand, is the aggregate of all the partial causes of the thing: "The *full cause* is the producer that is prior by nature to what is produced, that is, that which involves all the requisites that are sufficient (i.e. from which the remaining requisites follow)" (A VI 4, 564). Thus a state, according to Leibniz's causal theory, is not caused only by preceding states in the same monadic series, as Cover maintained when he attributed "real causation" to Leibniz. Rather the states of things external to the thing in question will also cooperate to produce a subsequent state. They do so not

51 See Silva (2016) for an alternative discussion of this question.

by the propagation of any causal influence, but by being necessary conditions for its production that are prior by nature to it, or more easily understood; and by being actual tendencies occurring in each monad.

The latter point is worth stressing against accounts that take Leibniz to be simply identifying the temporal order with the order of reasons, taking the causal theory to be an eliminative reduction of temporal succession to reason-inclusion. What such construals miss is Leibniz's *dynamism*, essential to which is his positing of active force or appetition, in addition to perception, as a defining characteristic of monads. Without appetition there would be no actual tendencies for states to proceed into subsequent ones, just reasons for them to do so. On the construal I am proposing, Leibniz is committed to the reality of becoming, even if this is tempered by his view that entities that are constantly becoming something different, like motions or bodies, cannot be taken to be substances. Substances are enduring entities, it is their states or accidents that are constantly changing: the law of the series is permanent, determining the succession of states, but the states actually come to be in succession.

This account is consonant with Leibniz's general views on causation. As is well known, he does not accept that any one substance has any real influence on any other, construing the phenomenon of causal action of one body on another in terms of the intelligibility of explanatory accounts. We saw above that *A*'s being "more easily understood" than *B* is explicated in terms of its serving better "to explain what happens in" *B*. That tallies with what Leibniz says elsewhere about identifying the cause of motion in physics. The following passage from "A Specimen of Discoveries" (from about 1686–89) is particularly clear:

> And that thing from whose state a reason for the changes is most readily provided is adjudged to be the cause. Thus if one person supposes that a solid moving in a fluid stirs up various waves, another can understand the same things to occur if, with the solid at rest in the middle of the fluid, one supposes certain equivalent motions of the fluid <in various waves>; indeed, the same phenomena can be explained in infinitely many ways. And granted that motion is really a respective thing, nonetheless that hypothesis which attributes motion to the solid, and from this deduces the waves in the liquid, is infinitely simpler than the others, and for this reason the solid is adjudged to be the cause of the motion. Causes are not derived from a real influence, but from the providing of a reason.
>
> (*Specimen inventorum*, A VI 4, 1620/LLC 311)

We will be discussing the relativity of motion in the third part of this book. There we will see that Leibniz ascribes relativity to motion insofar as it is conceived geometrically, that is, as change of situation. Nevertheless, it is possible to identify the cause of motion, Leibniz claims, by appeal to the most intelligible hypothesis. In the case of a boat, say, moving relative to the water, the water can equally be described as moving with respect to the boat, if that is held to be at rest instead. Still, it is 'infinitely simpler' to attribute the motion to the boat than to regard all the waves converging on the boat in just such a way as to cause it to move. So, by appeal to the most intelligible hypothesis, we identify the boat as the cause of the motion. Now, having identified the boat's being set in motion as the cause of the ripples in the water, we know that its states at some earlier time are not only earlier than its own later states, but also contain the reason for the water's later states.

This account of physical causation also fits with Leibniz's accounts of what it means for one thing to act on another. In the "Monadology" he explains that although "one created monad could never have a physical influence over the interior of another" (§51, 615/275), a "created thing is said to be active externally insofar as it has perfection, and to be passive towards another insofar as it is imperfect" (§49, 615/274). That is, a created thing can truly be said to be *active externally*, to *cause* something to occur in another, "insofar as what can clearly be understood in it serves to explain what happens in the other" (§52, 615/275).

The above considerations are relevant to Leibniz's defence of contingency as well as to his general theory of causation. Given existing conditions, he holds, a future state or event will be certain to happen, but that does not mean it is logically necessary that it should happen. Thus if a state *a* of a monad involves the reason for—or is a partial cause of—the existence of another of its states *b*, then by virtue of the monad's active force and the law of its series, *b* is certain to occur; but it is not logically necessary that it should occur. The same goes for states of even a composite thing like the hand of a clock, as Leibniz makes clear in the continuation of the passage from the *Definitiones notionum metaphysicarum atque logicarum* quoted from earlier:

> Next, from two contradictory states of the same thing, that is *earlier in time* which is prior by nature, i.e. which involves the reason for the other, or what amounts to the same thing, which is more easily understood. For example in a clock, in order to understand completely the present state of its hands, it is required that we understand its reason, which is contained in the preceding state; and so on. And it is the same in any other series of things, for there is always a certain connection, even though it is not always a necessary one.
>
> (A VI 4, 563/LLC 269–71)

The same thing may also be expressed in terms of final causation, since according to Leibniz final and efficient causal explanations are complementary ways of expressing the same phenomenon.[52] Thus a monad in some state *a* inclines

[52] Or, as Leibniz very succinctly puts it in the *Definitiones notionum* of 1685: "If A does B because he wants C, A will be the efficient cause, B the means, and C the final cause" (A VI4, 630). The teleological

towards a state *b* for the sake of some end, without this necessitating *b*. The action of the internal principle that brings about change does not always bring about the state to which it tends. Nonetheless, "it always obtains some part of it, and attains new perceptions [states]" ("Monadology"; GP VI 609/WFT 269). A state with a given tendency does not always bring about the result towards which it tends because there are other monads with possibly opposing tendencies, and these manifest as resistances to the monad's actions which become more clearly expressed in its future states. In fact, given the axiom of connection, the earlier state of every other monad in the same world involves the reason for any given state. Even so, among these states some may express more intelligibly the reason for this state, in which case we may identify them as its cause. Such requisites are *partial causes*, and all such requisites that are sufficient for the state to be produced constitute its *full cause*.

This notion of *full cause* is of the greatest importance for Leibniz's physics. For it is integral to the metaphysical principle that he takes as the foundation of his dynamics, the Principle of Equipollence, according to which "the entire effect must always be equal to the full cause." It is this principle, Leibniz claimed, that was his "Ariadne's thread" leading him to identify the correct notion of force involved in the conservation law, what we now call the conservation of energy. But it is important to note that the full cause involves the cooperation of other existents, whose states are the "concomitant requisites" constituting the cause.[53] They involve the conditions necessary for the production of the effect, including the initial conditions. Now, it is the sum total of compossible existents that makes a world. But this is a contingent world, because the laws governing the world are contingent. They are not necessitated, but depend on God's having selected the world for creation. The connection of cause to effect therefore involves the laws specific to each "series of things" or world of which they are constituents, laws which are contingent on God's choice. They are necessary only on the supposition of that choice, they are *hypothetically necessary*. It follows also that things that are necessary or absolute conditions of one another, or simultaneous, are so also on the hypothesis of God's having selected this world. Leibniz confirms this explicitly in the *Definitiones notionum metaphysicarum atque logicarum* of mid-1685:

> That is *earlier in time* which is incompatible with some position, and is simpler than it. The other position is called *later*. Those things are *simultaneous* which are by supposition co-necessary. I say, by supposition, i.e. with the series of things posited.

reading of cause is the one we naturally adopt from the inside, since we have access to our own reasons and our own desires and aversions; but this is compatible with the efficient causal reading, in which our endeavours (ideally) compete and cooperate externally with those of other substances.

[53] "And what we call causes are in metaphysical rigour only concomitant requisites" (*Principia Logico-Metaphysica*, A VI 4, 1647).

> ... clocks do not make earliness and lateness, they merely indicate it.... But the root of time is in the first cause, potentially containing in itself the succession of things, which makes everything either simultaneous, earlier or later.
>
> (A VI 4, 629/LLC 275)

It may be objected that in the preceding discussion I have freely passed back and forth between talking of monadic states, and of states of things in general, such as characterize composite bodies, phenomena. This would especially be objected by those who conceive of monads and their states as existing in a metaphysical realm, "the ontological ground-floor," distinct from the "phenomenal level." But, I would insist, although Leibniz conceives the changes of composites as arising from the changes among simples, he does not conceive them as existing on distinct ontic levels. What we find, rather is a conception where the states of monads are perceptions of the rest of the world from different points of view, and where the rest of the world consists in the phenomena external to the monad, ordered in space and time. As Leibniz writes in a manuscript essay from around 1712 first published by Couturat in 1903, translated by G. H. R. Parkinson as "Metaphysical Consequences of the Principle of Reason":

> every simple substance has an organic body which corresponds to it—otherwise it should have no orderly relation to other things in the universe.... there would be no order among these simple substances, which lack the interchange of mutual influx, unless they at least corresponded to each other mutually. Hence it is necessary that there is between such substances a certain relation of perceptions or phenomena, through which it can be discerned how much their modifications differ from one another in space or time; for in these two, time or place, consists the order of things which exist either simultaneously or successively. From this it follows that every simple substance represents an aggregate of external things, represented in diverse ways. (C 14/MP 175)

That is, it is not the states of the simple substances themselves that are perceived, but "the aggregate of external things," i.e. phenomena, and their changing relations; and this requires the monads in question all to be embodied. It is through its organic body that a monad is situated spatially, and this body is an (infinite) aggregate or composite of simple substances. Each of these represents the rest of the world from the point of view of its own body, from where that body is situated with respect to the other bodies at each given time.

In the "Monadology" we find Leibniz discussing how in matters concerning representation, "composites are analogous to simples" (§61/WFT 276). The state of each monad is a "representation of the details of the whole universe," but confusedly, and represents most distinctly "those that are either closest or largest in relation to each monad." (Note that this would make no sense unless he were talking about embodied monads.) Analogously, he writes, "each body is affected by the bodies which are in contact with it, and in some way feels the effect of everything that happens to them," and through the whole chain of bodies that are in mutual contact, "feels the effect of everything that happens in the universe."

But there is a problem with this analogy. For according to Leibniz's principles, all such action by contact should be elastic and take a certain time to occur. So there seems to be an incompatibility here with the idea that the mutual reflection of monadic states is instantaneous. Let me therefore say a little more about the notion in play here of "perceiving the rest of the universe."

In the above passage Leibniz says that such representation requires a mutual relation among coexisting perceptions. In the continuation of the essay Leibniz explains this with his famous image of monads as *living mirrors*, where each one reflects all the rest:

> ... Therefore, since every organic body is affected by the entire universe through relations which are determinate with respect to each part of the universe, it is not surprising that the soul, which represents to itself the rest in accordance with the relations of its body, is a kind of mirror of the universe, which represents the rest in accordance with (so to speak) its own point of view. (C 14/MP 175)

Here (as Leibniz explains elsewhere) this mirroring or representing does not require the presence of an image in the perceiving organs. Rather, "it is sufficient for the expression of one thing in another that there should be a certain constant relational law by which particulars in the one can be referred to corresponding particulars in the other" (C 15/MP 176–7). Perception, in this sense of representation or expression, does not require an action, but a relation of correspondence. This relation of expression, namely that of each state's "containing" or "involving" the other, is synonymous with that of being its condition absolutely, as we have seen.

This being so, I believe a distinction needs to be made: perception as it occurs phenomenally must be distinguished from perception as immediate expression or correspondence. When a body "in some way feels the effect of everything that happens to" other bodies, and thus the rest of the universe, this is an effect whose propagation takes time.[54] Insofar as we experience effects of other bodies nearby in one perception, this is still a perception that takes time, and depends on mediate requisites. In fact, as we shall see, Leibniz holds that any phenomenal perception takes time, and is further divided to infinity. If, on the other hand, each perception

[54] As Osvaldo Ottaviani has reminded me, the idea that perceptions take time is also implicit in Leibniz's correction of himself in the piece *De magnitudine* of 1676. Having defined simultaneous things as "those which can be sensed by one action of the mind," he notes that "since the action of the mind itself has an extent [*tractum*], it should be seen whether we should not call *simultaneous* those things such that if one exists, the other also exists" (A VI 3, 484/DSR 41).

or state is incompatible with a state occurring at a different time, and yet reflects every other compatible state, it must be of vanishing duration. States or perceptions in this sense are those that may be taken as absolute conditions of one another. As will be argued in detail in section 1.5, this is possible because of Leibniz's conception of a momentaneous state as one of a finite duration so short that no error will result from its being regarded as instantaneous. A perception as experienced is one in which any changes occurring within it are not perceived, because they are "either too small or too unvarying" to be noticed. They are nevertheless there. The states or perceptions that reflect or mutually condition one another, on the other hand, are momentaneous, and involve the fiction that no further change is happening within them. All this will be further discussed below.

In sum, Leibniz's theory of temporal order is a causal theory, based on his analysis of requisites. For one state to "involve the reason for another" is for it to be a mediate requisite for the other, and such mediate requisites are efficient causes, or "producers." Since these involve a series of intervening changes, cause and effect are precisely not simultaneous. Immediate requisites, on the other hand, are absolute conditions of one another; simultaneous states are of this kind, since each reflects or is in harmony with the other; whereas successive states are contradictory. Since a state is an immediate requisite precisely if it is not a mediate one, simultaneous states may be defined in terms of their being compatible or non-contradictory, as in the analysis given in section 1.1 above. A mediate requisite is defined as being "prior by nature" or "simpler by nature than that whose condition it is" (A VI 4, 628), i.e. such that its "possibility is more easily demonstrated." Such a partial cause is a necessary condition which will result in its effect provided various other conditions cooperate. These other conditions (such as initial conditions, non-interference from other things, etc.) are the "concomitant requisites" required for the effect to take place. All such requisites taken together constitute the full cause. A partial cause therefore only results in its effect contingently on these other conditions. But if all the conditions are in place, the effect will occur with certainty.

The preceding account explains the contingency of states in terms of their being partial causes, necessary conditions, requiring the cooperation of concomitant requisites to be sufficient conditions and to produce their effect. But given the full cause that all these conditions together comprise, the effect is certain. How, then is there any prospect for them to occur or have occurred otherwise than they actually do? Is Leibniz's theory of time not therefore necessitarian? These questions can be re-expressed as follows: once God has chosen to create the actual world, in what sense are other possible worlds really possible? And if all that happens occurs in the order it does because of the pre-established harmony, what sense can be made of temporal contingents and counterfactual statements, of things possibly occurring in time otherwise than they do? It is to these questions that I now turn.

1.3 Contingency, Compossibles, and Counterparts

From the time he met with Spinoza in December 1676 till the end of his life, Leibniz took great pains to distance himself from the Dutch philosopher's views on certain key issues. One of the most important of these was Spinoza's teaching that what exists, exists necessarily—a doctrine that Leibniz also considered to be implicit in Descartes' assertion that, in the course of time, all possibles would come into existence.[55] This was on his mind the day after one of his meetings with Spinoza, when he wrote:

> If all possibles existed, no reason for existing would be needed, and possibility alone would suffice. Therefore there would be no God, except insofar as he is possible. But if the opinion of those who hold that all possibles exist were true, such a God as the pious believe in would not be possible.
>
> (C 530; A VI 3, 582/DSR 105)[56]

The view that all possibles would be realized in time was regarded by Leibniz as dangerous to piety because it would make the actuality of the universe follow from its mere possibility, and thus deprive God of any choice in creating it—precisely as Spinoza maintained. Leibniz realized that he therefore needed to articulate a notion of possibility such that not all possible things are created: there must be unactualized possibles.[57] These would be things that were not impossible in themselves, but which did not fit in with the other things that God had decided

[55] "A famous philosopher of our century," Leibniz wrote in "On Freedom" [1689], "does not seem to have been far from such an opinion, for he expressly affirms somewhere that matter successively receives all the forms of which it is capable (Descartes, *Principles of Philosophy*, III, §47). This opinion cannot be defended, for it would obliterate all the beauty of the universe and all choice" (A VI 4, 1477). Likewise in "The Origin of Contingent Truths" [Summer 1689?], he had written "If everything that happens were necessary, then it would follow that only those things which existed at some time would be possible (as Hobbes and Spinoza want), and that matter would receive all possible forms (as Descartes wanted)" (A VI 4, 1663). See also his *Periculosa in Cartesio* (A VI 4, 1477–787). Leibniz subsequently had a controversy with the Cartesian Pierre-Sylvain Regis on just this point: see Lærke (2018) for an informative discussion.

[56] Cf. Couturat (1902, 12) and Loemker (1969, 169). Leibniz read three of Spinoza's letters to Oldenburg, probably while still in London prior to his leaving for Holland in November 1676, writing some initial comments in fair handwriting. As Ottaviani has argued (2016), he added further comments in a different hand later, very probably immediately after meeting Spinoza, as the correspondence with the above passage would suggest. Thus to Spinoza's statement in his Letter LXXV that "I conceive that all things follow with an unavoidable necessity from the nature of God," Leibniz then responds: "If all things emanate by a kind of necessity from the divine nature, and all possibles also exist, then the good and the bad, wrongly, will exist equally easily. And so moral philosophy will be destroyed" (GP I 123–4).

[57] As Leibniz wrote in "On Freedom and Contingency" of 1689, what drew him back from this precipice was "a consideration of those possibles which do not exist, have not existed, and never will exist" (A VI 4, 1653). For, he explains, "if there are possibles that never exist, then the things that exist cannot always be necessary." There exists a paper dated as 1677, "That not all possibles come into existence," in which we can see Leibniz arguing exactly this (A VI 4, 1352): "That not all possible come into existence; that is to say, there are certain possibles which neither are, nor were, nor will be."

are worthy of bringing into existence. As he wrote in a piece from the early 1680s, "On Freedom and Necessity,"

> The reason [*causa*] why some contingent thing exists rather than others should not be sought in its definition alone, but from a comparison with others. For, since there are infinitely many possibilities which nonetheless do not exist, the reason [*ratio*] why these exist rather than those should not be sought in their definition, (otherwise non-existence would imply a contradiction, and the other things would not be possible, contrary to hypothesis), but from an extrinsic principle, namely, from the fact that they are more perfect than the others.
>
> (A VI 4, 1445)

This was a crucial insight. Possible things could be defined as ones whose concept contains no contradiction, but only those possible things would be created which God judged to be best in comparison with others.

There are thus two aspects to contingency: (i) contingent things exist only if they are compatible with other things—infinitely many other things, in fact—and (ii) only those systems of possible existents are created that God deems the best. And such a judgement of comparative perfection would involve comparing alternative candidates for creation with each other through the whole course of their development. Only those things which are completely compatible with one another throughout their existence could be candidates for creation: that is, they would have to be *compossible*, to use Leibniz's term of art.[58] The aggregate of all such compossible things, substances in all respects compatible with one another throughout their histories, would then constitute a whole *possible world*.[59]

These correlative notions of compossibility and possible world are thus key ingredients in Leibniz's efforts to rescue philosophy from the impiety he saw in Spinoza's philosophy. For freedom of the will requires that certain things might have occurred otherwise. They could only do so, however, if the other things they co-existed with were accordingly different, and each of these would then have to

[58] The fact that Leibniz's notion of compossibility is essentially relational—and would indeed collapse if all relations had to be reduced to non-relational concepts—was first argued by Jaakko Hintikka (1972, esp. p. 161).

[59] As Osvaldo Ottaviani has argued (in an as yet unpublished paper, "Pure Possibles and Imaginary Roots"; private communication), this marks a change in Leibniz's conception of "compossible": during the Paris period he had characterized compossibles as those that are fully compatible with those that *actually exist*, whereas by the early 1680s he has reconceived them as those that are fully compatible with all other possibles systematically connected with them in a possible world. Before this, as Ottaviani convincingly argues, Leibniz's references to a plurality of worlds are cosmological: there are worlds within worlds, or worlds far away in the cosmos, but their denizens all have temporal and spatial relations to actual things in our world. See also Ottaviani's (2016, 35–41) for an argument for the importance of Nicolaus Steno's contribution to Leibniz's subsequent conception of different possible worlds, in comments he made for Leibniz on the latter's *Confessio Philosophi* in 1677.

have had a different history of development. That is, they would be denizens of another distinct possible world.

It is important to see that this notion of compossibility vastly delimits the scope of what are deemed to be possible existents. An object whose concept could be seen to contain no contradiction would not necessarily be a candidate for inclusion in a possible world that God could consider for creation. When in 1686 Arnauld brings up the example of a sphere as something we could conceive as existing independently of other things, and whose concept we would ordinarily consider without reference to how God might conceive it (GP II 33, A II 2, 33), Leibniz replies that this is not the kind of thing that would qualify as an existent. The concept of such an abstract kind or species is incomplete: it "contains only eternal or necessary truths, whereas the concept of an individual contains, considered as possible, what in fact exists or what is related to the existence of things and to time" (GP II 39, A II 2, 45). The concept of a concrete individual sphere, like the one on Alexander's tomb, would include reference to "the matter it is made from, the place, the time, and the other circumstances which by a continual sequence would finally envelop the whole succession of the universe, if one could pursue all that these notions entail" (GP II 39, A II 2, 45–6).

As this example illustrates, the question of whether a given concrete thing could be an existent necessarily involves a consideration of how other things are related to it *in time*, and indeed how it is related to all other existents in the same possible world. As Leibniz wrote in his *De natura veritatis* of *c.*1685–86, "all propositions that involve existence and time, by this very fact have the whole series of things included in them; for neither the 'now' nor the 'here' can be understood except through a relation to the others."[60] We saw in the previous section that any given state involves the states of other things occurring at the same time as its immediate requisites, and depends on preceding states as mediate requisites. Now, since between any two changes of state there are further changes, and causes for these changes, it follows that there is an actual infinity of efficient causes, or mediate requisites, for any state of an existing thing.[61] Correspondingly, there are reasons beyond reasons for every thing that happens. This inexhaustible infinitude of the reasons for contingent events is thus an essential component of Leibniz's characterization of contingency.

Leibniz explains this in terms of the indemonstrability of contingent truths. According to his theory of truth, the concept of the predicate is in some way

[60] (A VI 4, 1517). See Di Bella's very insightful analysis in his (2018), which includes a quotation of this passage (125).

[61] As Leibniz writes in a manuscript recently transcribed by Osvaldo Ottaviani, "On the Infinite" (LH 37, 5, Bl. 187–8 [ca. 1690–97]), "since time also is divided into actually infinitely many parts by varieties of changes, so that there is no particle of it in which some alteration does not occur, it follows that there is no assignable time, though finite at both ends, that is so small that there is not included in it an infinite series of causes."

contained in the concept of the subject in every true proposition, whether necessary of contingent.[62] In a necessary truth, it will be possible to demonstrate this a priori by a substitution of term for equivalent term until an identity is reached, in a finite number of steps. "In contingent propositions, however, the analysis proceeds to infinity, through reasons for reasons, so that there is never a full demonstration" (A VI 4, 1650).[63] This is analogous, Leibniz claims, to incommensurables in mathematics. For a commensurable proportion "is reduced to congruence with the same repeated measure" in a finite number of steps, but in an incommensurable one, no matter how far the analysis proceeds, there always "remains a new remainder that furnishes a [new] quotient" ("On the Origin of Contingent Truths," A VI 4, 1660, 1662). Analogously, a contingent truth "involves infinitely many reasons, but in such a way that there is always something that remains for which we must again give some reason" (A VI 4, 1662).

Leibniz gives the fullest explanation of this mathematical analogy in the *Generales Inquisitiones* (A VI 4, 776). But there, unlike in other texts, he cedes that a contingent proposition *can* be demonstrated "by its being shown by a continued gradual analysis that it approaches identities continuously, but never reaches them" (§134); this is similar to the case of an incommensurable ratio like $\pi/4$, which approaches a more and more accurate value without ever reaching a rational one. "So the distinction between necessary and contingent truths is the same as that between intersecting lines and asymptotes, or between commensurable and incommensurable numbers" (§135). After further consideration, though, he concludes: "But we can no more provide the full reason for contingent propositions than we can follow asymptotes forever and traverse infinite progressions of numbers" (§136). In other texts he is adamant that God knows contingent truths directly from the individual concepts of things, and not by reasoning from one contingent thing to an earlier one: "God does not need this transition from one contingent thing to another earlier or simpler contingent thing, a transition which has no exit, . . . but rather perceives in each individual substance the truth of all its accidents from its very concept" (A VI 4, 1517). The point is not that there is an a priori demonstration that requires an infinite number of steps; rather, there is

[62] An early example: "in every true affirmative proposition, universal or particular, necessary or contingent, the concept of the predicate is involved in some way in the concept of the subject" (A VI 4, 1654).

[63] Cf. what Leibniz writes in the first fragment we have where he explains contingency by comparison with infinite analysis, the "*De Natura Veritatis, Contingentiae et Indifferentiae . . .*," dated late 1685 to mid-1686: "In a contingent truth, even though the predicate is in fact in the subject, still by a resolution of each, though the terms be continued indefinitely, one never reaches a demonstration or identity . . ." (A VI 4, 1516). Leibniz is close to the infinite analysis criterion earlier, however, when he invokes considerations of imaginary roots and incommensurables to illustrate the difference between necessity and contingency in "On Freedom and Necessity," dated Summer 1680–Summer 1684 (A VI 4, 1444) (see Ottaviani 2021). It is in the *Generales Inquisitiones* (A VI 4, 776), however, that we see him first wrestling to give a clear exposition, as discussed below.

no such demonstration to be had at all.[64] There is no reduction to an identity, just as with incommensurables there is no reduction to a common measure: "just as a greater number indeed contains another incommensurable number, one never reaches a common measure, however far the resolution is continued into infinity, so in a contingent truth, [even though the predicate is in fact contained in the concept of the subject,] one never reaches a demonstration, however far notions are resolved" (A VI 4, 1516).[65]

But there is a second essential aspect to Leibniz's account of contingency. This is that only those possible things are actually created which God judges to be the best in comparison with others. So although all possibly existing things are contingent in that they depend on the existence of all others in the same world, this does not decide the question of which of these worlds is made actual: that depends also on God's choice, specifically, on his choosing among all possible worlds that world which contains the greatest perfection.[66] Consequently, whereas possibility and necessity are decided by the Principle of Contradiction, "all truths concerning contingent things or the existence of things rest on the principle of perfection" (A VI 4, 1445), where the perfection lies not in the thing taken in itself, but in its harmonious coexistence with other things in the same world, compared with what might have existed in other possible worlds.

Thus God freely chooses to create, out of all these possible universes or worlds, that one (and only that one) which is the best, or maximizes perfection. This is the *actual world*, W_A, incorporating everything that actually exists:

> [All] possibles are not compossible. Thus a universe is only a collection of a certain order of compossibles; and the actual universe is the collection of all the compossibles that exist, that is to say, those which form the richest composite. And since there are different combinations of possibilities, some of them better than others, there are many possible universes, each collection of compossibles making up one of them. (To Bourguet, Dec. 1714: GP III 573)[67]

[64] "Contingent truths . . . are known by God, not by demonstration (which implies a contradiction), but through his infallible vision, . . . by a kind of a priori cognition through the reasons for truths" (A VI 4, 1658). This contradicts Russell, for example, who was under the illusion that a truth of fact "requires an infinite analysis, which God alone can accomplish" (Russell 1903, in Frankfurt 1972, 373–4).

[65] See McDonough and Soysal (2019) and Arthur (2019) for further discussion of the indemonstrability of contingent truths.

[66] Bertrand Russell misconstrues Leibniz in this respect. Although he recognizes that the infinite analysis criterion should apply equally to possible worlds (Russell 1900, 61–2), he thinks the actual world is distinguished from merely possible ones by the contingency of the reasons connecting its terms, thus construing possible worlds as ones in which there is no contingency (25–39). For an analysis of the neo-Hegelian assumptions that lead him astray see Arthur (2017).

[67] Cf. what Leibniz wrote in the *Theodicy*: "By 'actual world' I mean the whole series and the whole collection of existing things, lest one might say that several worlds exist at different times and different places. For the whole collection must needs be reckoned together as one world . . . " (GP VI 107/H 128). Leibniz generally equates "world" with "universe,." I have denoted the collection of all possible worlds by U, the union of all possible worlds, so that any given world (the actual world, W_A, or any other

Each possible world is therefore an aggregate of compossible substances, "a chain of states or series of things whose aggregate constitutes the world."[68] In choosing to create one individual—for instance, Adam—God has chosen a whole world of substances compossible with this one. Now each of these possible worlds would contain substances with their own laws of development, constituting for that world a determinate concept in the divine mind. These are "laws of general order," as Leibniz explains in a letter to Arnauld of July 14, 1686: "each possible world depends on certain principal plans or aims on the part of God, which are proper to it," that is, on certain free decrees conceived as possible, which are the "laws of *general order* of that possible universe" (A II 2, 73/LAV 101). These laws of general order, or plans of God,

> enter into this individual concept of Adam, and determine the concept of this entire universe, and consequently both that of Adam and that of the other individual substances of this universe. For each individual substance expresses the whole universe of which it is a part according to a certain relation, through the interconnectedness which exists between things because of the linkages among God's resolutions or plans. (A II 2, 73/LAV 101–3)

Thus the laws of general order, constituted by God's free decrees, are the foundation for one state's involving the reason for all succeeding ones in the same world, and therefore for the interconnectedness of things. This has been expressed with admirable lucidity by Robert Sleigh, Jr.:

> According to Leibniz, a law of general order is analogous to an algebraic equation determining a line; it yields a sequence of events for the world to which it applies (DM §6). By definition it has no exceptions. We may think of it as a function assigning total states of its world to instants of time. Given that every possible world is fully characterized in terms of the properties and relations of its individual substances, the law of [general] order for a world may also be thought of as determining a set of laws of order—one for each substance in its world.
>
> (Sleigh 1990, 52–3)

particular possible world) is the quotient of this by the relation of compossibility. Since Leibniz denied infinite collections, however, strictly speaking such a union of all possible worlds (as occurs in Borges' *Ficciones*) would indeed have to be regarded as a fiction!

[68] From "On the Radical Origination of Things," November 1697 (GP VII 303/L 487). Note that, in keeping with this characterization of an individual as a "chain of states," I define compossibility as a relation holding between possible individuals, each characterized by its own series of monadic states—as contrasted, for instance with Mates (1986, 69–78), who portrays it as a relation among individual concepts. I believe that my construal is, on the one hand, consistent with Leibniz's statements, and on the other, avoids difficulties arising from construals of compossibility in terms of concepts, for instance, that of how there can be incompossibility when all purely positive terms are compatible with one another (Mates 1986, 76–7).

These considerations indicate a very close connection between the interconnectedness of things and Leibniz's notion of a possible world as an aggregate of compossibles. For the fact that the notion of a distinct law of general order is peculiar to each possible world implies that a state of one monadic series involves the reason for a state belonging to another distinct monadic series only if the two series are compossible, i.e. belong to the same possible world. Indeed, if we assume this as an axiom—I call it the Axiom of General Order—then it is possible to prove that the axiom of connection fails for non-compossible monadic series; and given this, that compossibility is an equivalence relation, partitioning the whole universe into distinct possible worlds.

I relegate details of the formal proof to Appendix 1c. We need to assume counterparts of Axioms 1 and 2 and Definitions 1 and 2, now redefined on states of all possible monadic series, in addition to the Axiom of General Order (Axiom 4), as defined above. On this basis it is provable that states of any two incompossible monads are compatible (i.e. not opposites) (Theorem 14); and therefore that the Axiom of Connection holds only for those monadic series which are compossible (Theorem 15). Now compossible series belong to the same world, and we have already proved that in such a world (i.e. if all the monadic series in question are compossible), simultaneity will be transitive, and therefore an equivalence relation. This entails that in any two compossible series, for each state of one series, there is a unique compatible state in the other (Theorem 17). This in turn entails that compossibility is an equivalence relation, as desired, so that it partitions the aggregate of all possible monadic series into separate possible worlds (Theorem 18).[69] Each possible world has its laws of series and interconnectedness determined by its law of general order, so that the states are ordered in temporal succession within each world by the reason-inclusion relation peculiar to that world. There is therefore a distinction here between two senses of time: there is the particular temporal order of each possible world, determined by the relations among particular states of that world—more accurately, this should be called its *history*; and there is time as the order of successives in general, in abstraction from the relations of reason-inclusion specific to any particular world. The general time concept orders successives not just in the actual world but in every possible world: this is *time as the order of possible successives.*

But the very tightness of this connection between the states of the compossible monadic series belonging to each world raises questions as to whether Leibniz's

[69] Cf. Arthur (1985, 274–5). Mates (1986, 77) expresses the same idea in terms of individual concepts: "In short, if two individual concepts belong to a single possible world, then they are present together or absent together in every possible world. Thus the relation of compossibility is transitive. Add to this the assumption that each individual concept is in itself capable of realization, that is, compossible with itself, and the obvious fact that the relation of compossibility between individual concepts is symmetrical, and one sees that this relation is an equivalence relation that partitions the totality of individual concepts into a set of mutually exclusive and jointly exhaustive equivalence classes."

possible worlds do not involve their own necessitarianism. For it would appear that once the first states of this world have been created, together with the laws of general order of that world, then everything else in that world would follow with necessity.

Leibniz denies, however, that any state follows another with metaphysical necessity. He cedes that given God's choice of the laws specific to a world, everything that happens in that world is predetermined, and happens with certainty. It is certain because God knows the complete concept of any individual, and this determines what happens to that individual at any time as a matter of fact. But the connection between cause and effect is contingent, precisely because the chain of causes between any two events that could exist is inexhaustible. Even if we grant this, however, Leibniz's concession of the *certainty* of what is to happen remains troubling. To use a concrete example, given God's choice to actualize the possible world that includes Adam, is it possible for Adam to have done anything differently?

Leibniz confronts this difficulty most notably in responding to the objections of Arnauld. The latter had objected to his claim that "the individual concept of each person contains once and for all everything that will ever happen to him," that it would commit Leibniz to asserting that once God had created Adam, his fate and posterity would "be bound to happen by a more than fatal necessity" (Arnauld to Leibniz, March 13, 1686; A II 2, 9/LAV 9). In his reply of April 12, 1686, Leibniz appeals to a version of the traditional distinction between *absolute* and *hypothetical necessity*: although God is absolutely free, what he does once he has already made certain decisions, he is obliged to do as a consequence of those decisions (A II 2, 17/LAV 21). So it is not a question of him creating some "vague Adam," and then deciding what he should do, but "a specific Adam" whose perfect notion includes all Adam's free actions, corresponding to God's free choice to create this particular Adam. "There is one possible Adam whose posterity is such and such, and an infinity of others whose posterity would be different; is it not the case that these possible Adams (if one may speak of them in this way) are different from one another, and that God has chosen only one of them, who is precisely our Adam?" (A II 2, 19–20/LAV 25).

Arnauld bridles at this talk of "possible Adams": "It is as if I were to conceive several possible 'me's, which is assuredly inconceivable" (Arnauld to Leibniz, May 13, 1686; A II 2, 35/LAV 47). The Arnauld that God has created "contains in its individual concept living in celibacy and being a theologian," whereas a possible Arnauld is a being "which has for one of its predicates having several children and being a doctor" (A II 2, 35/LAV 49). Arnauld concludes: "since it is impossible that I should not always have remained me, whether I had been married or had lived in celibacy, the individual concept of me involves neither the one nor the other of these two states."

Leibniz is happy to concede to Arnauld that, "taking Adam as an example of an individual nature, it is no more possible to conceive several possible Adams, than to conceive several 'me's" (A II 2, 36/LAV 51), although not that different contrary predicates might apply to him.[70] In his reply written the same month (July 14), Leibniz stresses that "all human events could not fail to happen as they actually have happened, once the choice of Adam is supposed made" (A II 2, 73/LAV 101). Leibniz is adamant, as Fabrizio Mondadori and Benson Mates have both stressed, that each individual substance can only exist in one world (Mates 1986, 138–40). For instance, in the manuscript summary of his philosophy *Specimen inventorum*, probably written in 1688, Leibniz writes:

> Hugo of St. Victor also saw this: when he was asked why God favoured Jacob and not Esau, he simply replied "Because Jacob is not Esau." That is to say, already contained in the perfect notion of an individual substance, considered by God in a pure state of possibility before every actual decree about what is to exist, there is whatever will happen to it if it should exist, and indeed the whole series of things of which it forms a part. Thus it should not be asked whether Adam will sin, but whether an Adam who will sin should be admitted into existence.
>
> (A VI 4, 1619; LLC)

Thus, he insists, once Adam is created, he is created along with a whole *order* (to Arnauld, July 14, 1686; A II 2, 73/LAV 101). Adam, and everything that will happen to him, is necessary on the hypothesis that the whole series of things of which he forms a part is to be created. This is the "hypothesis" in hypothetical necessity.[71] Since, in today's parlance, Adam is "world-bound," the Adam who sins does not exist in other possible worlds. It is therefore a contingent fact that Adam-who-sins exists, because there exist other possible worlds in which the sinning Adam does not exist, and God chooses to create not those worlds, but the one in which the Adam-who-sins exists. As Alan Nelson phrases this, "since God understands worlds such that if they existed, X would not exist, nothing that is actually true of X would be true in those worlds" (Nelson 2005b, 296). The

[70] Arnauld's position is evidently closer to Kripke's, where an individual, whatever its properties, is the same individual in any possible world.

[71] In his discussion of hypothetical necessity, Benson Mates charges that it depends on a modal fallacy which he calls the "fallacy of the slipped modal operator" (1986 chapter 6, sec. 2). He says Leibniz is guilty of it when he argues: "Granted that, if God foresees something, it will necessarily happen, yet this necessity, since it is only hypothetical, does not negate contingency and freedom." According to Mates, the first statement means "Necessarily, if God foresees something, it will happen": the necessity is not being conditionally predicated of the consequent, but unconditionally predicated of the whole conditional statement (Mates 1986, 117). I believe this is a misreading: hypothetical necessity is necessity given the hypothesis that God has created this world, not the necessity of a hypothetical statement. Here I follow the excellent treatment of these issues by Alan Nelson in his (2005b): something is hypothetically necessary if it follows from the concepts of the things that God would create if he created that world, so that, in short, "the hypothesis is that this world is created."

possibility of Adam's not sinning implies no contradiction just because there exist worlds, conceived by God, in which it is not true. This fact, moreover, is perfectly sufficient to underwrite the notion of real possibility that Leibniz requires: Adam's sinning is not necessary precisely because God could have created a world in which the sinning Adam did not exist.

But these considerations do not resolve the difficulty about events occurring counterfactually. For on the hypothesis that the sinning-Adam series of things is created, it is impossible that Adam not sin. There is no possible world, however, in which identically the same Adam decides not to take the apple and thereby avoid sinning. Now, on the one hand, Leibniz has satisfactorily answered this objection insofar as it concerns the contingency of Adam's action: it is no constraint on Adam's freedom that God foresees his free choice and decides to create this very apple-choosing Adam. But, on the other, Leibniz helps himself to our ordinary locutions about how Adam might have acted otherwise than he did, and it would seem incumbent on him to be able to give a satisfactory account of such locutions. As various commentators have observed, what is needed here is a way to specify, for any given individual such as Adam, who might have acted otherwise, a *counterpart* to that individual, say Adam*.[72] But if we understand these alternative Adams to be specified in terms of their complete concepts, then the individuals corresponding to their complete concepts would be incompossible. It would then follow that there would be no state of any counterpart-Adam simultaneous with any state of Adam in which he could do things differently than Adam might have at that time. Thus the different Adams cannot be identified by their complete concepts. In this connection Leibniz himself talks of the various Adams sharing a certain "finite number of predicates":

> in speaking of several Adams, I was not taking Adam as a determinate individual, but as a person conceived of in generality—that is, under circumstances which seem to us sufficient to determine Adam as an individual, but in truth do not sufficiently determine him as one: as when we mean by Adam the first man, whom God puts in a garden of pleasure, and which he leaves through sin, and from whose side God draws a woman.... But all of that does not determine him sufficiently, and so there would be several disjunctively possible Adams, or several individuals whom all of that would fit. That is true whatever finite number of predicates we may take which are incapable of determining all the rest. But what determines a particular Adam must involve absolutely all his predicates, and it is that complete concept which determines generality so that the individual is reached... (To Arnauld, July 14, 1686; A II 2, 77/LAV 107)

[72] Mates suggests that there are resources in counterpart theory to avoid this objection (1986, 146–8), and the same point was made explicitly by Fabrizio Mondadori (1975, 56), quoted by Jan Cover and John O'Leary Hawthorne (1999, n.9). See also Di Bella (2018, 131).

That is, the complete individual concept of Adam—what determines him as an individual in a given possible world—involves all the infinitely many predicates that apply to him at all the different times of his existence. These are not merely positive, simple attributes, but must instead include all predicates truly applicable to him at each time, including extrinsic (or relational) ones:[73]

> I say that the concept of the individual substance includes all its events and all its denominations, even those which are commonly called *extrinsic*, that is, those which pertain to it only by virtue of the general connection of things and from the fact that it expresses the whole universe in its own way.
>
> (To Arnauld, July 14, 1686; A II 2, 80/LAV 111)

What is important to appreciate here is that the "things that happen" to Adam must be properties that can be truly predicated of him *at a given time*, predicates whose applicability follows from the fact that he is in a particular state that reflects everything happening in the universe at that time.[74]

But here we are confronted with the difficulty that one cannot say that there is a state of Adam in which he (or someone very like him) does not take the apple in another possible world *at the same time as* he decides to take the apple in the actual world. For simultaneity, like individual substances, is world-bound. Leibniz makes this clear by using the same language of hypothetical necessity as he does about individuals in connection with the theory of time. In a passage we already quoted in the previous section, from a manuscript dating from mid-1685—a year or so prior to the exchange with Arnauld we have been considering—Leibniz defines simultaneous things as those which are "co-necessary" on the supposition that the series of things is created:

> That is *earlier in time* which is incompatible with some position, and is simpler than it. The other position is called *later*. Those things are *simultaneous* which are by supposition co-necessary; I say, by supposition, i.e. with the series of things posited. (A VI 4, 629/LLC 275)

[73] This has been insisted on by Massimo Mugnai: "on many occasions Leibniz repeats that the complete concept of each individual contains all the denominations—intrinsic and extrinsic (i.e. relational)—of such an individual" (Mugnai 1992, 133).

[74] Thus to take the example that has been prominent in the literature on the "lucky proof," if "Peter is a denier of Christ," then this is because he denied Christ in some particular set of concrete circumstances at a particular time. What caused Peter to do this, whether this is analysed in terms of his own reasons or the efficient causes of these circumstances, requires reasons beyond reasons to infinity. So I would say that it is not a question of demonstrating that the abstract predicate "is a denier of Christ" follows from the concept of the subject conceived as an infinite list of abstract predicates; only God knows all the concrete predicates that truly apply to an individual in a given state, but he knows this intuitively, not by demonstration. Obviously, though, we may know enough to demonstrate a contingent proposition from other contingent propositions.

But if simultaneous things are co-necessary on the supposition that a given series of things is created, this will restrict simultaneity to that series of things. Just as each individual substance is world-bound, so will be the aggregate of states simultaneous with a given state of that substance.

This is reflected in my reconstruction of Leibniz's theory of time, where for a state's to occur at an instant $t = I_x$ is for it to be a member of the equivalence class of states simultaneous with x. But in that case, what about states simultaneous with x that God did not actualize? As Jan Cover phrases the objection, "What then should we make of temporal counterfactual claims to the effect that s_n occurred at t, but might not have done so? If times are sets of simultaneous states, it is unclear how t—that very time—can fail to have s_n as a member" (Cover 1997, 313).

The answer, I suggest, depends on following up Leibniz's insistence to Arnauld that in talking of "vague Adams" he is not regarding them as completely specified by their complete individual concept—all the predicates that truly apply to them—but instead just by a finite collection of such predicates. This is consistent with Leibniz's view of how we think about things generally, since all we ever have when we think about or refer to existing things are incomplete concepts.[75] In talking about how Adam might have done otherwise, then, we are talking about how an "Adam" specified by only a finite cluster of predicates might have done so. The "real Adam," so to speak, the one specified definitively by his complete individual concept, is known only to God.

We can give substance to this conception as follows. Adam can have as many predicates in common with his counterpart-Adam as we wish, but an even finer discrimination showing the difference between Adam and Adam* will always be possible by way of more minute perceptions. So let us introduce the notion of two states being indistinguishable below a certain threshold Δ. We will suppose that two individuals m and n share a finite collection of predicates during a given time if and only if their corresponding states are indiscernible with respect to some threshold Δ during that time. Implicit in this proposal is the idea that the finer the threshold of discrimination under which the states are indiscernible, the more properties the two counterparts will share. We can make this more precise by introducing the idea of a *stage* of a monadic series S_m, which is some truncated sequence of states of that series running from state s to state t, which we denote $S_m|_s^t$. If for every state of $S_m|_s^t$ there is a unique corresponding state of another individual n that is indiscernible from it w.r.t. Δ, n is a *counterpart* of m during

[75] See Leibniz's discussion in the *New Essays* (A VI 6, 289–90), and the expositions of Leibniz's philosophy of natural language in Mugnai (2001, 244) and Arthur (2014, 48–53). See also the passages from the *New Essays* (A VI 6, 245) quoted by Mates (1986, 144–5) where Leibniz discusses similar people in other regions of the universe, and also his discussion of the different "Sextus's" in the *Théodicée* (GP VI 363), cited by Mates as showing that "There are many different possible persons who have all the properties *we* know of any given individual" (1986, 145). Finally, see also Leibniz's response to William Twisse (Grua, vol. 1: 358), quoted and discussed by Mugnai in his introduction to the *Generales Inquisitiones* (Mugnai, forthcoming).

that stage. The unique corresponding state that is indiscernible from some state x, with $s \leq x \leq t$, is *c-simultaneous* with x w.r.t. Δ.

Thus if we consider "Adam" as characterized only by a certain set of properties, such as "the first man, whom God puts in a garden of pleasure, and from whose side God draws a woman," various counterpart Adams are definable as above, who can be considered as having states indiscernible from his during a certain stage of his existence, but to whom other predicates not applicable to Adam may apply, such as "refuses to take the apple." A counterfactual such as "Adam might not have sinned" is then explicable in terms of there existing in God's consideration another possible world in which a counterpart-Adam—an individual sufficiently similar to Adam during some segment of his life as to be indiscernible from him within threshold Δ—does not sin. There are, I conclude, sufficient resources within Leibniz's theory of time to account for the same individual (identical within a certain threshold of discrimination Δ) acting otherwise than she or he in fact does at a given time.

Is this enough? Does it not only apply to counterparts and their possible worlds that are "close" to ours? Maybe, but this is all it needs to do. Worlds in which substances that are counterparts during a certain stage, afterwards diverge, are not such that we need to be talking about what happens at the same time. There can be radically different possible worlds having no mutual counterparts; but then there would be no sense in talking about what happened at the same time in them. Time as the order[76] of possible successives does not need to be, nor is it, an all-embracing container time; we need only the idea that the time between any two states in the actual world could have been, or could be, instantiated by different histories, might contain a different infinitely divided sequence of enduring states, and that time embraces these possibilities too, not just what actually happens in the actual world that God has created.

To summarize: I have argued that there are two essential aspects to Leibniz's account of contingency. First, not all possibles—things whose concepts contain no contradiction—are possible existents. The concept of an abstract object like a perfect pentagon contains no contradiction, but only those objects can exist that bear relations to all other existents in the same possible world, including in particular temporal relations. Second, only those possible things which God judges to be the best in comparison with others are actually created. So although all possibly existing things are contingent in that they depend on the existence of all others in the same world, which of them will actually exist depends also on God's choosing among all possible worlds that world which contains the greatest

[76] Heinrich Schepers (2018, 416) identifies Leibniz as using the notion of *order* in defining space and time for the first time in a fragment from June 1685–March 1686 (although possibly added later: "Place is the order of coexistents, time the order of changes" (A VI 4, 630.28, fn.20)). But in a fragment on analysis situs which Echeverría dates as from about 1682 (CG xvi 302), Leibniz writes "*Space* is a continuum in order of coexisting [***Spatium*** *est continuum in ordine coexistendi*]."

perfection. The temporal ordering of existents in a given possible world depends on the principles of general order specific to that world; this means that there is a different concrete temporal ordering or history in each different possible world. Although this means that there is no equivalence relation of simultaneity between events contained in different possible worlds, I argued that the meaning of counterfactual statements can be captured by appeal to counterpart individuals that are indiscernible from one another during a certain time within some threshold of discernibility Δ.

That explanation, however, depends on a distinction between distinct concrete temporal orderings, whose times are individuated by what happens at them—different histories—and time as an ideal entity ordering all possibles. To considerations of this sort, however, commentators have objected that time as the structure of all possibles is something merely ideal: it is a being of reason. As such it is on a different plane of being from the actual phenomena we experience as occurring in succession. And, they object, there is reason to suppose that it does not operate "on the monadic level," ordering the states of monads, since Leibniz characterizes relations as merely mental, and provides many examples of reductions of sentences involving relations. Although we have already given some discussion of these issues in the introduction, we need now to turn to a thorough investigation of the nature of time's ideality for Leibniz, and the nature of the reductions to which he is committed.

1.4 Reduction, Ideality, and the Homogeneity of Time

Time for Leibniz is not an independently existing entity, but an abstraction from existing things. What exist in reality are individual, enduring, concrete substances and their accidents or modifications; all the rest are abstractions, and therefore ideal. This is evidently in keeping with a broadly nominalist position, but there are different ways of interpreting this nominalism. On the most prevalent reading, as proposed by Bertrand Russell and endorsed by Mates, O'Leary-Hawthorne and Cover, Rutherford and Look, and Futch, Leibniz is seen as *eliminating time*:[77] time is an abstract entity, and since nominalists deny the reality of abstract entities, Leibniz is thought to reject the idea that time is part of fundamental reality. Moreover, that view is consistent with the received interpretation of Leibniz's philosophy of relations, which construes him as eliminating relations from his ontology. Thus there is wide agreement that Leibniz's intention to eliminate relations entails that there can be no relations among monads, straightforwardly

[77] See Russell (1900), Mates (1986, 227–35), Cover and O'Leary-Hawthorne (1999, 315–18), Rutherford and Look (LDB), and Futch (2008, 11).

entailing the ideality of time, and rendering a fundamental reality that is essentially timeless.

This eliminativist reading, however, is not the only interpretation of Leibniz on relations and their ideality. On an alternative reading proposed by Ishiguro, Hintikka and others, Leibniz never intended to eliminate relations in all senses.[78] His depiction of relations as ideal is understood rather as applying to relations conceived in abstraction from their relata, as abstract entities, but as in no way undermining the genuine relatedness of things. Thus although Leibniz denies reality to relations as entities in themselves, he does not thereby deny the reality of *relational facts*, such as the fact of two states of a substance occurring one after the other.[79] More controversially, these authors have held that Leibniz's thesis of the mind-dependence of relations does not apply to *relational properties*, such as the property of being the son of Adam. Other authors, notably McCullough and Plaisted, have claimed that Leibniz upheld the *reality of relational accidents*—the individual accidents of concrete substances which we would now call tropes—as the real foundations on which relations supervene.

The issue of the correct interpretation of Leibniz's philosophy of relations is very complex, though, and it would be too much of a distraction to try to enter into it fully here. Instead, I shall present a synopsis of some of the main issues, especially as they bear on space and time, in an appendix (Appendix 2). Since in the preceding sections I have committed myself strongly to a version of Leibniz's theory of time based on intermonadic relations, it will be no surprise that in the appendix I argue that the idea that Leibniz wanted to "eliminate" relations altogether is mistaken. As I explain there, I no longer subscribe to the relational properties interpretation (which I upheld in my (1985)), or the relational accidents as tropes reading (which I maintained in my (1994)). But it was no part of Leibniz's brief to deny the relatedness and interconnectedness of things by reducing relations to "monadic predicates," however enticing that connection to monads may seem to a modern eye. To be sure, relations taken in abstraction from things instantiating them are ideal, and not such as can qualify as existing things. Nevertheless, time as the abstract order of possible successives is not unreal, and obtains its reality from its foundation in the divine mind.

Here, however, I would like to take a different, but complementary, approach to the issue of the ideality of time, relating it more directly to the question of what type of *nominalism* Leibniz espouses, and what type of *reduction* is involved in his

[78] Ishiguro (1967), Hintikka (1969), Ishiguro (1972b), McCullough (1978), and Kulstad (1980).

[79] Here I should note that neither Mates nor Cover, at least, read Leibniz as eliminating such relational facts about monadic states. On their accounts, if it is true that one monadic state is before another, this will follow from statements attributing non-temporal properties to monads in these states. See Mates (1986, 227–35) and Cover (1997, 315–18). In this respect (to use Cover's terminology), their accounts are not "eliminativist," but "identificatory," notwithstanding their talk about there being no time in Leibniz's fundamental ontology.

thought on time. I already argued in section 1.2 that Leibniz is not proffering an eliminative reduction of temporal succession to reason-inclusion, since in addition to reasons for succeeding states, Leibniz posits an active force or appetition that takes a present state into succeeding ones. Without this dynamic element, there would be no temporal succession. But here I am concerned with a different claim for an eliminative reduction of time in Leibniz. For Benson Mates and Jan Cover interpret nominalism under the influence of Willard Quine as involving not just the elimination of space, time, number, etc. as abstract entities, but as eliminating them altogether as constituents of the world. I shall argue that this conception of nominalism is anachronistic, and is not adequate to Leibniz's notion of time as an ideal order to which actual things nonetheless conform, and which they actually instantiate. Moreover, although Leibniz countenanced "actual instants," instants identified by the changes occurring at them, he rejected the idea that time could be composed of instants, contrary to the claims of some commentators.

According to Benson Mates' rendition, Leibniz's nominalism "states that only concrete individuals exist and that there are no such things as abstract entities—no numbers, geometrical figures or other mathematical objects, nor any abstractions such as space, time, heat, light, justice, goodness, or beauty."[80] Thus he thinks that "Leibniz would agree wholeheartedly with that notorious pronouncement of present-day nominalism: 'We do not believe in abstract entities'" (Mates 1985, 173), referencing a classic paper by Nelson Goodman and Willard v. O. Quine (1947). Similarly, Jan Cover understands Leibniz to be committed by his nominalism to an "eliminative reduction" (Cover 1997, 300), where "apparent commitments to time, intervals of time, and instants" are "apparent only." On this basis Cover criticizes the account I gave of Leibniz's theory of time in my (1985) as deficient, in that it amounts to an "identificatory reduction" (Cover 1997, 299–302), whereas, he claims, Leibniz aimed to eliminate time (along with space, beauty, numbers and similar abstracta) from his ontology.

Now, both Mates and Cover draw attention to Leibniz's respect for the influential medieval English philosopher William of Ockham (*c.*1288–1348), revered by him as "the cream of the scholastic philosophers,"[81] and I agree with them that Leibniz's nominalism has much in common with Ockham's. But I am not sure

[80] This quotation is taken from the inside of the cover of Mates' book (1986), which gives an accurate synopsis of its contents that we may presume was written by Mates himself. Cf. "he did not believe in the existence of abstract entities of any sort" (Mates 1986, 171), "he does not believe in numbers, geometric figures, or other mathematical entities, nor does he accept abstractions like heat, light, justice, goodness, beauty, space or time..." (173).

[81] Thus Mates: "He has high praise for William of Ockham, whom he explicitly characterizes as a nominalist" (Mates 1986, 172); Cover: "He is, as Mates has argued recently, a nominalist about *abstracta* generally. This position is not unexpected, given Leibniz's willingness to retain more of the Aristotelian-Scholastic tradition than his contemporaries. He explicitly praises William of Ockham's nominalism in his preface to an edition of Nizolius" (Cover 1997, 304).

they have grasped either the full extent of the agreement in the two philosophers' positions, or the ways in which Leibniz went beyond Ockham's position (more accurately styled a kind of conceptualist-nominalism.) They claim, correctly, that Ockham denied the need to posit abstract entities or universals as independently existing; and yet that statements making apparent reference to them can still be understood using his theory of connotation. But it is not correct to say that he "denied abstract entities"; he denied that the abstract entities referred to by universal terms need to be posited as existents.[82] Thus Ockham denied any need to posit mathematical objects like numbers, points, lines, and surfaces, as really existing entities, claiming that by means of his theory of connotation, propositions involving them could always be reduced to ones involving only substances and qualities. Numbers are abstractions from groups of things numbered, surfaces abstractions from the bodies that are bounded by them, and so forth. In the same way, Ockham contended, time is an abstraction from the durations of enduring things, and there is no need "to posit a time distinct from enduring things (*rebus permanentibus*)" (496).

All of this is clearly consistent with the ontological parsimony mandated by Ockham's famous "razor": "To multiply beings according to the multiplicity of terms is erroneous, and leads far away from the truth."[83] But there are two salient points to be noted. One is that the individual, concrete things to which *abstracta* are reduced are things standing in certain relations to one another, such as bodies exceeding one another in length. And the other is that Ockham conceives the basic constituents of the world as *res permanentes*, enduring things, whose accidents follow one another in temporal succession. Thus, even though he denies the need to posit the reality of time or instants as independently existing entities, this does not entail a denial of the reality either of the enduring of things or of the succession of their accidents. On both these scores, I maintain, Leibniz is in complete agreement with him.

For his main conclusion that time is not an existing thing, Ockham appeals to a traditional argument which has its origins in the infancy of philosophy. Aristotle records it as follows:

> Some of it is past and no longer exists, and some is in the future and does not yet exist; these constitute both infinite time and the time that is with us at any

[82] See, for example, Spade and Panaccio (2019, §4). They claim that Ockham was *not* a nominalist in the sense that he sought to eliminate universal terms: his "program in no way requires that it should be possible to dispense altogether with terms from any of the ten Aristotelian categories (relational and quantitative terms in particular)." When he claims that the truth of "Socrates is similar to Plato" follows from the fact that they each possess a quality of the same species, such as whiteness, the idea is not that similarity is thereby eliminated, but that the two philosophers "are similar *without anything else added.*" There is no need to posit the abstract relation of similarity as an existing thing.

[83] Ockham, *Summa Logica* I, c. 51 in *Opera Philosophica* I, 171; cf. Loux (1974, 171).

> moment; but it would appear to be impossible for anything which consists of nonbeings to participate in being itself. (*Physics* 217b 34–218a 3)

A version of same argument was taken up by Augustine in his *Confessions*,[84] and is repeated by Aquinas and Averroës.[85] But it was Ockham who promoted this line most forcefully. In his *Quæstiones*, having raised the question "Whether time is something distinct from perduring things," he argues

> If one part of the whole does not exist, the whole does not exist; therefore, it is much more the case that if no part of the whole exists, the whole does not exist. But no part of time exists, as is clear inductively; therefore, etc.[86]

He then addresses the more particular question "Whether Aristotle or Averroës meant to posit a time distinct from enduring things (*rebus permanentibus*)" (Ockham 1984, 496), giving a *Contra* verdict by an explicit appeal to his razor in the form: "A plurality should not be posited without necessity."[87] Neither all the parts of time exist, nor even some, he argues, because all the parts are past or future, and so do not exist in the present:

> that which is composed of non-entities is not a positive entity; but time is composed of non-entities, because it is composed of the past which does not exist now, although it did exist, and of the future, which does not yet exist; therefore time does not exist. (Ockham 1984, 496.)

Augustine had concluded from this argument that whatever exists, exists in the present, a position now labelled "presentism."[88] So might one not then say that of

[84] Augustine, *Confessions* XI 11.13: "in the eternal, the whole is present; whereas no time is wholly present [*in aeterno,... totum esse praesens; nullum vero tempus totum esse praesens*]"; XI 20.26: "But what is now evident and clear is that neither future nor past exist [*quod autem nunc liquet et claret, nec futura sunt nec praeterita*]."

[85] Averroës: "Time is composed of past and future; but the past has already stopped being and the future does not yet exist.... It is the same for movement; no part of movement is in actuality" (quoted from Duhem 1985, 301).

[86] William Ockham, *Quæstiones*, 1984; Book IV, chapter 2, q. 7: "Whether time is something distinct from perduring things," 494; I am indebted to Emmaline Bexley 2007 for directing me to Ockham's views on time in the *Quæstiones*.

[87] "*Pluralitas non est ponenda sine necessitate.*" This is not exactly the usual formulation of Ockham's razor, "Entities should not be multiplied without necessity" ("*Entia non est multiplicanda sine necessitate.*"), but in these applications it clearly amounts to the same thing. See Brower (2005) for discussion.

[88] Augustine: "if nothing had become past, there would be no past time; and if nothing were yet to come, there would be no future time. How, then do these two times, past and future, exist, when the past does not now exist and the future does not yet exist?"; "For if future and past exist, I would like to know where they are? But if I cannot know that, I do know that they are not there as future or past things, but as present things. For if there they are also future, they are not yet there, and if there they are also past, they are not there now. Therefore, wherever anything is, it is only as present" (*Confessions* XI 14.17, XI 18.23; my translations).

time only the present exists? Ockham finds this equally objectionable: "of the present, nothing exists but the instant, which is neither time nor a part of time" (1984, 496). He supports this conclusion by his examination of Question 55: "Whether an instant implies any thing distinct from perduring things" (543), again applying his razor to answer in the negative. An instant is not a substance (an enduring thing) because it is always going out of existence; it is not an accident because there is no candidate for the substance of which it could be the accident (543). This does not prevent us from making true statements involving time, though. The statement that two processes last for the same time, for example, could be parsed in terms of a coincidence between the endpoints of their durations, these durations of processes being concrete particulars and the endpoints their modes. Thus Ockham denies the existence of either time or instants distinct from enduring things. But, of course, since the constituents of his world are enduring things, he certainly does not deny that things endure, nor that their accidents temporally succeed one another.

The similarity with Leibniz is quite striking. When Samuel Clarke insists on behalf of Newton's conception of absolute, self-existing time that duration exists eternally, Leibniz's reply reads like a paraphrase of Ockham:

> Everything which exists of time and duration, being successive, perishes continually. And how can a thing exist eternally if, to speak precisely, it never exists at all? For how can a thing exist if no part of it ever exists? Nothing of time ever exists except instants, and an instant is not even a part of time. Anyone who considers these observations will easily comprehend that time can only be an ideal thing. (To Clarke, V, §49: GP VII 402/LC 72–3)

Echoes of Ockham can also be seen in Leibniz's insistence earlier in the correspondence that "instants, considered without the things, are nothing at all, and that they consist only in the successive order of things" (Third Paper to Clarke, §6; GP VII 364/LC 27). This only applies to instants considered in themselves, however. Instants or intervals of time that are identified by what happens at them are not ideal: "the parts of time and place considered in themselves are ideal things... But it is not so with two concrete ones, or with two real times, or two spaces that are occupied, that is, truly actual" (Fifth Paper, GP VII 395/§27; LC 63).

The similarity of Leibniz's brand of nominalism to Ockham's extends to his philosophy of relations. He wants to eliminate the need to posit abstract entities such as instant, height, space and time, by showing how statements involving them can be rewritten in terms of expressions such as "is simultaneous with," "is congruent with," and so forth. Thus as he writes in a fragment "On the reality of accidents," written probably in 1688:

> Up to now I see no other way of avoiding these difficulties than by considering *abstracta* not as real things but as abbreviated ways of speaking... and to that extent I am a nominalist, at least provisionally.... It suffices to posit only substances as real things, and, to assert truths about these. Geometricians, too, do not use definitions of *abstracta* but reduce them to *concreta*: thus Euclid does not use his own definition of *ratio* but rather that in which he states when two quantities are said to have the same, greater, or lesser, ratio. (A VI 4, 996)

Leibniz gives the same example of Euclid's use of *ratio* almost thirty years later in his controversy with Clarke about the nature of space. Just as Euclid, "not being able to make his readers well understand what ratio is in the sense of the geometricians, defines what are the same ratios," so Leibniz, rather than posit places as existing entities which a body must occupy, is "content to define what is the same place" (Fifth Paper, §47: GP VII 402/LC 71).[89]

The salient point here is that Leibniz's reductions of these abstract entities is to *equivalence relations among concrete particulars*, such as sameness of place, sameness of ratio, and simultaneity. This is in marked contrast to the type of reductions adduced by Mates, or by Cover and O'Leary-Hawthorne. Mates claims that Leibniz should be interpreted as asserting that "every relational property of an individual is reducible, in his sense of 'reducible', to non-relational properties of that and other individuals" (Mates 1985, 219); and Cover and O'Leary-Hawthorne write of "inter-monadic relations or relational properties superven[ing] on monadic properties" (Cover and O'Leary-Hawthorne 1999, 83), to which they can be eliminatively reduced. Neglecting subtle differences between their interpretations (as well as difficulties as to what is meant by a "nonrelational" or "monadic" property—see Appendix 2), on both of these interpretations it is held that Leibniz would parse the relational fact represented by "Theaetetus is taller than Socrates" as (not equivalent to, but) resulting from facts such as "Theaetetus is six feet tall" and "Socrates is five feet tall." Logically, "is six feet tall" is a monadic predicate. But Theaetetus' *being six feet tall* is a relational accident of Theaetetus, since it implicitly involves a relation of comparison to some measure, so it cannot be said that this kind of reduction "eliminates relations" in any meaningful sense.[90] In fact, given what was said above, it seems that Leibniz would not have been inclined to take Theaetetus' height as

[89] An Ockhamist sentiment is clearly evident in what Leibniz wrote to Gabriel Wagner in 1698: "whoever requires a point, a line, or a surface to be something other than a limit, should adduce a reason. Beings should not be multiplied beyond their use. In all principles, everything should be assumed as simple as possible.... If something further is added, we will fall into absurdity—as usually happens whenever something is assumed without a reason" (A II 3b, 699).

[90] Cover and O'Leary-Hawthorne (1999) cede this point in the course of denying that it supports McCullough's view that Leibniz upheld the reality of relational accidents: "for us, but not for McCullough, this is but the first stage in a fully reductive analysis whereby only intrinsic accidents lie at the groundfloor" (72).

an irreducible accident, but would instead, like Ockham, have rewritten statements about his height in terms of relations of comparison between concrete particulars, the extended bodies of Theaetetus and Socrates.

The latter point is granted by Cover in his (1977). There he suggests a view of relational reductionism in which "apparently substantival commitments are understood by the relational reductivist as idiomatic, better rendered as his adjectival paraphrases have them—as committed only to material objects and events, bearing properties and standing in relations" (301). Significantly, he understands such reductions as "committed to" relations, even if these relations are themselves thought to be reduced to "non-relational" properties in a further stage of analysis. The import of the idea of "being committed to" is a Quinean understanding of nominalism, according to which "to be is to be the value of a variable." In Cover's words, "real commitments are to be read off paraphrases in the form '$\exists x(Fx)$' or '$\exists x \exists y(xRy)$', where now only material objects and events need to be reckoned as values of variables" (Cover 1997, 300–1). It is not clear that such a conception of existential commitment is appropriate for Leibniz's philosophy.[91] But we can ignore that for now, since it gives the correct result in this case: two things occurring at the same time or successively does not depend on reifying abstract times, but only on there being concrete events or states, existing in the requisite temporal relations.

As noted, however, Cover and O'Leary-Hawthorne go on to propose that such relations are eliminated in the last analysis, since statements ascribing relations can themselves be reduced to ones not involving them. Thus they propose that "inter-monadic relations or relational properties supervene on monadic properties" in the sense that

> for any relation **R** or relational property **R**(y), necessarily if x is **R** to y or has the property **R**(y), then there are monadic properties **F**, **G**...of x and y, and necessarily if any x and y have **F**, **G**...then x is **R** to y or has **R**(y).
>
> (Cover and O'Leary-Hawthorne 1999, 83)

[91] My misgivings about applying the Quinean account of existential commitment to Leibniz are as follows: First, Leibniz regarded mathematical entities (such as variables and their values) as ideal things, and not such as could actually exist: a *concretum* could not be the value of an *abstractum*. Second, there are no existential statements in Leibniz's logic, for the reason that he preferred an intensional approach (of concept inclusion), and this was precisely because it avoids questions of existence. Even on an extensional interpretation of his logic, Leibniz (correctly) regards universal statements as hypothetical. Although he does have a symbolism for universal quantification, the quantifier denotes, distributively, all those things to which what one says about them applies. But that does not entail existence, and not even possible existence: it is possible to quantify over non-existent things, such as square circles, if they are what one is discussing. A nominal definition can be sufficient to pick out what one is referring to without this entailing that such things are even possible, let alone the concepts of existing things. This is why Leibniz's relational reductions work. One does not have to posit the existence of ratios, or even decide whether there are such things, so long as one can identify things bearing the same ratio.

For the sake of argument, let us assume that this is correct, and that Leibniz did intend to reduce statements of this type to statements not involving the ascription of relational accidents to the things related. In the case of simultaneity, what would this involve? The one linguistic reduction Leibniz gives of "A is simultaneous with B" is that it would follow from "A exists today" and "B exists today."[92] But even if it is granted that the former would follow from the latter two facts, how could simultaneity supervene on these two facts without supposing that "today" is a concrete particular (albeit adverbial), so that times, in fact, would have to be presupposed? Here Cover and O'Leary-Hawthorne write vaguely about "more finely sliced monadic properties" (1999, 83) without enlightening us as to what they could be. So when it comes to the question of how we are supposed to derive an ontological conclusion from these premises concerning logic, we are left in the dark. In the final analysis, the kinds of logico-linguistic reductions suggested by Mates, Cover and O'Leary-Hawthorne for statements about times simply go in the wrong direction—from relations to supposed temporal particulars like "today"—whereas the reductions Leibniz champions that are relevant to his ontology of space and time obviate the need to posit times as existents by recourse to equivalence relations.

Another feature of Cover's description of relational reductions is worthy of comment, namely his invocation of "adjectival paraphrases." With regard to time, this seems to refer to the idea (which we also saw in Ockham above) that to compare two times is to compare the durations of two concrete processes. One of the examples Cover gives involves parsing "The duration of the homily was twice that of the race" as "$\exists x \exists y(x$ is a homily and y is a race and x is n times longer than y and $n = 2)$" (Cover 1997, 301). This returns us to the point I made earlier. Leibniz, like Ockham, posits enduring substances as existing: there is no question of him "eliminating time" in the sense of denying duration as an attribute of concrete existents. As Cover notes, duration is not the same thing as time (as it is for Newton). Like extension, it is an attribute of concrete existents.

Leibniz clearly articulated the latter distinction as early as 1676 in the fragment "On Magnitude":

[92] Leibniz gives this example in the course of distinguishing *relations of connection* from *relations of comparison*. Having resolved the relation of comparison "*A* is similar to *B*" into "*A* is red and *B* is red, and therefore *A* is similar (in this respect) to *B*," Leibniz writes "By *A* and *B* are understood things or individuals, not terms. But what shall we say about these: *A* exists today and *B* also exists today, ôr *A* and *B* exist simultaneously? Will this be a relation of comparison or of connection?" The implicit conclusion is: the latter. And "a relation of connection arises from the fact that *A* and *B* are in one and the same proposition *which cannot be resolved into a relation of comparison*" (A VI 4, 944; quoted in Mates 1986, 224; my italics). As we have seen in §1.1, simultaneity results from Leibniz's insistence that all coexisting things are interrelated; as embodied in the Axiom of Connection, this is a necessary condition for establishing simultaneity as an equivalence relation. (Leibniz adds: "It is the same regarding coexistence in the same place" (A VI 4, 944).)

> *Duration* is continuity of existing. Time is not duration, any more than space is collocation. And it would be inapt to say that a day is a duration, when we say rather that ephemerids endure for a day. Time is a certain continuum according to which something is said to endure. (A VI 3, 484/DSR 41)

He makes the same point many years later in his engagement with Malebranche's philosophy. In his dialogue, "Conversation of Philarète and Ariste" (1712, revised 1715), he demurs from Malebranche's Cartesian thesis that body is nothing but extension, objecting that this is to mistake something abstract for a concrete thing: "I deny that extension is a concrete thing," says Philarète (representing Leibniz) to Ariste (for Malebranche), "since it is an abstraction from the extended" (GP VI 582/L 620). *Something extended*, on the other hand, is a *concretum*, as is something enduring:

> extension is nothing but an abstraction, and demands something which is extended. It needs a subject; it is something relative to this subject as is duration. In this subject it even presupposes something prior to it. It implies some quality, some attribute, some nature in the subject that is extended.... One may say that in some way extension is to space as duration is to time. Duration and extension are attributes of things, but time and space are considered to be something outside of things, and serve to measure them. (GP VI 584/L 621–2)[93]

Similarly, Leibniz urges Clarke to recognize this distinction in their dispute over the ontic status of space and time:

> Things keep their extension, but they do not always keep their space. Each thing has its own extension, its own duration; but it does not have its own time, and does not keep its own space. (To Clarke, V, §46: GP VII 399/LC 69)

This is, of course, a traditional distinction, not one of Leibniz's own innovations. Leibniz can be understood to be appealing to Malebranche and the Newtonians in a common language—or at least, the one which comprised their common point of departure, that of the Cartesians. For even Descartes upheld the distinction between "the duration of the enduring thing," or "concrete time," and the abstract time insisted on by Gassendi.[94] In Leibniz's own deep metaphysics, however, the grounds of extension and duration are immensity and eternity, both of them

[93] Cf. also what Leibniz wrote to De Volder in July 1699: "Extension is an attribute; the extended, or matter, is not a substance, but substances. Moreover, duration, time, and the enduring thing, on the one hand, correspond proportionally to extension, place, and the placed thing, on the other" (GP II 184?/ LDV 104–7).

[94] See Arthur (1988) and Gorham (2007), (2008) for discussion of Descartes' views on time in relation to Gassendi's.

attributes of the Absolute, or God. These divine attributes are extension and duration insofar as they lack any limits, whereas created things, which "participate in" these divine attributes, are necessarily limited or bounded. The bottom line, however, is that every single created thing has its own extension and its own duration, and a bounded one in each case.

What Cover takes away from these distinctions is the idea that Leibniz is an "eliminativist" about time. Statements apparently committing us to "substantival idioms" are instead interpreted through "adjectival paraphrases" which "express spatial and temporal truths by quantifying over material bodies and events bearing properties and relations" (Cover 1997, 302). But it is surely contentious to suggest that this is to "eliminate" time *tout simple*. On the one hand, the relations in question are still temporal; and on the other, the "enduring" of substance is not a mere "adjective" of it, but part of its essential nature.[95] There is no deeper ontological level on which the enduring of substances is eliminated. Moreover, lest it be objected that permanence is one thing and succession another, we have Leibniz's own response to De Volder: "all individual things are successive ôr subject to succession.... Nor for me is there anything permanent in those things but that very law that involves the continued succession in individual things, corresponding to the law that is in the whole universe" (LDV 288/9).

To summarize: times and places do not have to be posited as independently existing. Considered in themselves ("outside of things") they are mere *abstracta*, mutually indistinguishable, and cannot therefore be individuated. Real times and places, on the other hand—"concrete" times and spaces that are "occupied," and therefore "truly actual"—are identified by what occupies them, what is happening at them, namely extended or enduring things.

All of this, it seems to me, agrees well enough with the exposition of Leibniz's theory of time given above in section 1.1 and in Appendix 1. Thus, suppose we are given two determinate states or events, such as the birth and death of Mohammed (b and d). Then the same relational fact can be expressed as b's being before d ($b <_t d$), and d's being after b ($d >_t b$); but the relation of temporal precedence itself, $<_t$, is something abstract, an *ens rationis*. Similarly, the time between b and d, both being states of the actual world ($b, d \in W_A$), is a time "occupied" by the durations of the things occurring in it, specifically in this case, the life-course of Mohammed. But considered as a *quantity* of time, a stretch of time within which any possible states might have occurred, such a time is purely ideal. Again, an *instant* considered as a simultaneity class of possible monadic states, that is, the class of states simultaneous with some arbitrary state, is *ideal*,

[95] It is possible that Cover's "adjectival" account is related to Edward Slowik's proposal to replace the absolute/relational dispute by the proposal of a "property theory of space and time." See Slowik (2019a).

a mathematical abstraction indiscernible from any other instant; and so is *time*, the family of these ideal instants successively ordered, $\{I_i, <_t\}$. An *actual instant*, on the other hand, can be identified only with respect to some determinate state or event happening at it.

Leibniz's talk of "real times" populated by "actual instants," however, is apt to lead to misunderstandings. Some have taken his occasional references to time and space as *phenomena bene fundata*[96] to be incompatible with the ideality of time. It is possible, though, that the intended referents are simply these "real times" and "occupied places," namely the concrete durations of things and places occupied by bodies, not time and space as such. But that is a different position from the Russellian one with which Cover finds fault. According to Russell, an instant of time is identified as a group of actual events; these are "such that no event outside the group is simultaneous with all of them, but all the events inside the group are simultaneous with each other," time being constituted by a continuum of such instants.[97] As Cover objects, however, this is not Leibniz's position, according to which time is an ideal notion, applying equally to actuals and to possibles, and not merely to actuals. As we shall see, Leibniz is adamant that time cannot be composed of instants, and this applies whether these are taken to be times at which changes actually occur, or abstract instants.

There is nevertheless a sense in which Leibniz wishes to hold that the ideal temporal order, governing actual as well as possible successions, is "real." Here "real" is being contrasted with what is merely imaginary, as opposed to how successive things are ordered in fact. Thus as he wrote in a long draft (not actually sent) of a letter to De Volder of July 6, 1699:

[96] A paradigm instance is in Leibniz's letter to Arnauld of October 9/19, 1687 (GP II 118–19; A II 2, 249/LA 253) where he refers to matter "taken as mass in itself" (i.e. without the entelechies in it) as "nothing but a pure phenomenon or well-founded appearance, as also are space and time." Cover thinks that this view of time as a phenomenon is an earlier position which Leibniz abandons in his maturity in favour of viewing time as ideal and as an order of all possible successives. He also notes (1997, 303) the passage translated by Loemker in "First Truths" (now dated as 1689): "Space, time, extension and motion are not things, but modes of consideration having a foundation" (A VI 4, 1648/L 270), observing that in fact Leibniz had deleted this passage. A similar passage occurs in LH 4, 3, 5e Bl. 24 of around 1695, where Leibniz refers (in the context of time) to space and motion as well-founded phenomena: see the quotation given below. Cf. also the passage in the unsent earlier draft of his letter to De Volder of January 19, 1706 (GP II 282/LDV 338–9): "Extension, like time and bulk, and the motion that consists of variations of these, disappears into the phenomena...." Finally, the supplementary study in Leibniz's letter to Des Bosses of February 15, 1712 suggests a different interpretation of the way in which they are phenomena: "And so the reality of bodies, of space, of motion, and of time, seems to consist in the fact that they are phenomena of God, that is, the object of his *scientia visionis*" (LDB 231–3).

[97] This is quoted from Russell's discussion in his *Our Knowledge of the External World* (1929, 126), cited by Cover at (1989, 299). Cover also cites similar conceptions of time as being identical with ordered sets of actual events from Nicholas Rescher (1979, 84), W. H. Newton-Smith (1980, 6–7), G. J. Whitrow (1980, 36), Hugh Lacey (1968), and Robin Le Poindevin (1990). See Cover (1989, 289 and 299).

> Time is no more nor less a being of reason than space. To co-exist, and also to exist before or after, is something real; although, I admit, they would not be real in the way substances or matter are commonly taken to be.
>
> (GP II 183/LDV 106–7)

This tallies with what he had written many years earlier (March–June 1681?), in a fragment he titled *The Origin of Souls and Minds*:

> That reality differs from the existence of a thing at a certain moment can perhaps be understood from the fact that time itself has a certain reality, yet it cannot ever be said that it exists now. But it is easier to show what these things are not, than to explain what they are in words, and demonstrate them through reasons.
>
> (A VI 4, 1461/LLC 263)

Leibniz's attempts to explain this in words elsewhere involved the appeal to space and time as "real relations." As detailed in Appendix 2, these are conceived as obtaining their reality through their being in the divine mind.[98] They do not, however, exist in the sense of existing at a certain time, on pain of an infinite regress. As Leibniz wrote in his dialogue of 1676, *Pacidius to Philalethes*: "Time itself ought not to be said to exist or not to exist at some time, otherwise time will be needed for time" (A VI 3, 565/LLC 209).

The difficulty Leibniz was grappling with is this. According to the traditional argument against the reality of time which he had inherited from Ockham and the Scholastics, time does not exist, because there is no time at which its parts exist together: "it cannot ever be said that it exists now." This will, of course, apply to any thing that exists successively, such as motion. Thus, for example, we find Leibniz arguing in the *Specimen Dynamicum* published in 1695 that "just like time, motion never really exists, if you take the matter in an exact sense [*ad ἀκρίβυιαν*], since a whole never exists when it does not have coexisting parts" (GM VI 235). From this he concludes that nothing exists of motion but what exists in it at each moment, namely the derivative force which is an accident of the enduring primitive force: "And so nothing in motion is real but that momentaneous something which must be constituted in a force striving towards change" (GM VI 235). This being so, motion seems to be reduced to a succession of momentaneous states, and time to a succession of instants.

But now the worry is that such a view should commit one to the composition of time from instants. Massimo Mugnai has published a fragment Leibniz

[98] Commenting on Leibniz's claim that a relation "outside the subjects," "being neither a substance nor an accident,... must be a purely ideal thing" (GP VII 401/LC 71), Cover says this is 'Leibniz's considered way of saying, *there is no such thing*' (Cover 1997, 305). Leibniz did indeed deny that time in abstraction from the things in it is an entity, a *thing*; but not that it has no reality at all.

wrote (perhaps in 1695) on a small piece of paper[99] in which he expresses just this worry:

> That a line is not composed of points is demonstrated by means of transverse lines; and by this argument it is shown that if you believe the contrary, then the whole will equal the part.
>
> It seems that time, however, must necessarily be composed from instants, because two instants cannot exist simultaneously; and so only the present instant exists; the past existed, the future will exist.
>
> But it seems possible to cut a line in proportion to time, so that there is a correspondence between anything that is supposed in the line and what is supposed in time. For example, in a uniform motion in a straight line at any moment a moving point exists in a new point of space. However, I see that it does not follow from this that a line is composed from an infinity of points, but only that there are infinitely many points in the line.
>
> Here indeed is the difficulty: If time is composed of instants, then there is a determinate multiplicity of instants in time. But is this therefore equal to the multiplicity of points in a line?

Some commentators have seen Leibniz as endorsing the view that time is composed of instants, citing this piece in partial support.[100] However, Mugnai's transcription left out the continuation and conclusion of the fragment on the other side of the slip of paper, in which Leibniz answers his question about whether time can be composed of instants firmly in the negative:

> I do not see how this can be denied. But then we are reduced into the same absurdities as arise if we compose a line from points.
>
> Therefore it must be said that time is not composed of a definite [*certi*] number of instants. From these and similar considerations it is concluded that space, body and motion are not things but phenomena, although well-founded ones.

In a manuscript written some time after 1693, first published by Luna Alcoba (1996), this same problem features very prominently. So the Mugnai manuscript perhaps constitutes a return to and resolution of the problem raised there, as

[99] Mugnai (2000, 135–7) gives the dating as "probably after 1695." Osvaldo Ottaviano has recently shared with me a full transcription of the fragment (LH 4, 3, 5e Bl. 24).

[100] Thus Vincenzo De Risi writes "That time must be constituted by instants is not something Leibniz can ever doubt" and that "time must be regarded as a set of instants, not a continuum with indistinct parts" (2007, 274, 275). He cites the Mugnai fragment on pp. 276–7. As we will see below, Carla Rita Palmerino interprets Leibniz as claiming "that time is composed of instants, as there is nothing real but instants" (2016, 77). She references De Risi (77–78), as well as Mugnai's "Two Leibniz Texts" (84, 97).

suggested by many points of correspondence. Carla Rita Palmerino (2016) has recently drawn attention to Luna Alcoba's neglected article, showing (by a very detailed and scholarly comparison) that the manuscript he transcribed is in fact Leibniz's reading notes and comments on Libert Froidmont's *Labyrinthus sive de compositione continui* (1631).[101] Thus we find in it the following passage:

> Argument: Time is composed of instants, for it is composed of those things which exist in it, but only instants exist in it, because only instants exist. In fact, there is no time, nor is time an entity, any more than is motion. Things move. Things are, will be, were. And just as many other things will be interjected. And this we estimate by uniform action, the continuous effect of which is similar to itself. (Luna Alcoba 1996, 185; Palmerino 2016, 94)

Here the "argument" given in the first sentence is not Leibniz's own, but an amplification of the following passage from Froidmont:

> Time and motion, they say, are composed of supposed pure moments (. . .) since in time nothing can exist except the present, for the past is no more, the future is not yet. (Froidmont 1631, 179; Palmerino 2016, 77)

As Palmerino notes, Froidmont is only reporting the argument, not endorsing it. On his own view, in Palmerino's words, "'the moment which we call present' does not add anything to the parts of past and future, but is rather 'the past and future at the same time, in the same way in which the point at which two lines touch does not add anything to their respective lengths'" (Palmerino 2016, 77). This is a nominalist position (consistent with Ockham's views, as noted above), whereas Leibniz's invocation of instants seems to contradict it. But (and here I disagree with Palmerino), Leibniz is not endorsing the composition of time from instants any more than had Froidmont. He does indeed emphasize that only instants exist in time, but this is perfectly compatible with his denials, in the Mugnai manuscript and elsewhere, that time is an entity capable of being composed.

Some light can be thrown on this by consideration of another manuscript, possibly from the same period (1695 or after), where Leibniz also compares time with a line—a manuscript that we will be examining in detail in the next section (§1.5). There he is adamant that continua "such as a line and time are ideal things like numbers," stressing that the comparison with numbers divisible into fractions is when place and time "are considered in themselves and abstracted from the

[101] In this piece Leibniz had abandoned his usual practice, when making reading notes on other authors, of clearly distinguishing his own comments from quoted text. This had led Alcoba to misinterpret the whole piece as presenting Leibniz's own views, involving many non-standard positions and sources. Palmerino has corrected this mistake, distinguishing the two by reference to Froidmont's book.

things existing in them."[102] Meanwhile, however, "in a spatial or ideal line, infinitely many actual points can be assigned, actually distinct from one another, and endowed with different motions." Yet "a line can no more be composed of points than a number from numerical minima." Because "there are no two actually assignable points between which another cannot be actually located," these actual points will be dense in the ideal continuum, but will not compose it:

> since in nature no two points can be assigned in the universe which have precisely the same motion, still, a continuous line will never therefore be resolved into points, nor into any finite division, nor will points compose a continuous line. Nor will two assignable points running into each other with different motions ever touch one another, since other points are always interposed.[103] The same holds for time, the actual instants of which contain just as many fulgurations of the divinity, ôr states of the universe.
>
> (*Locus et tempus sunt continua*, LH 37, 5, Bl. 134^{v})

Here the "actual points" are taken to be smallest portions of matter that can be individuated by their differing motions, yet that are still divided within by other parts with motions differing by smaller amounts still. They are dense in the continuum, but always separated from one another by indeterminate intervals of further matter corresponding to the differential of his calculus, as I shall argue in section 1.5. The difficulty of "how a whole body could be impelled if points are not touching each other immediately" is parried by the claim that mass is phenomenal, not substantial, otherwise the intervals between its actual points would be actual infinitesimals.

Similarly, in the Luna Alcoba manuscript Leibniz maintains that no two points can be immediately next to one another, and that it follows that points are "mere modes of thinking":

> Hence it will follow that two points cannot be immediately next to one another. What then? unless points are not things, but a mode of thinking.... But instead of points, lines and surfaces, are to be understood bodies whose depth, length and breadth is of an arbitrary minuteness, and thus are not determinate things, and this suffices in mathematics, that the error can always be made arbitrarily small.[104] (Luna Alcoba 1996, 189–90; Palmerino 2016, 96)

[102] This is the manuscript *Locus et tempus sunt continua*, LH 37, 5, Bl. 134, to be published in translation in a proposed volume by myself and Ottaviani, *Leibniz on the Metaphysics of the Infinite*.

[103] This notion that "other points are always interposed" perhaps explains his otherwise mystifying remark in the Alcoba piece that "just as many other things will be interjected" in any continuous stretch of time.

[104] Leibniz's reference here to his philosophy of infinitesimals is consistent with his reference in the same piece to his publication in the "*Journal des Sçavans*, 14 Sept. 1693, about the sum of infinite

The analogy between the points and instants therefore suggests that the "actual instants" are arbitrarily small parts of concrete time, individuated by changes happening in them, even though they are further divided into still smaller parts by actual changes within, so that in any time there are instants within instants to infinity. These instants are always separated from one another by indeterminate intervals of time corresponding to the differentials of his calculus. But they do not compose a continuous time.

To summarize: according to Leibniz, time is not an existing entity. It is an ideal order, an order governing the changes in successive things. What exists in it are these actual changes, separated by unassignable intervals, and derivatively the times or instants at which they occur. These instants are dense within any temporal interval, but do not compose into a continuum, which would contain not just the changes that do occur, but any possible changes. As a continuous order, time applies equally to actuals and to possibles.

Now this last consideration is precisely Leibniz's point in his controversy with Clarke over whether time could exist before the creation of the universe. The point is worth explicating a little, since as we have seen Leibniz's relational philosophy of time is often taken to be simply a structure of relations among actual events or states. This it is not: as we have seen, for Leibniz time and space are not just orders of actual existents, but of *all possible existents*. As Leibniz explained to Pierre Bayle in 1702,

> But Space and Time taken together comprise the order of possibilities of one whole Universe, so that these orders (Space and Time, that is) relate not only to what actually is, but also to whatever could be put in its place, just as numbers are indifferent to whatever the *res numerata* may be. And this involvement of the possible with the existent makes a continuity which is uniform and indifferent to every division. (Reply to Bayle, 1702: GP IV 568/L 583)

Consequently Leibniz could not follow the Scholastics in their rejection of "pre-mundial time" on the grounds that there was no actual change before the Creation: he was obliged to find a stronger argument, one pertaining to the *possibility* of change before the Creation.

Aristotle, of course, had no such problem. Having asserted the inseparability of time and change—a principle that Newton-Smith has appropriately dubbed "Aristotle's Principle" (1980, 14)—and noting that we cannot conceive of a time without conceiving an earlier time, he simply concluded that the Cosmos must be indefinitely old. Here the Christian theologians could not follow him, knowing

solicitations. And that nature never produces any action except by a veritable infinite multitude of concurrent causes" (Palmerino 2016, 78). The idea is that even an infinitely small element of motion, a *conatus* or endeavour, is resolved into infinitely many solicitations.

from the Bible that the world had been created only a finite time before. Faced with this dilemma, St. Thomas Aquinas, for instance, had argued that the time which could thus be *conceived* before the Creation must be just that, a conceptual or purely imaginary time. The alternative was to adopt the Stoics' solution of asserting the finite duration of the universe to be a mere interval of a self-existing infinite time, thus abandoning Aristotle's Principle altogether. But against this St. Augustine had already objected that it would entail a "falling back into a *loose indifference*" on the part of the Deity. It would mean, wrote Augustine, that "God was guided by chance when he created the world in that and no earlier time... although there was no difference by which one time could be chosen in preference to another." It would be as if God had "set the world in the very spot it occupies by accident rather than by divine reason".[105]

It hardly needs saying that this argument contains more than a hint of Leibniz's Principle of Sufficient Reason, and Leibniz himself does not miss the opportunity to avail himself of that principle in arguing against premundial time in his Third Letter to Clarke. Nevertheless, the argument from lack of sufficient reason is not adequate on its own to refute the idea of premundial time, as Leibniz appears to have realized. For Newton and Clarke did not share his and Augustine's conception of God as being outside of time: for them, time and space were coextensive with God, not the world; and according to their voluntarist theology, God's will by itself could be a sufficient reason for any of his actions. Thus Augustine's application of the principle of sufficient reason would be an unacceptable limitation on God's power of choosing. But Leibniz's argument against the possibility of premundial time, although it makes use of this principle, is subtly different from Augustine's, and altogether more potent. It is worth quoting in full:

> Supposing someone were to ask why God did not create everything *one year sooner*; and the same person should wish to infer from this that God has done something for which it is *not possible* that there should be a *reason*, why he did it *so* rather than *otherwise*: one would answer him that this inference would be true if time were something apart from temporal things, for it would be *impossible* that there should be *reasons* why things should have been applied to these *particular instants* rather than *others*, their succession remaining the same. But the same argument proves that *instants* apart from things are *nothing*, and that they consist only in the successive *order* of things; which order remaining the same, *one* of the two states, for instance that of the imagined anticipation, would not be any different from, and could not be discerned from, the *other* which is now. (To Clarke, III, §6: GP VII 364/LC 27)

[105] Augustine, *City of God*, quoted from Čapek (1976, 180).

The first half of this argument is similar to Augustine's, but not quite the same. Augustine had argued directly against the possibility that all things, events, etc., in the world might have been created earlier on the grounds of lack of sufficient reason. Leibniz, by contrast, does not just dismiss this possibility but uses it as a premise in a *reductio* argument against absolute time. He is prepared to provisionally grant the Newtonians' claim that the world might have been created sooner in order to demonstrate its absurdity. Thus he considers the case in which the date of the Creation and all subsequent events are displaced in this way as another possible state of affairs distinct from the actual one, and argues that if the temporal relations of a given event to all other events were still the same, the instant at which that event occurs would be indistinguishable in the two cases. Thus the supposedly earlier event in the hypothetical case is not in fact distinct from the corresponding event in our world, an obvious absurdity. It follows that the instants and indeed the entire temporal orders would be identical in the two cases.

Implicitly, Leibniz is here appealing to the Principle of the Identity of Indiscernibles, but in a context in which it is very difficult to deny. For if one posits temporal relations as existing among instants of absolute time, that is, independently of things and their states, and if the states of things are said to have their positions in time by virtue of occupying some such instants, then—since mathematical instants are in themselves all identical—there is no way of telling which instants they occupy; unless, of course, instants are individuated by the states or events occupying them, which is Leibniz's view.

It is important to appreciate the superiority of this position over Augustine's. For if Leibniz is interpreted as simply upholding the long tradition of the Schools that denied the existence of time in the absence of change—as Clarke interpreted him, and as many commentators still do—then his position is easily countered. For if one chooses to *define* premundial time as "imaginary," as had Aquinas, one does "not thereby *prove*," as Clarke rightly pointed out, "that [it] is *not real*." One might equally claim that all it means is "that we are wholly ignorant of *what kinds of things* [take place] in that [time]."[106] In other words, one cannot prove that the possibility of the world's being created sooner is not a *real* possibility by stipulation, since it is equally possible to stipulate the contrary, as did Gassendi, following the lead of the Stoics.[107]

[106] Clarke's argument (Third Reply, §2: GP VII 368/LC 31) was actually addressed to the case of "imaginary space," and I have taken the liberty of adapting it for the case of time.

[107] In the *Syntagma Philosophicum* Gassendi had argued: "Thus we say that the world could have been created a thousand years before its creation, not because there were years marked off by the repeated revolutions of the sun, but because Time flowed of which the appropriate measures, the revolving motions of the sun, *could* have existed then. And we do not say that all these times were imaginary..." (*Physicae, Sectio I, Liber I*, transl. Milič Čapek and Walter Emge, in Čapek 1976, 201).

For Leibniz, on the other hand, the impossibility of premundial time is not a consequence of an a priori identification of time with the duration of the actual universe. Indeed, as we have seen, for him time relates equally to all *possible* changes in the universe. However, in the absence of any possible change or event that might individuate it, an instant of this time of possible change is simply *ideal*. Thus Leibniz does not *assume* that premundial time is ideal, but, as he himself says, uses the argument from indiscernibility to "*demonstrate* that time without things is nothing more than a mere ideal possibility" (To Clarke, V, §55: GP VII 404/LC 75). For on his relational view, the possibility of a given event's occurring at a different time is represented by its bearing different temporal relations to other events in the temporal order. (This is precisely analogous to Leibniz's interpretation of change of place of a body as a change in its location within the order of situations.) But the Creation of the Universe is not such an event, since it cannot change place within the temporal order: no possible event could occur before the beginning of the world because it would then *be* the beginning of the world: "The beginning, *whenever* it was, is always the same thing" (IV, §15: GP VII 373–4/LC 38). Therefore

> if anyone should say that this same world which has actually been created could have been created *sooner*, without any other change, he would be saying *nothing intelligible*; for there would be no mark or difference by which it would be *possible* to know that it had been created *sooner*. (To Clarke, V, §55: GP VII 405/LC 75)

Thus we see that no elaborate reconstruction of Leibniz's position in terms of possible world semantics is necessary for an understanding of it. Indeed, the very point of the above argument is that a world in which the mutual temporal relations among all the states or events were the same, but in which any of the states or events occurred sooner, is *not* a possible one, since there is no criterion of individuation by which one could discern different instants within a given temporal order save by their correspondence to states or events occurring at them.

Without any actual things or processes to individuate them, that is, the parts of time are perfectly similar to one another: abstract time is perfectly homogeneous. Leibniz is explicit about this. In the *Initia Mathematica: De quantitate* of 1682 he writes:

> Those things are *homogeneous* which are either similar to each other or can be rendered similar to each other by a transformation, as a curve can be rendered similar to a straight line. Two straight lines are homogeneous, because similar; but a straight line and the arc of a circle are homogeneous things because the circle can be extended into a straight line. (GM VII 30)

and in the *Initia Rerum Mathematicarum Metaphysica*:

> *Time can be continued to infinity.* For since the whole of a time is similar to its part, it will be related to another time as its part is related to it, and so it can be understood to be continued into another greater time. (GM VII 22/L 667)

From these definitions it follows that the parts of any time are similar to one another and to the whole, so that time itself is homogeneous. Likewise, if the duration of some concrete process, like the rotation of the Earth on a given day, is equal to the duration of another contemporaneous process, such as the measure of time given by the motion of the hour hand of a clock through two full revolutions, and the duration of another day is equal to the same measure, or at least to one equal to it after a time translation, then the durations will be equal and similar, and therefore congruent. "Thus," Leibniz writes, "a pound of gold and a pound of lead are congruent; and so are today and yesterday" (GM VII 274).[108] (These latter considerations are key to his defining the measure of time.)

We can achieve further clarification of Leibniz's position on the homogeneity of time by comparing it with Newton's views on the same subject, particularly as expressed in his manuscript *De gravitatione*, first published in the 1960s. The apparent similarities of the views Newton expresses there to Leibniz's views (as well as his reference to space as an "order of situations" in the Scholium to the Definitions of the *Principia*) have moved Howard Stein to remark that "Newton is by no means so far as one might be led to suppose from Leibniz's view that the essence of space and time is in some sense relational" (1968, 194). For in *De gravitatione* Newton seems to anticipate Leibniz in his conception of time as an order, and also in his claim that instants are only individuated by their place in the order:

> For... the parts of duration are individuated by their order, so that (for example) if yesterday could change places with today and become the later of the two, it would lose its individuality and no longer be yesterday, but today... By their mutual order and position alone are the parts of duration and space understood to be just what they are in fact; nor have they any other principle of individuation besides that order and those positions, which therefore they cannot change.[109]

Commenting on this passage, Stein writes that "the idea formulated by Leibniz as the principle of indiscernibles is obviously a familiar one to Newton" (1967, 194).

[108] Here I must respectfully disagree with the learned De Risi, who insists that "congruence only applies to situational relations," and that "It naturally follows that, in order to be a continuum liable to be measured, time must [*per impossibile!*] arise from situational measures" (De Risi 2007, 274).

[109] Newton (2004, 25; translation somewhat modified in accordance with that of Stein, 1967, 194).

But a closer inspection shows that the similarities are not as great as they first appear. For although Newton's claim that "All things are placed in Time as to order of succession" (1687, 12) evokes Leibniz's definition of time as "the order of successive things," for Newton this does not constitute a definition. The "parts of duration," despite obtaining their individuality through their positions in the order, are not, for Newton, to be equated with the temporal locations of any *thing*, actual or possible: for duration is a mode of existence also of God, who exists independently of things. Accordingly there is a great deal of difference in the nature of the order of succession as conceived by the two thinkers. Newton conceives the order as obtaining among the parts of time considered *prior to any states of things* that might individuate them: they are individuated solely by their position in the order.[110] For Leibniz, on the contrary, the parts of time, taken in themselves, are indistinguishable. The Identity of Indiscernibles precisely does *not* apply to abstract things, only to existents. The order of the parts of time for him is parasitic on the order obtaining among the determinate states of things that individuate them.

Thus whereas for Leibniz the parts of time, considered in themselves, are completely indiscernible and wholly similar, and therefore *homogeneous*—each such interval being similar to any temporal continuum of which it is a part—for Newton, on the other hand, they are intrinsically distinguishable by their place in an order which exists absolutely (presumably in the mind of God), so that the parts of time are therefore *inhomogeneous*, even prior to their being filled by things or events. It is this very inhomogeneity or individuality of their parts prior to their being occupied by things or events that makes for the immutability and immovability of absolute space and time, according to Newton:

> As the order of the parts of Time is immutable, so also is the order of the parts of Space. Supposing these parts to be moved out of their places and they will be moved so to speak) out of themselves. For Times and Spaces are as it were their own places as well as the places of all things. All things are placed in Time as to order of succession, and in Space as to order of situation. It is from their Essence that they are places, and it is absurd for the primary places of things to be moved.
> (Scholium to the Definitions, Newton 1687, 12).

Now in modern physics the homogeneity of time is usually represented by the requirement of the invariance of all the laws of physics under the operation of *time translation*. That is, it is acknowledged that in a homogeneous time the translation of the dates and times of all events and states, say, backwards one year, can make

[110] This difference is noted by Stein: "Not identical, however: the relations that constitute space and give its parts their individuality are according to Newton *internal* relations" (1968, 194). The latter Bradleian term is not, however, used by Newton.

no real difference to our description of the world. Arguably, however, this cannot be so for Newton, since he claims that God must intervene from time to time in order to compensate for irregularities arising from the interaction of comets and planets, "which will be apt to increase till this system wants a reformation" (Query 31;Newton 2004, 138). For Newton this entails the non-conservation of quantity of motion, which must be compensated by the actions of God, since new motions require the agency of God or man or other living sources of active power—which is not otherwise conserved. This is made clear by Samuel Clarke on Newton's behalf in his correspondence with Leibniz. In his Fifth Letter, which Leibniz did not live long enough to see, Clarke (perhaps wilfully) failing to heed Leibniz's distinction between quantity of motion and quantity of force, writes:

> Indeed, all merely mechanical communications of motion, are not properly *Action*, but mere *Passiveness*, both in the bodies that *impell* and that are *impelled*. *Action*, is the beginning of a motion, where there was none before, from a Principle of Life or *Activity*: and if God or man, or *Any Living or Active Power*, ever *influences* any thing in the *material World*; and every thing be not *mere absolute Mechanism*; there must be a continual *Increase* and *Decrease* of the *Whole Quantity of Motion in the Universe*. Which this learned Gentleman frequently denies. (fifth reply to Leibniz, §§ 93–5; GP VII, 433/LC 110)

Leibniz denied that our world could be of this kind. He denied that God would have created a world so imperfect as to need occasional "restorations," and in any case rejected the idea that changes of motion would require a non-mechanical source. He argued that God would have equipped all the substances of the world with a source of active power of their own, and that without a constant measure of this active power in the universe, the world might either run down, contrary to what is observed; or the power might increase, making perpetual motion possible. Active power, for Leibniz, was not quantity of motion, but "force," the ability to do work, and he correctly identified its dimensions, so that it is equivalent both quantitatively and qualitatively to the modern concept of energy. I suggest that it is not outrageous to see in this linkage in Leibniz's thought between the homogeneity of time and the conservation of force a partial anticipation on his part of the relation between the translational symmetry of time and the conservation of energy encapsulated in the mathematical theorem discovered by Emmy Noether, a theorem that lies at the heart of modern physics.[111]

[111] According to Noether's Theorem, every differentiable symmetry of the action of a physical system has a corresponding conservation law. For example, if a system behaves in the same way however it is oriented in space (if its Lagrangian is invariant under a change of orientation), the theorem dictates that there will be a conserved quantity, namely the angular momentum. Similarly, if its Lagrangian is invariant under a translation in time, the corresponding conserved quantity is energy.

Of course, Newton did not in any case identify time with the order of succession, as did Leibniz. He regarded time as a precondition for existence, even for God's, and as consisting above all in *quantity*. This was a major motivation for Newton in making time fundamental to his mathematical physics, and was one of the main grounds on which the Newtonians attacked Leibniz's theory of time as inadequate. Thus (with Newton's tacit approval) Samuel Clarke objected to Leibniz's definition of time as an order of succession that if time is to be a quantity, it must be something more than just an order:

> The *Order of Things succeeding* each other in Time, is not Time itself. For they may succeed each other *faster* or *slower* in the same Order of Succession, but not in the same Time. (Clarke to Leibniz, IV, §41: GP VII 387–8/LC 52)

Leibniz replied by denying that the order of succession *could* remain the same if the time were greater or less:

> For if the time is greater, there will be more successive and like states interposed, and if it is smaller, there will be fewer, since there is no vacuum in times, nor, so to speak, any condensation or penetration, any more than there is in places.
> (To Clarke, V, §105: GP VII 415/LC 89–90)

This reply has puzzled Leibniz's modern interpreters. For if the states in question are monadic states, and one supposes these to be strictly instantaneous—i.e. of zero duration—then there would be infinitely many of them in any finite time, and no question of more or fewer of them composing a greater or smaller time. This has suggested to some that Leibniz's theory of time is at bottom *atomistic*, that each state is a temporal indivisible.[112] But there is no independent evidence for such a view, and it would fly in the face of his statements on the continuum problem.

A more prosaic interpretation would be that the states referred to are ordinary physical states. In fact, the reference to *like* states already points us in this direction. For this seems to indicate that we must already have made an abstraction to their relevant *similarities*; and if we have made abstractions, then we must be dealing with states of objects that appear among the phenomena,[113] such as the

The products of the angle and angular momentum, and likewise of time, and energy, or of (linear) position and momentum, all have the dimensions of *action*, the quantity first identified by Leibniz, who correctly defined its dimensions as well as stressing its importance for physics.

[112] J. E. McGuire, for example, argued in his (1976) that Leibniz had "an atomistic theory of duration and becoming [that] is incompatible with the unity of being that he attributes to real, substantial existence" (316), and also that for him "the perceptual continuum and the perception of becoming are . . . atomic in character" (314).

[113] Cf. Leibniz to de Volder, June 20, 1703: "Indeed, there is as much difference between substance and mass as there is between complete things, things as they are in themselves, and incomplete things,

successive swings of Huygens' pendulum clock. These oscillations would be different in themselves, involving different sequences of states of their constituent monads, but alike in relevant respects.

This reading is confirmed by Leibniz's discussion in the *Nouveaux essais*. To Locke's claim (*Essay*, II, ch. xiv, §16; 1690, 148) that "it is not then *motion*, but the constant train of *ideas* in our minds whilst we are waking, that furnishes us with the idea of duration," Leibniz replies:

> A train of perceptions arouses in us the idea of duration, but it does not create it. Our perceptions never have a sufficiently constant and regular train to correspond to that of time, which is a uniform and simple continuum, like a straight line. The change of perceptions gives us occasion to think of time, and we measure it by uniform changes; but if there were nothing uniform in nature, the time would still be determined, just as place is still determined even when there is no body that is fixed or immobile. The fact is that knowing the rules of non-uniform motions we can always relate them to intelligible uniform motions, and by this means we can predict what will happen when the various motions are combined together. In this sense time is the measure of motion, that is to say, uniform motion is the measure of non-uniform motion.
>
> (A VI 6, 152/RB 152)

Thus, in response to Locke's claim in §21 that "No two Parts of Duration can be certainly known to be equal" (Locke 1690, 150–1), Leibniz agrees with him that the inequality of the diurnal motions of the Sun—the natural days—has been revealed by the more exact measurements using the pendulum (A VI 6, 152–3/RB 152–3), as by Huygens in his *Horologium oscillatorium* (1673).[114] Locke had claimed that since "no two portions of succession can be brought together," all we can do in measuring time is "to take such as have continual successive appearances at seemingly equidistant periods"—"seemingly," because we have to measure them against "such as the train of our own ideas have lodged in our memories" (II, iv, §21; Locke 1690, 151). To such a naive empiricism Leibniz has correctly responded that the way we determine successive intervals of time to be equal is by appeal to uniform motion. So when in §22 Philalethes [Locke] remarks that it is very strange that time should be called (as by Aristotle) "the measure of motion," Leibniz is able to have to Theophilus respond that he has just explained how this should be understood, adding:

things as we grasp them through abstraction. In the phenomena we can define through abstraction whatever we want to ascribe to each part of mass, and everything can be distinguished and explained rationally, something that necessarily requires abstractions" (GP II 252–3/LDV 267).

[114] It is also consonant with claims Leibniz had made in earlier fragments on time from the mid-1680s: "The basis for measuring the duration of things is the agreement that obtains when different uniform motions are assumed (such as different accurate clocks)..." (A VI 4, 628–9/LLC 274–5).

> Indeed we could say that a duration is known by the number of equal periodic motions, each beginning when the preceding one ceases, for instance, by so many revolutions of the Earth or the stars. (§22; A VI 6, 153/RB 153)

The basis for regarding uniform motions as equal, moreover, is not merely empirical, but theoretical. In the *Horologium*, Huygens determines the equality of the swings of the pendulum by calculations based on the premises that in the absence of gravity and air resistance, a body will "continue its given motion with uniform velocity in a straight line" (Huygens 1986, 33), and that any composition of motions does not impede this. So an inertial motion acts as a theoretical surrogate for a uniform lapse of time, and underwrites the equality of successive intervals of the pendulum clock, or of some other uniform motion, such as that of Jupiter's moons.[115] The idea here is that two uniform straight-line motions that begin together and end together are congruent with one another, and having the same velocity they will trace equal lengths in equal times. The times will then be congruent for any two such motions tracing equal lines successively. On this basis the times of non-uniform motions may also then be determined, according to the known laws of motion. As Leibniz writes in the *Dynamica*,

> It is the same with velocities, for if two earlier velocities are in the same ratio as two later velocities, the later times will also be proportional to the earlier ones, since velocities (when equal lengths are traversed) are in reciprocal proportions to the times.
>
> It is impossible for another way of demonstrating these things to be found, since actions which differ in velocity cannot be reduced to *congruence*, as a little while before we reduced those things which have the same velocity to congruence, and they differ only by the magnitude of the moving body or length; and so a comparison of actions differing in velocity must be obtained by way of *similarity*.
>
> (*Dynamica* Part I, Book III, ch. 1, prop. 4; GM VI 350)[116]

The appeal to uniform motion, however, brings with it a suite of further difficulties. For we have to be able to represent such a motion in a space through time, a space with sufficient structure to support the idea of a uniform motion. Thus according to Leibniz's own definitions, we cannot define the measure of time

[115] Leibniz wisely concedes that measures of uniform motion such as given by the Earth's annual rotation around the Sun could change in the course of time, and the only way we would have to determine this is by "comparing this rotation with others–those of Jupiter's moons, for example—since it is unlikely that if they too underwent change, it would always be at a corresponding rate" (A VI 6, 152–3/RB 152–3).

[116] To this De Risi objects that the allusion to similarity here "remains an obscure one (because we do not understand how objects that have no *situs* can ever be similar)" (De Risi 2007, 269). As noted in Appendix A1.5, however, it is possible to define congruence and similarity in a one-dimensional continuum. See Hellman and Shapiro (2018, chs. 1 and 2).

without considering motion through space. Many modern interpreters have held that Newton's theory of time is clearly superior to Leibniz's on this score. But since a treatment of this issue involves us in a discussion of motion, space and space-time, I will defer it till later in this book (see §3.5).

Let me now summarize the conclusions of this section. We have seen that Leibniz did not identify time with what is in time: the instants of abstract time would be the times at which any possible changes could occur in it, whereas the actual instants in time designate the instants at which change actually occurs. This differs from the "eliminativist" interpretation of Cover, McGuire and others, according to which time does not apply to the actual: "if time is ideal, it cannot apply to the actual" (McGuire 1976, 312), it is part of an "ideal world accessible via abstraction or by thought alone" (Cover 1997, 303). But if temporal relations do not apply "on the monadic level," it is impossible to see how monads can be construed as substances whose modifications, such as derivative force, can come into being successively. I have argued that although Leibniz's nominalism agrees with modern nominalism in denying that *abstracta* like time have to be posited as existing things, considering them rather "as abbreviated ways of speaking," the idea is to reduce such entities to equivalence relations among concrete particulars, such as simultaneity among monadic states, or congruence among durations; and that although the equivalence class of simultaneous things will define an instant of abstract time, an actual instant is identified by reference to what happens at it.

Leibniz, as we have seen, holds that "space and time relate not only to what actually is, but also to whatever could be put in its place." But if what is in time, actual changes or the instants denoted by them, do not compose a continuum, then how can time, as something continuous, apply to them? How, as McGuire inquired, can the Law of Continuity apply to an actuality that is discrete? To answer these questions is the task of the next section.[117]

1.5 Change, Vague States, and Physical Continuity

These questions have caused puzzlement among commentators, and have prompted them to accuse Leibniz of contradicting himself.[118] On the one hand, we have his many proud assertions of his Law of Continuity, according to which

[117] I give a closely related treatment of the arguments that I propose in the following section in a chapter to appear in a Festschrift volume for Massimo Mugnai (Ademollo et al. 2021).

[118] Thus Bertrand Russell charged that "In spite of the law of continuity, Leibniz's philosophy may be described as a complete denial of the continuous" (Russell 1900, 111), a charge partially endorsed much later by J. E. McGuire: "the perceptual continuum is not truly continuous," and "Leibniz explicitly denies that actual substances are continuous in the sense of containing potential and indeterminate parts. On both these scores it seems that his philosophy is a denial of the continuous" (McGuire 1976, 311).

"Nature makes no leaps";[119] and as late as 1715 he was claiming that "*continuity* is found in time, extension, qualities, and motion—in fact, in all natural changes, which never take place by leaps" (GM VII 21/L 671). And yet in a well-known passage from a letter to De Volder 10 years earlier he had claimed (albeit without sending the bracketed last sentence) that

> In fact, matter is not continuous but discrete, and actually divided to infinity, even though no assignable part of space is devoid of matter. Yet space, like time, is something not substantial, but ideal, and consists in possibilities, that is, in an order of co-existents that is in some way possible. And so there are no divisions in it but those that the mind makes, and the part is posterior to the whole. In real things, on the contrary, unities are prior to multiplicity, and multiplicities exist only through unities. [It is the same with changes, which are not in fact [*revera*] continuous.]" (To De Volder, Oct 11, 1705; GP II 278/LDV 327)

But the analogy Leibniz presents here between *divisions in matter* and *changes* is worth pursuing further, since it provides the key to resolving the difficulty. It depends on Leibniz's distinction between true continua such as space and time, which consist only in the potential for division, and what are in space and time, namely matter and changes, which involve an actual division. According to this conception, matter is divided in such a way that every part of it is divided internally, as is each of these parts, and so on *in infinitum*. But no piece of matter is divided in all possible ways, where the points of division would correspond with all possible points in the continuum. Rather, there is an actual sequence of internal divisions that has no end. This is an example of Leibniz's "syncategorematic" interpretation of the actual infinite, according to which however great a number of divisions one might assign, there are in fact more than this, but there is no infinite number of them, and the divided parts are all extended, and never extensionless points. In a manuscript from 1689, Leibniz illustrates the conception by an idealized example of dividing every part, and part of a part, into halves, versus dividing them into thirds:

> A continuum is not divided into points, nor is it divided in all possible ways—not into points, since points are not parts but boundaries, and not in all possible ways, since not all creatures are in a given thing, but there is only a certain progression of them *ad infinitum*. Thus if you bisect a straight line and then any part of it, you will set up different divisions than if you trisect it. (A VI 4, 1648)[120]

[119] For instance, in the *Nouveaux essais* of 1704–5 Leibniz writes: "Nothing takes place suddenly, and it is one of my great and best confirmed axioms that nature never makes leaps. I call this the Law of Continuity..." (A VI 6, 56/RB 56).

[120] This important manuscript was first published by Couturat, and was interpreted by him as showing the logical foundation of Leibniz's metaphysics. It is usually known by its incipit "*Primae veritates* (Primary Truths)," but the Akademie editors give it the title *Principia Logico-Metaphysica*.

Accordingly, if "it is the same with changes, which are not in fact continuous," one might expect it to be similar for the temporal case. Indeed, Leibniz is quite explicit that "as a physical body is to space, so states or the series of things are to time" (To De Volder, June 30, 1704; GP II 269/LDV 305). This would then imply that, just as "in real things, that is, bodies, the parts are not indefinite . . . but actually assigned in a definite way" (268/303), so the same should apply to states. In that case the duration of any real thing should be divided into states—into a certain infinite order of states of finite length, each of them further subdivided. As we shall see in what follows, this does turn out to be Leibniz's position.

But is this not incompatible with the Law of Continuity? If every duration is divided into a certain ordering of discrete states, divided from one another by discontinuous changes, then surely we have no continuity, but a series of discontinuous leaps. This would set Leibniz at odds with himself, since in his *Theodicy* of 1710 he criticized the interpretations of continuous creation proposed by Pierre Bayle and Erhard Weigel (§384–93; GP VI 343) for doing just this. Similarly, he had written to De Volder in September 1699,

> Anyone who completely rejects continuity in things will have to say that motion is essentially nothing but successive leaps through intervals flowing forth not from the nature of the thing but as a result of the action of God, that is to say, reproductions in separate places, and would philosophize almost as if one were to compose matter from mere separate points. (GP II 193/LDV 127)

"This hypothesis of leaps cannot be refuted," Leibniz continues there (193/127), except by an appeal to the "principle of order." That refers to the "*principle of general order*" he articulated in his article in the *Nouvelles de la république des lettres* of July 1687, which he elsewhere calls his Law of Continuity. According to this principle, if the difference between two instances of a presupposed varying thing can be diminished until it becomes less than any given quantity whatever, then the corresponding difference in a co-varying quantity depending on it will also become less than any given quantity whatever. Applying this to states that differ over arbitrarily small intervals of time (as they will if they are actually infinitely divided in the way described above), then their difference will also become arbitrarily small as the intervals of time are diminished, yielding (by the Law of Continuity) a process of change that is continuous in time.

In this section, I will try to clarify this position of Leibniz's on change and temporal continuity by keeping a close eye on its historical development. Extremely important in this regard is his early dialogue, *Pacidius Philalethi*, written while waiting for calm weather to sail to Holland in late 1676.[121]

[121] This is translated in Arthur (2001). See also Samuel Levey's (2003), (2005), (2010), and (2012) for excellent further discussion of this dialogue, and Leibniz's treatment of time in it; for some rejoinders on the few points on which we disagree, see my discussion in Arthur (2018, ch. 5).

Although the conclusion of this dialogue is that all created things depend on the continual, direct action of God to keep them in existence—a conclusion that he will soon reject—this does not entail the rejection of the theory of change that he develops in the dialogue. Indeed, as we shall see, that theory remains a constant through his later writings. Leibniz rejects the occasionalist conclusion of the dialogue as soon as he is able to find a way to reposit principles of change in every created thing, something he achieves in 1678 by reconceiving force as an enduring attribute in created things. As we shall see, however, at the same time he models the subject of change—"that which by acting does not change"—on the human ego, and its states on the ego's perceptions. Since these developments can be traced back to problems articulated in the *Pacidius*, an analysis of their origins there will be very illuminating.

In that dialogue Leibniz gives an analysis of change according to which, because successive states are mutually contradictory, "there is no moment of change common to each of two states, and thus no state of change either, but only an aggregate of two states, old and new" (A VI 3, 566/LLC 211). In support of this he gives the analogy with the infinite dividedness of body:

> (NB. Just as bodies in space form an unbroken connection, and other smaller bodies are interposed inside them in their turn, so that there is no place void of bodies; so in time, while some things last through a momentaneous leap, others meanwhile undergo more subtle changes at some intermediate time, and others between them in their turn.... At any rate, it is necessary for states to endure for some time or be void of changes. As the endpoints of bodies, or points of contact, so the changes of states.... Nor is any time or place empty. During any state whatsoever some other things are changing.) (A VI 3, 559/LLC 195–7)

This is a concise note to himself that Leibniz wrote in the margins of his first draft in order to guide his rewriting of it. But it contains most of the main ingredients of his analysis of change. In this theory the only moments—strictly speaking, instants[122]—that are assigned are the endpoints (beginnings or ends) of finite temporal intervals, no more than two of which are ever next to one another. Thus concerning the spatial continuum, the interlocutors conclude that "the continuum can neither be dissolved into points nor composed of them, and that there is no fixed and determinate number (either finite or infinite) of points assignable in it" (A VI 3, 555/LLC 187). And the same applies to the temporal continuum. Moreover, because the only moments are endpoints of intervals, and no two

[122] In these passages Leibniz is mostly using 'moment' as synonymous with 'instant,' the temporal counterpart to 'point,' each being a mere endpoint, indivisible and lacking quantity. Sometimes, however, he will use the term 'moment' rather for an arbitrarily small duration, a usage that becomes more common later.

intervals have an endpoint in common, there is no moment of change, or state of change. In sum,

> at any moment that is actually assigned we will say that a moving thing is at a new point. And although the moments and points that are assigned are indeed infinite, there are never more than two of them immediately next to each other in the same line, since indivisibles are nothing but bounds. (A VI 3, 565/LLC 209) Nor is there any moment of time that is not actually assigned, or at which change does not occur, that is, which is not the end of an old or beginning of a new state in any body. This does not mean, however, either that a body or space is divided into points or time into moments, because indivisibles are not parts but extrema of parts. And this is why, even though all things are subdivided, they are still not resolved all the way down into minima. (A VI 3, 566/LLC 209–11)

There are several later references to this account of change, showing that Leibniz continued to uphold it in the years after 1676 and well beyond:

> Change is an aggregate of two opposite states in one stretch of time, with no moment of change existing, as I have demonstrated in a certain dialogue.
> ([Spring-Summer 1679] A VI 4, 307)
>
> Change is an aggregate of two contradictory states. These states, however, are understood to be necessarily immediately next to one another, since there is no third thing between contradictories. ([1683–85] A VI 4, 556)

And in his second last letter to De Volder of 10/11/1705, Leibniz writes:

> Endpoints of a line and unities of matter do not coincide. Three continuous points in the same straight line cannot be conceived. But two are conceived: the endpoint of one straight line [segment] and the endpoint of a second, out of which the same whole is constituted. In the same way in time there are two instants, the last instant of life and the first of death. (GP II 278/LDV 327)[123]

With this last remark about the last moment of life and the first of death, Leibniz implicitly alludes to his discussion of the paradox of the heap in the *Pacidius*, where he uses such riddles posed by Sextus Empiricus to introduce his novel analysis of change.

This is, to say the least, an original analysis; but it is in many respects quite baffling.[124] Two of the more puzzling things are as follows.

[123] See Levey (2002) for further discussion.

[124] See Samuel Levey's important essay (2010) for a penetrating analysis.

First there is the apparent contradiction in Leibniz's claims about states. Guided by the analogy with the actual division of bodies, we have depicted them as divided by changes occurring within them to infinity; yet we have seen him claim that "it is necessary for states to endure for some time or be void of changes." In the *Pacidius* Leibniz gestures at a solution to this discrepancy, comparing the leaps to the infinitesimals of his recently completed calculus. In a passage from the first draft that he subsequently deleted, he wrote:

> Whence you will understand that if it is a miracle for someone to be transferred from Paris to Rome in a moment, then it is a continual miracle, even if it is credible that the spaces through which these leaps occur are smaller than can be explicated by their ratio to magnitudes known by us. And these kinds of spaces are taken in geometry to be points or null spaces, so that motion, although metaphysically interrupted by rests, will be geometrically continuous—just as a regular polygon of infinitely many sides cannot be taken metaphysically for a circle, even though it is taken for a circle in geometry, on account of the error being smaller than can be expressed by us. (A VI 3, 569/LLC 409)

Here the idea is that even though the extended spaces through which the leaps occur are always finite, and take a finite time, such spaces and times are so small as to be "unassignable," smaller than any assignable. It is in this way that an infinite polygon can be taken for a circle. Thus even if motion is "metaphysically" interrupted by unassignable leaps (like the unassignable differences of his calculus), it will still be "geometrically continuous." We will come back to this analogy with the infinite polygon below, as well as this justification of the vanishing difference between how things change discontinuously in reality and a true geometrical continuity (which for Leibniz is purely ideal).

Secondly, however, Leibniz draws a startling conclusion from these premises in the *Pacidius* dialogue, namely that bodies *do not act*, and therefore *do not even exist*, between changes of state. For, he argues, given that "there is no moment of change common to each of two states, and thus no state of change either," then "if it is supposed that things do not exist unless they act, and do not act unless they change, the conclusion will follow that things exist only for a moment and do not exist at any intermediate time" (A VI 3, 557).[125] "Hence," he concludes, "it follows that proper and momentaneous actions belong to those things which by acting do not change" (566). Thus bodies exist at every assignable moment, but they do not

[125] On Levey's analysis, this constitutes Leibniz's argument that there is no precise shape in bodies. For a precise shape requires an enduring state during which it remains the same, but "no interval of time can be assigned during which an enduring state can persist the same and uniform throughout," since every state is further divided into states of lesser duration. See Levey (2010), esp. p. 163.

exist at the unassignable times between these moments. There is therefore *no temporal continuant to which the changes or actions can be ascribed.*

In the dialogue Leibniz uses this consequence to prove the necessity of "a superior cause which by acting does not change, which we call God" (567), "whose special operation is necessary for change among things" (568–9). Similarly, he wrote shortly afterwards in a piece probably from 1677:

> It is accepted, however, that nature is the work of God, and that whatever nature attempts comes about only through the will of God, for bodies in and through themselves are not the cause of their actions, since they are not even the same thing for longer than a moment.
>
> (*Demonstratio Quod Deus Omnia Possibilia Intelligit*, A VI 4, 1353)

Despite his innovations in the dialogue, then, Leibniz has concluded with a version of continuous creation not unlike those he will later criticize in the *Theodicy*.[126] As such it seems vulnerable to the same criticisms. There are no temporal continuants, just isolated instances of God's creating bodies here, there and everywhere, but only for an instant at a time.

In the dialogue, though, Leibniz refers actions to "those *things* which by acting do not change," and this hints at a different solution. This is in fact the view he wants to hold, as he believes that bodies contain their own individual principles of activity. He begins to develop this view in earnest after he has arrived in Hanover in 1677. Thus in what appears to be a draft of his intended book on physics written in 1678–79, he argues that "Substantial form ôr soul is the principle of unity and duration, whereas matter is the principle of multiplicity and change." For because of the actually infinite division of bodies into parts "each of which is agitated with a different motion," it follows that

> if we consider matter alone, no point will be assignable that remains together with another, nor any moment at which a body will remain identical with itself; and there will never be a reason for saying that a body is a unity over and above a point, and is the same for longer than a moment. (A VI 4, 1399/LLC 245)

That is, if a purely material body cannot retain its identity through time, it must contain within it some principle which does, which he identifies here as a "principle of unity and of duration" (A VI 4, 1399/LLC 245), or substantial

[126] This conclusion is anticipated in the piece *Infinite Numbers* from the spring of 1676. In fact, it is a return (after many detours) to more or less the position Leibniz had held in 1669. See Arthur (2009), where I trace his changing views on the infinitely small through four more or less distinct phases, the first of which is the one occurring in his original letter to Thomasius of 1669 and in manuscripts of that period, where he subscribes to a version of continuous creation, with bodies created at definite instants separated by actual infinitesimally small intervals, during which they do not exist.

form. Such a principle is modelled on the human Ego, which retains its self-identity while having a succession of different perceptions. In a passage discussing change from a fragment from 1683–85 Leibniz writes:

> The only difference that occurs when everything else remains the same, and makes there be no contradiction of any kind when the same things are said to be both contiguous and separate, is the difference of *time*. But whether those things are really the same that we think to be so is a matter for a more profound discussion. It is enough that there are some things that remain the same while they change, such as the Ego. (A VI 4, 562/LLC 267)

In this passage we see the Ego clearly identified as Leibniz's model for "those things which by acting do not change" (A VI 3, 566/LLC 211).

Deferring further discussion of that for the moment, what about the other difficulty concerning the enduring states that bodies are supposed to have between changes? In a piece dating from April–October 1686, ("*Dans les corps il n'y a point de figure parfaite*"), after alluding to his analysis of change in the *Pacidius* according to which a body exists only at the assignable moment it changes state, Leibniz notes that this analysis still presupposes enduring states between the changes. He then makes the intriguing suggestion that all such enduring states must be understood to be "vague":

> Now I believe that what exists only at a moment has no existence, since it begins and ends at the same time. I have proved elsewhere that there is no middle moment, or moment of change, but only the last moment of the preceding state and the first moment of the following state. But that supposes an enduring state. Now *all enduring states are vague*, and there is nothing precise about them. . . .
> (A VI 4, 1613–14/LLC 297; my emphasis)[127]

Not only do moving bodies of a definite shape not exist: strictly speaking, between changes they do not exist at all! Leibniz is interpreting "state" is its root sense of stasis, and arguing that only what acts, and thus changes, exists.

Of course, given Leibniz's identification of a substance's states with perceptions, this means that *perceptions* are also presupposed as enduring, rather than strictly instantaneous. This is confirmed in the continuation of the passage from the *Divisio terminorum* about change and the Ego quoted above:

[127] Cf. a marginal note Leibniz made to himself on the first draft of the *Pacidius*: "Why not say rather that the conclusion that things exist only at a moment, and do not exist at any intermediate time, will follow if it is supposed that things do not exist unless they act and do not act unless they change?" (A VI 3, 558/LLC 191).

> But if someone contended that not even I endure beyond a moment, he would not know whether he himself existed. For this he could know in no other way than by experiencing and perceiving himself. But *every perception requires time*, and so he either persists during the whole time of his perception, which suffices for us, or he himself does not perceive, otherwise he would persist only for a moment, namely, for that moment alone at which he exists.
>
> (A VI 4, 562/LLC 267, my italics).

This is an extremely interesting critique of the Cartesian *cogito*. Descartes had argued that from the fact that I perceive I now exist, it does not follow that I should exist a moment from now.[128] But since every perception requires time, my perceiving that I exist already presupposes my existing through a time, for if my existence were instantaneous, I would not be able to perceive it. Moreover, I need to be able to remember that I exist in order to form the thought. So what true substances have (and that Cartesian bodies lack), Leibniz claims, is *memory*. This is what links the perceptions together and forges the self-identity and persistence of substances through their changes. As he wrote in another fragment from 1685,

> Certainly those things which lack forms . . . do not persevere the same for longer than one moment, whereas true substances persist through changes; for we experience this in ourselves, for otherwise we would not be able even to perceive ourselves, since each of our perceptions involves a memory.
>
> (A VI 4, 627–8/LLC 273)

So we have a clear contrast between substances which persist through the time of their perception, and bodies and other "quasi-substances" which do not persist for longer than a moment. The perceptions of these substances, moreover, are such that not all the changes occurring within them are perceived. All this coheres with Leibniz's doctrine of *petites perceptions* in the *Nouveaux Essais*:

> There is at every moment[129] an infinity of perceptions within us, unattended by awareness or reflection, that is to say, changes in the soul itself, of which we are unaware, because these impressions are either too small and too numerous, or too unvarying, so that they are not sufficiently distinctive on their own. . . . but that does not prevent them from having their effect when they are combined with others, and from making themselves felt, at least confusedly, in the aggregate.
>
> (Preface; A VI 6, 53/RB 53)

[128] Descartes argued this in his Third Meditation and in his *Principles of Philosophy*, I, §31 (AT VIIIA, 13/CSM 200).

[129] Here a *moment* is to be understood as an arbitrarily short duration.

The idea is that a perception of however short a duration will still contain other perceptions in it. Even if a perception appears as one continuous state because of the limitations of sense—it is *vague*!—it is in fact infinitely divided into other smaller perceptions lasting for discrete durations. The instants that are actually assigned are those at which change occurs, change being the aggregate of its existence in one state at one instant and its existence in a contradictory state at the next. These perceptions, Leibniz says, *constitute* the individual's self-identity:

> These insensible perceptions also indicate and constitute the same individual, which is characterized by the traces or expressions they preserve of the previous states of this individual, thereby connecting these with its present state; and even when this individual itself has no sense of these traces of previous states.
> (*Nouveaux Essais*, A VI 6, 55/RB 55)

In this connection Leibniz refers to his Law of Continuity, according to which "any change always passes from the small to the large, and vice versa, through the intermediate, in respect of degrees as well as of parts" (*Nouveaux Essais*, A VI 6, 56/RB 56). "All of this supports the judgement that noticeable perceptions come by degrees from those that are too small to be noticed" (56–7). Since every perception of the same individual substance preserves traces or expressions of its previous states, it follows that the whole series of states forms a continuum, with each state or perception arising by degrees from the previous ones.

If this is a continuum, however, it is not the ideal continuum of mathematics. For it is constituted by the states or perceptions, which are actually divided from one another by actual changes of state, whereas in a mathematical continuum the instants would mark sites of possible changes. In fact, if the states are contiguous with one another, each having its own different endpoints, it would seem to constitute a *contiguum*, rather than a true continuum.[130] This is analogous to the case in the passage that I quoted from "Primary Truths," near the beginning of this section. There Leibniz claimed that

> a continuum is not divided into points, nor is it divided in all possible ways—not into points, since points are not parts but boundaries, and not in all possible ways, since not all creatures are in a given thing, but there is only a certain

[130] Here my analysis is in complete agreement with the account given by Douglas Bertrand Marshall in his (2011). He writes: "Points or surfaces only exist in the physically continuous quantity when there are parts which have those points or surfaces as boundaries. Moreover the parts of the physically continuous quantity have their own separate boundaries which merely touch one another. This is the sense in which they are not really continuous, for if they were really continuous, two points which were next to each other would not have separate boundaries.... In a physically continuous quantity such as a body, there can be distinct points p and p' whose distance from one another is zero. For instance, these points may lie on surfaces of two parts that are touching each other. In a mathematically continuous quantity, if p and p' are at zero distance from one another, then $p = p'$" (Marshall 2011, 13–14).

progression of them *ad infinitum*. Thus if you bisect a straight line and then any part of it, you will set up different divisions than if you trisect it.

("Primary Truths," [1689], A VI 4, 1648)[131]

In a previous publication I had suggested that a continuum in this latter sense of containing a particular infinite progression governed by a law of progression "is what Leibniz refers to on occasion as the *physical continuum*" (Arthur 2018, 279).[132] On this conception, a physical continuum consists in an infinite progression of determinate actual parts, united and determined by the law of progression, whereas a mathematical continuum consists only in potential parts. This is connected with the "real, physical identity" (A VI 6, 236/RB 236) that Leibniz claims (in the *Nouveaux essais*) underlies personal identity. The law of progression encoded in a monad is the basis of its continuation as numerically the same thing: "Organization or configuration alone, without an enduring principle of life which I call 'monad', would not suffice to make something remain numerically the same, i.e. the same individual" (231/231); "the identity of the same individual substance can be maintained only by conservation of the same soul, for the body is in a continual flux" (232/232).

Leibniz gives an account of physical continuity in distinction from mathematical continuity (the most comprehensive that I am aware of) in a fragment of uncertain date recently uncovered and transcribed by Osvaldo Ottaviani, *Locus et tempus sunt continua*.[133] It begins by contrasting the continuity of space and time with the mere contiguity of matter and of the incompatible states involved in change:

Place and time are continua, matter and change are contigua. A continuum does not have actual parts, except those that are assigned by an actual division. Nor

[131] Also in line with this definition of the physical continuum is the wonderful image Leibniz gives in the *Pacidius* when he compares its divisions with the folds in a sheet of paper or tunic: "Accordingly the division of the continuum must not be considered to be like the division of sand into grains, but like that of a sheet of paper or tunic into folds. And so although there occur some folds smaller than others infinite in number, a body is never thereby dissolved into points or minima" (A VI 3, 555/LLC 185).

[132] In support of this idea of physical continuity I quoted Leibniz's explanation to Des Bosses in his letter of January 24, 1713: "on the hypothesis of mere monads, the infinitude of the physical continuum would depend... on the principle of sufficient reason, since there is no reason for limiting or ending [the progression], or for its stopping anywhere; whereas the Mathematical Continuum consists in mere possibility, like numbers, and therefore necessarily contains infinitude in its very concept" (GP II 474/LDB 299).

[133] LH 37, 5, Bl. 134. Ottaviani and I are intending to publish a transcription and English translation of this manuscript together with other related pieces on the metaphysics of the infinite. There are no obvious external criteria for its dating. One could speculate a date of composition of around October 1705 on the basis of its content: the claim that the actual instants of time contain "just as many fulgurations of the divinity, ôr states of the universe" evokes what Leibniz wrote to De Volder in October 11, 1705: "Time is also resolved into unities of duration through actual changes, ôr into just as many creations infinite in number" (GP II 279/LDV 327)—as well as what he wrote to the Electress Sophie ten days later, quoted here below.

> does a line consist of points, or time of instants, even though there is no part of a line in which there is not an actual point, nor any part of time in which there is not an actual instant. A continuum, like a line and time, are ideal things like numbers; and in fact they are orders of possibles in which actuals are designated. Place or space is the order of possible co-existents; time is the order of possible changes or of incompatible states. Each order is continuous since nothing can be conceived as interposing that is not contained in it. (LH 37, 5, Bl. 134^{r})

A continuum is a mere ideal order, and has no points or instants in it except the actual points of matter (in the case of the spatial continuum), each distinguished from others by its having a different motion, or the actual instants of time designating changes of state. Thus the parts of the ideal continuum are *potential*, in contrast with the actual divisions into parts that occur in actuality. In this respect the parts of ideal place or time are analogous to the fractions into which we can divide unity in infinitely many different ways, but from which it cannot be regarded as composed. But "the comparison of number with place or time is when they are considered in themselves and abstracted from the things existing in them." The parts of matter, by contrast, are actual parts that are individuated by their differing motions:

> These parts [of unity] are therefore potential not actual, and so also are the parts of a line, and a line can no more be composed of points than a number from numerical minima. And in fact to every section of a line into parts there corresponds proportionally a section of unity into parts. Meanwhile in a spatial or ideal line infinitely many actual points can be assigned, actually distinct from one another, and endowed with different motions, which do not compose the line. And these points will be extremities of the parts into which the line is actually divided by the variety of motions in matter.
>
> (*Locus et tempus sunt continua*, LH 37, 5, Bl. 134^{r})

So here the "actual points" are described as extremities of the parts into which matter is actually divided. This suggests that, as in the *Pacidius*, the extremities of two contiguous parts of matter are touching but distinct, because they belong to parts of matter distinguished by their differing motions.

But in the immediate continuation of this passage Leibniz proceeds to discuss the division of a given line *AB* into two equal parts by "an actual point *C*," with *AC* and *BC* likewise to be divided into two equal parts by "actual points *D* and *E*, and so on, always taking actual points in the middle of the part by division." This clearly suggests that the "actual points," while still distinguished by their differing motions, are now to be regarded as being precisely *not* contiguous. For, assuming such a continued bipartition of the line as proceeding to infinity—"as it certainly can in fact proceed, since in nature no two points can be assigned in the universe which have precisely the same motion"—it will follow that between any two such actual points—distinguished by their different motions—there will always be another:

> Nor will two assignable points running into each other with different motions ever touch one another, since other points are always interposed. The same holds for time, the actual instants of which contain just as many fulgurations of the divinity, ôr states of the universe. (LH 37, 5, Bl. 134^{r})

Leibniz then appeals to the same example of infinite bisection or trisection that he had given in "Primary Truths," quoted near the beginning of this section:

> For since actual divisions can be instituted in actually infinitely many different ways, e.g. by infinite tripartitions or mixed bipartitions and tripartitions, or by any other partitions whatever, just as if every line were cut in the extreme and middle ratio, it is clear that the simplest, which proceeds by perpetual bipartitions, is not determined as unique. Hence it could be understood that what is actual of place and of time is made up of [*conflatum*] points or instants ôr is an aggregate of them; but not place itself and time itself, which are continuous and potential things. Namely, just as place could be divided in a different way, as by tripartitions, so also could time; which are no more the aggregates of points or instants, or of the lines into which they can be resolved, than number is the aggregate of the fractions into which it can be broken down.
>
> (*Locus et tempus sunt continua*, LH 37, 5, Bl. 134^{v})

Here it seems that "what is actual" of place are the assignable points individuated by their differing motions (more accurately, endeavours), separated by unassignable gaps. This, Leibniz notes, invites a difficulty, that of "how a whole body could be impelled if points are not touching each other immediately." It is here that he introduces a distinction between physical and mathematical continuity. But it is interestingly different from the way I had phrased it above, and this is worth a sizable digression for the further light it throws on his views on the mereology of change:

> But that must be said to be equivalent to a continuum [in] which there are no two actually assignable points between which another cannot be actually located. This is physical continuity, not geometrical;[134] geometrical points are endpoints of the continuum, not elements of it; but physical points are the elements of mass. Both

[134] After this declaration, the rest of Leibniz's explication is written in a darker ink, as are many interpolations and clarifications he has interjected. His marginal comment, however, that "The whole of the latter is an intelligible explanation of the phenomena, not of substances..." is written in the original fainter ink, so the darker ink commentary and corrections would seem to have been added afterwards.

> are indivisibles. But mass is not something continuous, but a conjunction; for there is no separation of its parts, ôr no interval can be assigned between two of its parts in which there is not some part of it. But if mass were something substantial, not phenomenal, we would arrive at infinitely small real things, intervals of nearest points or instants.
>
> (*Locus et tempus sunt continua*, LH 37, 5, Bl. 134^{v})

A physical continuum, according to the definition of the first sentence, is essentially a dense aggregate of physical points or "elements of matter"; and likewise, a physical continuum of duration will be an aggregate of what is actual in time, namely the changes in it. But this claim that there are no two points between which there are no others appears to contradict the account of the physical continuum given above, where the points bounding the intervals *are* in fact immediately next to one another, the last of one with the first of the next, with these intervals forming a contiguum, not a continuum. Leibniz notes this objection himself at the end of the fragment: "It can be objected that in every order where there is prior and posterior, there are two things immediately next to one another; the issue comes down to an examination of this rule. If it is accepted, it will turn out that the latter determinations [concerning the physical points not touching] would not be brought into play" (LH 37, 5, Bl. 134^{v}).

Vincenzo De Risi (2019, 127, n. 30) has noted a similar apparent discrepancy in views Leibniz expressed on other occasions. For example, in a letter to Bernoulli in the Fall of 1698 Leibniz stated that "even if in motion all the points are gone through, it does not however follow that there are two points infinitely close to one another, and even less that there are two next to one another" (GM III 536). Yet in his letter to De Volder of October 11, 1705 he asserts that "we cannot conceive three continuous points in the same straight line; but two may be conceived: the endpoint of one straight line [segment] and the endpoint of a second, out of which the same whole is constituted" (GP II 278/LDV 327). But we need to take into account the immediately preceding sentence in that letter: "Endpoints of a line and unities of matter do not coincide." This is the same distinction that we see in the above passage from *Locus et tempus sunt continua*: the points we can conceive as next to one another are "endpoints of the continuum," i.e. those of contiguous line segments, "geometric points." But they are not physical points, and it is these that are held to be "elements of mass" or "unities of matter."[135]

Unfortunately, though, this answer only spawns further difficulties.[136] What is the status of the distinct endpoints of contiguous segments? If these *geometric*

[135] De Risi wisely notes, however, that "The apparent oscillation in his views here may perhaps be accounted for by saying that the passage addressed to Bernoulli concerns ideal objects, while the passage to De Volder deals with the real world" (De Risi 2019, 127, n.30).

[136] Here I am indebted to Vincenzo De Risi for suggesting I needed to describe these difficulties explicitly.

points are points in the mathematical continuum, they should be identical, not separate and next to one another. And what, on the other hand, are these physical points that Leibniz describes as "elements of mass" and "indivisible"? If these are the physical points he refers to elsewhere, such as in his *Système Nouveau*, then they cannot be indivisible in the strict sense. For those physical points "are indivisible only in appearance, whereas mathematical points are exact, but are nothing but modalities" (GP IV 483/WFT 149). A physical point is not strictly indivisible; rather it is an arbitrarily small portion of matter, distinguished by its differing motion (i.e. endeavour). Since every such point is distinguished from any other, no matter how close, by a differing endeavour, no two of these physical points can be actually next to each other.

Leibniz sidesteps the first difficulty in *Locus et tempus sunt continua* by abandoning the appeal to endpoints and instead concentrating the analysis on *midpoints* of successive intervals. He argues that if one bisects a line and each part of the line, "always taking actual points in the middle of the part by division," then such an "actual simultaneous division proceeds to infinity, since in nature no points can be assigned which have precisely the same motion" (LH 37, 5, Bl. 134^{r}).[137] The midpoints of each of these subdivided lines will be individuated by a different motion (strictly speaking, by a different *conatus* ôr endeavour), but between any two such physical points there will always be further points differentiated from them by their own motions; so "no two of [these] assignable points running into each other with different motions will ever touch one another, since other points are always interposed" (LH 37, 5, Bl. 134^{v}).[138] Analogously, the "actual instants" of time marking actual changes would also always have further instants interposed. "The same holds for time," Leibniz writes, "the actual instants of which contain just as many fulgurations of the divinity, ôr states of the universe" (LH 37, 5, Bl. 134^{v}).[139]

[137] Leibniz stresses that there is nothing unique about taking midpoints or "perpetual bipartitions into two equal parts." "For since actual divisions can be instituted in actually infinitely many different ways, e.g. by infinite tripartitions or mixed bipartitions and tripartitions, or by any other partitions whatever, just as if every line were cut in the extreme and middle ratio, it is clear that the simplest, which proceeds by perpetual bipartitions, is not determined as unique." This echoes his demonstration that, in finding quadratures, it is not necessary to decompose the area under a curve into equal subintervals. See Knobloch (2002) for details. That one should nevertheless adopt the simplest partition is an example of an argument form we will consider in detail in §3.5 below.

[138] This agrees with Marshall's analysis in his (2011, 19): "even if the physically continuous quantity does not always contain the midpoint for any pair of points in it, the physically continuous quantity does contain points within ϵ of the midpoints. I believe this is a reasonably straightforward sense in which the difference between physical and mathematical continuity is less than any given. Since ϵ can always be chosen to be far finer than the level of acuity of human sensation, it follows that the physically continuous quantity will be indiscernible from the mathematically continuous quantity as far as all appearances are concerned."

[139] —here we might have expected Leibniz to have said "The same holds for time, *between any two actual instants of which are contained* just as many fulgurations...": I will come back to that below.

But what about the fact that Leibniz describes these physical points as "elements of mass" or "unities of matter"? In his October 1705 letter to De Volder, Leibniz also describes the "unities of matter" as not touching, but explains how their denseness is related to the "order of changes":

> One unity is not touched by another, but there is a perpetual transcreation in motion. This is, namely, that when a thing is in such a state that by continuing its changes through an assignable time there would have to be a penetration at the next time afterwards, each and every point would be in another place, as the avoidance of penetration and the order of changes demands. (GP II 278/LDV 327)

This mention of "transcreation" refers us back to Leibniz's Paris notes, where the notion first occurs. Leibniz introduces it (and its synonym, "transproduction") in the manuscript *Infinite Numbers* of c.April 10, 1676, and then makes it the centrepiece of the dialogue *Pacidius Philalethi* with which I began my analysis.[140] Strictly continuous motion, he declares in *Infinite Numbers*, is impossible, and must occur "by a leap," so that "motion is nothing but *transcreation*" (A VI 3, 500/LLC 93). The fact that in all three places—in the dialogue, in *Locus et tempus sunt continua*, and in his October 1705 letter to De Volder—the transcreation or denseness account occurs *alongside* the other account of the continuum in which there are contiguous sections, strongly suggests that Leibniz regards the two accounts as compatible. My suggestion is going to be that the account in terms of contiguous sections or states pertains to how the continuum appears in the imagination: successive states are really distinct, because contradictory, but when we represent them in the imagination, we do so using geometry, the science of the imagination. We use the image of contiguous line segments to represent successive states when in fact the states are vague: in reality the states no more have precise boundaries than bodies have precise shapes.[141] This also relates closely to the idea that we represent differences in the calculus as if they are indivisible elements, when in fact they are unassignable.

We can gain further insight into his reasoning by returning to the context in which he formulated his idea of transcreation. In "On the Secrets of the Sublime" of February 1676, Leibniz speculates that "a perfectly fluid matter is nothing but a multiplicity of infinitely many points, ôr bodies smaller than can be assigned," leaving a "metaphysical vacuum" (A VI 3, 473/LLC 47). "A metaphysical vacuum is an empty place however small, only true and real," constituting the difference between a true continuum and a perfect fluid. A perfect fluid is "not a true

[140] In "Infinite Numbers" of *c.*April 10, 1676, Leibniz concludes that continuous motion is impossible, and must occur "by a leap," so that "motion is nothing but *transcreation*" (A VI 3, 500/LLC 93).

[141] For an informative discussion of Leibniz's treatment of vagueness, see Samuel Levey's (2002), and for application to the idea of precise boundaries in matter see the same author's (2010) and (2012).

continuum, even if space is a true continuum"; rather it is "discrete, ôr a multiplicity of points." This supposes that both the points ("bodies smaller than can be assigned") and the "metaphysical vacuum" are actual infinitesimals. But in that case, "any part of matter would be commensurable with any other," as would the circle be to the square (LLC 49). Leibniz suggests he should look into this more carefully. This he does two months later in "Infinite Numbers," which begins with a thorough examination of this issue. But there he comes to the conclusion that unassignable things—like a line smaller than any assignable, or the "metaphysical vacuum"—are *not* actual infinitesimals, but *fictions*. There are no such things *in rerum natura*, even though they express "real truths": "these fictitious entities," he writes, "are excellent abbreviations of expressions, and for this reason extremely useful" (A VI 3, 499/LLC 89–91).

In other words, even though it is "really true" that matter is discrete—it is a multiplicity of physical points, each distinguished from the others by a different endeavour—the difference between it and a true continuity is infinitesimal. The "metaphysical vacua" separating these points are so small that, however small one takes them to be, they are smaller still. Clearly these "metaphysical vacua" evoke the "metaphysical" interruption of motion by rests referred to in the *Pacidius*, referenced earlier, an interruption which still leaves the motion "geometrically continuous—just as a regular polygon of infinitely many sides cannot be taken metaphysically for a circle, even though it is taken for a circle in geometry, on account of the error being smaller than can be expressed by us" (A VI 3, 569/LLC 409). Leibniz had elaborated on this topic in "Infinite Numbers":

> The circle—as a polygon greater than any assignable, as if that were possible—and other things of that kind, are fictive entities. So when something is said about the circle, we understand it to be true of any polygon such that there is some polygon in which the error is less than any assigned amount *a*, and another polygon in which the assignable error is less than any definite assigned amount *b*. But there will be no polygon in which it is less than all assignable errors *a* and *b*, even if it can be said that polygons somehow approach such an entity in order.... And even though this ultimate polygon does not exist in the nature of things, one can still give an expression for it, for the sake of abbreviation of expressions.... For entities of this kind, i.e. polygons whose sides do not appear distinctly, are made apparent to us by the imagination, whence there arises in us afterwards the suspicion of an entity having no sides. (A VI 3, 498, 499/LLC 89, 91)

It is by extending this analysis to motion that Leibniz arrives at his idea of transcreation (or transproduction):

> For something to become another thing is for something to remain which pertains to it rather than to the other thing. But this is not always matter. It

> can be the mind, understanding a certain relation; for instance, in *transproduction*, even though everything is new, still, by the very fact that this transproduction happens by a certain law, continuous motion is imitated in a way, just as polygons imitate the circle. And hence one is said to come out of the other, by a similar abuse, as it were, of the imagination. (A VI 3, 503/LLC 99)

Thus it is the imagination that fills in the gaps between the points or instants, giving rise to the illusion of a perfect uniformity and true continuity: "in the mind there is thought of uniformity, yet no image of a perfect circle: instead we apply uniformity to the image afterwards, a uniformity we forget we have applied after sensing the irregularities" (A VI 3, 499/LLC 91).[142] Almost thirty years later, Leibniz gives a striking illustration in a letter to Sophia, Electress of Hanover:

> Matter appears to us as a continuum, but it only appears so, as does actual motion. It is similar to how alabaster powder appears to make a continuous fluid when one boils it over the fire, or how a spoked wheel appears continuously translucent when it turns with great enough speed; without one being able to distinguish the locations of the spokes from the empty spaces between them, our perception unites the separate places and times.
> (To Electress Sophia, October 21 (NS), 1705; GP VII 564)

This evokes the doctrine of *petites perceptions*, where the irregularities in a given perception are unnoticed and it is sensed as smooth and uninterrupted. Either we do not sense the gaps (because of the limitations of sense perception), or we forget we have sensed them, and apply uniformity to the image afterwards.[143] Thus not only the image of a perfect circle, but also the geometric representation of states as smooth and uninterrupted between changes, are products of the imagination. This helps to clarify Leibniz's discussion of physical continuity in the *Locus et tempus sunt continua*. It is because there is always further matter between any two physical points that "no interval can be assigned between two of its parts in which there is not some part of it," so that between any two such assignable points there is always another. But the interval between any two physical points is

[142] Again, see Sam Levey's publications, in particular his (2003), (2005), and (2010), for analyses of Leibniz's infinitely divided continuum, the role of the imagination in filling in the gaps, and the connection with the idea of "vagueness," and also for his critiques of the idealist readings of Sleigh and Adams of the "no precise shape" doctrine. See also the above-cited paper by Marshall (2011) for an excellent discussion of the applicability of the geometry of the continuous to infinitely divided matter.

[143] For a thorough discussion of the relation between sensation and imagination in Leibniz's thought, I refer the reader to Lucia Oliveri's forthcoming book. She explains (*inter alia*) how minute perceptions make available to the mind the infinite actual modifications of discrete matter that are first unified into distinguished sensations, then synthesized by the imagination into perceptible wholes, with shapes and sizes determined by confused sensations.

unassignable, and cannot be something real.[144] If, on the other hand, "mass were something substantial, not phenomenal," he argues there, "we would arrive at infinitely small real things, intervals of nearest points or instants." Analogously with time, the actual instants at which change is taking place are separated by "vague" states in which further changes are occurring, but of which we are not conscious. As Leibniz wrote in "Infinite Numbers," "they are not on this account any less sensed by our consciousness [*a nobis consciis*]. Rather we forget them, just as we are oblivious of the things we dream" (A VI 3, 499/LLC 91).

There is therefore a very strong correlation in Leibniz's thinking between the fictional nature of infinitesimals and the insubstantiality of matter (and the phenomenality of states). In the *Locus et tempus sunt continua* this prompts him to add:

> So here is a popular way of explaining the issue, suitable for those who do not grasp that material things are only phenomenal. For if the things are real, we will resolve space into a multiplicity of points, and time into a multiplicity of instants. Should we not therefore say that in fact between any two actual instants or points another must be interposed, and that we never arrive at two of them that are unassignably distant, between which nothing actual is interposed? this could be said on the subject of the continuum. (LH 37, 5, Bl. 134^{v})

This is consistent with the marginal note Leibniz had made on the first sheet, before entering into the discussion of physical continuity: "The whole of the latter is an intelligible explanation of the phenomena, not of substances. That is, everything is offered to the mind as if this were so; and one monad makes up the deficit of the phenomena of the other" (LH 37, 5, Bl. 134^{r}).[145]

Coming back to the apparent discrepancy between Leibniz's statements about the denseness of points, then, we can say this. Leibniz presents two different models of physical continuity, without acknowledging any incompatibility between them. On one model, the continuum is actually divided into contiguous parts. Having a line stand for the continuum, it is divided into contiguous line

[144] This discussion seems to undermine De Risi's contention (2019, 128) that "it is possible that these unassignable spaces are to be regarded as actually infinitesimal *vacua* between parts of matter. They would thus tend to push Leibniz toward a non-Archimedean geometrical system." In support of that contention, De Risi construes a passage from Leibniz's reading notes on Froidmont as "clearly represent[ing] an opening toward non-Archimedean considerations" (128, n.33), when in my opinion it is a view that Leibniz finds in Froidmont and quotes for the purpose of rejecting it, which he then does. But see the whole footnote; indeed, see the whole of De Risi's excellent article for an authoritative treatment of Leibniz's views on the continuity of space.

[145] We might also relate this claim to Leibniz's assertion in a piece from the late 1670s, "A Body is not a Substance," that "if we only say this, that bodies are coherent appearances, this puts an end to enquiry about the infinitely small, which cannot be perceived. But this is also a good place for that Herculean argument of mine, that all those things which are such that it is impossible for anyone to perceive whether they exist or not, are nothing" (A VI 4, 1637/LLC 261).

segments each of which has a different endeavour at each different moment. Here the points in the continuum are "extremities of the parts into which the line is actually divided by the variety of motions in matter"; they are *geometric points*, indivisibles as the endpoints of the subintervals into which the continuum is divided, and no more than two of them will be next to one another. On the second model, the points in the continuum are *physical points*, "bodies smaller than can be assigned," each distinguished from the others by a different endeavour; in this case, between any two such points there will always be others: they will be dense.[146] I suggest the following interpretation of how it is that Leibniz regards these two models as compatible. If the intervals (the contiguous line segments) of the first model are regarded as real, then because of the actually infinite division Leibniz takes as definitive, they will have to be actual infinitesimals. The assumption of actual infinitesimals, however, leads to contradiction. In fact the intervals are unassignable, so small that no error is derivable from neglecting their extension. This corresponds with the fact that we perceive bodies and changes as continuous, even if what is actual in them is discrete.[147]

This still leaves moot the status of the "geometric points" of the first model, as pointed out by De Risi. They subsist in a kind of untenable middle ground between the physical points and truly mathematical points. A point in the mathematical continuum marks the place where it can be divided. If a division is actually made, there will be two disjoint actual parts, each with its own endpoints, so the rightmost of one is contiguous with the leftmost of the other. These boundaries are limitations of the extended, they are modifications of each extended part, and specific to it. If on the other hand no such division is made, but only considered as possible, the two endpoints of those potential parts mark the same point of possible division of what is still continuous—the actual, contiguous endpoints occur at a single point of the mathematical continuum. This accords with Leibniz's definitions (in the *Specimen geometriae luciferae*, the *In Euclidis Πρωτα*, and also in earlier manuscripts from the 1680s), where two co-integrant parts of a continuous line have a point in common. The contiguous endpoints, I suggest, must therefore be regarded as artifacts of the representation. Insofar as a change is represented as occurring at an actual instant, this is a boundary between the two contradictory successive states; but since these states are vague, so too will be the boundary. In *Locus et tempus sunt continua* Leibniz finesses this by taking intermediate points within the vanishing intervals.

[146] Cf. De Risi (2019, 131): "Leibniz, in fact, was obliged to attempt to characterize an infinite sequence of elements (i.e. the boundaries of bodies) within the *continuum* of space, which is dense in this latter but does not exhaust it (since, in addition to the boundaries, there is also matter in space)."

[147] Ted McGuire, in his perceptive essay (1976), drew attention to the non-continuity of phenomenal extension and change (see esp. pp. 306–7), but equated it with what he calls the "perceptual continuum," which "comes into existence in discrete chunks" (311; see also 306, 310–11, 314). This contradicts Leibniz's claims that matter and change are perceived as continuous, with the irregularities smoothed out by the imagination.

This side-stepping of the contiguous points is consistent with his claims in the *Pacidius* and beyond that "there is no moment of change common to each of two states, and thus no state of change either"; instead, as we saw, change is defined as "an aggregate of two states, old and new" (A VI 3, 566/LLC 211). Yet "all enduring states are vague, and there is nothing precise about them" (A VI 4, 1613–14/ LLC 297).

So the analogy between matter and change comes down to this. There is no assignable point in space at which there is not a unit [*unitas*][148] of matter that is not actually moved with a different motion (endeavour): this is a body of arbitrary smallness, a physical point. Similarly, there is no assignable state in which there is no actual change occurring. So change occurs not at the endpoint of one interval or the next, but within a state of arbitrary smallness. Perceptions are always of a finite duration (even though we represent them as bounded segments of extended duration in our imagination); but because meanwhile other things are changing, and these changes must be reflected in every monadic state, these states must in fact be of vanishingly small duration, so that they are momentaneous. The duration of any created thing must be an aggregate of such momentaneous states, produced by the changes of state at each actual instant. As Leibniz explained further to Electress Sophia in his letter of October 1705:

> And one can thus conclude that a cluster [*un amas*] of matter is not a true substance, that its unity is only ideal, and that (setting aside extension) it is only an aggregate, a cluster, a multitude of an infinity of true substances, a well-founded phenomenon, without ever violating the rules of pure mathematics, but always containing something besides. And one can conclude also that the duration of things, or the multitude of momentaneous states, is the cluster of an infinity of flashes [*d'eclats*] of Divinity, each of which at each instant is a creation or reproduction of all things, having no continual passage, strictly speaking, from one state to another. (GP VII 564)

This echoes the "just as many fulgurations of the divinity, ôr states of the universe" of *Locus et tempus* (LH 37, 5, Bl. 134^{v}), and is in turn echoed by Leibniz in his essay "Monadology,"

> all the created or derivative monads are productions of God, and arise, so to speak, by the continual fulgurations of the Divinity from moment to moment (§47, GP VI 614)

[148] Leibniz's use of the term *unitas* here is certainly problematic, since it confuses the unities that are in matter—the monads—with bodies of arbitrary smallness, which monads certainly are not. There is an analogous problem with his describing the momentaneous states as *unitates*.

It should not be forgotten, however, that transproduction "happens by a certain law" (A VI 3, 503/LLC 99), this being a necessary condition for actually dense states to be perceived as a continuous motion, or for a polygon of arbitrarily many sides to be depicted by the imagination as a true circle. Leibniz makes precisely this point in his *Specimen Geometriae luciferae* of around 1695, one of his most important and original essays in geometry. In a passage noted already by Ernst Cassirer in his *Leibniz' System* (1902, 183), and more recently praised by De Risi (2019, 131–3) for its prescience in distinguishing denseness from true continuity, Leibniz writes:

> From these considerations the nature of continuous change can also be understood: it does not truly suffice for it that between any states that you choose an intermediate one is found, for other progressions can be thought of in which such an interpolation may be made perpetually, so that [this state of affairs] cannot be conflated with something continuous; instead, it is necessary that a continuous cause can be understood that is operating at every moment…And such changes can be understood in respect of place, species, magnitude, velocity, and indeed also of other qualities that do not belong to this consideration, like heat and light… (*Specimen Geometriae luciferae*, GM VII 287)[149]

The mention here of the "continuous cause operating at every moment" takes us back to the idea of continuous creation. As I argued above, in the *Pacidius* Leibniz argued that the discontinuity in motion and change demonstrated the necessity of "a superior cause which by acting does not change, which we call God" (567), who produces the changes among things by his "special operation" (568–9). Subsequently, the simple substances that he posits in things are supposed to perform that role, with each of these substances producing its changes from its own store, and consequently the changes in composite things too. But the discreteness of actual changes creates a profound difficulty for this notion, one that is well stated by McGuire: "For if change in the temporal nature of substantial existence can only be characterized by means of a discontinuous series of momentary states, simple substances cannot truly endure" (McGuire 1976, 316).

If I am right, the key to resolving this difficulty is the recognition that Leibniz is modelling the continuity of substance on his (successful) rendering of continuous transitions in his calculus. The law of the series of monadic states is an extremely complex encoding of the future behaviour of the substance, giving its state at any instant. But this law (equivalent to the complete concept of the individual substance) remains the same, while the substance corresponding to it produces its

[149] The dating of the *Specimen* is also difficult. De Risi suggests that internal clues suggest that it is closely connected with other manuscripts datable as from the mid-1690s, and certainly after 1693. See De Risi (2019, 131) for discussion.

states in accordance with the law. (The law is what corresponds to the physical memory in a substance, underlying its physical identity.) The individual substance is the subject of change and the source of its own momentaneous states, and is what remains the same through these changes. It is "the thing which by acting does not change." As Leibniz writes to De Volder:

> The succeeding substance will be considered the same as the preceding substance as long as the same law of the series or of simple continuous transition persists, which makes us believe in the same subject of change, or monad.
>
> (Jan 21, 1704/LDV 291)

This can usefully be compared with the discussion in the *Specimen Geometriae luciferae* in which the above distinction between density and continuity occurred. There Leibniz discusses how a solid figure can be constituted as an "aggregate of all states of a certain continuous transformation":

> We may understand something continuous not only in simultaneously existing things, and in fact not only in space and time, but also in a transformation [*mutatione*] and in the aggregate of all states of a certain continuous transformation. For instance, if we assume that a circle is continuously transformed, and goes through all the species of ellipses preserving its magnitude, then the aggregate of all these states, that is to say of all the ellipses, may be conceived as continuous even if the ellipses are not in contact and do not exist together but one of them is produced by the other. (GM VII 285; De Risi 2019, 132)

We may even form a solid figure from all these ellipses, says Leibniz, "that is to say, a solid whose sections parallel to the base are all those ellipses taken in order."[150] Likewise, one might say, when a moving body is at a new physical point at each different assignable instant, its motion may be regarded as continuous—even though the points are not touching—because its existence at each point is produced by a continuous cause, the monads internal to the body. In the same way, too, so may a continuous duration be aggregated from momentary states. These "momentary states" or intervals between changes correspond to the differences between successive values of variables, the differentials of Leibniz's calculus. In forming an integral, one has a variable stand for a value (such as the location of a point) at each assignable instant, and the particular functional relationship allows one to calculate the quantity of the aggregate that results from summing all the differences. This, I suggest, is why Leibniz feels entitled to describe physical continuity as constituted by the dense physical points or actual instants, and

[150] As De Risi remarks, 'From a mathematical point of view, Leibniz seems to be remarkably close to the modern idea of a continuous fibration of a manifold' (De Risi 2019, 132).

equally as the aggregate of the intervals between these points or instants. The actual instants at which changes are occurring are densely ordered, just as are the physical points on an infinitely divided line. But when you calculate an integral as a sum of differences, you do so by means of a function that gives values at each assignable instant. As Leibniz explains to the Electress Sophia,

> Thus although matter consists in a cluster [*Amas*] of simple substances without number, and although the duration of creatures, as well as actual motion, consists in a cluster of momentaneous states, nevertheless it must be said that space is not composed of points, nor time of instants, nor mathematical motion of moments, nor intension of extreme degrees. It is just that matter, the course of things, and finally any actual composite, is a discrete quantity, whereas space, time, mathematical motion, intension or the continuous increase that we conceive in speed and in other qualities, and finally everything that gives an estimate that extends to possibilities, is a quantity that is continuous and indeterminate in itself, or indifferent to the parts that we can take in it, and those that are actually taken in nature. (GP VII 562)

Nevertheless, despite the lack of uniformity in actual changes, the mathematics of the continuous is still applicable, as Leibniz explains to Bayle in 1702, replying to the latter's article 'Rorarius' in his *Critical Dictionary*:

> This inclusion of the possible with the existent makes a continuity which is uniform and indifferent to every division. It is true that in nature we never find perfectly uniform changes, such as are required by the idea of motion that mathematics gives us, any more than we find actual figures that are precisely like those we learn about in geometry, because the actual world does not remain in an indifference of possibilities, but arises from actual divisions or multiplicities, whose results are the phenomena that present themselves, and which are varied in their smallest parts. Yet the actual phenomena of nature are ordered, and must be so, in such a way that one never encounters anything in which the law of continuity, or any of the other most exact rules of mathematics, is violated. (GP IV 568)

As we have seen, the imagination "supplies the deficit" between physical and ideal continuity, filling in the gaps between the physical points. That is the continuum as perceived. If we agree to call this the "perceptual continuum", then this is perfectly continuous. But this is not the same as the series of perceptions that is generated in each substance. According to Leibniz this is a physically continuous series of perceptions, consisting in successive finite perceptions, each apparently uniform, although divided within by changes of which the perceiver is unaware, "because these impressions are either too small and too numerous, or too unvarying [*unies*]" (A VI 6, 53/RB 53), with this division proceeding internally without limit. But the continuity of these perceptions, what makes them perceptions of the same individual, is provided by the law of the series, which provides the basis for each state containing traces of the previous states of that individual, its physical memory. For this, as we have seen, "it is necessary that a continuous cause can be understood that is operating at every moment" (GM VII 287).

Now, one might object that if all the states are of finite duration, this still only gives us a discontinuous transition. But the idea is that the states are of an assignable duration only on a limited level of discrimination—they are vague! Increasing the resolution reveals further subdivisions, without limit, so that they can be treated as "unassignable". This has a parallel with how Leibniz conceives the transition from an infinite series of terms differing by finite amounts to a truly continuous sum or integral. In a letter to Bernoulli in 1702, he writes:

> For instance, 1/3 + 1/8 +1/15 + 1/24 + 1/35 etc. or $dx/(xx-1)$, with x equal to 2, 3, 4, etc., is a series which, taken wholly to infinity, can be summed, and dx is here 1. For in the case of numbers the differences are assignable. (GM V 356)

He then contrasts it with an integral of the same form, where

> x and y are not discrete terms but continuous ones, i.e. not numbers that differ by an assignable interval but the abscissas of a straight line, increasing continuously or by elements, that is, by unassignable intervals, so that the series of terms constitutes the figure... (GM V 357)

On his syncategorematic interpretation of infinitesimals,[151] such "unassignable intervals" correspond to assignable ones that may be made so small that the resultant error is smaller than any that is pre-assigned. Leibniz calls these 'incomparable' to finite ones.

> We must at the same time consider that these incomparable common magnitudes themselves, being not at all fixed or determined, can be taken to be as small as we wish in our geometrical reasoning, and so have the effect of the rigorously infinitely small. For if any opponent tries to contradict this statement, it follows from our calculus that the error will be less than any error that could be assigned, since it is in our power to make this incomparably small magnitude small enough for that purpose, inasmuch as we can always take a magnitude as small as we wish. (To Pierre Varignon, 1702; GM IV 92)

[151] For accounts of Leibniz's syncategorematic approach to infinitesimals, see Ishiguro (1990), Arthur (2008), Levey (2008), and especially Arthur (2013a), Rabouin (2015), and Rabouin and Arthur (2020).

Analogously, then, this analysis can be applied to any continuous transition: all changes in such a transition are in actual fact discrete, as are the intervals between these changes. But since there are always changes in every interval at a further level of discrimination, there are no intervals so small that further change is not occurring within them. Thus, taking an "actual instant" to be one at which change occurs, then although we never arrive at two instants separated by an actually unassignably distance, the intervals between them "can be taken to be as small as we wish," and have the effect of being rigorously infinitesimally small. The resulting duration will be the integral of such momentaneous states. It does not matter that the partition of the duration is into a particular progression of instants, and not into all possible instants, because the difference of this physical continuity from a rigorous geometrical continuity can be rendered smaller than any pre-assigned difference, and will consequently be (mathematically, if not metaphysically) null.

This analysis, I believe, resolves the difficulties noted by Russell and other commentators about the apparent incompatibility of Leibniz's assertions of the discreteness of the actual and the universal applicability of his Law of Continuity. Let me now summarize its main conclusions:

- Every duration of an enduring thing is a series of successive states that are opposite (i.e. mutually contradictory) to one another, and each of which contains the reason for those following it.
- Change is the aggregate of two opposite states in the same succession of states.
- When represented geometrically, immediately successive states are contiguous to one another.
- (Geometric) instants are the endpoints of such states, so that there are never more than two such instants immediately next to one another in the same time.
- All enduring states are vague. This means that no change is discernible within them, at a certain level of discrimination; and that they have no precisely determined boundaries.
- But during any state, further changes are actually occurring, even if they are not discernible on a given level of discrimination.
- Any given state is actually infinitely divided by these changes within it into further states (where the infinity is understood syncategorematically).
- This entails that the changes are dense within any interval. But they are not continuous, since they are separated by unassignably small states.
- The states, understood as intervals between changes, correspond to the infinitesimal differences (differentials) of Leibniz's calculus: they are "unassignable," momentaneous, that is, of a duration that is not so small that a smaller one cannot always be assigned, and which is therefore null.

- These unassignables, however, cannot be understood as actual infinitesimals, since the notion of the actually infinitely small contains a contradiction. They are rather, fictions, abbreviated ways of speaking, which are nevertheless of the greatest utility.
- The duration of any thing is therefore divided into an actual infinity of such momentaneous states, separated by actual changes.
- Provided these states and their changes are generated by a law of progression, they constitute a physical continuum, consisting in an infinite progression of momentaneous states separated by changes.

I show the coherence of the theory of change constituted by these theses by giving a mathematical exposition based on them in the last section of Appendix 1, A 1.5.

This account of change also gives us fascinating insight into Leibniz's deep metaphysics and some of his motivations for it. First, there is the claim in the *Pacidius* that "things exist only when they act, and do not act if they do not change." This analysis of change implies that bodies, understood geometrically à la Descartes, *do not act*, and therefore *do not even exist*, between changes of state. Understood in this way, bodies are not temporal continuants: they exist at every assignable instant, but they do not exist at the unassignable times between these instants. So if the world consisted only of material bodies (as it did for Hobbes, for instance), they (and indeed the whole universe) would stand in need of being continuously created by God at each assignable instant. If, on the other hand, there is to be action in bodies, they would have to contain certain other "things which by acting do not change," producing their changes. As we saw, Leibniz took the human ego or self as his model for such active temporal continuants when he rehabilitated the derided substantial forms or primary entelechies of the Scholastics. Such an entelechy is a continuous cause within body, encoding in the "law of its series" all subsequent states of the substance, and actively producing its series of states through appetition. The states are conceived by Leibniz as perceptions, that is, representations of the rest of the universe. The perceptions of any one substance are connected by (not necessarily conscious) memory, so that each perception contains traces of every preceding one, and emerges from them "in degrees of the infinitely small" in a continuous process of change governed by the law of succession. As far as they can be perceived, these states are vague, and involve abstraction from the changes that are actually produced during them. But we are oblivious of these internal changes, the imagination filling in the gaps, so that the states are perceived as ideally continuous. Even so, these changes of state are always occurring, discernible only at a more profound level of discrimination, "one monad making up the deficit of the phenomena of the other" (LH 37, 5, Bl. 134^{r}). For even if there is no change apparent in a thing in a given state, nevertheless, since other things change all around it while it is in this state, it modifies its relations with them, and these changes are reflected in its minute subliminal perceptions. Since each state reflects the changes happening in all the others with which it is compatible, each extended state is further divided without end. It follows that the states are of vanishingly small duration: they are instantaneous. There are so many states that the difference from a true mathematical continuum can be made smaller than any assignable error by increasing the resolution of the analysis. Monadic duration is thus constituted by an actually infinite aggregate or cluster of momentary states, each produced by successive "fulgurations of the divinity" in accordance with the law of the series of each monad.

1.6 Conclusion

In his *Geometry and Monadology* Vincenzo De Risi described Leibniz's philosophy of time as "incomplete, uncertain, and definitely obscure," and the attempt to reconstruct it as "presumably an impossible task" (2007, 270). What I have done in this chapter may thus be regarded as an attempt to meet the implicit challenge in those words by giving a coherent reconstruction of his theory of time. In summarizing some of my main conclusions in that construction, I shall also attempt to bring out their contemporary relevance where appropriate.

One of the main obstacles that De Risi saw stemmed from his conviction—shared with other commentators—that Leibniz regarded monads as lying "outside any temporal order, while only their phenomenal manifestations occur in time" (271). Contrary to this widely accepted view, I have endeavoured to show how Leibniz's theory of time unfolds very naturally once it is recognized that it is based on relations among the states of substances, simple as well as composite, where the earlier states are mediate requisites of later ones, thereby involving the reasons for them. States that succeed one another in time are contradictory to one another, and those that do not contradict one another are compatible: insofar as they are states of things coexisting in the same world, they are therefore simultaneous. Moreover, this compatibility reflects the fact that each state of a simple substance represents the same universe according to its point of view. It follows from Leibniz's axiom of connection, according to which an earlier state of any one substance involves the reason for states simultaneous with any of its later states, that this harmony of simultaneous states, once established at the beginning, remains in force: "it is there that the whole cause of the *harmony* is to be found" (to Clarke, V §91; GP VII 412/LC 85).

A major reason why such an account has been resisted, of course, is the dominant interpretation of Leibniz's thesis of the ideality of relations as precluding their applicability to the actual, and therefore to monads. In my (1985) I let the strength of the relational theory I reconstructed on the basis of Leibniz's own statements serve as a rebuttal to that claim. Here, in response to criticisms, I have expanded that treatment to show more fully how it fits with Leibniz's accounts of causation, contingency, possible worlds, and continuity, presenting a formal exposition of these aspects of the theory in Appendix 1.

Some have objected that a theory of time based on relations among monadic states could not be Leibniz's fundamental position, since he held relations to be ideal, and reducible to non-relational predicates. There are a number of issues here. I have treated the question of reduction and the ideality of time in §1.4, where I argue that the ideality of temporal relations applies to them in abstraction from the things they relate, whether these are monadic states or the phenomenal states aggregated from them, but it does not prevent these relations from nonetheless truly applying to those states. The kind of reduction involved is that of reducing statements making reference to times to ones involving only what event is *at the same time as* another: that is, it involves a reduction to an equivalence class of states containing changes that occur at the same time. Relations of temporal precedence, on the other hand, are based on causal precedence, but neither order of precedence is itself reducible. Time, consequently, is an ordering of successive equivalence classes of simultaneous states. As an order, it is an ideal thing, rather than something existing in its own right; nevertheless, it is the order in which any successions of states must occur. It therefore has a reality deriving from the possible successions God could create; it is, like space, a *real relation*.

This interpretation involves a reading of Leibniz's philosophy of relations that extends also to space. But because of the complexity of the issues involved, I have deferred a detailed defence of that to Appendix 2. There I present my stance on the various issues in relation to the main lines of interpretation proposed by other commentators. On the one hand, Leibniz subscribed to the nominalist tradition of analysis—more accurately, the nominalist-conceptualist position advocated by Ockham, which denies that the truth of a relation holding between two things requires there to be a corresponding relational accident existing in one or both of the things. On the other hand, since the time of his *Dissertation on the Combinatorial Art* in 1666, Leibniz had been deeply committed to the view he inherited from the Herborn encyclopaedists, and Johann Heinrich Bisterfeld in particular, according to which there is a universal harmony or connection among all created things, "the principles of which are relations."[152] These two strands in his thinking are not incompatible, but they definitely seem to pull him in different directions. With the nominalists he denies that referring to things as happening at a given time requires the positing of self-existing times, and classifies relations as "mental," supervening on the things that stand in these relationships; but with the realists he maintains that relations of connection do have reality outside the minds

[152] The quotation is from the *Dissertatio* (A VI 1, 199). See my discussion in Arthur (2014, 36ff.) and Mugnai's introduction for the forthcoming edition of the *Dissertatio*.

of perceivers, and ascribes the reality of the spatial, temporal, and causal orders to the divine mind: to the extent that we correctly discern the reasons for one thing in another, we partake in a limited way of divine omniscience, and similarly things standing in certain relations of mutual situation or temporal precedence do so insofar as they partake in the divine attributes of immensity or eternity.

In these respects time, as the order of possibly successive things, is entirely analogous to space, the order of possibly simultaneous things. But there are also some striking disanalogies, which De Risi regards as obstacles to the construction of a coherent Leibnizian theory of time. One of these stems from Leibniz's insistence that quantity can only be discerned in things that are compresent, and thus simultaneous, not successive. Related to this is the fact that Leibniz's theory of space is built upon the foundation of congruence of situations, and time is not an order of situations. A third disanalogy is that successive states, and, in the limit, actual instants, are contradictory to one another, which De Risi sees as entailing a denial of the homogeneity of time, as well as a denial of its continuity.

In opposition to the last of these objections, I have construed Leibniz's time as intrinsically homogeneous, and therefore as not differing from space in this respect. It is true that in the spatial case there is nothing corresponding to *appetition*, where each state of a monad is oriented towards the future: it involves the reason for those after it. This is the basis of the asymmetric ordering of states, and thereby the asymmetry of the temporal order of succession. But asymmetry does not entail inhomogeneity. Time is homogeneous if the laws and the phenomena are invariant under a translation of time: the state of affairs where the universe is created a year earlier with everything in it retaining the same temporal relations is indiscernible from the actual state of affairs, and therefore identical with it. This corresponds to Leibniz's characterization of homogeneity in terms of the similarity to one another of all the parts of time. Actual times are identified by reference to the events happening at them, and in this respect analogous to actual places, which are distinguished by the bodies occupying them. The parts of time in themselves, however, are indistinguishable from one another, as are those of space; the fact that each part of time is oriented from past to future does not in the least detract from this.

Nor does this intrinsic asymmetry contradict the symmetry of physical laws. The direction of time, as I have argued elsewhere,[153] is constituted by the fact that a process is a transition from an initial to a final state, not by temporal symmetries or asymmetries in the types of process that may occur. It does not matter whether a type of process can occur only one way with respect to the direction of time, individual processes (process-tokens) all occur with an orientation from initial to final state, where the former is earlier than the later. Time, in short, is the order of

[153] See Arthur (2019), chapter 4: "Classical Physics and Becoming".

all possible successions. That the actual succession of states constituting the created world tends towards perfection, as Leibniz claims, is a property characterizing world history, not the temporal order.[154]

The fact that successive states contradict one another certainly constitutes a difference between the temporal order of changes and the spatial order of situations. But, I have argued, for Leibniz the fact that actual changes divide any state into successive opposed substates does not prevent the Law of Continuity applying to a given process. This is because every state or perception is subdivided by changes occurring within it at a higher order of discernibility. No matter how short the duration of a state is taken to be, there are in fact further shorter substates within, rendering the interval between successive changes infinitesimal. In this way, although any sequence of actual changes will be dense, no difference from true continuity can be assigned: from the fact that for any difference that might be assigned, there will in actuality be a smaller one, it follows that there is no assignable difference. Meanwhile, a state in which no change is discernible beneath a given threshold of discernibility is treated as void of changes. Such an enduring state, however, is "vague": the absence of boundaries in it is a feature of how it is represented (either mathematically, or in perception), so that such homogenous enduring states are artefacts of the imagination, rather than reality, which is infinitely variegated.

Here there are interesting resonances with contemporary quantum physics. For according to quantum theory, an isolated system undergoes a continuous development of its state in time (Schrödinger representation) or remains in the same state through that time (Heisenberg representation), although no actual change occurs until there is a discrete observable event, the probability of which is computable from the theory. When such an event is detected, of course, this will be the result of some interaction with a detector, so the original system will no longer be isolated, and there will be a change of state. This was interpreted by Werner Heisenberg and Niels Bohr as showing that physical quantities have no independent reality prior to being brought into being by an observer making a measurement—indeed, on some extreme interpretations of this so-called Copenhagen Interpretation, quantum states represent the observer's knowledge rather than objective states of affairs. Modern physicists, however, are more inclined to follow Einstein and Schrödinger in rejecting any positivistic reduction of reality to what human beings can observe or measure. Still, even if we resist any temptation to see Leibniz's monadic perceptions as an anticipation of this idea of

[154] The target of that remark is De Risi: "Nor could we maintain that the moving backwards of time would be indiscernible from its moving forwards, as elsewhere Leibniz seems inclined to accept the hypothesis that cosmic perfection increases (rather than diminishes) in the course of history.... In any case, it is not clear whether continuity and the architectonic hypothesis of an increase in perfection are enough to grant the uniqueness of the temporal order, or if considerations on causality also need to be applied" (2007, 272).

quantum states representing only an observer's knowledge,[155] one might still see a resonance between his idea that the enduring states between actual changes are "vague," i.e. involve no discernible changes, and the enduring states of quantum theory, to which there correspond only probabilities, but no actual values of certain physical quantities between detectable events. For instance, Leibniz's claim (made on the basis of his reasoning about the continuum) that a body has no definite location or shape at any given instant, is upheld in quantum theory, albeit with a very different connotation of vagueness. More profoundly, perhaps, his contentions that questions of discernibility should enter at the most fundamental level of description of physical reality, and that qualities that are manifested by macroscopic bodes (such as extension and shape) might be emergent and scale-dependent, can be seen as showing remarkable foresight.

One charge that has often been made against Leibniz's philosophy is that, despite his best efforts, it collapses into necessitarianism. This seems to be laid bare in his theory of time and causation, where each state not only involves the reason for those that follow it in the same substance, but also all subsequent states in the same possible world, resulting in a completely deterministic system. Leibniz's response is to embrace the determinism, but to deny that this entails logical or metaphysical necessity. One state involves the reason for another through an infinity of intermediate reasons—this is why it is a mediate requisite. The state that is the effect is the consequent of the state involving the reason for it, but between any two such states there is a limitless sequence of intermediate reasons connecting them. The infinitude that is implicit in such involvement is such that there can be no demonstration of the consequence; if there could be such a demonstration, its opposite would entail a contradiction, and the consequence would be necessary.

Here I am correcting some recent analyses of Leibniz's notion of contingency, which suppose that there *is* a demonstration of a contingent proposition (for God), only not one obtainable by finite calculation. This criterion of the impossibility of an infinite demonstration for a contingent truth, I argue, is a symptom for Leibniz of the fact that the complete individual concept of anything that could qualify as a possible existent involves all other things in the same world, and is therefore infinite. So its candidacy for actual existence depends on God's choosing to create the whole world in which it exists, that world being the one that contains the greatest perfection.

One might wonder how physical laws are possible if the link between any two states is contingent. Here it is important to recognize that our knowledge is always

[155] Monads, of course, are not necessarily human, and perceptions not necessarily conscious. According to Leibniz there are monads even in the smallest portion of matter, and the vast majority of them are incapable of knowledge: their perceptions are representations not of human knowledge, but of the whole of the rest of the universe, which is assumed to exist objectively.

about *kinds of things*, rather than individuals in all their infinite glory. That is, in Leibniz's terms, we are dealing with incomplete concepts: we make abstraction from things themselves, conceptualizing them in terms of a finite set of properties, and find regularities among their properties which we codify as laws. The laws are specific to this particular possible world, and thus contingent. They are hypothetical: if such and such conditions obtain, this consequence will follow. But assuming the truth of the laws (of which our knowledge is fallible), and assuming we have correctly identified the initial and boundary conditions, what follows is certain. But it is not *necessary*, precisely because we have abstracted away from the infinite complexity of the actual conditions that obtain. This is important because it shows that Leibniz's concept of cosmic determinism is quite distinct from Laplace's, with which it is usually confused. Laplace assumed that it would be possible in principle to determine all the positions and motions of all the bodies in the universe at a given time, so that, given a knowledge of the correct laws (which he thought that Newton had made a good start in identifying), all subsequent states of the universe would logically follow. Leibniz, on the other hand, held that it would be in principle impossible to determine all the positions and motions of all the bodies in the universe at a given time. God would know the state of each and every substance at every time, since he would know the law of the series of each substance. But knowledge of the individual laws or complete concepts of each individual involves knowledge of the infinite, which is feasible only for an omniscient being, and even then, intuitively, not analytically.[156]

Determinism, of course, has been under fire ever since the establishment of quantum theory in the 1920s and 30s. For the objective indeterminacy implied by the theory is incompatible with Leibniz's Principle of Sufficient Reason. According to orthodox quantum theory there is, for example, no reason sufficient to explain why a radioactive decay event occurs exactly when it does; moreover, if that shortcoming were only a feature of our lack of knowledge of what determines the time of decay, this would violate the predictions of quantum theory. Since Laplace explicitly follows Leibniz in appealing to the Principle of Sufficient Reason as a justification for cosmic determinism, this objective indeterminacy would appear to undermine determinism on either of their conceptions of it. On further reflection, though, the implications of quantum theory are less alien to Leibniz's philosophy of determinism than one might think. For on Leibniz's conception, although not on Laplace's, there is an objective chance element that is not reducible to human ignorance. This is due to the infinitude of the conditions actually pertaining to the situations where the laws are applied, as well as to the contingency of the laws themselves (which are not logically necessary, but only those necessary to the best possible world). There is some analogy here to the

[156] For an analysis of the continuing relevance of Leibniz's views on determinism and causation, see Arthur (2019b).

natural philosophy of David Bohm, who also insisted that "causal laws must be corrected by taking into account contingencies," with these objective contingencies arising, because of the "quantitative infinity of nature," from their "complex, multifold and interconnected character" (Bohm 1957, 139, 141, n.*). Thus according to his "hidden variables" interpretation, particles have well-defined trajectories in space, but undergo "random fluctuations originating at a deeper level" (111–16), constrained by the requirement that their average values conform to the probability distributions given by the Born Rule of the standard theory.[157]

The idea of states "involving reasons" for one another, of course, is apt to sound quaint to a modern ear, and the construal of causes in terms of reasons excessively rationalistic. In this regard a proposal by John Collier is particularly interesting, namely to treat causation as the transfer of information.[158] On this proposal, physical information is defined through Schrödinger's and Brillouin's Negentropy Principle of Information as the difference between the maximum possible entropy of a given system and its actual entropy, so that "the information in a specific state of a physical system is a measure of the capacity of the system in that state to do work." The transfer of information is always in a specific direction, even though the capacity for work (as in Leibniz) is undirected. A causal process can then be conceived as "a computation (though perhaps not a Turing computation or equivalent) in which the information in the initial state determines information in the final state" (Collier 1999, 11). Collier claims that this entails a distinction between the certainty of one physical state's following another and the necessity of this happening, a distinction almost identical to Leibniz's. The causal connection between two physical states is such that the information in one determines the information in the other, but it is not a necessary connection, since the conditions determining the initial state are highly contingent.[159]

A more straightforward connection between Leibniz's views on matters temporal and modern views is provided by the very idea of a causal theory of time. On the one hand, there is a historical lineage traceable through Kant to the causal

[157] In direct opposition to Leibniz's conception of nature, however, Bohm does not "suppose that the same pattern of things is necessarily repeated at all levels," nor "that the general pattern of levels that has been so widely found in nature thus far must necessarily continue without limit" (1957, 139). Indeed, David Layzer (1990) has argued that the discreteness of the quantum of action, in contradiction to Leibniz's s Law of Continuity applied to levels of discernment, is crucial to the understanding of order through statistical mechanics. Consequently, he argues, "irreducible or objective randomness—randomness that does not arise from human ignorance—can exist only in a universe whose most fundamental laws are quantal rather than classical" (1990, 43).

[158] See in particular Collier (1999). I give an account of Collier's theory and its connection with Leibniz's ideas in Arthur (2021).

[159] "Causal connection is necessary in the same way that computation or deduction is necessary, but it is not necessary in the sense that it is impossible for things to be otherwise. The necessity depends on contingent conditions analogous to the premises of a valid argument" (Collier 1999, 2, fn.2). "Although the causal relation is necessary, its conditions are contingent, so it is necessary only in the sense that given the relata it cannot be false that it holds, not that it must hold..." (11).

theories of time of Georges Lechalas and Hans Reichenbach.[160] On the other, there is the independent advocacy of this approach by Alfred Robb, E. C. Zeeman and John Winnie to construct a causal theory of Minkowski spacetime, and also the more recent causal network theories of spacetime. Soon after Einstein had published his special theory of relativity and Minkowski had reformulated it in spacetime terms, Robb proposed an alternative foundation to Einstein's based on absolute relations of before and after. His premise is that one instant *A* is *absolutely before* or *chronologically precedes* another, *B*, if a physical influence can be propagated from *A* to *B*.[161] This is essentially identical to Leibniz's definition of temporal precedence in terms of *A*'s being a mediate requisite of *B*. The difference is solely that (1) Robb adopts as a premise that a physical influence cannot exceed the speed of light, *c*; and (2) that he does not avail himself of anything corresponding to Leibniz's Axiom of Connection. The result is that the temporal order is no longer a serial ordering, but instead a strict partial ordering, what Robb calls a "conical order": it is no longer the case that any two points in spacetime are either before, after or simultaneous with one another. Only points such that a possible influence can connect them are in an absolute relation of before and after, and the only simultaneous events are ones happening at the same point in spacetime. On this basis, as Robb showed and as was subsequently more widely recognized, the whole structure of Minkowski spacetime could be derived.

Combining such considerations about the causal basis of time with Leibniz's espousal of the relativity of motion led Reichenbach to see him as having largely anticipated Einstein's relativity. This is now acknowledged to be a case of over-interpretation.[162] The kind of relativity of motion Leibniz envisages is not unconnected with Einstein's, as we shall see in detail in chapter 3. But Einstein did not subscribe to a relational theory of space like Leibniz's, and no one before him had remotely anticipated that measures of temporal intervals might depend on the motion of the reference frame from which they are represented, or that the time elapsed in going from one place *and time* to another might vary with the space-time path taken from one to the other. Also, as I have explained in this chapter, the notion of mediate requisite is one of two notions of requisite in Leibniz's causal theory. In addition to the mediate requisites (those linking the states of any causal process connecting two events), there are the immediate requisites responsible for the mutual involvement of simultaneous states, the "connection of all things." These, we have seen, are derivable from mediate requisites on the assumption of the Axiom of Connection. So had Leibniz eschewed that axiom, and restricted his

[160] See Henryk Mehlberg's account in his (1935/1937).

[161] "Thus if I can send out any influence or material particle from a particle P at the instant A so as to reach a distant particle Q at the instant B, then this is sufficient to show that B is after and therefore distinct from A" (Robb 1921, 11). For a discussion of Robb's approach and its continuation by Zeeman and Winnie, see my (2019).

[162] See, for example, De Risi (2012) and Arthur (2019c).

theory to mediate requisites propagated with a finite speed, he would have anticipated Special Relativity. But the eschewal of instantaneous connections would have been ruinous to his whole metaphysical system. For without them there would be no possibility of the pre-established harmony that lay at its heart, at least as he conceived it.[163]

Turning, finally, to the question of the quantity of time, it has been objected (beginning with Locke and Clarke) that order is one thing, and measure another. In the case of space, as we shall see in chapter 2, relations of situation are taken to involve distance and thus measure (even if only a proportional metric), and equality of measure is expounded in terms of congruence.[164] But time, in De Risi's words, "is just *not* a situational order" (2007, 272), and congruence would appear to be inapplicable, both "because motion and superposition are not applicable to time,"[165] and "because quantity can only be known through the consideration of simultaneity" (273). To this it must be replied: Leibniz is explicit that although situation is not applicable to time, position in an order is.[166] Secondly, congruence is indeed applicable to successive intervals of time, since substitutability of congruents does not involve actual motion, any more than it needs to in space, just the similarity and equality of intervals. The equality of the intervals into which a given time may be resolved, on the other hand, is something that needs to be established, and this is done by means of equal spatial intervals being traversed by a uniform motion.

These considerations bring into play both space and motion, however. So let us now turn to an account of the way Leibniz lays the foundation for his theory of space, before investigating how he treats motion in space in the last section of chapter 3.

[163] Having said that, however, one might see some place for Leibniz's idea of a pre-established harmony in quantum physics. This has been suggested by Jürgen Jost in a forthcoming paper, "Leibniz and Modern Physics" (forthcoming; private communication). The states of two subsystems that have split from an original system and become spatially separated are, according to quantum theory, "entangled." Even though they can be situated so that there can be no communication between them, the statistics of observations on similarly prepared systems show correlations that cannot be explained by their common history and standard probability theory. In Leibniz's terms, such pairs of states have no mediate requisites in common; yet, at least statistically, they exhibit a kind of harmony.

[164] Cf. "*Equal* things are those which can be resolved into their different individual parts that are congruent to the different individual parts of another thing" (*Specimen Ratiocinatorium Mathematicarum, sine calculo et figuris*, C 563).

[165] Compare Locke's claim in his *Essay*: "In the measuring of extension, there is nothing more required but the application of the standard or measure we make use of to the thing of whose extension we would be informed. But in the measuring of duration this cannot be done, because no two different parts of succession can be put together to measure one another" (II, ch. 14, §18; Locke 1690, 148).

[166] Cf. this fragment of uncertain date, transcribed by Couturat: "*Position* is a mode of discerning <even> those things that are indiscernible in themselves, as for instance two points have nothing in themselves by which they can be distinguished [*discernantur*], but are distinguished by position. *Situs* will be position of coexisting, it is therefore a species of position. There is also position of instants, not situation" (LH 35, 3, Bl. 19 r; C 540–1).

2

De loco: Leibniz's Theory of Space

> "I do not say, then, that space is an order or situation, but an *order of situations*, or an order according to which situations are disposed, and that abstract space is that order of situations when they are conceived as possible."
>
> (to Clarke, V, §104; GP VII 415)

Introduction

If Leibniz's contribution to the theory of space is measured by his influence on subsequent developments in mathematics and physics, it is a tale of brilliant innovation, incisive criticisms, and inspired suggestions. First and foremost, Leibniz is celebrated as the most powerful and influential protagonist of the relational theory of space, according to which space consists in relations, and is not (as Newton claimed) an entity existing in its own right. His criticisms of Newton's absolutism and defence of a relationalist alternative have reverberated down three centuries, helping to inspire Mach's philosophy of space and time, Einstein's theories of relativity, and now a new generation of researchers in quantum gravity.[1] In mathematics, his idea of *analysis situs*, a generalized geometry of situation not involving co-ordinates, is celebrated as having inspired Hermann Grassmann's invention of vector algebra as well as the modern mathematical science of topology. If we judge Leibniz's contribution by what he himself achieved, however, it is a tale of largely unfulfilled promise and enigma, concerning which there is scant interpretive consensus. His manuscripts on *analysis situs* are episodic attempts in an ongoing enterprise that he never managed to bring to completion, and recent historians of mathematics have rejected earlier identifications of this project with modern topology. It has not been clear, moreover, how this mathematical work connects with his relational theory of space, nor how the latter can be made adequate to the description of motion necessary for Leibnizian dynamics. These difficulties are accentuated by the standard interpretation of Leibniz's metaphysics, according to which his doctrine of the ideality of space is supposed to follow from his "denial" of relations. Thus on the one hand it is widely

[1] See, for instance, Julian Barbour's account in Barbour (1999, 40) of the influence of Leibniz on his own and Lee Smolin's work on the theory of time and space. See also Smolin (1997).

Leibniz on Time, Space, and Relativity. Richard T. W. Arthur, Oxford University Press. © Richard T. W. Arthur 2021.
DOI: 10.1093/oso/9780192849076.003.0003

held that fundamental reality for Leibniz consists only in monads (simple substances), which are not located in space. This is problematic, since phenomenal bodies, which Leibniz describes as aggregates of simple substances, *are* located in space. On the other hand, as I have already noted in chapter 1, the idea that Leibniz excluded relations from his fundamental ontology sits very uneasily with his relational theories of space, time and motion, and would appear to undermine the very foundations on which they are based.

In this chapter I shall try to resolve some of these enigmas. Although the incomplete nature of Leibniz's theory of space cannot be remedied, I will attempt to show that there is considerably more coherence to his views than is usually credited. In earlier papers I had argued that the connection between the relational theory and mathematical analysis situs requires a recognition that the elements of his mature theory are not existing bodies (as on a standard relationalist view) nor bare points (as on an absolutist view), but *situations*, which are representations of systematic relations among points, abstracted from the individual, relational accidents of bodies.[2] This requires a rethinking of his place in the relationist tradition, where he is usually portrayed as a champion of what I have termed body-relativism, thought to extend from Aristotle through him to Mach and Einstein. In §2.1 I examine the consequences of this reconceptualizing of his relationism. By situating Leibniz's thought in relation to the views of his contemporaries, I show the probable influence of Huygens, and how relativists were able to distinguish true from apparent motion without referring motion to absolute space, as did Borelli and Newton. I examine this issue in regard to its bearing on the debates over the nature of spacetime, and respond to some recent papers on this topic, especially Slowik's claim that Leibniz espoused a property theory of space.

I then proceed in §2.2 to a consideration of the genesis of Leibniz's thinking on space.[3] I trace the development of his views from his earliest writings of 1669, through the period where he studies Spinoza's views in Paris, to the fragment *De tempore locoque*... probably written in Rome in 1689. Without attempting to give a complete account of his sources, I argue that his earliest views not only bear marks of the probable influence of Gassendi and Patrizi, but also owe something to Averroist precedents for the conception of space as involving indeterminate magnitude and figure prior to its acquiring boundaries through the bodies in it. While engaging with Spinoza's views Leibniz develops the idea that the divine attribute of immensity is the basis of space, which persists unchanged and indivisible through the changes of the network of internal boundaries induced in it by the motions of the bodies it contains. This view remains intact while he develops his theory of time and space as "real relations."

[2] I first argued this in Arthur (1987) and in more detail in Arthur (1994).
[3] This is a reworking and amplification of the first section of Arthur (2013b).

In §2.3 I turn to his *analysis situs*, showing the underappreciated importance of Hobbes's treatment of situation for Leibniz's own. Hobbes offers two ways of characterizing a situation: either as to all intents and purposes a figure, with lines joining its vertices; or as given by the angles and distances of these vertices to an external point. The former links with Leibniz's view of space as a changing network of figures, the latter with his conception of the point of view of a substance, a perception of the rest of the universe from the organs of sense of its body, as if from a point. I then show (following De Risi) how Leibniz develops his new "geometrical characteristic" or *analysis situs* to derive various properties of space, including uniformity, isotropy, homogeneity, and arcwise connectedness.

In §2.4 I turn to the metaphysics of space. Here I am mainly concerned with De Risi's contention that late developments in Leibniz's thinking on *analysis situs* were intimately connected with an alleged "phenonemenalist turn" in the metaphysics of his final years. De Risi gives a complex and closely argued phenomenological interpretation based on the premise that Leibniz sought to reduce all phenomena to situational relations, while connecting Leibniz's distinction between quantity and quality with his theory of monadic expression. On his interpretation, space for Leibniz is "transcendentally determined by conditions on the possibility of sensible experience" (De Risi 2007, 427), phenomena—*all* phenomena—are mappings of situational relations, and substances are determined by their phenomenal representations. I argue that this leaves out of account the crucial role played by appetitions in producing the external phenomena to which the internal ones are correlated, as well as Leibniz's continued commitment to divine immensity as the basis of space. Also, while agreeing with Edward Slowik that the foundation provided by divine immensity is stressed by Leibniz in his later writings, I argue that this does not constitute a late change of position and abandonment of relationalism for a "property theory" of space; nor, on the other hand, is the diffusion of situation that De Risi identifies as the key to Leibniz's mature expositions of *analysis situs* indicative of a change of view: rather, it is the mathematical expression of the possibility of diffusion of divine presence (immensity) that Leibniz had maintained throughout.

One further issue arising from this discussion is the delicate question of whether Leibniz recognizes more than one type of space, and this is the topic of §2.5. It is standardly assumed that the abstract space of his geometry underpins the actual or phenomenal space of his physics. But it has remained unclear whether these are two different spaces, and if so what exactly the difference is between them. Some authors have claimed that this opens up the possibility of a difference of structure, so that physical space might have different properties in different possible worlds, perhaps lacking continuity, for instance, or connectedness, or flatness. Against this, I side with De Risi, and with Debuiche and Rabouin, in holding that there is only one space for Leibniz, the Euclidean, three-dimensional, homogeneous, isotropic, continuous, space articulated in his *analysis situs*. This does not dispose completely of the question, since it has also been contended by myself, previously, and by De Risi, that the system of boundaries actually instantiated from one instant to another can nevertheless be taken as constituting a phenomenal space, in distinction from the continuous mathematical space that encompasses all possible boundaries. This is connected with Leibniz's way of distinguishing continuity from contiguity, and De Risi's charge that the latter notion hamstrings Leibniz's understanding of space until the last. In response, I argue that although the parts of bodies in space have their own boundaries, and are thus contiguous, we abstract from this in representing them mathematically, where the parts are merely potential, and can therefore share extrema. Moreover, the actual phenomena—physical points and changes—are not contiguous but dense, and separated by (fictional) intervals of space and time that can be rendered as small as desired, and therefore null.

I will return to a discussion of Leibniz's theory of space in connection with his philosophy of motion in the last section of the next chapter.

2.1 Relative Space and the Order of Situations

Space, for Leibniz, is "something merely relative," and not an independently existing entity. His clear statement of this position, particularly in his controversy with Newton and Clarke,[4] is seen as a vital link in the chain of relationalist thought stretching before him from Aristotle to Descartes, and after him through Mach to Einstein. But although his place in this tradition is undeniable, the brand of relationalism he advocates is, I have argued (Arthur 1987; Arthur 1994) crucially different in certain respects from that of Aristotle and Mach: his is not a body-relative space, one defined by reference to existing bodies. In particular, Leibniz's innovations in this regard are intimately related to his founding work in generalized geometry (*analysis situs*), whose formulation was accompanied by highly original reflections on the nature of space. Since I wrote those papers very much more has been learned about the relation of *analysis situs* to Leibniz's theory of space, especially with the publication of Leibniz's *Characteristica geometrica* and related studies by Javier Echeverría and Marc Parmentier in Leibniz (1995,

[4] Thus, as he wrote in his Third Paper for Clarke, "I have indicated more than once that I hold space to be something purely relative, as time is" (§4, GP VII 363/LC 25). This had been his view since 1677, but one might wonder where he expected Clarke to have seen it. In the published version of the *Correspondence*, Clarke included a helpful Appendix of relevant quotations taken from Leibniz's published works (LC 127–139), almost all from the *Specimen Dynamicum* (1995), the *New System* (1698) and the *Theodicy* (1710). None of these concerns the relational nature of space. One concerns the relativity of motion, the example of the two similar perfect concentric spheres, one revolving in the other, from *On Nature Itself* (LC 136), although the even more relevant passage from the *New System* is missing: "As for absolute motion, nothing can determine it with mathematical rigour, since everything terminates in relations" (GP IV 487/L 459).

abbreviated CG), and De Risi's magisterial study (2007). Although I took some of this into account in my (2013b), in the following pages I will attempt a more thorough reappraisal, especially in light of subsequent critiques of the traditional body-relativist interpretation, notably those of Edward Slowik (2016) and Valérie Debuiche and David Rabouin (2019). Slowik, indeed, goes so far as to contest the idea that Leibniz was a relationalist about space at all, seeing him instead as advocating a version of a property theory of space. So in this section I will outline my position more carefully, as well as explaining what bearing it has on the modern debate over the nature of spacetime, all of which, I agree with Slowik, is highly anachronistic with respect to the views of Newton and Leibniz.

According to the traditional relationalist view advocated by Aristotle, Descartes and Mach (I shall ignore the idiosyncrasies of their views and some of the profound differences that exist among them), a place or spatial location is not a position in some absolute or container space, but rather a position *relative to some body or bodies*. For Aristotle, "the place of a thing is the innermost boundary of what contains it";[5] according to Descartes, "when we say that a thing is in a given place, all we mean is that it occupies such and such a position relative to other bodies; but when we go on to say that it fills up a given space or place, we mean in addition that it has precisely the size and shape of the space in question" (Descartes 1644, II, §14; AT VIII 48/Cottingham et al. 1985, 229); for Mach, "All our principles of mechanics are, as we have shown in detail, experimental knowledge concerning the relative positions and motions of bodies" (Mach 1919, 229). Space, on this perspective, is a structure that depends on the things existing in it, and places consist precisely in the positions of bodies *relative to* a given body or set of existing reference bodies (such as the "fixed stars"). I call this kind of relational theory of space *body-relativism*. On such a view, the motion of a body consists in its occupying different positions with respect to those reference bodies, taken as immobile. On the basis of such statements as that "space, taken apart from things, has nothing in itself to distinguish it, and indeed has nothing actual about it,"[6] Leibniz has usually been taken as one of the chief representatives of this position.

The idea that bodies' positions are determined relative to other bodies is, of course, one of the guiding intuitions behind the principle of relativity, as the history of thought on this topic from Descartes to Einstein amply demonstrates. Here I am referring to the principle of the *universal relativity of motion*, according to which *all motion is relative to which body or system of bodies is hypothesized to be at rest*, and the phenomena (although not their description) and laws of physics are invariant under change of hypothesis, called by Leibniz the "Equivalence of

[5] Aristotle, *Physics*, IV, 232a 4–5; quoted from Huggett (1999, 59).
[6] 5th Letter to Clarke, §67; (GP VII 407/LLC 79).

Hypotheses." The snug way in which body-relativism and universal relativity fit together can be seen in the writings of Descartes:[7]

> To determine the position, we have to look at various other bodies which we regard as immobile; and in relation to different bodies we may say that the same thing is both changing and not changing its place at the same time.
>
> (Descartes 1644, II, §13; 1985, 228)

In the continuation of this passage, Descartes acknowledged that one might well determine the motions of terrestrial bodies by reference to "certain fixed points in the heavens." But, he argued, "if we suppose that there are no such genuinely fixed points to be found in the universe (a supposition which will be shown below to be probable), we shall conclude that nothing has a permanent place, except as determined by our thought" (228).

The young Christiaan Huygens adopted this idea of the relativity of motion to other bodies regarded as immobile, at least for the case of linear motions. (For circular motion, as we shall see, he thought for many years that he had a criterion for true motion, namely that it would produce centrifugal force.[8]) In a manuscript dating from 1654 he wrote:

> It does not seem possible to understand what rest or motion there is in bodies except in respect of other bodies. For concerning motion we can imagine nothing but what changes the mutual distance and disposition of bodies among themselves. And so a body that moves is said to move in respect of other bodies with which it changes situation, and to rest in respect of those with which it conserves its situation.
>
> (*De motu corporum ex percussion*, *Appendix* 1; Huygens [1654] 1929, 111)

This was still Huygens' position (with circular motion excepted, as mentioned) when he took Leibniz under his wing during the latter's sojourn in Paris in

[7] In this section I shall ignore the issue, much discussed in the recent literature, of whether Leibniz's commitment to the Equivalence of Hypotheses is in tension with his use of mv^2 as the measure of force. This will be a central issue in the third chapter. Also I shall not try to impose modern distinctions between 'relational' and 'relative' on this material, since I think it would distort the reporting of the history of the subject. Thus spacetime could be supposed as absolute, as an existing thing, although constructed from spatiotemporal relations; or, the issue of how it is constructed could be regarded as distinct from the question of whether all motion is relative. Moreover (here with Reichenbach's claims in mind that Leibniz anticipated Einstein's relativity), no one before Einstein had conceived that velocity might be relative to the reference frame adopted.

[8] For a reconstruction of Huygens' thought in the period from 1668/9 to 1687, and the identification of centrifugal force as the probable referent of the κριτήριον Huygens cited for the reality of circular motion, see Mormino (1993, 57–65). Cf. Huygens' remark in Codex 7A, Fragment 9: "For a long time I thought that in circular motion one had a criterion [κριτηριον] for true motion, from centrifugal force" (Mormino 1993, 236).

1672–76, although after further reflection in the 1680s he came to regard all motion, including circular, as relative.

By the time of Leibniz's stay in Paris, however, several treatises on motion were circulating that distinguished relative motion from motion as change of place in a space conceived as absolute. For example, Giovanni Borelli, in his *De vi percussionis* (1667) had written:

> Therefore local motion will be the successive passage from one place to another in some determinate time, extending, by successive contacts, through all the consecutive parts of place ôr of the space traversed... In addition, the passage of local motion either occurs from one place of mundane space to another, or in the relative space of some container; the former shall be called real and physical motion, the latter we will call relative motion, although often it does not involve a change of its situation in place, or the space of the universe. (Borelli 1667, 1–2, 3)

Here Borelli assumes that it is unproblematic for bodies to be located in an antecedently existing absolute space. Similarly, Ignace-Gaston Pardies wrote in his 1670 that "the absolute speed of a body must be distinguished from its respective speed. I call *absolute speed* that which is considered in a body compared with the space in which it moves, while *respective speed* is that which is considered in two bodies compared together, by which speed the two bodies mutually approach or recede from one another."[9]

In his notes on his contemporaries recorded in the *Codex Hugeniorum 7A*, Huygens showed increasing exasperation with such explanations of motion. For example, in Fragment 3 from about 1687–89, he writes:

> Local motion, says Borelli, is the successive passage from one place to another, etc. But what is place? Is it immobile space? But how can we call a thing immobile when we are still seeking the definition of motion? Is a place then a space capable of accommodating a body and defined by its position? Then it is defined with respect to other bodies. (Mormino 1993, 146)[10]

[9] "il faut distinguer la vîtesse absoluë d'un corps, & sa vîtesse respective. J'appelle *Vîtesse Absoluë*, celle qui se considere dans un corps comparé avec l'espace dans lequel il se meut: & *Vîtesse Respective*, celle qui se considere dans deux corps comparez ensemble, par laquel vîtesse ces deux corps s'approchent ou s'éloignent mutuellement l'un de l'autre" (Pardies 1670, 44). Similarly, Edme Mariotte wrote in his (1673, 2–3) that "Vistesse Respective de deux corps, est celle avec laquelle ils s'approchent, ou s'esloignent l'un de l'autre, quelles que soient leurs vistesses propres. [The respective speed of two bodies is that by which they mutually approach or recede from one another, which are their proper speeds]." Huygens later regarded this as a usurpation of his notion of respective velocity (see Mormino 1993, 31, n.63).

[10] Similarly, he wrote in Fragment 5 (in 1687 or later): "Motion cannot be understood except as relative. What is place? Is it immobile space? But how can we call something immobile when we are still

Huygens' scepticism over how location with respect to space could be achieved without reference to bodies had been piqued much earlier when he was asked by Henry Oldenburg in August 1669 for his opinion on William Neile's question,[11] "if there were but a single body in the whole immense mundane space," and yet all motion were respective, how then could God move it? Huygens' reply was uncompromising:

> Concerning his [Neile's] supposition of one single body in all the world, I say that one could not consider there to be either motion or rest in this body, because there would be nothing to which they could be referred. And if he does not wish to agree with me that motion and rest can only be considered relatively, I beg him to tell me and define what it is to take one or the other of them as absolute and without relation.[12]

It is not a stretch, I believe, to see Leibniz as having become convinced of the correctness of such a relational view under Huygens' influence.[13] Although he believed that there must be something in body that explains why one body is able to rebound from another in such a way that the collision laws of Huygens, Wren and Wallis are satisfied, he was already convinced in April 1676 that the conservation of quantity of motion "must be asserted of the action ôr respective motion by which one body is referred to another, or acts on the other" (A VI 3, 493/LLC 76–9)—a formulation in complete conformity with Huygens' position noted above that "a body that moves is said to move in respect of other bodies with which it changes situation." That is, the conservation law underpinning the correct laws of collision depends only on the relative motions of the bodies concerned, on their velocity differences. This is in contrast to the laws of

seeking the definition of motion?" (Mormino 1993, 174). For Huygens' further objections to Borelli's account of motion, see the rest of Fragment 3 and Mormino's commentary in his footnotes (Mormino 1993, 146–53).

[11] For a discussion of this episode involving William Neile (1637–70) and Oldenburg, see Dana Jalobeanu's (2011, 114 ff.).

[12] Huygens to Oldenburg, October 30, 1669; quoted from Mormino (1993, 57). This response is reflected in the *Codex Hugeniorum 7A* (in a passage also discussed by Ed Slowik in his (2016, 24)), where Huygens writes: "since I hold that motion is nothing but a relation to another body, we say (with no offence to God) that he could not make the relation be to what does not exist, that is, make one body be two or more. Similarly, I say, nor can one solitary body be constituted by God as being at rest, since rest, just like motion, is relative to something else, and neither can be predicated of a single body" (Mormino 1993, 186–88). This is just the kind of reasoning Leibniz will later employ against Clarke in their correspondence. God could not create the world at a different time, for instance, since the time of Creation is identified as being the time at which Creation occurred: "that particular time, considered without reference to things [sc. the creation event itself], being an impossible fiction" (Fifth Paper, §58; GP VII 405/LC 77).

[13] Leibniz even went as far as to claim in a letter to Thomas Burnett in 1697 that in Paris (and until Newton's *Principia* appeared) he had agreed with Huygens that centrifugal force is a criterion for absolute motion (GP III, 205). As we shall see in chapter 3, this claim is contradicted by the evidence of his writings on relativity.

Descartes, Pardies, and Leibniz's own collision laws from his *Theoria Motus Abstracti* of 1672. Leibniz explains the link with the relational theory of space in a manuscript he wrote shortly after his return to Hanover in early 1677:

> If *space* is a certain thing supposed in pure extension, whilst the nature of *matter* is to fill space, and *motion* is change of space, then motion will be something absolute; and so when two bodies are approaching one another, it will be possible to tell which of them is in motion and which at rest; or, if both are moving, with what speed they are moving. And from this will follow those conclusions which I once showed in the *Theory of motion abstractly considered.* But in reality space is not such a thing, and motion is not something absolute, but consists in relation. (A VI 4, 1968/LLC 225)

The consistency of Leibniz's position with Huygens' (on linear motion)[14] is still evident in the unpublished second part of the *Specimen dynamicum* of 1695, where he writes:

> motion considered apart from force—that is, insofar as only the geometric notions of size, shape and their variation are considered in it—is really nothing other than change of situation. Therefore *motion, as far as the phenomena are concerned, consists in mere relation* [*respectu*]...It must therefore be maintained that, if several bodies are in motion, it cannot be inferred from the phenomena which of them is in absolute, determinate motion or at rest; rather, rest can be attributed to any of them you choose and the same phenomena will still be produced. (GM VI 246–7)

Given this background, it seems plausible to suppose that Leibniz's doubts about whether there is such a thing as a background space against which motions could be judged was intimately tied up with what he had learned from Huygens about the laws of collision. (There is good evidence for this, as we will see later, in §3.1). And this suggests that he, like Huygens, regarded place as defined by relation to bodies.

But there is a second, less historical, motivation for seeing Leibniz as a relationalist of the traditional kind, which I think is much more influential. This is that it provides exactly the expected contrast with the views of Newton. For, as Newton wrote in the *Principia* (1687, 12/1999, 410) a body-relative space is one

[14] Huygens' mature position on circular motion is pithily summarized in Fragment 8 of the *Codex Hugeniorum 7A*: "There is no motion except with respect to something else. But relative motion consists in the change of distance either of two or more bodies among themselves, or of the parts of one to the parts of the other (as when two spheres revolve around one another without changing their distance from their centres)" (Mormino 1993, 224). For a lucid discussion of this, see Howard Stein's (1977).

determined "on the basis of the positions and distances of things from some body we regard as immobile," having earlier declared that what he calls *relative space* is defined "by our senses through its situation to bodies" (11/409). Thus according to this almost irresistible contrast, we have Newton the absolutist, who holds that a class of absolute motions can be physically distinguished, and that these define a preferred frame of reference for all motion, *absolute space*, and Leibniz the relativist, who holds all motion to be relative to which body is taken to be at rest, places to be relative in the same way, and space to be any of the *relative spaces* so determined.[15]

As enticing as such a dichotomy may be, however, it cannot be sustained as an accurate reading of the historical state of affairs. In the first place, as I have argued elsewhere (Arthur 1994), it is anachronistic to read Newton's relative spaces as reference frames, and absolute space as a specially distinguished relative space. What Newton had in mind was more in keeping with Borelli's view, as quoted above. A relative space is a circumscribed "movable dimension" of absolute space, that is, a region of space determined by the bodies occupying it, like the "subterraneous space" or, to give a more modern example, the stratosphere. This is how it is described in the *Principia*:

> Absolute space, by its own nature without relation to anything external, always remains similar and immobile. Relative space is any mobile measure or dimension of it, which is defined by our senses through its situation with respect to bodies, and is commonly taken for immobile space: such as the dimension of a subterraneous, an aerial or a celestial space, defined by its situation with respect to the Earth. Absolute and relative space are the same in species and in magnitude, but they do not remain always numerically the same. For if the Earth moves, for example, the space of our air, which relatively and with respect to the Earth always remains the same, will now be one part of the absolute space into which the air passes, and now another part of it, and thus absolutely will be perpetually changing.
>
> (Newton, Scholium to the Definitions 1687, 12/1999, 408–9)

Thus the space occupied by the Earth is a relative space: it is a finite region of space, coincident with now one, now another, region of absolute space as the Earth describes its path around the Sun. But such a delimited region of space is not an infinite reference frame in the post-Einsteinian sense, which coordinatizes the whole infinite space (and also time), nor is there a class of equivalent such frames,

[15] See for example, Nick Huggett's characterization of Leibniz's position as involving "relative reference frames," each consisting in "the relative locations of all bodies from some reference body," so that "this view is the opposite of Newton's position that space is a substance separate from matter" (Huggett 1999, 160–161). This informally summarizes what is presented more carefully and formally in, for example, Friedman (1983), Torretti (1983), or Earman (1989).

with one privileged as absolute.[16] In practice, as of course Newton recognized, we take a space whose coordinates are determined relative to some existing bodies, like the fixed stars; their being taken to be at rest is only an empirical hypothesis, however, and one that is potentially refutable. This does not alter (what Newton took to be) the fact that there is one fixed background space in which the places of bodies are defined, and in which they have their true motions.[17]

Regarding Leibniz, on the other hand, there are reasons for questioning whether body-relativism is the correct reading of his mature position. In the first place, according to his considered definition of space as "the order of existence of states which are simultaneous" (*Initia rerum mathematicarum metaphysica* [April 1715]; GM VII 17), it comes out as something *instantaneous*; that is, it is the ordering of all those states occurring at the same instant. Thus space for Leibniz, at least on this definition, is not an enduring space, as it would be if it were a Newtonian relative space, which is a space relative to some body or system of bodies considered as remaining at rest through some duration.[18] Moreover, it is a relation that extends to possibles as well as actuals. As Leibniz wrote in the *Nouveaux essais* in answer to Locke's professed nescience about whether a void space would be a substance or an accident, "[space] is a relation: it is an order, not only among existents, but also among possibles as though they existed" (A VI 6, 149).

Unfortunately, though, recognition of the instantaneity of space for Leibniz appears to make for intractable difficulties for his position. These were clearly articulated by Howard Stein in his classic paper, "Newtonian Space-time" (1967). Stein argued that when Newton supposed the existence of points of space that retained their self-identity and mutual relations through time, he was effectively positing spacetime structure, including an affine connection that was necessary in order to be able to represent inertial motions through time and space. Moreover,

[16] This reading of Newton's relative space (which I proposed in my (1994)) has since been endorsed by Ori Belkind in his (2013, 473). It demands a reconsideration of what Newton says about the relativity of motion, especially in relation to his Fifth and Sixth Corollaries to his Laws of Motion in the *Principia*. This I have given in Arthur (1994), and more recently in Arthur (2015b), where I present a comparison of the two authors' readings of relativity. I argue that what Corollary 5 says, for instance, is "that these bodies' motions 'are the same among themselves', i.e. their *relative motions* remain the same, whether their space is moving inertially or at rest. But this does not alter the fact that the *true motions* of these bodies will be different in the two cases, since it would require a force to move these bodies and their accompanying relative space from rest to a given velocity in absolute space" (1994, 226).

[17] Recent work by De Risi has even cast doubt on whether Newton thought of space itself as having a structure, as opposed to consisting in the potential for figures with structure to be embedded in it. De Risi argues that although Newton, following Patrizi, saw figures and geometrical objects as "sections and parts of an all-encompassing absolute space," "these metaphysical conceptions of space still regarded it as a geometrically amorphous background which is a mere ontological precondition of the possibility of geometrical figures (which are parts and pieces of it)" (De Risi 2016, 42). This may seem contradicted by Newton's claim above that "Absolute and relative space are the same in species and in magnitude" ("in *figure* and magnitude" according to Cajori's older translation); but I will not pursue this issue further here.

[18] As we shall see, however, Leibniz does have a concept of space corresponding to Newton's relative space, a space taken as an order of situations relative to a system of bodies taken to be at rest. This space, however, is a construction of the imagination rather than an actually enduring entity.

on the basis of his analysis of Newton's criticisms of Descartes' treatment of motion in the undated paper *De gravitatione*..., Stein argued that Newton had shown himself to be aware that, in addition to the spatial relations holding among bodies at an instant, it is necessary to posit such a structure in order to be able to represent motions.

This has led Stein and other modern interpreters to regard Leibniz's treatment of motion as similarly defective.[19] The criticism is this: in characterizing space as the order of relations among simultaneous things, and time as the order of succession of the non-simultaneous, Leibniz can be taken to have described a spacetime with a certain structure that Jürgen Ehlers (1973) and Howard Stein (1977) have called "*Leibnizian spacetime*."[20] This amounts to absolute simultaneity + a time metric + a Euclidean metric structure for the instantaneous spaces, but with no basis for the identification of the same point of space at successive instants—no affine connection.[21] But without such a connection of sameness of points through time, the spacetime implicit in Leibniz's philosophy is inadequate for representing inertial motions. Stein summarizes this argument in a later essay:

> The force of Newton's argument is great. In a famous polemic that came to a head late in the lives of both men, Leibniz took up the cudgels against Newton on behalf of a "relational" view of space and motion—not, indeed, that of Descartes, but one that was still open to Newton's criticism that on that view neither straightness of a path of motion nor constancy of speed is a concept that makes sense. (Stein 2002, 229)

This is a serious criticism. For in his physics Leibniz, like Huygens, presumes the existence of bodies moving in a straight line at a constant velocity; that is, he takes for granted that he can treat inertial motions as a privileged class of motions.[22] But, from a modern perspective, this means that in his physics he is presupposing *Galilean spacetime*. This is what we get if we add to *Leibnizian spacetime* a "standard of rotation...by choosing a rigid frame and declaring by fiat that it is non-rotating" (Earman 1989, 31), so that there will be a family of preferred frames that are not rotating with respect to it; and if we then single out one of the preferred affine connections, so that the preferred frames will be "the inertial

[19] "The point of central philosophical interest is that such a 'connection'... *cannot*—as Newton quite clearly indicates—be *defined in terms simply of the spatial relations of bodies*... And this point affects the program of Leibniz quite as much as the fanciful cosmology of Descartes" (Stein 1967, 187).

[20] As John Earman reports, Julian Barbour (1974) and Barbour and Bertotti (1977), (1982) called a different structure by this name: it lacks a time metric. Earman calls their spacetime (more accurately) "Machian Spacetime" (Earman 1989, 27–8).

[21] For a formal description and details, see Earman (1989, 27–31).

[22] Cf. Earman (1989, 72): "And to the modern reader the most glaring inconsistency in Leibniz's analysis is that it assumes an absolute or invariant notion of straight-line motion, a notion that is simply unavailable in the relationally acceptable Machian and Leibnizian spacetimes."

frames, frames whose world lines are straight lines in space-time" (33). Newtonian spacetime is then effectively what results from singling out one of these inertial frames as the unique frame in which motions actually occur.

Contra Stein, I have argued that Newton's argument begs the question. He claimed that in Descartes' cosmology (and the same applies to Leibniz's) bodies are always in motion; so if a body *B* is said to move relative to a body *A* taken to be at rest, then, since the body *A* is in fact moving, its place will have changed by the time *B* has moved anywhere. But this presupposes Newton's own understanding of motion as being the change of place in absolute space. On a relational understanding, however, the particular place of a body is defined by reference to other bodies: if the place of *B* is defined relative to the body *A*, which is hypothesized to be at rest, then, by definition, *A itself always remains in the same place*! Descartes, of course, cannot be extricated from the mess of his definitions, since he holds that there is only a distinction of reason between body and (generic) space.[23] But this would not apply to Huygens or Leibniz, who distinguished them in fact.

The same considerations apply to Slowik's charge that in his response to Locke's definition of being "in the *same place*," Leibniz helped himself to a surrogate of Newton's absolute space. Locke gave the example of "a company of chess-men, standing on the same squares of the chess-board where we left them" (Locke 1690, 132–3; Bk. II, ch. xiii, §8). Since "the distance from certain parts of the board [is] that which determines the place of the chess-men," he maintained, we would say they were still in the same places if they occupied the same squares even after the board had been moved; and that the board was in the same place in the cabin of a ship even if the ship had sailed, and so forth. Responding to this in the *Nouveaux essais* (Bk. II, ch. xiii, §8), Leibniz wrote:

> 'Place' is either *particular*, as considered in relation to this or that body, or *universal*; the latter is related to everything, and in terms of it all changes of every body whatsoever are taken into account. If there were nothing fixed in the universe, the place of each thing would still be determined by reasoning, if there were a means of keeping a record of all the changes or if the memory of a created being were adequate to retain them—as the Arabs are said to play chess on horseback by memory. However, what we cannot grasp is nevertheless determinate in the truth of things. (A VI 6, 149)

Slowik charges that "Leibniz's claim that 'if there were nothing fixed in the universe, the place of each thing would still be determined by reasoning' is deeply antithetical to relationism, needless to say, and mimics Newton's use of absolute

[23] Cf. Ori Belkind's analysis of Newton's argument and its inapplicability to Leibniz's views: "His argument here is directed at Cartesian physics, not relational definitions of true motion in general" (2013, 485).

place in the *Principia*" (2016, 64). In one sense, it is true, this idea of a universal place or space in which all the mutual positions can be determined mimics Newton's absolute space, since all the places of bodies relative to one another could be calculated in such a space. But this does not entail that such a space is an existent: it could be a mental construction, like the Arabs' mental picture of the chess game. As Leibniz wrote in a fragment from February 1677, "The absolute motion we imagine to ourselves, however, is nothing but an affection of our soul while we consider ourselves or other things as immobile, since we are able to understand everything more easily when these things are considered as immobile" (A VI 4, 1970/LLC 229). Slowik thinks that this is not an option available to a relativist:

> Put simply, a relationist must define the notion of, for instance, "same place" by means of a material reference frame: they cannot, as does Leibniz, countenance the possibility that there may be no fixed material frames at all while simultaneously insisting that "same place" is still "determinate in the truth of things"—determinate with respect to what? (Slowik 2016, 64)

It is true (as we shall see in detail in what follows) that Leibniz does explain sameness of place with respect to a set of "fixed existents," these being a set of coexisting things whose mutual situations remain the same, by hypothesis. Things are in the same place if they have equivalent relations of situation to such fixed existents, so that place is an equivalence class of situations. The fact that the fixed existents would not actually remain in the same mutual situations does not matter if we could calculate how they all moved relative to one another from that starting point. As Leibniz wrote in a further fragment, probably written in Vienna in 1688,

> Absolute space is no more a thing than time is, even though it is pleasing to the imagination; indeed it can be demonstrated that such entities are not things, but merely relations of the mind trying to refer everything to intelligible hypotheses—that is, to uniform motions and immobile places—and to values deduced on this basis.
>
> ("Motion is not Something Absolute"; A VI 4, 1638, LLC 333)

Thus the idea is that motions and subsequent locations can be calculated with respect to a set of existing bodies assumed as remaining in fixed mutual spatial relations, even if those bodies themselves move in the meantime—just as the chess moves are not conducted with respect to some absolute space, but with respect to their positions on the board; and even if the "board" is carried only in the imaginations of the horseback riders, the moves can be determined or calculated. Here there is an important proviso: that the relative positions of the moving bodies should also take into account their relative distances. As we shall see, this is one important respect in which the relations of situation countenanced by Leibniz differ from modern relational conceptions. These are still relative situations, however, and (contrary to Slowik's claim) there is nothing here antithetical to relationism (unless this is taken in a modern, anachronistic sense), or to commit Leibniz to regarding spatial location as a monadic property.

We still have to meet Stein's objection, of course, that in order to be able to calculate all the positions of bodies to one another in such a universal space, we would need to presuppose a spacetime structure adequate to the description of motion involved, including a kinematical connection (affine structure) in addition to instantaneous relational spaces. But again it can't be assumed that this is a property of a presupposed existing spacetime (a space lasting through time), as opposed to a space constructed in the imagination, without also begging the question. It is true that there must be some ontological foundation for the connectedness through time of this motion, but it does not follow that this must be provided by perduring points of space.

What, then, provides the ontological foundation for the connectedness through time necessary for inertial motion? Huygens, for his part, showed little interest in such questions, preferring to stick with established empirical facts, such as the established fact of the existence of inertial motion, without worrying about providing a foundation for it.[24] But for Leibniz the case is quite the contrary. He recognized the need for a kinematical connection for motions, but he did not ascribe this to space, since space is not a thing existing through time. Nor did he ascribe it to bodies, at least not directly, since he did not regard bodies, taken merely geometrically, as remaining self-identical through time; yet he recognized that there must be something in them to perform that role, on pain of everything reducing to ephemeral phenomena. The provision of such "genidentity," in fact, is one of the key roles he sees as performed by the substances within bodies. Very briefly,[25] according to his "reformed philosophy," bodies are aggregates of substances, while substances are constituted by an infinite series of states or perceptions, and a force or appetition by which each of these states tends to subsequent ones in the monadic series in accordance with the individual law of the series. And these perceptions or states are, as we have seen, representations of the rest of the universe from the point of view of the monad's organic body. As such they encode all the information for locating this body in relation to all the others coexisting with it at each instant, and also the tendencies for changes of location. That is, the

[24] Cf. what he wrote to Leibniz in November, 1691: "But this study [sc. improving geometry] should not prevent us from working in physics, for which I believe we know enough and more geometry than is needed; but one must reason with method on experiments, and amass more of them, almost following Bacon's project" (GM II 112).

[25] As I explain in the Introduction above, I gave a book-length argument for this reading in Arthur (2018).

ontological basis for the kinematic connection, for Leibniz, is in the substances, not in space. As he explained to De Volder in his letter of June 20, 1703,

> But in the phenomena ôr aggregates, every new change is derived from collision according to laws prescribed partly by metaphysics and partly by geometry, for one needs abstractions in order to explain things scientifically. Thus in a mass we regard the individual parts as incomplete things, each contributing something of its own, but we regard the whole as completed by the running together [*concursu*] of all of them. And so we understand any body whatever as in itself tending in a straight line along a tangent, even if by the continued impressions of other bodies its motion in fact follows a curve. But in the substance itself, which is intrinsically complete and involves everything, the construction[26] of the curved line itself is contained and expressed, since every future state is also predetermined in the present state of the substance. (GP II 252/LDV 267)

Of course, there still remains the question of whether Leibniz provided a theoretical underpinning for the representation of motions in space. And here it must be said that the above-mentioned critiques of his relational view have been conducted with scant reference to how Leibniz himself set about representing motions in space and time. Partly this is a result of the unavailability in translation of many key texts; but a bigger obstacle has been the assumption that Leibniz understood spatial relations in the modern sense. For as we shall see, a major shortcoming of the customary readings of Leibniz's relationalism is the failure to recognize that the spatial relations he takes as basic are *relations of situation*; and these are the subject of his numerous and multifaceted writings over a forty-year period on his new approach to the foundations of geometry, *analysis situs*.

And this brings us to another reason for rejecting the usual reading of Leibniz as a body-relativist. For in this connection Leibniz explicitly rejects the pivotal assumption of traditional body-relativism, namely that space depends on some particular arrangement of bodies. For in the Fourth Letter of his correspondence with Samuel Clarke, §41, Leibniz disclaims any direct dependence of space on bodies. He writes:

> The author contends that space does not depend on the situation of bodies. I answer that it is true that it does not depend on such and such a situation of bodies, but it is that order which renders bodies capable of being situated, and by which they have a situation to one another by virtue of existing together.
> (GP VII 376–7/LC 42)

[26] Paul Lodge says Gerhardt's "*constructio*" is a misreading for "*conservatio*," which he translates as "maintenance"; but it seems to me that *constructio* makes better sense. I thank Jeffrey Elawani for bringing this passage to my attention in support of my interpretation.

So Leibniz's term *situation* is not synonymous with a *spatial ordering of bodies*, as it would be on a traditional body-relativist reading. For if 'order' and 'situation' are synonyms, then, as Clarke objected, Leibniz's assertion that space is "that order which renders bodies capable of being situated" would amount to the nonsensical claim "that *Situation* is the cause of *Situation*" (Clarke, Fourth Reply, §41, GP VII 387/LC 52). In response Leibniz agrees that to say that "space is an order or situation which makes things capable of being situated . . . would be nonsense," but denies that this is what he said:

> I don't say that space is an order or situation, but an *order of situations*, or an order according to which situations are organized [*rangés*], and that abstract space is that order of situations conceived as possible.
>
> (Fifth Letter, §104; GP VII 415/LC 89)

Thus the primitives of Leibniz's theory are neither simple *points*, as they would be on an absolutist view, nor are they merely *bodies + spatial relations*, as they would be on a conventional relationalist view. Instead, they are *situations* of bodies, and space is an ordering of equivalent such situations. When these situations are thus conceived in abstraction from any bodies that might be situated, their order constitutes *abstract space*, the order of all situations conceived as possible. This is the continuous mathematical space he treats in his *Analysis Situs*. Thus, as was already recognized by Ernst Cassirer long ago, the situations in Leibniz's theory of space are the *sitûs* of his *Analysis Situs*, and it is to his writings on this subject that we must turn in order to understand his relational theory of space.[27]

We will explore in the following sections how to understand precisely what these situations are, and Leibniz's efforts to build a theory of space out of them. But this is an appropriate place to make some preliminary remarks concerning the implications of this fact—that the primitives of his theory are situations—for understanding the type of relationalism Leibniz embraced.

Perhaps the first thing to note is that relations of situation, as Leibniz conceived them, involve the mutual distances of the things so related. This is by no means as idiosyncratic as it appears to the modern reader, used to the twentieth-century understanding of relations that Russell, Whitehead and others built upon the foundations built by Schröder and Peirce and others of the late nineteenth century. But it was not unheard of in the seventeenth century. As is shown by Locke's example of the chess-pieces whose places are entirely relational, and yet determined by their "distance from certain parts of the board," it was not a bone of

[27] I argued this connection between the theory of space and *analysis situs* in Arthur (1986, 1987, 1988 and 1994a)—unaware that it was common knowledge to those who had read Cassirer's *Leibniz* (Cassirer 1902). The significance of this connection has become much clearer with the publication of Leibniz's *Characteristica geometrica* by Echeverría and Parmentier in Leibniz (1995), and Vincenzo De Risi's ground-breaking study (De Risi 2007).

contention between him and Leibniz. And as late as the closing years of the nineteenth century, Russell regarded spatial relations as involving distance, and had in fact tried to build a relational theory which derived distance from foundations provided by Projective Geometry. But the fact remains, as observed by Slowik, that this "collapses the difference assumed in modern relationism between relative order and spatial distance in favor of the former notion, order of situations" (Slowik 2016, 76).

However, Slowik claims, secondly, that Leibniz "then treats order of situations as an internal property of each body, an internal property which when idealized apart from bodies becomes space" (76).[28] Let me explain why he is moved to make this assertion, before explaining why I think it is in error.

As a first item of evidence for his interpretation of space as a property, Slowik cites Leibniz's reply to Samuel Clarke's claim in his Third Reply of their correspondence. There Clarke had said "Space is not a being, an eternal and infinite being, but a property or a consequence of the existence of an eternal and infinite being. Infinite space is immensity, but immensity is not God" (GP VII 368/LC 31). To this, Leibniz objects in his Fourth Letter, if space is a property, then "what substance will that bounded empty space be an affection or property of" that Newton and Clarke suppose could exist between two bodies?

> §9. If infinite space is immensity, finite space will be the opposite to immensity, that is to say, mensurability, or bounded extension [*l'étendue*]. Now extension must be the affection of something extended [*un étendu*]. But if that space is empty, it will be an attribute without a subject, an extension without anything extended. That is why, in making space a property, the author falls in with my opinion, which makes it an order of things, and not something absolute.
>
> (GP VII 372–3/LC 37)

This is a hypothetical argument. *If* space is a property, then an empty space will be an attribute or property without a subject, an extension with nothing to be the subject of that extension. But then it could not be something absolute, but would have to be more like an order. So what "falls in with [Leibniz's] opinion" is not that space is a property, as Slowik suggests, but the consequence that it is thereby not absolute.[29] This is made clear in the subsequent exchanges. Clarke recognizes that it is his space-as-property view that is under attack, and replies by denying that void space would lack a substance just because it lacked body: "in all void space, God is certainly present," as are possibly other intangible beings that we cannot

[28] John Earman defines the property view as follows: "it would agree with the relationist in rejecting a substantival substratum for events while joining with the absolutist in recognizing monadic properties of spatiotemporal location" (1989, 14)

[29] Slowik (2016) reads this remark as Leibniz committing himself to a property theory of space: see *inter alia*, pp. 72, and 76.

sense. In his Fifth Letter, Leibniz responds with a broadside of objections to Clarke's space-as-property view (GP VII 398–9/LC 67–9). If immensity is infinite extension, then "finite space will be the extension or mensurability of something finite," with the absurd consequence that in changing place a body would leave its own extension (§37). It doesn't seem reasonable to say that the volume of space in an exhausted receiver ("round or square") is a property of God, so then it would have to be the property of the "immaterial, extended, imaginary substances that the author seems to fancy in the imaginary spaces" (§38). But in that case "subjects will leave off their accidents, like clothes, so that other subjects may put them on" (§39). And in the following paragraphs of the Fifth Letter (§§40–6), Leibniz offers a series of further arguments against what he calls "this strange imagination that space is a property of God" (GP VII 399).

But Slowik thinks Leibniz has a different kind of property view, where the order of situations is an internal property of each body which, when "idealized apart from bodies becomes space" (2016, 76). His argument for this depends on an interpretation of Leibniz's philosophy of relations à la Russell in terms of *internal relations*, an interpretation I have critiqued elsewhere.[30] Slowik quotes the passage from Leibniz's short essay on "how people come to form the notion of space" that constitutes §47 of his Fifth Letter to Clarke:

> And here it would be well to consider the difference there is between place and the relation of situation of the body that occupies the place. For the place of *A* and *B* is the same, whereas the relation of *A* to the fixed bodies is not precisely and individually the same as the relation that *B* (which is to take its place) will have to the same fixed bodies; and these relations [of situation] agree only. For two different subjects, such as *A* and *B*, cannot have precisely the same individual affection, since it is impossible that the same individual accident should be found in two subjects, or pass from one subject to another. But the mind, not contented with an agreement, looks for an identity, for something that should be truly the same, and conceives it as being extrinsic to the subjects; and this is what we call *place* and *space*. But this can only be an ideal being, containing a certain order, in which the mind conceives the application of relations. (GP VII 400–1/LC 70)

What does it mean for a relation of situation to belong to "the body that occupies the place"?[31] As Slowik notes, this must be understood in terms of Leibniz's metaphysics. Each state of a substance, as noted above, is a representation or expression of the rest of the universe from the point of view of that substance's

[30] In Arthur (unpublished). I have deferred discussion of Leibniz's philosophy of relations to Appendix 2, to which I shall make reference as we proceed.

[31] Clarke had translated Leibniz's "le rapport de Situation du corps qui occupe la place" as "the relation of situation in the body that fills up the place," which (possibly) contributed to Slowik's notion that Leibniz thought spatial relations, and space itself, are somehow internal to the body.

organic body. The situation of a body, then, must be expressed in the states of the organic bodies contained in it. Slowik quotes the passages concerning place as an extrinsic denomination from the 1696 essay, "On the Principle of Indiscernibles" (see Appendix 2). Place is not a purely extrinsic denomination; rather, "that which has a place must express place in itself," "so that in fact, situation really involves a degree of expressions" (C 9).

Slowik reads this as an instance of Leibniz's "tendency to view place, position, etc., as akin to an internal property."[32] But that is precisely *not* what Leibniz is saying in these passages! First, place must be distinguished from situation. Place is an external denomination, conceived as if it is purely extrinsic to its subjects; whereas relations of situation are specific to bodies. As Leibniz says just before the passage quoted from his Fifth Letter, "*place* is that which is the same in different moments to different existents, when their relations of coexistence with certain other existents which are supposed to continue fixed from one of those moments to the other, agree entirely together" (GP VII 400/LC 70): the place is assumed the same, while "the relations of situation agree only" (401/70). It is the situations that are affections or individual accidents of bodies; but even these cannot be precisely the same, since they are relational accidents of different subjects. Moreover, as relational accidents they must, on Leibniz's account of relations, have a foundation in the relata on which they are based; and this foundation, as I have argued (see Appendix 2), is the degree of expressions in the individual states expressing the rest of the universe. So what is internal to each substance is a state expressing the relations of situation of its organic body to everything else in the universe at that time. Two bodies which have wholly equivalent situations to other coexistents, but not (*per impossibile*) individually the same accidents, are in the same place. Then "space is that which results from the places taken together" (400/70). Thus, as I wrote in a previous publication, Leibniz is defining place "in terms of an equivalence: it is the equivalence class of all things that bear the same situation to our (fictitious) fixed existents. And when we take all possible situations relative to these fixed existents, we have a manifold of places, or abstract space" (Arthur 1994, 237).[33]

Slowik notes that in §47 Leibniz proceeds to liken this to the case of the three possible ways of considering *ratio* or proportion, adducing it as "another example of the use of the mind to form, on the occasion of accidents that are in subjects, something corresponding to them outside the subjects" (401/71). There is the ratio of L to M, conceived as a relational accident of L, so that L is "the subject of

[32] Slowik has a third argument for ascribing the property view to Leibniz: "Third, the statement that bodily extension 'is at it is only by virtue of the abstract one [space]', signifies that even bodily extension cannot be separated from the order of situations (space), a conclusion that is also in line with his view that both bodily extension and space are internal to each body" (Slowik 2016, 76). I will return to that claim in §2.4 below.

[33] This passage is quoted approvingly by both Slowik (2016) and Debuiche and Rabouin (2019, 182).

that accident that the philosophers call a relation or *rapport*"; or of M to L, conceived as an accident of M; or, conceived in abstraction from both, when, "being neither a substance nor an accident, it must be a purely ideal thing, the consideration of which is nevertheless useful" (401/71). According to Slowik, this is another instance of the above-mentioned "tendency to view place . . . as akin to an internal property" (Slowik 2016, 73).[34] Not only are relations presented as subject-based accidents, but "the abstracted third way of considering the relation is not put forward as negating the first and second interpretations, rather, it is just presented as a different manner of understanding the relation" (2016, 74). But as we concluded in Appendix 2, the contrast here is between the relational accidents, which are constantly changing as a result of the changes in what is expressed in the states of substances at different instants, and relation in the sense of something abstracted from them. Here that abstract relation is that of "*being in the same place*," with abstract space being a system of such (equivalence) relations of being in the same place. In fact, Leibniz presents the point of this example of the ratios as follows:

> Moreover, I have here done much like Euclid, who, not being able to make understood what *ratio* is absolutely in the sense of geometricians, defines rather what are *the same ratios*. And it is in this way that, in order to explain what *place* is, I have desired only to define what is *the same place*. (GP VII 401–2/71)

As Debuiche and Rabouin comment on this passage, Leibniz himself makes clear that his intention is to construct space "as a system of what we would now call equivalence classes" by his comparison with Euclid's definition of proportion (2019, 181). "Places and space are henceforth not conceived in terms of a particular situation of bodies (which, strictly speaking, makes no sense), but in terms of an equivalence relation between situational relations" (181). "Which shows," Leibniz writes, "that in order to have an idea of place, and consequently of space, it is sufficient to consider these relations and the rules of their changes, without needing to fancy here any absolute reality outside the things whose situation we consider" (GP VII 400/LC 69).

In sum, insofar as situation is internal to monads, it is a feature of their states, comprised in their particular points of view, which change from one instant to another. These states and points of view are specific to a given monad at a given time, and are changeable accidents, not properties. Nor are these situations places: places are what we would term equivalence classes of situations, with respect to a

[34] He adds: "Yet, as all Leibniz scholars know, the price to be paid for such a scheme is astonishingly high: it apparently renders the whole of space internal to each body/substance, with phenomenalist or idealist consequences difficult to evade: e.g., "situation really involves a degree of expressions" in (e) (more on this below)" (Slowik 2016, 77).

set of bodies fictionally taken as immobile. Space, finally, is a system of such places, and thus denotes an order of situations conceived as possible. It is not a property of monads, and does not presuppose the actual existence of bodies, merely the relations they might have and the rules governing their changes.

In conclusion, we have seen that Leibniz's view of space is uncompromisingly relational. Although it is indeed the case that his brand of relationalism is radically different from that presupposed by modern writers concerned about the foundations of spacetime, the idea that it is a property theory in disguise is profoundly mistaken. But in order to fully appreciate the implications of Leibniz's construal of space in terms of situational relations, and indeed to address the criticisms about the inadequacy of his account for representing motion, we need first to give a full account of the development of Leibniz's thinking about the nature of space, beginning with the origin of his account of space as an order of situations. Once we have accomplished this, we can return to the question of how he sought to provide for what we would call spacetime structure in §3.5 below.

2.2 On the Genesis of Leibniz's Theory of Space

In Leibniz's earliest studies, space is presented as something quasi-substantial, from which the primary matter that fills it receives its extension. Thus in letters to his former professor Jacob Thomasius he describes it as "almost more substantial even than body" (October 6, 1668),[35] and as "the primarily extended entity, ôr mathematical body, which contains nothing but three dimensions, and is the universal place of all things" (April 30, 1669; A II 1, 34). In the latter letter he declares that "Primary matter is mass itself, in which there is nothing but *extension* and antitypy, i.e. impenetrability. It has extension from the space it fills."[36] And in his *Confession of Nature Against Atheists* of 1668, he argues that "a body is defined as that which exists in space" (GP IV 106/L 110). "A body has at once the same magnitude and figure as the space which it fills" (106/110). However, a body cannot derive a determinate shape and size from space, since "although the term 'space' involves some magnitude and figure, it does not involve a determinate magnitude and figure" (GP IV 107/L 111).

[35] "Indeed, space is almost more substantial even than body; for if body is eliminated, space and its dimension remain, and when no other body comes into it this is called a vacuum, but on the contrary body does not remain when space is eliminated" (A II 1, 19).

[36] (A II 1, 26/L 95; LLC 337). Leibniz published a revised version of this letter at the end of the preface of his edition of Nizolius in 1670. My citations in this book will be to the letter actually received by Thomasius, as reconstructed by the Akademie editors on the basis of Kortholt's notes on the differences between the published texts and the letter Thomasius received (A II 1, 23–38). Leibniz's main aim in the letter is to persuade Thomasius that the philosophy contained in the eight books of Aristotle's physics—as opposed to the Aristotle of then recent Scholastic philosophy—is perfectly compatible with the basic tenet of the "reformed philosophy" that he is siding with, "that only magnitude, figure and motion are to be used in explaining corporeal properties" (A II 1b, 25/L 94).

There are several traditions that could be seen as sources for such conceptions. There are strong indications of an influence of Patrizi in the claim that body derives its extension from space. According to Patrizi, God first emanated the extension of space independently of matter, and then emanated unextended prime matter, which becomes extended through being diffused in space.[37] On the other hand, the idea of space as that which is filled by matter agrees with Descartes' equation of space with the scholastic notion of 'internal place' (*Principles*, II, §10; AT VIIIA 45/CSM 227). But, like Gassendi, Leibniz insists on impenetrability (resistance to being penetrated or compressed) as essential to matter, and thus agrees with him in rejecting both Descartes' equation of extension with matter and his denial of a real distinction between extension and corporeal substance (*Principles*, II, §11).[38] Whether he also agreed with Gassendi's conception of space as absolute and self-existing is not clear, as he does not mention it here; Henry More, clearly, is another possible source for regarding space as something substantial. But, without attempting to pin down exactly who among the moderns were Leibniz's most important influences, we can say that Patrizi and Gassendi were probably preeminent among them.

On Aristotle's own view, magnitude is an accident of a given individual substance. Geometry, as the science of magnitude, is therefore not a science of space, but a study of the properties of figures conceived as abstracted from individual bodies. Under the influence of the Neoplatonists, however, magnitude came to be conceived as an accident not of substance, but rather of matter itself.[39] Thus extended matter could be regarded as a material substrate, in itself devoid of any geometrical properties. In the Averroist tradition, this matter came to be regarded as "intelligible matter" or "imaginative matter," so that (according to some in this tradition) geometrical objects could be regarded as abstractions from corporeal matter, and yet as constructed in the imagination. On such an interpretation, extension and quantity derive not from substance itself, but from the bodies of which figures are abstractions or *phantasms*. Hobbes can be regarded as

[37] Compare with Leibniz's explanation in the April 30, 1669 letter to Thomasius: "Space is the primarily extended entity [*Ens primo-extensum*], ôr mathematical body, which contains nothing but three dimensions, and is the universal place of all things. Matter is the secondarily extended [*ens secundo-extensum*], ôr what has something apart from extension ôr mathematical body and physical body, that is, resistance" (A II 1, 34/L 110). I am indebted to Vincenzo De Risi for drawing my attention to this correlation with Patrizi's views. See his (2016) for discussion of Patrizi's contributions to the theory of space.

[38] "But the essence of matter ôr the very form of corporeity consists in ἀντιτυπία ôr impenetrability" (A II 1b, 26/L 95; LLC 337). This point is reiterated in criticisms of Descartes penned in 1675: "He says the nature of body consists in extension, for everything else can be taken away from body except, so to speak, corporeality. To me it seems that there is a certain quality besides extension that cannot be taken away from body, namely impenetrability, i.e. what makes one body yield to another; and I do not see how this could be derived from extension" (A VI 3, 215/LLC 25).

[39] My remarks here are deeply indebted to De Risi's work, especially his wonderfully succinct and illuminating potted history of the concept of space in his introduction to (De Risi (ed.) 2015, 1–13).

heir to this tradition.[40] Leibniz was immersed in study of Hobbes at the time he wrote the pieces quoted above. But in them he also displays knowledge of Averroist precedents of his own views. As we have seen, in his *Confession of Nature Against Atheists* he held that although space involves magnitude and figure, "it does not involve a determinate magnitude and figure"; since, he explains, "what pertains to the term 'existing in that space' is motion, for when a body begins to exist in a different space than it did before, by that very fact it is moved" (GP IV 107/L 111). This is consistent with what he wrote about *primary matter* in his letter to Thomasius:

> . . . Now this continuous mass which fills the world is, as long as all its parts are at rest, primary matter; everything is produced out of it through motion, and everything is dissolved back into it through rest. For regarded in itself there is no diversity in it, but only homogeneity, except as a result of motion. . . . Matter has *quantity* too, but this is *interminate*, as the Averroists call it, or *indefinite*. For so long as matter is continuous, it is not cut into parts, and therefore does not actually have boundaries in it (I am not speaking of the inside boundaries of the world or whole mass, but those of its parts), although it does have extension or quantity in it. (A II 1b, 27; A VI 2, 435/LLC 337)

Here primary matter is conceived as something extended yet indeterminate, containing no inner boundaries and thus no definite parts. The boundaries of the objects within it, the parts of matter, are the result of the differing motions by which they are distinguished. Clearly, here we see prefigured Leibniz's mature distinction between primary matter as something indeterminate, and secondary matter as something actually divided into determinate parts, i.e. into bodies of various shapes and sizes (as in, for instance, the discussion of matter in the preface to his *Nouveaux Essais*). In explanation of how this could be so without a void separating the parts, Leibniz invokes Aristotle's distinction between continuity and contiguity: the parts of a thing whose adjacent extremities are together are *contiguous*, i.e. merely touching; whereas those whose adjacent extremities are one and the same boundary are *continuous*.[41] On this reading, even the Cartesian

[40] Cf. De Risi (ed.) (2015, 7): In the Averroist tradition, imaginary space "was a creation of the imagination, which again hinted toward a connection of it with the ontology of mathematical objects as proper products of the *phantasia*. . . . Even Hobbes, well into the seventeenth century, was still claiming that while geometry may well have a substrate in imaginary space, this only comes from the corporeal origin of imaginary space itself—since extension and quantity are nothing but bodily features." Cf. Hobbes: "SPACE *is the phantasm of a thing existing without the mind simply*; that is to say, that phantasm, in which we consider no other accident, but only that it appears without us" (Hobbes, *De corpore*, chapter vii, 1839b, 83/1839a, 45).

[41] See Aristotle's discussion in his *Physics*, Books 5 and 6 (i.e. *E* and *Z*), and my translations and discussion in Appendix 2a of LLC, pp. 347–8.

plenum is discrete insofar as it is comprised of contiguous parts, notwithstanding the continuity of extension conceived prior to any division.

The same goes for space. In fact, in the early 70s the *only* difference Leibniz conceives between space and primary matter (regarded as something indeterminate, yet possessing extension and quantity) is the *resistance to change* of matter. Thus in December 1675 Leibniz argues

> If I imagine in a space, instead of something extended, a perfect fluid which is at rest, but which, when another body is floating in it, moves to keep the place filled, then I mean nothing other than empty space. It would be matter if the motion of the body were retarded by its motion. (A VI 3, 466/LLC 31)

Therefore a space is something extended—an *extensum*[42]—which, so long as it is conceived as continuous, is itself "interminate" in the sense that it "does not actually have boundaries in it." But by the same token, insofar as it contains bodies that are moving within it, it will be divided into parts along with the matter it contains. This takes for granted the Cartesian view that the parts of extension are individuated by their differing motions; in his letter to Thomasius, Leibniz is concerned to show how this is compatible with his (Averroistic) understanding of Aristotle.[43] What this view entails, although Leibniz does not spell this out just yet, is that while space in the abstract includes all possible divisions into parts, space conceived through time is a constantly changing partition, whose ever-changing boundaries are determined by the motions of the matter within it. Consistently with this, in March 1676, in his last year in Paris, we find Leibniz arguing:

> Supposing space to have parts—that is to say, so long as it is divided by bodies into empty and full parts of various shapes—it follows that space itself is a whole or entity accidentally, that it is continuously changing and becoming something different: namely, when its parts change, and are extinguished and supplanted by others. (A VI 3, 391/LLC 53)

By the time Leibniz writes this, however, there have been substantial developments in his thinking. In particular, his engagement with the philosophy of Spinoza, perhaps refracted through the prism of his friend Tschirnhaus's understanding of it, is evident in some of the themes and terminology of the Paris notes in 1676. For in these manuscripts Leibniz distinguishes between the *extended* per

[42] As explained in the Glossary below, I generally leave the term *extensum* untranslated, in recognition of its function as a technical term on a par with *continuum*. In Arthur (2001), I translated it as "the extended," "extended thing," etc.

[43] Similarly, in a fragment from the same period, Leibniz writes: "Aristotle's primary matter is the same as Descartes' subtle matter. Each is divisible to infinity. Each lacks forms and motion in itself, each acquires forms through motion" (A VI 2, 279–80/LLC 343–4)

se, which, although actually infinite, is not divisible into parts and exists eternally, and *extension*, which is divided into different parts at different times. This corresponds to Spinoza's distinction between God as a substance, one, indivisible, and absolutely infinite, whose attribute of being *extended per se*, like his other attributes, "expresses an infinite and eternal essence and is thus immense [*immensum*, immeasurable]";[44] and *extension as conceived abstractly*, which has finite parts that can be enumerated in the imagination, and which is therefore measurable (A VI 3, 276–8; LLC 103–7). Now although it is true that the idea of God as immense is theologically traditional, it is perhaps significant that the term *immensum* makes its first appearance in Leibniz's meditations on space just after he begins discussions of Spinoza's thought with Tschirnhaus in February 1676. In any case, the theme is taken up in earnest in the continuation of the above-quoted passage written in March:

> But there is something in space which remains through the changes, and this is eternal: it is nothing other than the immensity of God, namely an attribute that is one and indivisible, and at the same time immense. Space is only a consequence of this, as a property is of an essence. It can easily be demonstrated that matter itself is perpetually being extinguished, or becoming one thing after another. In the same way it can be demonstrated that mind also continuously changes, excepting that which is divine in us, or which comes from without. In a word, just as in space there is something divine, the immensity of God itself, so in mind there is something divine, which Aristotle used to call the active intellect, and this is the same as God's omniscience; just as what is divine and eternal in space is the same as God's immensity, and that which is divine and eternal in body, ôr a movable being, is the same as God's omnipotence, and what is divine in time is the same as eternity. (A VI 3, 391–2/LLC 55)

What is curious here is that (in a manuscript dated March 18, 1676) Leibniz appears to be grappling with Spinozan theses to which he is not supposed to have had direct access until after receiving them from Schuller over a month afterwards.[45] This is

[44] See Leibniz's copy of the excerpts from Spinoza's *Ethics* sent him by Schuller, probably in the second half of April 1676 (A VI 3, 275–82/LLC 101–17). Intriguingly, the phrase "and is thus immense" is missing from Definition 6 in the canonical version; it is not clear whether it is an addition by Leibniz, or a feature of a variant manuscript he was copying from—perhaps the original, as suggested by Gebhardt (SO IV 390). See LLC 399. Mogens Lærke confidently claims that we know the phrase was added by Leibniz (Lærke 2009, 212), but gives no evidence for his claim. He suggests that the *immensum* is alien to Spinoza's thought; yet for Spinoza extended substance itself is essentially immeasurable, measure being "nothing but a mode of thinking, or rather, of imagining" (A VI 3, 279/LLC 109).

[45] I am not sure what the explanation of this is. It could just be that Tschirnhaus, without yet letting Leibniz see the Letter on the Infinite (a copy of which was in his possession in Paris), had since apprised him of more of Spinoza's views than was contained in Leibniz's record of their conversation in February.

especially evidenced by his distinction of divisible space from the extended per se (translating *extensum per se*), and of duration from eternity. Previously, in a note appended to his record of what Tschirnhaus had told him of Spinoza's views in February, he had written of the whole of space as "the maximum of all extended things [*extensa*]" and eternity as the "the maximum of all successive things" (A VI 3, 385/LLC 43). In the continuation of the March 18 passage, however, like Spinoza, he denies succession or duration to eternity ("the necessity of existing"), and denies divisibility to both eternity and immensity (divine ubiquity):[46]

> One attribute serves admirably to disclose another; for eternity is something indivisible, since it is the necessity of existing, which does not express succession, duration or divisibility. In the same way omnipresence or ubiquity is not divisible, as space is; and omnipotence is not subject to variation like the forms of bodies (A VI 3, 391–2/LLC 55)

This may not represent a change of position, however. For in a fragment ("On Magnitude") that probably dates from after Leibniz made his comments on the material he received from Schuller,[47] he explains that eternity may be taken in either of two senses, either as something unbounded (whether duration or time), or as the necessity of existing:[48]

> *Eternity*, if it is conceived as something which is homogeneous with time, will be an unbounded time; if as is an attribute of something eternal, it will be duration through an unbounded time. But the true origin and inmost nature of eternity is the very necessity of existing, which does not in itself indicate succession, even if it should happen that what is eternal should coexist with all things.
>
> (A VI 3, 484/DSR 41)

What is clear, however, is that "Immensity corresponds to eternity, and that just as eternity does not indicate succession, so immensity does not indicate extension or parts" (A VI 3, 484/DSR 43). Thus Leibniz follows Spinoza in distinguishing the divine basis of what is extended from extension itself (and its division into parts), and in conceiving this basis as the divine attribute of immensity: that is, the

[46] This echoes doctrines contained in Spinoza's Letter on the Infinite sent him by Schuller: "the source of the difference between *eternity* and *duration*," Spinoza writes there, is that "we conceive the existence of a substance to be of a wholly different kind than the existence of modes" (A VI 3, 277/LLC 105–7). Only the existence of modes is explained by means of duration, "whereas that of substance is explained by eternity, that is, by the infinite fruition of being."

[47] "On Magnitude" is dated by George Parkinson as "Early ? 1676" on the basis of its contents. But Leibniz's remark there that "colour is not something intelligible per se, even though it is conceived per se, because it is not conceived distinctly" (A VI 3, 483/DSR 41) seems to me to postdate his discussion of this point in his notes on Schuller's excerpts from Spinoza (see his note L_2, A VI 3, 277/LLC 105). See (LLC 398, n.18; 399, n.2; and 400, n.11) for further discussion of the dating of this piece.

[48] This was noted by Parkinson in his introduction to DSR, xxxvii.

attribute of being extended per se, which is indivisible, infinite and unbounded. But he goes beyond Spinoza in making an analogous case for three other divine attributes, eternity, omnipotence, and omniscience. Thus (i) the divine basis of duration is the attribute of eternity or duration per se, which is also *indivisible, infinite* and *unbounded*; and these properties are shared by (ii) the divine attribute of omnipotence that undergirds the variation in forms of bodies, and also (iii) the divine attribute that is the basis of our minds or knowledge, omniscience.[49]

Leibniz expands on this analogy in another piece ("On the Origin of Things from Forms") from the same period (April 15–30 or so). Here again (but as if for the first time![50]) he contrasts changeable space with its unchanging basis, identifying the latter as the indivisible "extended per se," "the *immensum*," which has modes but no parts:

> Our mind differs from God as the absolute *extensum*, which is a maximum and indivisible, differs from space or place; that is, as the extended per se differs from place. Space is the whole of place. Space has parts, but the extended per se does not have parts, although it does have certain modes. Space, by the very fact that it is dissected into parts, is changeable, and variously dissected; indeed, it is continuously one thing after another. But the basis of space, the extended per se, is indivisible, and remains during changes; it does not change, since it pervades everything. Therefore place is not a part of the extended per se, but a modification of it arising from the addition of matter, that is, it is something resulting from it and matter. Plainly, the divine mind is to ours as the space they call imaginary (whereas it is maximally real, for it is God himself insofar as he is considered to be everywhere ôr immense) is to place, and to the various shapes that are produced in the *immensum*.... And so it is the *immensum* which persists during continuous change of space; this, therefore, neither has nor can have bounds, and is one and indivisible. (A VI 3, 519/LLC 119–21)

Places are not parts but modifications of the *immensum*, arising from the addition of matter, i.e. "bulk, ôr mass." When this is added, "there result spaces, places and intervals, whose aggregates give Universal Space" (A VI 3, 519/LLC 121). Thus, Leibniz adds, universal space is the aggregate of places, as the Republic of Minds is the aggregate of individual minds, and the divine mind is to ours as the *immensum*, or "real space," is to universal space (519/121). But there is a disanalogy

[49] For an analysis of the respects in which Leibniz departs from Spinoza regarding the analogy with mind, see Lærke (2009). He notes Koyré's observation that Spinoza was strictly opposed to the Aristotelian notion of the divine mind as *intellectus agens*; but argues that Leibniz, understanding Spinoza through the prism of Tschirnhaus, might not have realized that at this time. See Lærke (2009, 215) for discussion and references.

[50] In "On the Origin of Things from Forms" Leibniz writes "So, in order to distinguish it from space, it is best I should call it 'the Immensum'" (A VI 3, 519/LLC 121). But he had already called it this is "On Secrets of the Sublime" of February 11 (A VI 3, 475/LLC 51), and in the pieces already quoted.

between aggregates of minds and aggregates of places, Leibniz notes, in that whereas the soul or mind "cannot be destroyed," places and shapes are continually destroyed, "for when what occupies a place is eliminated, so is its place" (121/521). In sum,

> this universal space is an entity by aggregation, and is continuously variable; that is, it is a composite of spaces empty and full, like a net, and this net continually receives different forms, and thus changes; but what persists through these changes is the *immensum* itself. For this *immensum* is God himself insofar as he is thought to be everywhere, that is, insofar as he contains that perfection or absolute affirmative form which is attributed to things when they are said to be somewhere. (A VI 3, 519/LLC 121)

In these reflections we see Leibniz criticizing the "imaginary space" of the Scholastics, claiming that on the contrary "it is maximally real"; but the "it" in that sentence connotes not *space* but the *immensum*, and it is this that he insists "is God himself insofar as he is considered to be everywhere ôr immense." In fact, he is perfectly sympathetic to the idea that extension and duration, taken in abstraction from the things in them, are imaginary or beings of reason. In his annotations on Spinoza's "Letter on the Infinite" he picks up on Spinoza's claim that "Time is for limiting duration," commenting:

> That is, I suppose, by conceiving duration as an entity per se abstracted from its subject, in which case it would be, as he says, imaginary or a being of reason. Which considerations strongly agree with Hobbes's. For Hobbes calls place the phantasm of existence and time the phantasm of motion. (A VI 3, 279/LLC 109)

This is consistent with how imaginary space is conceived in the Averroist tradition. As De Risi observes, in this tradition space "is a product of the imagination," pointing the way towards a conception of mathematical objects "as proper products of the *phantasia*" (De Risi (ed.) 2015, 7). For scholars in this tradition, the notion of *spatium imaginarium* "hinted at the fact that the imagination cannot help but conceive further extension beyond the bounds of heaven" (7). Perhaps significantly, Leibniz feels free to use this as an argument for the infinitude of space.[51]

Indeed, what Leibniz finds congenial in Spinoza's thought on the extended is this idea of the basis of space—in contrast to the space we conceive—being "maximally real." Space, taken as the divine attribute of being extended per se,

[51] See, e.g., *A Chain of Wonderful Demonstrations about the Universe* of December 12, 1676: "That space is infinite is demonstrated from the fact that however [large] it is supposed to be, there is no reason why it should not have been made larger" (A VI 3, 585/DSR 109); and the very similar argument he gave a year earlier (A VI 3, 470)/LLC 39. He repeats the argument in 1678, applying his Principle of Sufficient Reason (A VI 4, 1987/LLC 233).

is *maximal*: it is infinite in the sense of being "the greatest in its own kind."[52] Thus the one comment he is moved to make on his record of what Tschirnhaus had told him of Spinoza's views in February is precisely how he himself "is accustomed to say" that this is the middle of three "degrees of infinity," situated between the absolutely infinite, "which is in God," and the "mere infinite," which lies between bounds, like the asymptote of a hyperbola.[53] This second, intermediate degree of infinity, Leibniz writes, "is that which is greatest in its own kind, as for example the greatest of all extended things is the whole of space, the greatest of all successives is eternity" (A VI 3, 385/LLC 43). Here "the greatest of all successives" must be eternity in the sense of being "duration of something eternal through an unbounded time," rather than its basis in "the very necessity of existing, which does not in itself indicate succession" (A VI 3, 484/DSR 41). Similarly, space as the greatest of all extended things is unbounded extendedness, as opposed to the "basis of space," divine omnipresence.[54]

What Leibniz would *not* have found congenial, of course, is Spinoza's identification of God with Nature, and his concomitant denial that God is a person. "God is not something metaphysical, imaginary, incapable of thought, will or action, as some have made him out to be, so that it would be the same as saying that God is nature, fate, fortune, necessity, the world; rather God is a certain substance and a person," Leibniz writes (A VI 3, 475/DSR 27), before setting himself the task of establishing exactly that in these meditations, "On the Secrets of the Sublime." This thought puts him in mind of the (heretical) views of Vorstius (Conrad von dem Vorst, 1569–1622), for he immediately continues:

> The excessive abstractions of the imaginings[55] of philosophers, by which they have reduced God to some imperceptible nothing, were the reason why Vorstius, indignant with these chimerical opinions that are inimical to divine honour, made God corporeal and enclosed in a certain place, in order to show that he is, on the contrary, a certain substance and a person. (A VI 3, 475/DSR 27)

Of note here is the sympathy Leibniz shows for Vorstius' motivations. Later (in his notes of March 18) he shows a similar sympathy for the motivations of "those who

[52] See Definition 6 of Spinoza's *Ethics*, reproduced by Leibniz in his annotations on what he received from Schuller (A VI 3, 276/LLC 103).

[53] See also his note L24 on Spinoza's Letter on the Infinite, "I have always distinguished...," where he makes the same distinction of the three degrees of infinity. He maintains this distinction later, as we shall see.

[54] This distinction will acquire profound significance in my criticisms of De Risi's phenomenological interpretation of Leibniz on space (in §3.3).

[55] The Academy edition has "abstractiones Philosophorum imaginariorum," which Parkinson translates as "the abstractions of those who are supposed to be philosophers"; instead I have taken the liberty of interpreting it as a slip of the pen for "abstractiones Philosophorum imaginationum," which makes more sense. (Osvaldo Ottaviani assures me that Leibniz does indeed seem to have written 'imaginariorum'.) It could perhaps be translated as "abstractions of philosophers of imaginary things."

believed God himself to be the matter of things," saying that they seem to have had some inkling that "there is in matter, as also in space, something eternal and indivisible" (A VI 3, 392/LLC 55). The reference here though, is probably not to Vorstius. For Leibniz had begun his April 1669 letter to Thomasius by agreeing with his mentor's dismissal of the "utterly monstrous opinions" of the obscure Baghemin of Stettin, among which was precisely the view that "God is primary matter" (GP I 15; GP IV 162; A II 1b, 22/L 94). Leibniz traces the source of this view to Julius Caesar Scaliger, and following him, Sennert and Sperling. Again, however, he is sympathetic to their motivations, saying that they wanted forms to be produced by God through his own active power, rather than "from the objective and, so to speak passive power of nothing" (A II 1b, 24/L 94).

Now, Leibniz himself also wants to attribute the production of forms to what is active in matter, whose internal motions bring about divisions and boundaries, and thus places. For these places are the places of the things located, and their figures—which Leibniz calls "forms" in his letter to Thomasius—are distinguished from one another by their boundaries, which are in turn produced by their differing motions. The places produced, however, are modifications of the divine attribute of immensity, the extended per se, already presupposed as a basis of space providing magnitude and figure, although not determinate figures and magnitudes. It is the latter that are produced by the addition of matter and its motions. And the extended per se is not itself primary matter, since the latter also includes antitypy. Nevertheless, although Leibniz does not identify God with primary matter, for him the places to which the addition of forms in matter give rise are modifications of the divine attribute of immensity, which "is God himself insofar as he is considered to be everywhere ôr immense" (A VI 3, 519/LLC 121).

Two notable features of what Leibniz in 1676 calls universal space are that it is a mere aggregate of places, and that it is continually changing through time. Both of these are features he will later use to characterize body as phenomenal: body is phenomenal, he will argue, both because it is a mere aggregate of many things (despite appearing as one), and because it is constantly changing, and never the same thing from one moment to another.[56] Universal space in this sense, then, appears to be a phenomenon, although Leibniz does not actually call it phenomenal on these grounds.[57] Space as it is in itself, however, is free of determinations. In fact, already in this period Leibniz is making a sharp distinction between the continuity of space and the discreteness of the matter contained in it. As we have seen, space as a continuous entity is in itself "interminate," containing no boundaries, even if it is variously divided into portions by the matter moving within it.

[56] See Arthur (2018), especially chapter 2, 56–9.

[57] In my (2013b) I described this conception of universal space as a constantly changing aggregate of its parts as "phenomenal space"—a conception which has since been criticized by Debuiche and Rabouin (2019). I will return to this topic in §3.4 below.

Thus in "On the Secrets of the Sublime" Leibniz contrasts a physical plenum that is "divided into a multitude of points" or "into all the parts into which it can be divided" and "a true continuum":

> A perfect fluid is not a continuum, but discrete, that is, a multitude of points. It does not therefore follow from this that the continuum is composed of points, since liquid matter will not be a true continuum, even though space is a true continuum; from which it is again clear how great a difference there is between space and matter. (A VI 3, 473/LLC 47)

That the continuum cannot be composed of points, or even determinate parts, is a constant refrain of Leibniz's throughout his career. Apparently in contradiction to this, though, we find him arguing near the end of "On the Origin of Things from Forms" that a continuum can be composed out of an aggregate of places. In the course of stressing a second qualification on his analogy between the universal republic of minds and universal space, he states that "a continuum cannot be composed of minds as it can of spaces." In an effort to explain, he writes that "the soul is not an entity by aggregation, but universal space is an entity by aggregation" (A VI 3, 521/LLC 123). This suggests that universal space, formed by the aggregation of places, can compose into a continuous space. If so, however, it goes against his fast-developing ideas about the continuum in this period. The hugely important essay "Infinite Numbers" (ca. April 10) is an investigation precisely of whether continuous quantities in geometry can be composed of determinate parts, whether they be finite or infinitely small. Leibniz concludes that they cannot. He then comments:

> The reason why the unbounded, that is, that which is greater than anything finite, is something and the infinitely small is not, is that in the continuum the maximum is something, and the minimum is not.... In the continuum, the whole is prior to its parts; the absolute is prior to the limited; and so the unbounded is prior to that having a bound, since a bound is a kind of addition. There is no maximum number, and no minimum line. (A VI 3, 502/LLC 97)

These conclusions are repeated in "On the Origin of Things from Forms," prior to the above-mentioned passage on composing a continuum from spaces:

> There cannot be such a thing as the fastest motion or the greatest number, for number is something discrete, where the whole is not prior to the parts, but the opposite.... Whenever a whole is prior to its parts then it is a maximum, as in space and the continuum. If matter is, just like figure, that which makes a modification, then it seems that matter is not a kind of whole.
> (A VI 3, 5202/LLC 121)

These considerations concerning continuity and entities by aggregation are connected with Leibniz's reconsideration in these months of what magnitude consists in. Thus in the essay "On Magnitude" we find him abandoning his former conception of magnitude simply as the multitude of parts: "I once used to define magnitude as the number of parts, but later I considered that to be worthless unless it is established that the parts are equal to each other, or in a given ratio" (A VI 3, 482/DSR 37). Instead, he prefers a definition according to which "magnitude is that through which it is recognized whether some thing is a whole."[58] This repeats his claim in "Infinite Numbers" of ca. April 10, 1676 that "Magnitude is that constitution of a thing by the recognition of which it can be held to be a whole." The point is that the emphasis is now on an aggregate being recognized as one, continuous, or having magnitude, by a mind understanding the relation among the components, not by a simple addition of them as units:

> I do not know whether what is really divided, that is, an aggregate, can be called one; it seems to be, though, since there are names invented for this.... For something to become another thing is for something to remain which pertains to it rather than the other thing. But this is not always matter. It can be mind itself, understanding a certain relation. (A VI 3, 503/LLC 99)

On such an understanding, then, universal space as the aggregate of places would no longer be understood as having a magnitude deriving from the multitude or sum of its parts—since a continuous quantity is not composed of parts. Its magnitude would instead derive from the relations among the parts. This tendency towards a relational understanding of space (and time) can be seen in what Leibniz writes about them in "On Magnitude." In his discussion of duration and time, for example, he states that "if two things are of such a kind that it is impossible for one to be understood without the other, they are *simultaneous*" (A VI 3, 484, DSR 41), thus beginning his journey to the theory of time in terms of relations among requisites that we explored in chapter 1 above. And on extension and place he writes:

> Therefore that according to which things are said to be extended, to which extension alone is applicable per se, but in an order with respect to those things which contain something else besides, is called *Place*, and absolutely, *Space*. And it is according to space that we conceive things as situated, as simultaneously perceptible, as distant from one another, and as having figures.
>
> (A VI 3, 484/DSR 41)

[58] He goes on to define magnitude in terms of *congruence*, or being "capable of being brought within the same boundaries" (A VI 3, 482/DSR 37). As I noted in LLC 398, this more sophisticated definition is good evidence that "On Magnitude" was composed later than "Infinite Numbers."

Thus we are already on the threshold of his conception of space as determined by relations of situation. In the summer of 1676, as we saw in §2.1 above, Leibniz had defined a situation as "a mode according to which any body can be found, even though we recognize nothing in it specifically by which it can be distinguished from the others," ("Mechanical Principles," A VI 3, 103).[59] And by early 1677—only a few months later—Leibniz was already firm in his conviction that space is not "a certain thing supposed in pure extension," but rather, like motion, "consists in relation" (A VI 4, 1968/LLC 225). Then in a manuscript written in about 1679, he describes time as a "relation of things with each other," and as being, like space, "generic, and comprising all that there is."[60]

Now it might be thought that with the advent of this relational understanding of space and time, Leibniz's prior commitment to a space whose basis is "maximally real" would simply evaporate. In this regard, it is very interesting to compare the above quoted passages from "On the Origin of Things from Forms" of April 1676 with a short fragment written some thirteen years later, "On Time and Place, Duration and Space" (*De tempore locoque*...).[61] The latter begins by taking for granted the relational theory, time and space being "real relations, or orders of existing." As real relations, however, they must have a foundation in reality,[62] and Leibniz declares, just as he had in the papers of 1676, that this foundation of time and space is divine magnitude, that is, eternity and immensity, respectively:

> Time and place, or duration and space, are real relations, i.e. orders of existing. Their foundation in reality is divine magnitude, to wit, eternity and immensity. For if to space or magnitude is added appetite, or, what comes to the same thing, endeavour, and consequently action too, already something substantial is introduced, which is not in anything other than God or the primary unity. That is to say, real space in itself is something that is one, indivisible, immutable; and it

[59] Leibniz builds upon this definition of situation in the conspectus to his "little book on the *Elements of Physics*" in 1678: "The extended is what has magnitude and situation. But magnitude is a way [*modus*] of determining all the parts of a thing, that is, all the parts by means of which the thing could be understood; *situation* is a way of determining those by which the thing could be perceived" (A VI 4, 1987/LLC 233).

[60] This is in "Metaphysical Definitions and Reflections," probably a draft of the introduction to the *Elements of Physics*: "And this relation of things with each other is called *time*, which is also [like space] generic, and comprises all there is, for nothing can occur which is not before, after or simultaneous with any other given thing" (A VI 4, 1397/LLC 243).

[61] Although in LLC I date this piece as possibly from as early as 1686, it is attributed by the Akademie editors to Leibniz's Italian journey on the basis of the watermark (March 1689–March 1690). The doctrine of space and time as real relations is also found in *Specimen Inventorum*, probably written while he was in Vienna in 1688 (A VI 4, 1621/LLC 313); also, in a piece that can confidently be dated as mid-1685, we find Leibniz affirming that "whatever is real in space and time consists in God comprising everything" (A VI 4, 629/LLC 275).

[62] See Appendix 2 for a treatment of Leibniz's highly ramified position on relations. For a comprehensive review and interpretation, see Massimo Mugnai's (1992), (2012) and (2018).

> contains not only existences but also possibilities, since in itself, with appetite removed, it is indifferent to different ways of being dissected. But if appetite is added to space, it makes existing substances, and thus matter, ôr the aggregate of infinite unities. (A VI 4, 1641/LLC 335)

Here we see that, just as before, immensity or "real space in itself" is characterized as "one, indivisible, immutable," and also as "indifferent to different ways of being dissected": it is only divided into parts by the addition of something substantial. Only now what has to be added is not matter directly, but endeavour, equated with appetite: when this is added to space, "it makes existing substances, and thus matter, ôr the aggregate of infinite unities." Space, considered in itself, "contains not only existences but possibilities"; there is only a particular order of existing things in it when it has been divided by the motions or endeavours within it.

It is also noteworthy that space is still equated with magnitude; previously, as we have seen, Leibniz had regarded space as involving magnitude and figure, but not "determinate magnitude and figure" (GP IV 107/L 111). This is consistent with what he writes here, where space is "indifferent to different ways of being dissected" prior to its being actually divided by the motions in it.[63]

Of course, there are also some significant differences between the two pieces, due to the profound changes in Leibniz's ontology that have occurred in the intervening years. One, clearly, is the identification of endeavour not simply as a neo-Hobbesian element of motion, but as *appetite*. It is still the case that Leibniz will insist that the parts of space are identified by their containing matter moving with different motions at any instant, and thus with different endeavours; but now the endeavours are interpreted as modifications of substances: they are results of the tendencies substances have to change state, their *appetitions* or *appetites*. This is, of course, a reflection of the fact that between composing the Paris manuscript and penning the above one in Italy, Leibniz has rehabilitated substantial forms. A substantial form, as he explains in a manuscript written in Hanover some time between 1683 and early 1686, "is a principle of action, ôr primitive force of acting" (*De mundo praesenti*, A VI 4, 1508/LLC 287). In it there is "a kind of cognition, that is, an expression ôr representation of external things in a certain individual thing," conjoined with a principle of action, "i.e. an endeavour ôr appetite in accordance with this cognition of acting" (1508/287). It is according to its representation of external things in its substantial form that the body is a

[63] In between, in a set of definitions from 1680, Leibniz writes: "*The infinite* is that which has magnitude absolutely, *the finite* involves the negation of certain things of the same kind. But the latter seems to hold in continuous things, in [discrete] things it seems that quantity cannot be conceived absolutely, as some whole. As magnitude is the number of parts, so the infinite, which involves something beyond the parts of which there is a number, is greater than any assignable" (*Definitiones*, A VI 4a, 406).

unity per se.[64] Now, instead of the differing motions of the parts of matter being instituted directly by God (as in the dialogue *Pacidius Philalethi*), they are the result of the actions of substances within matter, and from which bodies result. As Leibniz writes in a list of definitions from the same period (1683-early 1686), unless a body "contains within itself a certain substance that is one, corresponding to the soul, which they call a substantial form or first entelechy, is it no more a substance than a pile of logs, and will only be a real phenomenon, like a rainbow"; for every body or part of a body is subdivided into parts, whereas "there is no real substance that is not indivisible." Consequently, he concludes,

> mathematical things like space, time, a sphere, an hour, are only phenomena, which are conceived by us on the model of substances.... And indeed, it could be demonstrated that those things that are divisible and consist in magnitude, such as space time and bulk [*moles*], are not complete things, but something must be superadded to them, which involves all those things that can be attributed to this space, this time, this bulk.
> (*Divisio terminorum ac enumeratio attributorum*, A VI 4, 559–60/LLC 265–7)

Here Leibniz's assertion of "something needing to be superadded" to space evokes what he had written in "On the Origin of Things from Forms" of 1676, where he claimed that the modifications of the *immensum* "do not occur by any change in it, but by the superaddition of something else, namely bulk ôr mass. From the addition of bulk and mass there result spaces, places and intervals, whose aggregates give Universal Space" (A VI 3, 519/LLC 121). But in thee writings from 1688 to 1689, "those things that must be superadded" to space in order to make its parts determinate are, as already mentioned, the "endeavours ôr appetites" in the forms of the substances contained in them. It is the addition of appetite (or endeavour) to space that results in matter being divided into actual parts ("this space, this time, this bulk"); this is because the endeavours are now modifications of substances, and these produce the discriminations and boundaries of the parts of matter.

In the Italian fragment *De tempore locoque*..., Leibniz is careful to say that the modifications are of something *in* God, not modifications of God. Nonetheless, when he says that "something substantial is introduced, which is not in anything other than God or the primary unity," it might seem to follow from this "that all created things are in God," as he notes in another important fragment written in Vienna in 1688. To this objection "it must be responded," he writes, "that the

[64] Here is the full passage: "*Substantial form* is the principle of action, ôr primitive force of acting. There is moreover in every substantial form a kind of cognition, that is, expression ôr representation of external things in a certain individual thing, according to which the body is a per se unity, namely in the substantial form itself. This representation is conjoined with a reaction ôr endeavour ôr appetite, according to this cognition of acting" (A VI 4, 1508/LLC 335).

reality of created things is not the very thing which is absolute in God, but is limited, for it is the essence of created things to be limited" (*De Abstracto et Concreto*; A VI 4, 990). This can be illustrated, he explains, "by the image of space and body":

> the extension of space and that of a created thing are different because the extension of space is in itself absolute, unbounded, impartible, devoid of any change; briefly, whatever is real in space is the very omnipresence of God. But the extension of body is limited in all ways. Indeed, however, it seem impossible to deny the reality of what we attribute to the space immediately required for bodies, and the absolute reality immediately required for limiting them.

Thus where Leibniz had previously described universal space as "an entity by aggregation, and continuously variable," "like a net which continuously receives a different form, and thus changes," he now insists on the extension of space as being devoid of any change. But that applies to "the extension of space in itself," that is, the extended per se,[65] which formerly he had also characterized as immutable. And it is still the case that Leibniz conceives space insofar as it is occupied by bodies—what we might call space conceived concretely—as differently partitioned by the boundaries of the bodies in it, and that they are really so divided by "the absolute reality immediately required for limiting them."

Returning to the Italian fragment, the other major change over the Parisian one from 1676 is, of course, the characterization of time and space as "real relations, i.e. orders of existing." But what does Leibniz mean by a "real relation"? He is clear that he does not mean a relation that involves "real accidents," as in the scholastic doctrine, which he consistently opposes. According to that doctrine, "a *real accident* is a being that is supposed to arise and be destroyed by the very fact that the substance is supposed to change" (*De Abstracto et Concreto*; A VI 4, 989): it is a transitory real being in one or more of the related substances that is the foundation of the changing relations between them. As I discuss further in Appendix 2, for Leibniz a real relation is rather "a relation of connection" as opposed to a mere "relation of comparison" that would arise from considering the related things together; as a "relation of existence" it is irreducible to a mere relation of comparison because it involves "the universal connection of things"—as is the case with "coexistence in the same place" (A VI 4, 944).

But how is this relational conception compatible with the idea of space as divided into cells? This is where the idea of space as an *order of situations* comes in, and that is the topic of our next section.

[65] Cf. the fragment from 1685 given by De Risi: "*Space*, namely absolute [space], [is that] in which nothing else can be considered but extension" (LH XXXV, I, 5, Bl. 49; De Risi 2007, 624).

2.3 The Analysis of Situation

As part of a continuing campaign to gain admittance to the *Académie Française* in Paris,[66] Leibniz wrote to Huygens in September, 1679, informing him of a project for a new mathematical science he had conceived, attaching a specimen. The appended essay (GM II 20–7) is a sketch he had made of this "New Characteristic," which he also calls the Analysis of Situation (*analysis situs*). Describing it to his Dutch mentor, he writes:

> I am still not satisfied with algebra, because it does not give the shortest methods or the most beautiful constructions in geometry. This is why I believe that, so far as geometry is concerned, we need still another analysis which is distinctly geometrical or linear, and which will express *situation* [*situs*] directly as algebra expresses *magnitude* directly. And I believe I have found the way and that we can represent figures and even machines and motions by characters, as algebra represents numbers of magnitudes. (GM II 17–20/L 248–9)

Typically, Leibniz has overstated the case for his new science, and it is not too surprising that Huygens reacted with some severity[67] to Leibniz's claims that his new science could supersede two thousand years of classical geometry—not to mention the Cartesian algebraic geometry of which Huygens himself was probably the most gifted practitioner alive!—and could even allow the symbolic representation of machines and their motions. Nonetheless, the very idea that there could be a more fundamental approach to geometry than the algebraic, expressing properties of figures directly, has proved inspirational for a number of thinkers down the years, ever since the letter first came to light as a result of the unfortunate Samuel König's dispute with Maupertuis in 1752.[68]

Leibniz had, in fact, been developing his ideas on this subject since his last year in Paris, and had already drawn them up on his arrival in Hanover in January 1677 into a manuscript he titled *Characteristica geometrica*—not to be confused with his 1673 critical reading notes on Honoré Fabry's *Synopsis Geometrica* (1669) to which he gave the same title, nor with the substantial treatise bearing the same title that he produced in August 1679 as a summary of his considerable labours on

[66] As part of his campaign, Leibniz also wrote about his project to Father Jean Bertet in September 1677 (A II 1b, 571–4) and to the Abbé Jean Gallois in December 1678 (GM I 182–90; A II 1b, 669–70). There is also a very informative letter to Gallois from September 1677 (A II 1b, 566–71) that he decided not to send. See De Risi (2007, 62) for an enlightening discussion.

[67] The relevant portion of Huygens' response is given (in both the original French and English translation) by De Risi in his (2007, 71–2).

[68] See Vincenzo De Risi's lyrical account of the circumstances of the letter's publication (2007, 104–5).

the subject that year, of which his letter to Huygens was a summary.[69] I will not recount here the details of the progress of his thought over the years, since this has already been delivered by Vincenzo De Risi in his masterwork *Geometry and Monadology* (2007). This includes a comprehensive analysis of the development of Leibniz's thinking on *analysis situs* (2007, 40–101), of the fortunes of the project after Leibniz's death (101–16), and also an invaluable assessment of the dating of all the relevant texts, the "Tentative Chronology" (117–26). I will restrict my discussion to a few salient points.

As De Risi notes, the germs of some of the leading ideas may be found in texts written even prior to Leibniz's becoming fully acquainted with higher mathematics in Paris. *Situs* or situation occurs as a technical term even in his earliest writings. Initially *situs* denotes a disposition of smallest parts or unities in relation to the whole. Thus in his *Dissertatio de arte combinatoria* (1666) Leibniz writes concerning *complexions* that "the disposition of the smallest parts, or of the parts assumed to be the smallest (that is, the unities) in relation to each other and to the whole can itself also be varied. Such a disposition is called a *situs*." In his definitions, he says "*Situs* is the location of parts." Absolute situation is that of the parts with respect to the whole, relative situation that of parts to parts. (GM V 14; A VI 1, 172/L 77).[70] Five years later, in the *Theoria motus abstracti* of 1671, Leibniz denies the existence of Euclid's "partless point"[71] on the grounds that "such a thing has no situation (*situs*), since whatever is situated somewhere can be touched by several things simultaneously that are not touching each other, and would thus have several faces." This conception of a point as 3-dimensional, I suggest, betrays the influence of Hobbes, whom Leibniz was studying assiduously at this time, as mentioned above.[72] For in opposition to Euclid, Hobbes defined a point as a body whose magnitude "is not considered (even though there always is

[69] The 1677 and 1679 versions of the *Characteristica geometrica* are included in Echeverría's valuable selection with French translations and notes by Marc Parmentier, *La caractéristique géométrique* (Leibniz 1995b, abbreviated CG), along with four drafts of the latter, the sample sent to Huygens, and many other fragments from the same period. For his notes on Fabry see De Risi (2007, 51), and for a discussion of the *Characteristica geometrica*, see pp. 57–72.

[70] "In the former much consideration is given to priority and posteriority; in the latter, none. It will be best, therefore, to call the former *order* and the latter *vicinity*; The former is disposition; the latter, composition.... in vicinity there can be no variation, but only situation ...." Leibniz gives the examples of the arrangement of elements in a straight line for absolute situation, and in a circle for relative situation. See the quotation and discussion in De Risi (2007, 41–4).

[71] "*A point is that which has no part, or has no magnitude*" (Euclid, *Elements* Book 1, Def. 1; Toddhunter edition, 1933, p. 1).

[72] De Risi (2007, 45) draws attention to another text written in the same period, where Leibniz writes: "A geometry must be written without motion, but only with situation, i.e. locus or distance" (*Elementa de Mente et Corpore* (1671), A VI 2, 282). He also mentions the possibility that Leibniz was later inspired by his reading of the projective geometry of Desargues, La Hire and Pascal while in Paris (2007, 28), but then he discounts this on p. 141, concluding in his long footnote 14 that: "Therefore one must adapt to the evidence that Leibniz's theory of similarity does not anticipate projective geometry in the least, nor does it owe anything to Desargues or Pascal.". For an insightful treatment of the influence of projective geometry on his thought, see Débuiche (2013).

some magnitude)."[73] Leibniz follows him in rejecting Euclid's definition of a point as lacking parts or magnitude, although he will also reject Hobbes's definition, instead defining a point as "that which has no extension, i.e. whose parts are indistant, whose magnitude is inconsiderable, unassignable, i.e. smaller than can be expressed by a ratio to another sensible magnitude unless the ratio is infinite, smaller than any ratio than can be given";[74] and indeed the conception of the infinitely small as possessing parts and magnitude survives in his mature characterization of infinitesimals (although not points) as "incomparable" with finite quantities, but as having an assignable ratio to other magnitudes of the same order of infinity.[75]

But it is in Hobbes's definitions of situation and figure that his influence on Leibniz can be seen most clearly, and it is here, I suggest, that we can see the origin of Leibniz's idea for his analysis of situation.[76] In *De Corpore* Hobbes defines situation as follows:

> §20. *SITUATION is the relation of one place to another.* Now, the situation of several places is determined by four things: *by their distances from one another*; *by their individual distances from a given place*; *by the order of the straight lines drawn from a given place to the remaining places*; and *by the angles which are made by the lines so drawn.* For if the distances, order, and angles are given, that is, are certain, then their individual places will also be certain, as other places cannot be.
>
> §21. Points, however many they may be, have a *similar situation* with an equal number of other points, when all the straight lines drawn from some one point to all of the latter have individually the same ratio to all those drawn at equal angles from the same single point to all of the former.
>
> (*De Corpore*, II, chapter XIV; Hobbes 1839b, 172–3)

[73] "If the magnitude of a body that moves be not considered (even though there always is some magnitude), the path along which it passes is called a *line*, or *one single dimension*, the space through which it passes, a *length*, and the body itself, a *point*" (Hobbes, *De Corpore*, II, chapter VIII, §12; Hobbes 1839b, 98–9/LLC 359); "by a point is not to be understood that which has no quantity or which cannot be divided by any means (for there is nothing of this kind in nature); but that whose quantity is not considered, that is, of which neither quantity not part is computed in a demonstration" (*De Corpore*, III, chapter XV, 2; Hobbes 1839b, 177/De Risi 2007, 168).

[74] Even after rejecting indivisibles, Leibniz will define a point in *De Minima et Maxima* of November 1672–January 1673 as being "of a length, breadth and depth infinitely smaller than any given sensible thing" (A VI 3, 99/LLC 15).

[75] Cf., for instance, his *Defense du calcul:* "This is what makes me speak on other occasions of incomparables, because what I say of them has its place whether one understands infinitely small magnitudes or one employs magnitudes of a smallness that cannot be considered and is sufficient to make the error less than that which is given" (Pasini 1988, 708). For a recent treatment of Leibniz on infinitesimals see Rabouin and Arthur (2020).

[76] Leibniz's access while in Paris to the papers of Pascal, and also his interaction with Desargues, were likely factors helping to stimulate his conceiving a different approach to geometry: see Echeverría (CG 27), De Risi (2007, 28, 49) and Débuiche (2013, 426). But none of these authors comments on the possibility of a debt to Hobbes.

Hobbes then proves that if straight lines are drawn from some point D to three points A, B, and C, and then three other lines are drawn from a point H to the points E, F, and G, in such a way that the angles ADB and EHF are equal, and likewise BDC and FHG, and the lines DA, DB, and DC are proportional to HE, HF, and HG, respectively, then the two situations ABC and EFG are *similar*, since "those things that differ only in magnitude are similar." Interestingly, the proof is achieved by laying the smaller figure on top of the larger one. Hobbes then proceeds to define *Figure*:

> §22. *FIGURE is quantity determined by the situation, or by the placing of all its extreme points.* Now I call those points extreme which are contiguous to the place which is outside the figure. In lines and surfaces, therefore, all points may be called extreme, but in solids only those which are in the surface that includes them.
>
> And *similar figures* are those for which the extreme points in one of them are all placed similarly to all the extreme points in the other; for such figures differ in nothing but magnitude. (*De Corpore*, II, chapter XIV; Hobbes 1839b, 174)

Now, the concept of situation was freely used by other mathematicians of the time, and (as Mattia Brancato (2015) has argued), Leibniz may well have been persuaded of its importance by his exposure to Weigel's views when he studied with him in Jena in 1663.[77] Weigel's definitions of *situs*, however, are (typically for him) extremely wide-ranging, intended for application to all fields of study.[78] There is a much closer correspondence with what Hobbes had written in *De corpore*, and it was likely under Weigel's influence that Leibniz had been moved to study Hobbes (Brancato 2015). At any rate, I propose that we can see in Hobbes's account certain features that might well have been salient for Leibniz. First, there is the definition of situation as a relation of places to one another ("*situs est relatio loci ad locum*"). Second, there is the idea of having points stand in for the places. (It is the combination of these two features that we find in Leibniz—he will later define

[77] Brancato argues that Leibniz's first exposure to Hobbes was probably conditioned by Weigel's rather metaphysical reinterpretation of Hobbes's philosophy, prior to his studying Hobbes's own writings: see especially his discussion in Brancato (2015, 48ff).

[78] Weigel published his *Corporis pansophici pantologia* in 1673, and Leibniz made very detailed excerpts from and notes on it in 1683 (A VI 4b, 1187–1200). There, Weigel gives a definition of *situs* in general—"*Situation* is a terminative mode by which something for a certain reason [*ratione*] stands in an association and complex with other things"—remarking in the subsequent Scholium that "Situation is demonstrated by the prepositions 'to', 'at', 'within', 'without' [*ad, apud, intra, extra*]." He then proceeds to differentiate it into subcategories: *extensive situation* (either *abstract* or *concrete*), *simple* (either *relative* or *correlative*) or *continued*, extensive situation being "regarded either abstractly or concretely. Regarded abstractly it is either formal or terminal," and so forth. Notably, he applies such conceptual divisions to the figures of geometry. (A VI 4b, 1191–4). The *Corporis pansophici* includes a reference to Leibniz's *Theoria motus abstracti*, and Leibniz mentions a book by Weigel in 1670 that was probably a draft of this work. For analysis of Weigel's influence on Leibniz, see Brancato (2015, esp. 40–68, 137–41, 168–70).

places in terms of situations). Third, the idea that situation reduces to distances, either those of the placed points to one another, or of each of them to an external point of reference; and that this gives an order, which also involves the angles of the straight lines joining them to the external point. Some other features are standard features of classical geometry: (4) the idea that if these angles are the same in two different situations of points, but the corresponding lines are proportional, then the situations are *similar*; (5) the idea of a figure as determined by all its extreme points; and (6) the idea that if two figures are similar, then the only difference between them is one of magnitude.

Regarding the first feature, Leibniz will constantly attempt to provide a more general and philosophical definition of situation. For example, in a fragment published by Echeverría and Parmentier from January 1680, he writes: "A situation is nothing other than that state of a thing by which it happens that it is understood to exist in a certain way simultaneously with [*simul*] other extended things" (Fragment 14, CG 276).[79] In geometrical practice, however, Leibniz will always avail himself of Hobbes's characterization of it, in which it is points that are situated, and the situations of things are represented by lines between points: "Any two points whatever have a situation to one another; that is to say, there is a path from one to the other" (Fragment 11, September 1679; CG 246); "The situation of the parts of a thing to each other is called *figure*" (*Conspectus*..., 1678; A VI 4, 1987/LLC 233).

Second, with respect to this representation of places by points, De Risi sees a break-through in Leibniz's maturity (1712–16), where he finally understands "a point in purely abstract terms, according to its functional definition, thus completely disengaging it from geometrical intuition" (De Risi 2007, 173). "A point," De Risi asserts, "will rather become a place that does not contain any other place (a definition made more concrete, but also simpler, by the introduction of the *locus* concept)" (171), citing definitions such as that in *Recta definio* of 1715: "A *point* is the simplest place [*locus*], ôr that in which there is no other place. And so if *B* is in *A*, and for that reason *B*∞*A* [i.e. *B* coincides with *A*], this is the same as that *A* is a *point*." (614). Of course, the locus concept is traditional in geometry, and Leibniz certainly makes use of it in his early studies. Indeed, as De Risi acknowledges, there are definitions in Leibniz's earlier work according to which

[79] De Risi notes that "*situs* is a notion that Leibniz always tends to assume as primitive" (2007, 132), and that "in the case of situation we are left at a loss—it is just intuition that provides us with this primitive notion" (133). What he means is that Leibniz's attempts to define it are metaphysical (i.e. transcendental, according to him; see below), not geometrical, and that in AS all the work is done not by defining it directly, but by defining "what having *one and the same situation* means" (133). On the one hand, this is exactly right: according to Leibniz's provisional nominalism this is precisely the way forward, to avoid questions of existence in favour of a reduction to equivalence relations (see Appendix 2); but on the other, it neglects the fact that Leibniz is building on Hobbes's characterization of situation, where it is nearly synonymous with *figure*, as in the quotation I give from the *Conspectus*.

"a point contains only itself" (171).[80] His claim is only that this abstract definition of a point is made more precise in later years by defining it in terms of the locus concept. But I am not convinced that this change has the significance that De Risi imputes to it.

Now this conception of *point*, as coinciding with anything that is in it, follows from Leibniz's definition of congruence. As he writes in the essay he sent to Huygens, the "New Characteristic," two figures are "congruent with respect to the order of their points" means that "they can occupy exactly the same place, and that one can be applied to or placed on the other without changing anything in the two figures except their place" ("New Characteristic," A III 2, 851; CG 260/L 251). A point is, as it were, a degenerate figure, since it only has one point to be superposed. It follows that "all points in the world are congruent to one another" (A III 2, 851; CG 261/L 252).[81] Now, even though it is perfectly traditional to define congruence by the overlaying of one figure on another so that their boundaries coincide, and to regard the congruence of figures as amounting to their similarity together with equality of magnitude, Leibniz would have found both ideas in Hobbes, in connection with his conception of situation.

Of course, the fact that magnitude or distance is included in the concept of a situation (the third of the above noted points of correspondence with Hobbes) means that Leibniz's space is a metric space—not a topological or affine space, as some have suggested.[82] As we shall see, Leibniz will define space as the *order of situations*, abstractly conceived, so that it is a system of relations among points (standing for abstract places). In its final version, Leibniz's absolute space is, as De Risi explains, *constituted* by (although not composed of) situated points (2007, 173–5); we will return to this point below.

As to the fifth point of correspondence noted, the idea of a figure being determined by all its extreme points, De Risi notes that "because of the uniformity of space, any determination of a figure integrally depends upon its *ambitus*, i.e. (it being a figure) on its boundary" (2007, 213) But this is difficult to define precisely, as Leibniz's continuing attempts at definition of figure testify.[83]

[80] Compare with this definition from his middle period: "A point is what has situation to anything else placed in the extensum, but does not itself have extension" ("*Scheda* on situation and extension" [1695]; De Risi 2007, 588; in Appendix 3.7 below).

[81] Cf. also "*any two points whatever can be congruent to each other*, that is, similar and equal to one another" (Fragment V, August 11, 1679; CG 82).

[82] De Risi (2007, 176): "Space is in itself a metric space. The system of the relations that give structure to the set of points constituting space is not a system of topological, or affine, or however else defined relations, but is simply a set of metric relations." The same point was stressed by Graham Solomon in his PhD thesis, *Leibniz's Analysis Situs in Mathematical Context* (London: University of Western Ontario, 1989).

[83] De Risi argues that "Leibniz was to find his most adequate definition through the properly dimensional determination, which is grounded in the fact that a manifold with boundary is bounded by a non-necessarily connected manifold (of a lesser dimension) with no boundary" (2007, 212). Leibniz gives a kind of summary of his previous attempts to define figure in answer to Locke's own

Not all these ideas about situation are unique to Hobbes, of course, as we have already noted. But one point of correspondence that seems to me to strongly suggest a debt to him, and one that becomes seminal for Leibniz, is this idea that there are two equivalent ways of conceiving situation: (1) as determined by the relations of distance among the points demarcating the situated thing, so that to all intents and purposes situation is synonymous with *figure*; and (2) by reference to an external point, so that, in Hobbes's words, it is determined "*by their individual distances from a given place; by the order of the straight lines drawn from a given place to the remaining places*" (Hobbes 1839b, 172). For while the first of these seems closely related to Leibniz's conception of space being divided into determinate parts, the second seems equally closely related to the idea of similarity of perspectives and thus points of view.

Consider, say, a tetrahedron. Anything at the apex of the figure has a determinate situation with respect to the other three vertices at its base, the situation being determined by the angles at each vertex, the proportions among its edges, and the distance between any two vertices. Thus a situation of bodies is a given geometrical arrangement of bodies, modelled on the arrangement of the vertices in a geometrical figure.[84] Again, the same situation between the four vertices could be represented by drawing lines from them to some arbitrary reference point, and then taking the angles between these lines, the proportions between them, and the distance from any one of them to the reference point. Thus a situation may be regarded as determined either (i) by its own boundaries—a three-dimensional *extensum* by its surrounding surface, a plane figure by various lines, and a line by its terminating points—or (ii) by lines drawn from its vertices to a given external point.

According to the first of these ways of regarding situation, space can be regarded as an aggregate of places, where the places, its parts, are determined at any given time by the boundaries of the bodies in it: "Supposing space to have parts—that is to say, so long as it is divided by bodies into empty and full parts of various shapes—it follows that space itself is a whole or entity accidentally, that it is continuously changing and becoming something different..." (A VI 3, 391/ LLC 53). On this conception, a figure is determined by the situation of its vertices, and similarity of situation, as on Hobbes's account, requires putatively similar

rather feeble attempt in the *Nouveaux essais* of 1704–5: "To say that it is an extended thing [*étendu*] bounded by an extended thing is not sufficiently general, for a whole spherical surface is a figure and yet it is not bounded by anything extended. Again, one might say that a *figure* is a bounded extensum, in which there is an infinity of paths from one point to another. This includes surfaces lacking boundary lines that were not included in the preceding definition, and it excludes lines because from one point to another in a line there is only one path or a determined number of paths. But it would be better still to say that figure is a bounded extensum that can receive an extended cross-section, or simply that it has *breadth*, another term for which up till now no one has given a definition" (A VI 6, 148; De Risi 2007, 212–13).

[84] Cf. also the remark in "On Analysis Situs" which equates situations with figures in geometry: "But similarity is seen best of all in the situations ôr figures of geometry" (GM V 179/L 255).

figures to be bought into a superposition, so one can see that the only remaining difference in the situations is one of magnitude. Determination of comparative magnitude, that is, requires that the figures be perceived at the same time. That insight can be seen clearly in what Leibniz wrote in his *Conspectus Libelli Elementorum Physicae* of 1678:

> The situation of the parts of a thing to each other is called *figure*. This is the source of *similars*, which cannot be discerned unless they are simultaneously perceived. (A VI 4, 1987/LLC 233; L 277–8)

In the second way of conceiving a situation, on the other hand, it can be taken as a complex[85] of distances and angles of the vertices of such a figure to an arbitrary point; and indeed with respect to such a point one can take a whole order of situations of all the bodies coexisting with it. The situation of a body with respect to all other bodies simultaneous with it is thus given by all the angles and distances from the body (now thought of as a point) to all the other bodies coexistent with it, giving a spatial representation of all those coexistents at such a point. This is how Leibniz described situation in the *Principia mechanica* of 1676, as described in §2.1, where after characterizing it as "a mode according to which any body can be found, even though we recognize nothing in it specifically by which it can be distinguished from the others," he adds that "this way of finding the body depends on knowing its distance from other bodies, and also on knowing the angle, that is, the figure which it makes with another body" (A VI 3, 103).[86]

Now this way of conceiving situation aligns very nicely with the way in which Leibniz describes the difference between the different perspectives offered in perception and the intuited conception of space as it is in itself. It is manifested in his famous figure of the town or city, occurring in both the *Discourse On Metaphysics* (§14) and the *Monadology* (§56–7), but already presented by Leibniz in his letter to Thomasius of April 30, 1669 as an illustration of the difference between how "an essence differs in its qualities only in relation to sense":

[85] I use the term 'complex' advisedly. On the one hand, this is a technical term occurring in Combinatorial Topology, and so signals the close correspondence between the approach taken by early practitioners of that science and Leibniz's own approach in his *analysis situs*, which I have explored elsewhere (Arthur 1986, 1987); and on the other, it is a term Leibniz himself uses in this connection. Thus in a passage on the meaning of 'being in', he says: "Thus, an accident is in no other place or time than in the subject, nor is a part in another thing different from the whole, nor is a thing placed anywhere other than in place, and something ordered is not in an order different from the complex of the ordered things" (LH IV, 7 B, 3, Bl. 56v; quoted from an unpublished paper of Massimo Mugnai's.)

[86] In a recently transcribed fragment *Situs puncti (II)*, (MT #50, (39955)), dated from watermark evidence as 1679–81, Leibniz defines the situation of a point as "the mode of determining its distance from any others whose distance from one another is determinate." As an example he gives a point *e* lying outside a triangular pyramid *abcd*; the situation of the pyramid is determined by lines joining its vertices to an internal point *f*, so that the situation of *e* is now given by the lines joining it to the vertices of the pyramid.

> For just as the same city presents one aspect of itself if you look down on it laid out in ground-plan [*in Grund gelegt*] from a tower in its centre, which is exactly as if you were to intuit it, so it appears otherwise if you approach it from outside, which is exactly as if you were to perceive the qualities of body. And just as the external aspect of the city varies depending on whether you approach it from the east or the west, so similarly the qualities vary according to the variety of the sense organs. (GP I 19; A II 1b, 28–9/L 97)

Leibniz presents the same image forty-five years later in the *Monadology*:

> And just as the same town looked at from different sides will appear quite different and is, as it were, multiplied in perspective, in the same way it happens that with the infinite multitude of simple substances it is as if there are just as many different universes, but these are only perspectives of a single one, according to the different points of view of each monad. (§57; GP VI 616)

The significance of this is as follows. The different perspectives on the city/town when approached from different directions correspond to the differing *points of view* of the perceivers. Here we need to interpret "point of view" literally and spatially: monads have a point of view from the perspective of the body whose substantial form they constitute, as perceived in the sense organs of that body.[87] This clearly corresponds spatially to the situation of each individual's body in relation to all the others it coexists with, yielding representations that will change as the body moves. Each such representation can thus be regarded as a perception,[88] provided the individual is equipped with organs of sense in which these representations are presented with varying degrees of clarity and confusedness. Supposing now that the individuals are also equipped with appetites that take them through the changes of representation in an autonomous fashion, according to some internal law, then one has all the ingredients of Leibniz's substantial forms. He argues just this in a piece written in about 1681:

> Insofar as God relates the universe to some particular body, and regards the whole of it as if from this body, or what is the same thing, thinks of all the appearances or relations of things to this body considered as immobile, there

[87] In my paper (2015a), "The relativity of motion as a motivation for Leibnizian substantial forms," I argue that monads need to be where their bodies are through time in order to represent motions, their changing relations to other bodies. See also Jen Nguyen's (2018), in which she argues, inter alia, that "points of view serve as the metaphysical foundation for situation, which, in turn, serves as the metaphysical foundation for extension" (62).

[88] "Each substance," as Leibniz writes in his *Specimen Inventorum* of 1688, "expresses the whole universe according to its own situation and point of view [*aspectum*], inasmuch as everything else is related to it, and hence it is necessary that some of our perceptions, even though clear, are nonetheless confused, since they involve infinitely many things" (A VI 4, 1618/LLC 309).

> results from this the substantial form or soul of this body, which is completed by a certain sensation and appetite.
>
> (*The Origin of Souls and Minds*; A VI 4, 1460; LLC 261)

Similarly, five years later he argues in the *Discourse on Metaphysics* that God produces substances as results of considering the world from all its infinite points of view:

> For God, so to speak, turns on all sides and in all ways the general system of phenomena which he has seen fit to produce in order to manifest his glory; and as he regards all the faces of the world in all possible manners, since there is no relation that escapes his omniscience, the result of each view of the universe, as regarded from a certain place, is a substance which expresses the universe according to that view, if God should see fit to actualize his thought and produce that substance. (§18, A VI 4, 1549–50/WFT 66)

Clearly, then, there is a significant correlation between Leibniz's rehabilitation of substantial forms in 1678 and his emerging new science of *analysis situs* in the same period. The idea that situation is a relation among coexistents, i.e. things that can be perceived simultaneously,[89] the close correspondence of situation with the point of view of a substance, and the construal of perception as the representation or expression of the whole universe, all remain constants of Leibniz's thought into his maturity. In fact, the above considerations cast Leibniz's much-quoted remarks from his *New System* in a more revealing light. The substances that God creates with appetition so that "they can be the source of their own actions," together with perception, which gives them the ability to represent the whole universe, are "what might be called *metaphysical points*;...and *mathematical points* are their *points of view* for expressing the universe" (GP IV 477/WFT 149). Each perception is a representation of everything that coexists with a substance at a given time; and God creates each substance as something actualizing such successive representations, as we saw above. The substances are *metaphysical points*, and the *mathematical points* from which the situations of all the other bodies are represented in the sense organs of these substances, their *points of view*, are precisely the points from which the situations of all these other bodies are considered, as in the second way of conceiving situation that Leibniz inherits from Hobbes.

I shall return to a discussion of the significance of this figure of the city for understanding Leibniz's metaphysics of space in the next section. But first I should give a brief description of Leibniz's new approach to geometry, sketching his

[89] Cf. §108 of the 1679 *Characteristica geometrica*: "If A and B are perceived simultaneously, and then C and D are perceived simultaneously, there is a distinction of situation" (CG ix 228).

intentions and some of his results. I do not pretend to give anything like a comprehensive account—something that even Vincenzo De Risi did not pretend to give in his marvellously rich study (2007) of the philosophical implications of *analysis situs*, on which I shall continue to depend heavily.[90] But I will try to sketch some of the main developments in Leibniz's thought on *analysis situs*, especially insofar as they bear on an understanding of his metaphysics of space.

According to his descriptions of the project, Leibniz's aim with his *analysis situs* is to represent the relations among geometrical figures directly, without recourse to the Cartesian co-ordinates and equations of ordinary analysis.[91] In explaining his name for this new type of calculus in 1693, he says he calls it this "because it explains situation directly and immediately, so that even figures that are not delineated are depicted in the mind through symbols; and whatever the empirical imagination understands from figures, this calculus derives by a certain demonstration from the symbols."[92] He denotes the relation of situation of A to B by 'A. B'; since the relation is mutual, this is equally the relation of B to A. He uses A, B, etc. to represent given points, and the last letters of the alphabet, X, Y, and Z, to represent variable points (CG 241/251). Among the relations he defines are *coincidence*: "If there are two lines AB and CD and the two points A and C are one and the same, we designate it thus: A ∞ C, that is, A and C coincide" (CG 170); *equality*, denoted A *Π* B, A equals B (CG 180); and *similarity*:

> (31) *Similars* are those which cannot be discerned when considered by themselves... For we can discover no attribute, no property, in one that we cannot also find in the other; and calling one of these A and the other B, we denote similarity: A ~ B. (CG 182)

The most important basic relation between two situations, though—at least in his efforts of 1679—is that of *congruence*, since the relation of situation is based on it. Leibniz denotes "is congruent to" by the symbol '$\cong$',[93] so A.B is congruent to C.D is written A.B $\cong$ C.D. Thus "A.B $\cong$ C.D signifies that the situation between the

[90] As De Risi notes, Leibniz's writings on the subject in one year alone—the *annus mirabilis* of 1679—take up much of the 350-page bilingual edition by Echeverría and Parmentier (CG).

[91] Together with the 1679 paper on the new characteristic that Leibniz sent to Huygens (A III 2, N347/L 248–53), Loemker translates an undated paper that he ascribes to the same period "On Analysis Situs" (GM V 178–83/L 254–8). As De Risi mentions, however, Eric Aiton has dated that second paper as belonging to 1693 (De Risi 2007, 85, 125).

[92] ("On Analysis Situs," GM V 182–3/L 257). Similarly, he told Huygens, that "its chief utility consists in the inferences and reasonings that can be done by operations without characters, which could not be expressed by figures (and still less by models) without multiplying them too much, or without becoming embroiled in too great a number of points and lines" (CG 259/L 250).

[93] Leibniz's symbol in his 1679 manuscripts is actually 'γ', but in a footnote he urges himself to use rather the symbol '∞' (CG 260, n.27) (which he later uses for coincidence). '$\cong$' is my approximation to the symbol for congruence he uses later. He first listed these four relations together with their symbols in the third draft of the *Characteristica*, dated August 1–11, 1679, including also symbols for 'greater than' and 'less than' (CG 116).

points A and B is the same as the situation between C and D" (CG 236). Any two points A and B are congruent: "Every point is congruent to every other point; that is, A ≅ B, or Y ≅ Z" (CG 246). Moreover, as we have seen, the situation of two arbitrary points to one another constitutes a *path* from one to the other: "Any two points whatever have a situation to one another, that is to say, there is a path from one to the other" (CG 246). As we have seen, two figures are congruent if "one can be applied to or placed on the other without changing anything in the two figures but their place" (CG 260/L 251). But since the paths can be understood as connecting any two points, this is the same thing as having the vertices of the two figures coincide. Thus a triangle can be represented as A.B.C, and it is congruent with another triangle E.F.D if one can transfer the points A, B, and C, while conserving their mutual situation, into the place of the points D, E, and F in order, so that A falls on D, B on E, and C on F, where these pairs of points are joined by rigid lines (no matter whether these are straight or curved) (CG 260/L 251).

On this basis, in his writings of 1679 Leibniz proceeds to demonstrate several interesting results, all achieved combinatorially.[94] Some examples:

> the reflexivity of congruence, A.B ≅ B.A;[95]
> "A.B ≅ D.E signifies nothing other than A ≅ D and B ≅ E at the same time, with the situations A.B and D.E unaltered" (§44, CG 192);
> "If A.B.C ≅ B.C.A, it follows that A.B ≅ B.C ≅ A.C, (ôr that the triangle is equilateral)" (§47, CG 194);
> "If because A.B.C. ≅ A.D.C. we have B ∞ D, then C is said to be *in the same straight line* as A.B." (CG 134)
> "If three points A. B. C. are said to be *situated in a straight line*, then supposing A. B.C ≅ A.B.Y, we will have C ∞ Y." (§50, CG 196)
> If A.B.C ≅ A.B.Y, the locus of all Y's will be a *circle*. (§96, CG 222; CG 263/L 252)
> "If AY ≅ BY ≅ CY then the locus of all the Y's, ôr Ῡ, is called a *straight line*." (§97, CG 222; CG264/L 253)

Thus Leibniz aims to abstract from the relations of situation basic relationships that do not depend on anything concerning the lineal depiction of the figures involved, but only coincidence of the points marking their boundaries. Once these

[94] Cf. Echeverría, who having identified four phases in Leibniz's methodology, characterizes his aim in the fourth phase as "to obtain *purely combinatorial* definitions, theorems, and finally demonstrations" (CG, Introduction, 18).

[95] "The proposition A.B ≅ B.A is always true, that is to say, the situation of A to B is the same as that of B to A; for their situation is a mutual relation which does not involve any distinction between the points themselves, and points themselves will already have no distinction per se, because they are always congruent" ("Latin fragment of a letter to Huygens," September 1679; CG 236).

abstract relations are represented algebraically, they can be regarded as embodying these relationships independently of the Euclidean lines and surfaces in which they had been depicted, so that the Euclidean constructions of figures would not be necessary to achieve geometric results. Proceeding in this way, Leibniz told Huygens, he could "see the possibility of extending the characteristic to things which are not subject to the imagination."[96]

One is tempted to say that Leibniz presupposes Euclidean space, and then, having abstracted the relations and properties of relations by means of these combinatorial operations, dispenses with this background Euclidean space, so that it acts as a kind of scaffolding enabling the construction of the abstract theory, rather than a necessary presupposition of the theory's content.[97] But this way of describing things involves projecting back into Leibniz's time our current understanding of Euclidean geometry as defining Euclidean space. As De Risi has made clear in recent publications, to do so is anachronistic, and seriously underestimates the originality of Leibniz's achievement. For Leibniz was the first to conceive Euclidean geometry as concerned with the properties, not of figures *in* space, but *of* space: the properties of the relations he derives are *properties of space itself*. Let me illustrate this point by reference to some specific examples of such properties, all of which depend upon the characterization of space as an order of situations.

One of these is the *uniformity* (or *isotropy*) of space. In §9 of the *Characteristica geometrica* of 1679 Leibniz writes:

> Now if we are to treat things in order it should be recognized that the first consideration is that of *Space* itself, that is, absolute pure Extensum. By *pure* I mean free of matter and motion, and by *absolute* that it is unbounded and contains every extension. Thus all points are in the same space and can be related to one another. But whether this space is some thing distinct from matter, or whether it is only a constant appearance or phenomenon does not matter here.
>
> (CG 150; GM V 144)

Clearly this evokes what he wrote in the pieces we examined in the preceding section from 1676 and 1689, where space, considered in itself, is "indifferent to different ways of being dissected" prior to its being actually divided by the motions in it, and "contains not only existences but possibilities." It also corresponds to his later definitions from the Fall of 1714:

[96] (CG 265/L 253). Similarly, Leibniz wrote to Gallois in December 1678: "If this characteristic of geometry were established, as I see it could be, it would lead one infallibly to a point where one can go farther, insofar as this is possible, just as well as algebra does" (A III 2, 566; CG 12).

[97] I said just that in my (2012, 20). In earlier publications I had pointed out that a similar strategy was followed by practitioners of Combinatorial Topology in the early twentieth century.

> *Absolute* unbounded *space* is that which is greatest [*amplissimum*] in situation. Therefore every place is in absolute space. Absolute space is a continuum, otherwise some place could be interposed which would not be in it. Absolute space is uniform, for variety arises from determinations, which are not consistent with its being the greatest.
>
> Absolute space is *uniform*, ôr everywhere similar (congruent) to itself. Thus if one uses a surface to cut a solid out of it, this will be uniform inside.[98]

As De Risi points out, these characterizations of the uniformity of space are supported by the properties Leibniz derives in the *Characteristica geometrica*, properties we would regard as characterizing *isotropy*:

> §60. Between any two congruent things infinitely many other congruent things can be assumed, for one cannot pass into the place of the other while preserving its own form, unless it passes through other congruent things.
>
> §61. Hence from any point whatever a line can be drawn to any other point. For one point is congruent to another.
>
> §62. Hence also from any point whatever a line can be drawn through any other point.
>
> §63. A line can be drawn which passes through however many given points.
>
> §64. In the same way it may be shown that a surface can go through however many given lines. For if they are congruent, it is clear that the generating line can be in every one of them successively. If they are not congruent, it is clear that during the motion the generating line can be so increased, diminished and transformed, that when it comes to this, it is made congruent.
>
> §65. Anything whatever can be placed in space while preserving its own form, that is to say, to anything existing in space infinitely many other congruent things can be assigned.
>
> §66. Anything whatever preserving its own form can be moved in infinitely many ways. (CG 202–4)

Putting these components together—the possibility of congruence everywhere in space, and the definition of motion in terms of continuous successive congruence—we have the possibility of rigid motion in space (De Risi 2007, 180).[99]

[98] These are from the manuscript De Risi has published as *Spatium absolutum*; (2007, 609 and 608, resp.) See also his discussion on p. 178.

[99] Louis Couturat had already drawn attention to §§60–7, dubbing them "axiomes de libre mobilité" (axioms of free mobility) (*La Logique*, 417–18). Echeverría notes that "these propositions also have a relationship with the modern theory of arcwise connection, thus they lead Leibniz on the way to topology and, more generally, to the search for the most general axioms and structures for founding geometry" (CG 38–9).

The second of property of space I wish to consider here is *homogeneity*. Leibniz attempts to capture it in the *Characteristica geometrica* (CG 174), but cancels his attempt (De Risi 2007, 160, n.31).[100] The reason for his hesitations, according to De Risi's analysis, is the difficulty he has in formulating an account of the equality of measure independently of congruence. Leibniz was well aware, of course, that two figures whose situations are similar and also equal are thereby congruent, but if either similarity or equality is defined in terms that still rely on measure, and measure itself is defined in terms of congruence, he cannot escape circularity (De Risi 2007, 140–7). The beginnings of a way out of this dilemma, De Risi observes, can be seen in his analysis of equality in the *Specimen geometricae luciferae* of 1695. Using the notion of a continuous transformation, he is able to define two magnitudes as homogeneous if there is a third magnitude equal to the first and similar to the second (160):

> For, just as those things are *equal* which are either congruent or can be rendered congruent by a transformation, so those things are *homogeneous* which are either similar (whose homogeneity is evident in itself, like two squares to each other, or two circles), or can be rendered similar at least by a transformation; which transformation happens, however, if nothing is subtracted or added and yet it becomes different....
>
> It is also clear that those things are homogeneous which are generated by a continuous increase or decrease of the same thing, with the exception, that is, of minima and maxima, i.e. extrema. And since those things are equal one of which can become the other by a transformation, it is also clear that those things are *homogeneous* with one another which are themselves similar, or at least similar to those things to which they are equal. (GM VII 282, 83)[101]

Here Leibniz is very close to the idea of an invertible continuous transformation that preserves neither quantity not quality—that is, the notion of a bi-continuous transformation or homeomorphism that is basic to topology. But probably because he is conceiving space in terms of situations that are defined in terms of congruence—and thus depend on measure—and also because he is conceiving continuity in terms of possible relations of situation, such a result is beyond his purview.[102]

[100] In the cancelled passage Leibniz writes: "I understand those things to be *homogeneous* which have something in common, as do a point, a line, a body, and their parts" (§24; CG 174). But he gives an improved definition a few pages later, where he defines two things as homogeneous "if any parts whatever in one may be assumed equal to parts in the other, and the same can always happen in the remaining ones" (§30; CG 180).

[101] None of these ideas seem to be original to the 1690s, however. As we shall see, the defining of homogeneity in terms of a transformation, the characterizing of such a transformation in terms of continuity, and the notion of co-integrant parts, can all be found in papers from the early 1680s.

[102] Again, see De Risi's lucid discussion (2007, 166–7).

This prompts a discussion of the continuity of space. This is, in fact, a large topic, because Leibniz's thought on continuity is brimming with innovation, as De Risi has discussed in a superb article (De Risi 2019). Since Leibniz regards any continuum with something conceivable outside it to have a bound, he is able to define continuity in the sense of *connectedness*, as follows. It depends on the idea of "co-integrant parts" [*cointegrantes*], defined in "*Hic memorabilia*..." (dated as 1695) as parts which "taken together, equal the whole" (De Risi 2007, 587). This allows Leibniz to define continuity (in the sense of connectedness) in the *Specimen geometricae luciferae* as follows:

> A continuum is a whole any two of whose co-integrant parts (that is, parts which taken together coincide with the whole) have something in common, and indeed if they are not redundant (i.e. have no part in common), i.e. if the aggregate of their magnitudes is equal to the aggregate magnitude of the whole, they have at least a boundary in common. (GM VII 284)

Leibniz is proud of this notion of continuity because, as he says in a note he wrote on top of the *Hic memorabilia*..., "it does not presuppose the notions of similarity, transformation or motion" (De Risi 2007, 586). It is therefore applicable also to time and other non-spatial magnitudes, as De Risi remarks (2019, 139). Nevertheless, it clearly applies to space as a whole, since space is defined by Leibniz as "a continuum consisting of all places [*locis*]" ("*Hic memorabilia*...," De Risi 2007, 587), or in the "*Spatium absolutum*..." of 1714, as "*an extensum* possessing continuity and coexistence of parts" (De Risi 2007, 608). The place of an extensum, we recall, is "that in which it is situated," as defined by its boundaries, and therefore a determinate part of space. But space is the order of all such situations taken abstractly, and it is this that is continuous, because all pairs of possible co-integrant parts have a boundary in common. The significance of Leibniz's achievement in defining continuity of space (in this sense of connectedness) is well summarized by De Risi. It was Leibniz's definition of space as *an order of situational relations* which "opened up the way for him to [give] an account of continuity along the same lines," making it "possible, for the first time, to conceive of continuity not as a monadic predicate or as a binary relation, but rather as a system of such relations" (2019, 153).[103]

[103] There is, on a modern understanding, no single property of being continuous. Instead there are various distinct properties, none of which is perhaps a perfect representation of the intuitive concept. Besides connectedness, and the arcwise connectedness approximately captured by §§60–8 of the *Characteristica geometrica*, there is also continuity in the sense of Dedekind completeness, the property necessary for determining, for example, whether there are precisely two points of intersection when the extended radius of a circle is drawn through its circumference. As De Risi explains (in his 2019), Leibniz gives definitions that come quite close to this concept, without, however, any inclination to establish a correspondence with the real numbers.

However, De Risi qualifies these remarks with a caution. Although the above definition is articulated in terms of relations, these are not *situational* relations, but rather *mereological* ones, ones depending on the relation of part to whole. In fact, De Risi maintains, Leibniz was never able to derive continuity using situational reasoning, because "one cannot compose any extension from unextended things" (2019, 154). In fact, he "did not attempt to reduce the elements of geometry to points and situational relations alone, taking continuity to be a derivative notion that could be defined from these latter. Rather he permitted extension itself, or continuity, to count among the basic elements of geometry: on a par with situational relations" (2019, 154). The reason for this, De Risi maintains, was his continuing commitment to a neo-Aristotelian conception of continuity that was still common in the seventeenth century and "that Leibniz himself could not get rid of for many years" (De Risi 2007, 183). This refers to the opposition of continuity proper to the contiguity of the determinate parts of the extended, a mereological conception that was inadequate to his goal of giving all geometric properties real, as opposed to merely nominal definitions. For example, Euclid's definition of a circle as "the figure described by motion of a straight line in a plane about a fixed end" (G VII 294/L 230)—represented in Leibniz's characteristic (CG 263) by the locus of all the Y's, ôr $\bar{Y}$, such that $A.B.C \cong A.B.Y$—is a real definition, since it proves the possibility of the circle a priori by giving the means for its construction.[104] But this does not *establish* the continuity (in the sense of completeness) of the locus so produced. Although the Euclidean definition depends on the uniformity of space, which Leibniz has established by means of his situational analysis, continuity, on the other hand "is defined rather in terms of parthood relations, and therefore depends on the (independent) mereological structure of the same space" (De Risi 2019, 162).

The root of the difficulty, in De Risi's estimation, is that there is no means of transition from points and boundaries of the extended to its continuity: "The extended is continuous, but it is perfectly heterogeneous to a point, which is only its boundary. Thus there can be no passage between the continuum and a set of points, which belong to two different orders of consideration" (De Risi 2007, 183).[105] According to De Risi's analysis, Leibniz addressed this difficulty to some extent by introducing the notion of *homogony*. Two entities, for instance, a point

[104] For this distinction see for example Leibniz's discussion in the *Discours* (§24; L 319).

[105] Marc Parmentier says something very similar: "For Leibniz in 1679 geometry has two poles, extension and *situs*, and one clearly perceives his temptation to recover this duality by that of space and point.... Space and point oppose one another in a dual relation, as extension and as *situs*, as two terms that are too abstract or too simple. Between them, no dialectic nor conciliation is possible" (CG 343). Cf. De Risi: "at this stage of his thought [*c.*1685], Leibniz is unable to reduce the notion of space (or extension) to that of situation, and thus he confines himself to oppose the two pairs: point and space, situation and extension" (2007, 166–7); "Let it be recalled that he has always doubted whether he should give priority to the notion of situation, which we have accepted so far, or rather to the concept of extension" (2007, 350, n.40).

and a line, are homogonous "when one can disappear into the other by a process of continuous change" (GM VII 20/L 668). This is a weaker notion of continuous transformation than one occurring only between homogeneous things, which is a special case of homogony. For the kind of continuous change involved in homogony may also result in a change of dimension by 1.[106] As we will see in detail in section 5 of chapter 3, this notion has its origins in classical geometry, where (especially under the influence of Proclus) a line was conceived as generated by the flux of a point, a surface by that of a line, and so on. This is important, as it gives Leibniz a way of characterizing the relationship between a figure and its section, and of the trace generated by the motion of a figure. De Risi sees Leibniz as having formalized this notion as homogony in the mid-1690s,[107] as a result of which it assumes a dominant position in his *Initia Rerum* of 1715—at the culmination of Leibniz's efforts in *analysis situs*—while the relation of congruence finally takes a back seat to those of similarity and equality, and measure conceived independently of congruence.

This story of the development of Leibniz's thought on space is actually in need of substantial amendment as a consequence of the appearance of further mathematical papers of Leibniz, some dating from the 1680s, which have recently been transcribed by De Risi and other scholars. As we shall see below, the notion of *congenitas* (later to be come homogony), its connection with the notion of a transformation in connection with the Law of Continuity, as well as the definition of continuity in terms of "co-integrant parts," can all be found in papers dating from the 1680s. I will defer discussion of the implications of that till later. For now I will forgo further discussion of the mathematical achievements and shortcomings of Leibniz's *analysis situs* as an approach to establishing geometry, for which I refer the reader to De Risi's excellent analyses.

In sum, the significance of Leibniz's efforts in connection with these properties of space (uniformity or isotropy, homogeneity, and continuity)[108] is that, even if he has not succeeded in demonstrating them by our standards, or perhaps even his own, he is the first to have actually conceived these as properties *of space*, and not

[106] For Leibniz's exposition of homogony as relating homogeneous as well as heterogeneous things, see the piece *De homogeneis et de homogonis* in Appendix 3.6. As De Risi has explained (see his 2007, 163, fn. 34), with this notion of homogony, Leibniz is close to the idea of a continuous one-parameter family of continuous transformations. While bi-continuous transformations between homogeneous things are basically homeomorphisms, the result of the totality of these transformations is *not* necessarily a homeomorphism. In particular, an object transformed through a continuous succession of transformations through a continuously parametrized interval may change dimension. This cannot happen with a direct, bi-continuous transformation (homeomorphism).

[107] in his (2007) De Risi maintains that the concept of being homogonous appears first as *συγγενὴς* in the *Specimen geometriae luciferae*, and in the roughly contemporaneous "*Scheda* on situation and extension" of 1695(?) (which I have translated in Appendix 3.7 of this book below), as "congeneous": "A point is congeneous with an extensum. For a point is made from an extensum by a continuous change, as demonstrated in proposition [12]" (De Risi 2007, 589). See De Risi's discussion of homogeneity and homogony (2007, 160–71) for further details.

[108] To these De Risi adds the three-dimensionality of space; see his analysis (2007, ch. 2).

of the figures contained in space. This is because, by characterizing space in terms of possible situations and transformations among them, in abstraction from the determinations that actually give them their measures, that is, the boundaries of the bodies which would determine them, he could attempt to determine properties of any possible figures that could be co-perceived, and thus of space itself.

2.4 The Metaphysics of Space

We have seen some of the independent developments in his understanding of mathematics that have led Leibniz to this point. But now I want to turn to considerations of how these developments are related to his wider metaphysics, in particular his metaphysics of space and its relationship to his theory of substance. These concern the way Leibniz distinguishes quality and quantity, and how this relates to his metaphysics of substance; but they also concern the significance of his construal of spatial extension in terms of the possibility of coexistence, and even the possibility of being co-perceived. In particular, I will be concerned to address the interpretation of De Risi, who sees Leibniz as providing a new, phenomenological grounding for his theory of space in the last years of his life (roughly, 1712–16).

According to De Risi, these new developments depend upon a characterization of similarity that is deeply rooted in Leibniz's theory of monadic expression. Whereas for him in 1677 the idea of defining similars through co-perception was "perhaps just a brilliant trick," in his mature thought it "becomes the only truly logical foundation of Leibniz's metaphysics as phenomenology" (2007, 403, n.79). The intrusion of phenomenological considerations, moreover, "enters into the foundations of geometry especially in relation to continuity" (2019, 163). Thus, De Risi explains, construing space as "in a certain way, a product of the mind (as an abstraction from physical extension, or perhaps as a form of intuition in an almost Kantian sense)" (2019, 163), allowed Leibniz to pursue a different approach to establishing the continuity of space, where an a priori proof of continuity will involve "determining continuity as a transcendental condition of space. This seems to me the only true deduction about continuity of space that can be found, although obscurely, in Leibniz's work" (De Risi 2007, 183, n.54).

In order to assess this interpretation, we should begin with similarity and quality. For a crucial distinction, and one that is fundamental to Leibniz's whole project, is that between *quality*, which can be conceived in things individually, and *quantity*, a determination of which can only be made by comparison. As Leibniz explains in the 1693 essay "On Analysis Situs":

> Thus it does not suffice to call things similar whose form is the same, unless one gives a further general notion of *form*. But in undertaking an explanation of

> quality or form, I have learned that the matter finally comes down to this: those things are *similar* which cannot be distinguished when observed individually. For the quantity of things can be grasped only by their compresence, or by the actual application of an intervening thing; whereas quality presents something to the mind which you can know in a thing individually and can then apply to the comparison of two things even though there is no actual application of an intervening thing as a measure by which one thing may be compared to another, either immediately, or through the mediation of a third thing. (GM V 180/L 255)[109]

According to De Risi's analysis, however, this constitutes only a "negative definition" (2007, 146–7): it says what quantity is not. Leibniz never succeeded in showing how one could define a positive concept of quantity, and thus of equality, without presupposing measure—as, indeed, this definition does. But in the context of his mature epistemology, De Risi maintains, this defect is turned into an advantage: for now quantity is given by co-perception of phenomena, which is essentially spatial. And, given the mutual exclusivity of quantity and quality, this means that the qualities a monad has in itself, while expressible in spatial terms, are never fully disclosed by the phenomena, and are therefore expressed confusedly. "What is distinctly apprehended in a phenomenon pertains to its expressivity, and manifests itself as the qualitative aspect of the representation. What is confusedly apprehended," on the other hand, "manifests itself as the quantitative aspect of the representation" (2007, 382–3). Thus what preserves quality is similarity, and what preserves quantity is precisely equality. In this way, De Risi maintains, Leibniz is able "to deduce the operational geometrical determination from the phenomenological one. *Geometria situs* and metaphysics of space have finally merged" (383).

On this reading, it is not the distinction between quality and quantity itself that leads to the reconfiguring of the *analysis situs* project in Leibniz's final years, and indeed to its reinvigoration, but its tie-in with what De Risi sees as Leibniz's "late phenomenalism." In fact, as he is well aware, the quality/quantity distinction is no innovation of his last years, but is coeval with the *situs* project itself. Leibniz had already clearly articulated it in his unsent letter to Gallois of 1677:

> After thoroughly investigating, I have found that two things are perfectly similar when they cannot be distinguished except *by compresence*, for example, two unequal circles of the same matter cannot be distinguished except by seeing them together, for then you can see that one is bigger than the other. You might say: I will measure one today and the other tomorrow, and thus I will distinguish them even without seeing them together. But I maintain that this is still to

[109] Leibniz gives precisely the same characterization of quantity and quality in the *Initia Rerum* of 1715 (GM VII 18–19).

> distinguish them, *not by memory, but by compresence*, because you have the measure of the first one present, not in your memory, for *magnitudes cannot be retained in memory*, but in a material measure engraved in a ruler or some other thing. For if all the things in the world affecting us were diminished in the same proportion, it is evident that not a single person could notice the change.
> (GM I 180; quoted in De Risi 2007, 58–9)

In this remarkable passage we see the first known occurrence of the famous "Leibniz shift" argument, an argument which in due course would inspire Hermann Weyl's theory of gravitation, and consequently be seminal for the development of gauge theory, in the guise of gauge symmetry.[110] In the continuation of the same passage, moreover, we see again the figure of the town/city seen from different sides, reinforcing the close link we discussed above between Leibniz's emerging analysis situs and his rehabilitation of substantial forms:

> Through this definition I easily demonstrate very beautiful and general propositions, for example, that when two things are similar according to one operation or consideration they are also similar according to the others; for example, supposing two towns are unequal in size but appear perfectly similar when looked at from the eastern side, I say that they will also appear perfectly similar when looked at from the western side, provided the whole town is revealed in each view. (GM I 180)

Noteworthy here, too, is the mathematical conclusion Leibniz derives from this, that two situations that are similar from one reference point (point of view), are similar from all points of view. Robert McRae has suggested that the idea of representation or expression corresponds to the idea of an *isomorphism* (1976, 23, 42, and *passim*). Formally, such a one–one mapping would require a complete similarity of structure; in the above example, each perception or representation from a given point of view would have to represent all the relations in the city, and vice versa. In an excellent article on the topic, Chris Swoyer (1995) has argued that such a characterization of expression as isomorphism is generally much too strong to capture Leibniz's notion of expression in all the contexts in which he uses it (although he allows that McRae intends a similarity of structure of some vaguer sort): "expression is a species of representation, and representations typically select, abridge, and distill" (Swoyer 1995, 87). A circle, for example, "has many

[110] The same figure is repeated in the *Characteristica geometrica* of 1679: "If indeed we imagine that God should diminish the appearance of everything in us and around us in some cubicle while conserving the same proportion, everything would appear the same, nor could the earlier state be distinguished from the later unless we stepped out of the sphere of things that had been proportionally diminished, namely our cubicle; for then a distinction would appear when the things presented were coperceived with those that had not been diminished" (CG 182).

features that are not preserved by its perspectival projections onto an ellipse, parabola, or hyperbola" (81). On Swoyer's interpretation of Leibniz's general account of expression, "one thing *expresses* a second just in case there is a structure-preserving mapping from either to the other" (82). He proposes that the analogy with perspectival projections suggests that "just as different projections of a circle onto a plane from different points of perspective give rise to separate correlating relations between the circle and each of its projections, the expressions of the world by different monads give rise to separate correlating relations between the world and each monad" (96). Moreover, "the relationships that are preserved by *each* of the correlating relations or transformations between monads are those that are objectively real," and this common core is what accounts for the similarity of perceptions from different points of view.[111] As Leibniz writes in the *Monadology*, "this representation of the details of the whole universe is confused, and can only be distinct with respect to a small part of things, namely those which are either closest to or largest in relation to each monad; otherwise every monad would be divine" (§60; GP VI 617).[112]

De Risi agrees with Swoyer's analysis, although he argues that monadic perceptions should instead be regarded as different partial isomorphisms—that is, *homomorphisms*—of the same isomorphism, since they are all (confusedly) representing the same thing, the whole universe. He quotes the continuation of §60 of the *Monadology*: "It is not in the object of knowledge, but in the modification of knowledge of the object, that the monads are bounded. They all go to infinity, to the whole [*au tout*], but they are limited and distinguished by the degrees of distinct perception" (GP VI 617). Here "the whole," De Risi claims, "means all the relations preserved by perceptual isomorphism" (2007, 321, n.17).[113]

[111] Swoyer's paper is very rich, and he deals with many other species of expression than the perspectival expression considered here: expression in analytic as well as in projective geometry, linguistic expression and surrogative reasoning, expression through models, and expression in measurement. I should also add that in his treatment of monadic expression, he offers a version of what he takes to be the "reductive phenomenalism common in Leibniz's later writings," where "a mind-independent external world drops out of the picture," so that "correlating relations between the items in it and the minute perceptions of individual monads are merely apparent" (Swoyer 1995, 96). In this case, the remaining correlating relations among the monads "give rise to an objective conception of an external world, since it is simply the features that are invariant under these transformations (i.e. that are common to all monads) that are objective" (96). This is the interpretation that De Risi takes for granted as representing Leibniz in his final system. See Daniel Garber's (2009) for a book-length treatment of the thesis that Leibniz abandoned realism for such a reductive phenomenalism in his final years.

[112] The reader should also consult Valérie Debuiche's (2009) for a thorough exploration of the mathematical roots of Leibniz's notion of expression. She concludes that it "cannot be reduced to a mere mathematical notion."

[113] Leibniz's idea that the whole world is represented in every perception of each substance becomes seminal for the future development of idealism, where it takes the form of the *doctrine of internal relations*. Bertrand Russell, enlightened by his discovery of that doctrine in his reading of Leibniz, identified it as the Achilles heel of idealism, which he then sought to expunge. As I have recently argued, however, Russell's identification of the holism implicit in this doctrine with the brand of Neo-Hegelian

Also, of course, the lateral perspectives on (representations of) the city need to be compared to the view from the tower above. Each substance, that is, expresses the whole universe from its own point of view, but the view from above, the ground plan, corresponds to God's understanding, and is the basis of the correlations among the various perspectives. As Leibniz writes in the *Discourse on Metaphysics*,

> It is only God—from whom all individuals continually emanate, and who sees the universe not only as they see it, but also quite otherwise from them all—who is the cause of this correspondence among their phenomena and who brings it about that what is particular to one is public to all; otherwise there would be no point of connection. (A VI 4, I550–1/WFT 66)

De Risi understands the difference between God's "view" of the universe and that of created substances as residing in the "limited aesthetic expressivity" of a created monad, in contrast to God's representation. On his reading (contrarily to what we have just quoted), God does not see the universe as created beings see it *at all*; rather, "God's representation must be the noumenal world itself, thus the set of monads" (De Risi 2007, 336).[114] This implies "that *no monad*... ever achieves the situational isomorphism" (338): not God, since that would imply that his representation would be the ideal limit of finite representations, whereas his representation in fact *is* the world: "no difference exists between God's ideas of the world and the world itself" (338); and not finite beings, since they have limited aesthetic comprehension: the only infinite they can aspire to is a syncategorematic one, and thus one that is necessarily incomplete. Therefore "The identity between representations and things-in-themselves is only a prerogative of divine apperception, i.e. of hypercategorematic infinity" (339).

The perceptions of created substances, on the other hand, being always to some extent confused representations of this divine representation, will thus be representations of a representation. Indeed, as we shall see shortly, since "each monad is

absolutism he found in his contemporaries (especially Bradley and Joachim), and also with the peculiar doctrine of the reduction of relations to non-relational properties that he inherited from Bradley and Lotze, led him to seriously misrepresent Leibniz's metaphysics (Arthur, unpublished).

[114] De Risi (2007, 335, n.28) quotes another passage that runs against his interpretation that not even God can represent things to himself either as they really are or as humans see them, from a preparatory note Leibniz made in February 1712 for his reply to Des Bosses: "In fact, God sees things exactly as they are according to geometrical truth, although he also knows how each thing appears to every other, and so all other appearances are contained in him eminently" (GP II 438). This is said by Leibniz right after again contrasting the scenographic views (lateral perspectives) with the ichnographic, or ground plan. He makes the contrast in the same terms in his letter to the Landgrave of Hesse-Rheinfelds (GP II 19), and in his *Specimen inventorum* (A VI 4, 1618/LLC 309). Leibniz is also very clear in a passage from December 1676: "There is no doubt that God understands how we perceive things; just as if someone who wants to give a perfect conception of a town, he will represent it in various ways.... For God... understands things in infinitely many ways, whereas we do only in one" (A VI 3, 400/DSR 115).

determined by its own situational homomorphism (i.e. by its own perception)" (488), each substance will be determined by its representation as a phenomenon. This reading, in appearing to reduce substances to mere representations, seems to turn Leibniz's philosophy into a variant of a position that he determinedly opposed. What is missing (as Leibniz explains in *On Nature Itself*) is his central idea that "the very substance of things consists in the force of acting and being acted upon" (§8; GP IV 508/WFT 214). In order for God to produce substances it is not enough for him simply to create their *perceptions*, their differing perspectives on the universe from different points of view at different times. In addition, as we have seen, he needs to create an *appetition* in each one that will serve to produce these perceptions and thread them through time. This is the substantial principle inherent in things, their *force*:

> No enduring thing can be produced if no permanent force can be impressed in things by divine power. Failing that, it would follow that no created substance, no soul, would remain numerically the same, and nothing would be conserved by God. Then all things would be reduced to mere transitory, evanescent modifications or phantasms, so to speak, of one permanent divine substance; and, what comes to the same thing, nature itself, or the substance of all things, would be God—a doctrine of very ill repute that a subtle, though profane, author has recently imported into the world, or perhaps revived. (GP IV 508–9/WFT 214)

Similarly, Leibniz had claimed in *De tempore locoque*... that if space is conceived independently of appetition, then "it is indifferent to different ways of being dissected. But if appetite is added to space, it makes existing substances, and thus matter, ôr the aggregate of infinite unities" (A VI 4, 1641/LLC 335).

Now De Risi, as we shall see, does not deny that appetition is a crucial part of the make-up of substances. Moreover, the profound changes he sees as occurring in Leibniz's philosophy are imputed to his last years, whereas *De tempore locoque*... was written in the late 1680s, and *On Nature Itself* was published in 1698. De Risi's intention is not to give an account that would hold for Leibniz's construal of space throughout its whole development, but only one that would link his supposed adoption of a phenomenalist metaphysics in his final years with developments in his theory of space in those years. Nevertheless, he sees the change over to this supposed "late monadological metaphysics" as being precipitated by Leibniz's mathematical studies of the mid-nineties—a claim in tension with the somewhat later appearance of the quoted passages from *On Nature Itself*.[115] I have

[115] "And, as the determination of space as a situational order first appears in the mathematical studies of the nineties, what we have here is that the most remarkable advancement between Leibniz's *Discours* and his late monadological metaphysics might be accounted for by his geometrical results of those years" (De Risi 2007, 300–1, n.2) Against this suggestion that the change began in the

argued elsewhere at book length against the idea that there is any such profound change in Leibniz's philosophy over to the very type of phenomenalism he had consistently opposed, and I will not attempt to repeat all those arguments here.[116] Nor can I enter into an extended critique of De Risi's very scholarly phenomenological reading in all its intricate detail. Here I will restrict myself to pointing out some of its consequences with respect to the understanding of space, and related aspects of the change in the metaphysics of space and substance that he imputes to Leibniz.

De Risi sees Leibniz as giving a *synthesis* of space: it is "transcendentally determined by conditions on the possibility of sensible experience" (De Risi 2007, 427). Even in his *Characteristica geometrica* of 1679, Leibniz is investigating "the conditions laid upon the possibility of experience" (406), "the ways in which consciousness (still not an expressive one) could produce spatial representations"—although this is from a time "when he had not yet formulated his monadological theory of the phenomenon" (406). What the latter consists in, for De Risi, is Leibniz's articulation of a theory of expression, where any given phenomenon is produced in a given monad as a confused perception, that is, is expressed as a particular perceptual homomorphism. (Here the Principle of Pre-established Harmony is simply a statement of the fact that the differing homomorphisms are partial mappings of the one isomorphism, mapping onto the world as a whole.) Quantity, however, cannot reside on the side of the noumenon, since all that could be discerned in the thing-in-itself would be quality, as determined by similarity, and in themselves, monads have no situation in space. Thus it is "the material side of the isomorphism that determines *space as such*, and actually also *situs* as such," with the result that "the possibility of being situated is . . . a universal property of representation" (341). The purely material side of an expressive morphism, being independent of the external world must, accordingly, "be a product of the spontaneity of consciousness," and the cause of material properties "will need to be found in the Transcendental Ego" (406).

Evidently, this reading is suffused with Kantian categories and terminology. De Risi does not, however, identify Leibniz's transcendentalism with that of Kant. Monads are certainly not unknowable things-in-themselves: that would be "a total misunderstanding of Leibniz's metaphysics, as in fact monads in themselves are perceived as phenomena" (De Risi 2007, 328 n.22). Although on his reading the representation of things in space is "produced from a simple synthetic act of

mid-nineties, however, De Risi elsewhere suggests that it began only a few months before his death in November 1716: "As Leibniz died a few months after he had begun this revolution in his spatial metaphysics, we have been left with his few suggestions and our poor conjectures . . ." (428). In private communication, De Risi has clarified to me that he thinks that the chronology is not precise, but that in his (2007) he was concentrating on the years 1712–16, even if some of the advances Leibniz built on in these last years had occurred earlier.

[116] See Arthur (2018) for a book-length treatment.

consciousness" (407), "the genuinely expressive character of sensibility causes a phenomenon adequately to represent, at least in its qualitative aspects, the thing itself" (405). On the other hand, however, the thing-in-itself or monad is identified with a set of situational relations: "Every phenomenon expressing a noumenon expresses (non-situated) monadic relations as situational relations" (341). Thus "the object of the different perceptual homomorphisms of each monad (i.e., all in all, the external object to which the content of representation of each substance refers) is in turn nothing other than the set of all these homomorphisms" (De Risi 2007, 328–9); "each monad is determined by its own situational homomorphism (i.e. by its own perception)" (488). In thus reducing substances to sets of relations, De Risi is giving an interpretation that is close to Ernst Cassirer's, as he expressly admits.[117] As a result, he sees Leibniz as more radical than Kant: he "has actually phenomenized substances (not just dogmatically assumed phenomena as things-in-themselves), because according to him substances are actually reduced to a set of relations" (405, n.81).[118]

Here I think De Risi has overstepped his mark. For (as he has subsequently explained to me) he did not intend to maintain that substances are reduced to sets of relations. Monads, he agrees, assuredly exist externally to the perceptions of any perceiving monad, and are what the perceptions are representations of. Moreover, they do not consist solely in sheaves of perceptions—the representations in each "metaphysical point" of the rest of the universe from its point of view—but also in *appetitions*, by the agency of which each one *actively produces* these perceptions successively through time according to its individual law, and also by virtue of which the physical phenomena of bodies, derivative forces, and motions are produced. Without the substances threading the successive perceptions (and producing them!), we have a mere phenomenalism of the kind Leibniz was trying to oppose. In De Risi's view, however, appetition, being a mere internal principle which brings about the passage from one perception to another (*Monadology* §15, *Principles NG* §2), is entirely determined by the content of the perception in question. Although substances exist in addition to perceptions (as a result of the necessary substance upon which they depend for their existence), they are entirely determined by their own representations. Moreover, according to De Risi,

[117] "I think I have throughout favoured a transcendental-phenomenological reading of Leibniz's philosophy, in many respects similar to the one that guided Cassirer more than a century ago. I think, in other words, that the core of Leibniz's late philosophy of space resides in his theory of a phenomenon and its knowledge" (De Risi 2007, xix). Cf. also 320, n.17.

[118] He writes: "we can also see that the key step of passing from the concept of substance to the concept of function, that Cassirer has pointed out as the dawn of the Critical perspective, is somehow more radically taken by Leibniz than by Kantian transcendentalism" (De Risi 2007, 405, n.81). As we will see in 3.4, this influence of Cassirer is also evident in De Risi's construal of Leibniz's notion of corporeal substance as the monad itself, insofar as it is taken with the further determination given by its relations of domination and subordination to other substances.

these representations are all essentially spatial.[119] Whatever is non-spatial in a phenomenon (motion, force, time) may be grounded in spatial features, in the sense that the complete knowledge of the spatial properties of a phenomenon would entail the knowledge of its past and future states, and therefore its motions, the forces acting upon it, the development of the phenomenon over time.[120]

On the reading I favour, by contrast, what is real for Leibniz is the *dynamism* of substance. The appetition of substances gives rise to forces, the active forces being what is real in motion, the passive forces being what is real in extension, where their diffusion constitutes extended matter. Although space derives its reality from the divine attribute of immensity, its extension derives from that of matter, and that in turn is constituted by the diffusion of force within it. As Leibniz wrote in the second (unpublished) part of the *Specimen Dynamicum*,

> It must be recognized, above all, that force is something utterly real in substances, even created ones; but space, time, and motion are to some extent beings of reason, and are true and real, not in themselves, but insofar as they involve the divine attributes of immensity, eternity, and operation or force in created substances. (GM VI 247/AG 130)

That space and time are real to the extent that they participate in the divine attributes of immensity and eternity was a feature also of Leibniz's early thought, as we saw in §2.2; that motion is real to the extent that it depends on force, participation in the divine attribute of operation, is intimately connected with Leibniz's reconceptualization of substantial form as (primitive) force, the very being of created substances.[121] On the phenomenalist reading favoured by De Risi, on the other hand, monads cannot be said to *be in* matter in a literal sense, so that extension can no longer be conceived as a diffusion of the passive force of resistance (antitypy) in a body, and motions must be understood as reduced only to a succession of representations. What is being diffused, according to De Risi, "is not any objective (expressly given) material property, but only a formal property of the possibility of experience" (2007, 415).

As noted, the key to De Risi's interpretation is to see Leibniz as having given (already in 1679) what amounts to a transcendental synthesis of space in terms of

[119] This spatialization of Leibniz's philosophy is pervasive in De Risi's interpretation, e.g.: "Every phenomenon expressing a noumenon expresses (non-situated) monadic relations as situational relations" (2007, 341); "no monadic property exists that cannot be in some way expressed as a precise situational relation" (401); "the total system of such boundaries determines a specific configuration in space (a set of figures), which finally expresses all the properties of the things-in-themselves" (493). I confess I do not see support for these claims in Leibniz's own words.

[120] For this formulation I am indebted to De Risi himself (private communication).

[121] See my discussion in Arthur (2018), esp. ch. 5, section 5.3. For a detailed treatment of the connection between Leibniz's views on the relativity of motion and his conception of substantial forms, see Arthur (2015a).

"conditions on the possibility of sensible experience," and to see him, on returning to this in his last years armed with a theory of the continuity of space based on monadic expression, as giving a phenomenological derivation not only of space itself, but even of the spatial representations constituting phenomena and determining substances themselves.

So before turning to the theory of expression, let us first examine the implications of Leibniz's metaphysical grounding of space in 1679. For one can well imagine how Leibniz's construal of situations in terms of the possibility of co-perception (in the *Initia Rerum* of 1715) would appear to the likes of Edmund Husserl and Ernst Cassirer as a phenomenological deduction.[122] But Leibniz describes space in terms of the possibility of co-perception already in the last, remarkable, passages of the *Characteristica geometrica* of 1679, §108, and this is where De Risi sees Leibniz as giving a transcendental deduction of magnitude, path, and continuity. Leibniz writes:

> §108. If A.B are perceived simultaneously, and in turn C.D are perceived simultaneously, there is a distinction of situation. But what is perceived when A.B are co-perceived and what is perceived when C.D are co-perceived, are similar: for it is clear that nothing can be observed in them individually which will not be observed in both. Thus if what is perceived when A.B are co-perceived and what is perceived when C.D are co-perceived can be distinguished from each other, they are distinguished solely by magnitude. Thus when A.B are simultaneously perceived they are simultaneously perceived as having some magnitude.
>
> (CG ix 228)

Magnitude, that is, cannot be perceived in situated things taken in themselves, but requires a comparison, and this requires them to be perceived together, *simul*, at the same time. But here Leibniz argues in the opposite direction, that the very possibility of co-perception is not just a necessary, but a sufficient condition for the perception of magnitude. What Leibniz is doing here, as De Risi explains, is giving a synthesis of elements occurring simultaneously in a given perception, that is, a synthesis of coexistents. But since space is the order of coexistence, and situation is the "mode of co-existing," this is a spatial synthesis (De Risi 2007, 408). This is a totally abstract synthesis, characterized by simultaneity alone, realizable "through a situational relation, i.e. through the representation of a (still undetermined) distance between two terms" (410). The above passage from §108 continues with Leibniz's deduction of path:

[122] As De Risi notes, even Hans Reichenbach, in the process of *resisting* a transcendentalist reading of Leibniz in favour of his own conventionalist one, saw him as having "required consciousness productively to participate in organizing the representation of a phenomenon" (De Risi 2007, 428, n.99).

> When two things are perceived to exist simultaneously in space, by that very fact a path is perceived from one to the other. And since they are congruent, by that very fact is conceived the path from one into the place of the other. Now, two points are congruent to one another. So what is perceived when two points are simultaneously perceived is a Line, that is, the path of a point.
>
> (CG 228; modified after De Risi 2007, 411)

This is then followed by a deduction of the continuity of space—not continuity in the sense of linear completeness, nor connection in general, but the arcwise connectedness that we observed above (De Risi 2007, 411). As De Risi acutely observes, this depends on a crucial assumption that Leibniz makes, namely that any situational order can be decomposed into (and therefore constructed out of) a multiplicity of two-term syntheses (2017, 411). This is because of a prior commitment to the more general thesis that "any set of relations can be reduced to simply binary relations variously combined [with one another]" (408). The passage continues:

> When we conceive two points as simultaneously existing, and ask why we say they are simultaneously existing, we will think the reason is that they are simultaneously perceived, or at least that they can be simultaneously perceived. When we perceive something as existing, by that very fact we perceive it to be in space, that is, that there can exist indefinitely many other things absolutely indiscernible from it. Or, what is the same thing, we perceive that it can move, ôr that it can be first in one place and then in another, and because it cannot be in several places at the same time nor move in an instant, we thereby perceive this place as continuous. (CG 228–30)

Here Leibniz is dealing with space in the abstract, that is, abstracting from the particular contents of any perception. Thus the only determination occurring is that of situation; and what "has situation and does not have extension" is a point. He is therefore able to define abstract or universal space as the locus of all such abstract points congruent to a given point: "Universal Space is the locus of all points."[123] Thus, in his notation, if A is that arbitrary point and Y a variable point, absolute space is all points Y for which $A \cong Y$. De Risi characterizes this as space being *constituted by points*. As he observes,

> After a century of endless controversies between sorites and infinitesimals, limits and minima, the indivisible and the undivided, here comes Leibniz with this amazingly simple result—space is, somehow, *constituted by points* ("*constitui,*

[123] *Scheda on situation and extension*, [1695], definition (LH XXXV, 1, 14, Bl. 90; see Appendix 3.7 below).

> *dico, non componi*"). This apparently most anti-geometrical and anti-philosophical consideration comes to him from the more complex definition of a point he has used for some time now, that of a locus regarded as unextended but situated.... Space turns out for him to be a set of relations between unextended (but situated) elements. (De Risi 2007, 173–4)

This, in my opinion, is an observation of profound importance for the history of geometry, and De Risi is to be commended for identifying it, and bringing it to the attention of scholars. One may wonder, however, why De Risi sees it as a discovery of Leibniz's last years. It is true that in the fragment *Rectam definio*..., which De Risi dates as 1715, we find that formula expressed: "If $Y \cong A$, then $\bar{Y}$ will be space" (De Risi 2007, 614), and we find a similar formulation in *Magnitudinis nomine*... of January 1712 (614). But we can find the same idea in the fragment "On Space and Point" of 1685, "Space is the locus of all points" (624), and indeed a wholly equivalent formula occurs in the *Characteristica geometrica* of 1679:

> If $Y \cong (Y)$, then the locus of all Y, ôr $\bar{Y}$, will be the absolute extensum, ôr *Space*. (CG ix §89, 220)

It may well be that Leibniz, armed with a superior conception of continuity formulated in the mid-90s, is able to return to this conception with renewed confidence in his mature years. But it still seems to me that this conception of space as the locus of all points bearing situational relations to one another is already in place in 1679. This appears clearly articulated, for example, in the fragment that Echeverría titles "New Characteristic for expressing position and motion," with a probable date of September 1679:

> Any two points whatsoever have a *situation* to one another, ôr a path is given from one to the other. The situation of the first point *A* to the second point *B*, I denote *A.B*. This is the same as saying that the two proposed points can be understood as the extrema of the extensum connecting them.
>
> Hence it is clear that all points can be understood to be in one and the same extensum, which is the unbounded ôr absolute extensum, namely space; in which space, that is, there is nothing but extension, and therefore other things can be understood as having an assignable situation in it, but it itself cannot be understood as situated anywhere. A point, on the contrary, is that in which nothing can be understood to be situated, ôr a point is the simplest thing that can be understood in the extensum.
>
> Since therefore all the points in the universe are congruent with one another, and all possible points are in unbounded space, it follows that unbounded space will be the locus of all points congruent to a given point, that is, the locus of all points

> taken absolutely. To express this symbolically [*speciose*], let there be the congruents Y ≅ A, and the locus of all Y will be unbounded space.
>
> (CG xi 246–7; Appendix 3.2)

Moreover, as we have just seen, insofar as Leibniz can be held to have successfully derived the ideas of a path in space and also its arcwise connectedness through synthesis, this is already in place in 1679. It is true that he does not yet have the mature definition of a straight line that he adumbrates in the *Rectam definio*... and the *Initia rerum* both of 1715, where it is determined as the simplest path between two points by an appeal to the Principle of Reason; and that in the latter are also showcased the notion of section by means of his notion of homogony, and much more. But the one crucial element that is missing earlier, for De Risi, is a conception of *continuity* in the sense of what we would call *completeness*, a result of his investigations of the mid-90s, which I shall return to below. For without this, continuity in the form of arcwise connectedness would still be parasitic on a conception of continuous extension that had not properly been explicated in terms of points. This reintegration is only achieved, according to De Risi, when Leibniz reconceives the constitution of space by points in the light of his phenomenological theory of expression in his last years.

Turning now to that matter, we may observe that in that last passage from CG ix §108, the very fact that Leibniz is there discussing not just what can be perceived in space but "indefinitely many other things absolutely indiscernible from it," indicates that he is there dealing with figures or situations in abstraction from anything instantiating them. He is treating *indiscernible* points being in the same relations of situation, and thus absolute or mathematical space (cf. De Risi 2007, 412, 415). He is discussing what all perceptions must have in common, but not the specific contents of the perceptions of actual phenomena. To discuss the latter, we need to move over to a discussion of Leibniz's theory of expression.

For the actual situations of phenomena are expressed more or less confusedly in the perception of each monad. Thus, although the representation of a plurality of coexistent objects regarded as indiscernible gives rise to the perception of a homogeneous extension (which is a purely extrinsic interrelation of positions), the relations among the actual phenomena must be expressed in the perceptions.[124] As Leibniz writes in an important little essay written in around 1700, "On the Principle of Indiscernibles" (C 8–10),

[124] Cf. De Risi: "from the mathematical point of view the concept of order of situations only leads to the representation of a *spatium absolutum*, which, being a uniform one, is inadequate to diversify among monads in perceptual terms. Therefore in order for a real expressivity of space to be possible, the addition of the phenomenal distinction of each monad's organic body is necessary" (2007, 492).

> To be in a place seems, abstractly at any rate, to imply nothing but position. But in actuality, that which has a place ought to express place in itself; so that distance and the degree of distance involves also a degree of expressing in itself the remote thing, either a degree of affecting it or of receiving an affection from it. So, in fact, situation really involves a degree of expressions. (C 9/Leibniz 1995b, 133)

Leibniz gives this example as an illustration of his doctrine "that there are no purely extrinsic denominations" (8/133). Phenomena cannot exist in a relation of situation to one another as if this were purely extrinsic to them. Instead, this relation of situation must be expressed as a particular quality in *A* representing the degree of remoteness of *B*, and a quality in *B* representing the degree of remoteness of *A*. As Leibniz explains, this is why place and position, like quantity and number, "are mere results, which do not constitute any intrinsic denomination per se, and so they are merely relations which demand a foundation from the category of quality, that is from an intrinsic accidental denomination" (9/134). Thus we falsely "conceive position as something extrinsic, which adds nothing to the thing posited, whereas in fact it adds the way in which that thing is affected by other things" (9/134).

This doctrine concerning extrinsic denominations, of course, is appealed to by those commentators who claim that Leibniz "denied relations"—the issue we have treated in Appendix 2. The idea is supposed to be that Leibniz accepted only monads and their monadic accidents as real, and accordingly ruled out relations. De Risi does not so interpret it, obviously, since he not only analyses space in terms of relations, but even goes so far as to claim that "substances are actually reduced to a set of relations," as we saw. Nevertheless, the notion of expressivity is central to his interpretation. He interprets the above doctrine that there are no purely extrinsic denominations as implying that expressivity, since it involves the qualitative aspect of the representation, is what is distinctly perceived, whereas "what is confusedly apprehended manifests itself as the quantitative aspect of the representation" (2007, 382–3). That is, the relations of situation, since they involve quantity, are simply "the non-expressive residue of the ideal isomorphism" (341).

I would interpret the state of affairs regarding situations and extrinsic denominations, rather, as follows. Abstractly conceived, situations are indeed merely extrinsic. The situations of actually existing phenomena, on the other hand, must correspond to the degrees of expressivity in the substances which are representing these phenomena. Just as there is an intrinsic denomination corresponding to what is doing the expressing, so there is an extrinsic denomination corresponding to what is expressed. The first is the degree of expression within the monad, according to which monadic point of view is expressed or represented in the situation of its body; the second is the situation as it occurs in the phenomena. That is, the actual situation is manifested as the lines and surfaces marking out the boundary of the phenomenal body of the monad. The abstract relations are ideal;

but the situations of the actual phenomena will be marked out by the boundaries of the phenomenal bodies, the shapes and figures which appear in perception.[125] These lines and surfaces are relational accidents of the bodies in question, modes or modifications of these bodies. But relational accidents are not entities that should be posited as independently existing; rather, they are results of the intrinsic modifications of substances, modifications that change from one instant to another. Thus, on the one hand, relations of situation require a fundament in the monadic perceptions themselves, and are results of the harmony among these perceptions; and yet they nonetheless exist (transiently) as modifications of actual phenomena, constituting the ever-changing boundaries of existing bodies depicted in the imagination.

De Risi, with an eye to the subsequent history of phenomenalism initiated by the intervention of Kant, sees Leibniz as giving a reduction of phenomena to the internal perceptions of monads. There is certainly textual evidence for such a reading,[126] and perhaps evidence of this kind inspired Kant to read the world of phenomena as a world of mere appearances, and to read that back into Leibniz. As I have argued elsewhere (Arthur 2018), however, a better reading (i.e. one that coheres with a much greater range of texts, and also with his solution to the continuum problem) is that with his notion of pre-established harmony Leibniz envisioned a one–one mapping (homomorphism) between the content of perceptions internal to the monad—"internal phenomena"—and what those are perceptions of—"external phenomena," namely the bodies and their motions that are produced by the forces of the substances contained in them.[127] The move from saying that all we know of phenomena is contained in our perceptions and the conditions for their possible occurrence, to saying that this is all that phenomena are, is (in my opinion) the classic fallacy of idealism.[128]

[125] Cf. Leibniz's "Remarks on M. Foucher's Objections" [1695]: "Properly speaking, the number ½ in the abstract is a simple relation, by no means formed by the composition of other fractions, even though in numbered things there is equality between two quarters and one half. And the same can be said of an *abstract* line, for there is composition only in *concrete* things, or *masses*, of which these lines mark the relations" (GM 4 491/WFT 184; Russell 1900, 246). Until I understood Leibniz's analysis of situations, I was puzzled by Leibniz's claim here that lines "mark the relations" in concrete things.

[126] Well known instances are Leibniz's assertion in the *Discourse* §14 that "In reality nothing can ever happen to us other than thoughts and perceptions," and his claims in the last letter to De Volder (January 19, 1706) that "I do not see what argument could prove that there is anything in extension, bulk, or motion beyond the phenomena, i.e. beyond the perceptions of simple substances," and that "the reality of sensible things consists in nothing other than the agreement of phenomena" (LDV 334).

[127] By "external phenomena" I mean ones external to any particular monad: these will be aggregates of the monads external to it, and the accidents and relations resulting from them. De Risi, by contrast, reads Leibniz's distinction between a subjective phenomenon and an objective one as that between the "expressive limit of the apprehension of a single monad" and "the ideal situational isomorphism," which is unattainable because of the infinitude of its object—namely, the whole world as known by God (2007, 370). I agree with him, however, when he says that Leibniz seems habitually to exploit the ambiguity between objective and subjective senses of the term "by assuming now one and now the other possible meaning of phenomenon" (xix).

[128] Perhaps De Risi does not commit this fallacy, since he acknowledges the existence of the substances represented in the perceptions (as well as their ultimate cause, God). Nevertheless his

These general considerations about Leibniz's metaphysics of expression, however, do not engage with other aspects of De Risi's case for a late change in Leibniz's views on space and continuity, which I now want to examine. He contends that in the years 1712–16 Leibniz finally threw off the shackles of his commitment to the distinction between contiguity and continuity, a development that was made possible by his new theory of diffusion. This is the idea of "the infinite repetition of the congruence relation," from which "Leibniz will later derive the possibility for any figure whatsoever to be taken through a continuous congruence from one place to another of extended space" (De Risi 2007, 413). It is the process by which "a homogeneous extension originates from the repetition *partes extra partes* of the identical," and "in his mature years," De Risi maintains, "Leibniz will actually find a very precise and technical name to describe it," namely *diffusion* (413). This development thus allows Leibniz finally to derive extension through the process of diffusion, whereas until then there had simply been an opposition between the pairs point and space, and situation and extension (167); and also to deduce continuity phenomenologically.[129]

That is, De Risi interprets extension as deriving from the diffusion of situs itself, "the formula according to which *extensio* is *diffusio* of *situs*" (415). But according to him, because diffusion is "a process that occurs only in a phenomenon," "what is being diffused (in this instance!) is . . . only a formal property of the possibility of experience" (415). Curiously, De Risi does not present any text of Leibniz's where he is supposed to have presented this mature theory of diffusion of *situs*. Marc Parmentier does, however, in a passage already quoted by Ernst Cassirer in his discussion of analysis situs (1902, 146). The text in question is in Leibniz's *Entrétien*, his dialogue on Malebranche, and it runs as follows:

> As for myself, even though I would always distinguish the notion of the extended from that of body, I do not cease to believe that there is no substance which could be called space, that is to say, that there is no subject which has nothing but extension in it.[130] However, were I to admit such a substance, I would always

reading is idealistic insofar as he claims that for Leibniz's God there is an "identity between representations and things-in-themselves" (2007, 339), and that "in Leibniz the identity of object and concept in an intellectual act makes sense, because for him the only concrete objects are spiritual substances, i.e. nothing but simply representational contents" (365, n.50). And in general, on his "transcendental-phenomenological reading" De Risi wishes to present Leibniz's ontology as residing in "his theory of a phenomenon and its knowledge" (xix), thereby deriving ontology from the theory of knowledge.

129 According to De Risi, what happened in Leibniz's last writings was that "space changed into an abstract and unprecedented system of relations (points), while the extended went on being defined according to the former metaphysical usage, i.e. as a coexistent continuum. Hence the problem would [a]rise of demonstrating that space is extended" (167). In the late writings the "production of extension through diffusion is closely related to the deduction of continuity that Leibniz seeks to obtain from his phenomenological attempts" (415).

130 Here I follow the French quoted by Parmentier from Erdmann, an earlier draft which Leibniz subsequently cancelled. Where the cancelled draft has "je ne laisse pas de croire qu'il n'y a point de substance qui puisse être appellée espace; c'est à dire qu'il n'y a point du sujet qui n'ait rien que

> distinguish between the extended or extension and that attribute to which the extended or diffusion (a relative notion) is related, which would be situation or locality. Thus the diffusion of place would form space, which would be like the πρῶτον δεκτικόν or primary subject of the extended, and through which it would apply to the other things that are in space. Thus the extended, when it is the attribute of space, is the diffusion or continuation of situation or locality, just as the extension of body is the diffusion of antitypy or materiality. For there is place in a point as well as in space, and consequently place can be without extension or diffusion, but diffusion simply in length makes a local line endowed with extension. It is the same with matter: there is matter in a point as well as in body, and diffusion simply in length makes a material line. The other continuations or diffusions in breadth and depth form the surface and solid of the geometers, and in a word, form space with respect to place and body with respect to matter. (GP VI 585/AG 262)

Since this text was written in 1712 (and revised in 1715) it might be thought to support De Risi's case for a late change of view. And indeed Parmentier cites the passage in reference to "the definitive theory," which "will thus suppose a distinction in extension itself, rejecting space from the side of situation and revealing the duality between extension and situation as a false opposition" (CG 343). De Risi notes Parmentier's argument, but thinks that the opposition between situation and extension lasts beyond Leibniz's reading of Malebranche, and that in any case the mere "word '*diffusio*' does not solve the problem" (2007, 78, n.86). Both authors note the necessity for Leibniz to solve other problems first, a major one being determining a satisfactory and less schematic definition of a straight line (CG 346–8, De Risi 2007, 68–9, 226–64).

All this is more than a little puzzling to me. As we have already noted, the ideas of the generation of a line as the path of a point, with surfaces similarly generated from lines, and solids from surfaces, are all in place in his earlier work, as is the idea of continuity in the guise of arcwise connectedness. So while Leibniz's progress in the foundations of geometry between 1679 and his death is certainly remarkable (including his attempts on the parallel postulate, the three-dimensionality of space, his improved definitions of homogeneity, and of section, and so forth), the question we are trying to settle is whether there is a fundamental change in his understanding of space associated with the concept of diffusion of place.

The concept of diffusion (or repetition) constituting extension is not new in 1712. Leibniz had already published a paper in the *Journal des Sçavans* of January

l'étendue" (CG 343), Gerhardt gives the corrected version: "je ne laisse pas de croire qu'il n'y a point de vuide, et même qu'il n'y a point de substance qui puisse estre appellée espace" (GP VI 585; Ariew and Garber follow Gerhardt, AG 262).

1693 where he wrote that "the extended signifies only a repetition or continued multiplicity of that which is spread out; a plurality, continuity and coexistence of parts" (De Risi 2007, 517). Similarly, in "On Body and Force: Against the Cartesians" of May 1702,[131] and again in his letter to De Volder of June 30, 1704 (LDV 305), he had described how extension is not absolute, but relative to the "nature" that is extended. That "nature" is identified by him there as *το δυναμικὸν*, the force of acting and being acted upon that constitutes substance. The extension of bodies, that is, is construed in terms of each substance's resistance to a change in the boundaries of its organic body, this resistance being the *antitypia* which, as Leibniz had always believed, is what distinguishes matter from mere extension. Moreover, even though the precise formula for extension—"the continuous, simultaneous diffusion or repetition of a certain nature"—is a product of his maturity, the idea that the force or action of every finite substance is "always diffused or to some extent impeded" occurs already in the "Metaphysical Definitions and Reflections" of 1678–79;[132] as does also the conception of matter as "the principle of being acted upon (*principium passionis*)," which, since it "must necessarily effectually contain multiplicity in itself," is "a continuum containing a plurality of things at the same time, that is, an *extensum*" (1400/246). Thus already in 1679 Leibniz has the concept of a diffusion of force, where force is equated with form (or the principle of action, or appetite), in addition to the concept of matter as merely passive, an extensum, that is, a simultaneous, continuous multiplicity.

In the above quotation from the *Entrétien*, Leibniz presents the constitution of space by the diffusion of situation as analogous to the constitution of matter through the diffusion of antitypy. But at the same time he cautions that "there is no subject which has nothing but extension in it."[133] So his meaning appears to be that, taken in itself, the extension of space can be understood as a continuous diffusion of situation, but that the situations only have meaning relative to what could instantiate them. De Risi interprets this to mean that "what is being diffused... is not any objective (expressly given) material property, but only a formal property of the possibility of experience" (415).[134] For, while recognizing

[131] This piece is included by Gerhardt in 1860 as a supplement to Leibniz's much earlier letter to Honoré Fabry (GM VI 81–98). The title is that supplied by Ariew and Garber (AG 250). A new edition of the Latin has been produced by Henrik Wels (https://www.uni-muenster.de/Leibniz/DatenVI5/op2642.pdf) and will be published in A VI 5. Wels notes that it must have been addressed "to the Jesuit Giovanni Battista Tolomei, and not, as previously assumed, Johann I Bernoulli."

[132] (A VI 4, 1398/LLC 245). As I mention in my (2018), this important piece is not very happily titled; it is likely that it consists of drafts for his projected little book on physics that he was proposing at that time.

[133] This can perhaps be seen as a direct repudiation of the view he had expressed to Thomasius in 1669, when he claimed that "space is a substance" on the grounds that it is the subject of geometry.

[134] This should be compared with Cassirer's comment on Leibniz's Characteristic: "In this way, the sensual residue that still sticks to the given elements is stripped off.... The distinction of forms, which might seem to be accepted simply as a sensual fact, is thus recognized here as an achievement of pure thought. [*So wird der sinnliche Rest, der an den gegebenen Elementen noch haftet, abgestreift.... Die*

that Leibniz had defined extension as a coexistent continuum early on (2007, 415), De Risi contends that he does not satisfactorily connect extension with situation until he explains "the theory of diffusion through congruence": "the purpose of the *diffusio* process is that of generating extension" (414, n.89). But, as I have already argued, the formal notion of the constitution of space by points congruent to an arbitrary point is already in place in 1679. De Risi contends that "the diffusion of *situs* is a one-of-a-kind *a priori* phenomenological production, and it generates space as a continuous set of all possible syntheses" (416, n.90), it generates a "manifold" (414, n.89). On the contrary, I shall argue, for Leibniz extension is not the production of a finite consciousness. Abstractly, space consists in the "diffusion of place" (GP VI 585), where place consists in an equivalence class of situations, but the extension of this space is the abstract counterpart of the extension of the matter it contains. The extension of matter is produced by the forces or appetitions of the substances contained in it, forces which are themselves limitations of divine omnipotence; likewise, the figures of their bodies are produced in the unbounded extensum by limitations of divine omnipresence, with the result that abstract space may be conceived as the locus of all their mutual situations conceived as possible.

For the basis of space is divine immensity. Leibniz is unshakable in this conviction. This is what accounts for spatial extension's not being merely "a formal property of the possibility of experience," but something with a reality founded in a divine attribute that is independent of human consciousness. We traced this view in §2.2 from the *De Origine rerum ex Formis* of 1676 through to the short tract probably written in Rome in 1689. But it remains in place in all Leibniz's mature discussions as well.[135] It is there in the *Specimen Dynamicum II* of April 1695 (in a passage we already quoted in §2.3), which is worth quoting again to illustrate how, after creating his dynamics, Leibniz conceives the divine attribute of omnipotence (which in 1676 he had said was analogous to immensity and eternity) as manifested in created substances as force:

> It must be recognized, above all, that force is something utterly real in substances, even created ones; but space, time, and motion are to some extent beings of reason, and are true and real, not in themselves, but insofar as they involve the divine attributes of immensity, eternity, and operation or force in created substances. (GM VI 247/AG 130)

Unterscheidung der Formen, von der es scheinen könnte, dass sie als einfach sinnliche Thatsache hingenommen warden muss, wird also hier als Leistung des reinen Denkens erkannt.]" (Cassirer 1902, 150).

[135] Cf. Slowik: "One might think that these Spinozistic or Neoplatonic tendencies would have been long since abandoned by the time of Leibniz' mature monadic writings (post mid-1690s), yet there are a number of discussions in this later period that are strongly reminiscent of the immensum" (Slowik 2016, 69)

Particularly illuminating, however is the essay on the metaphysics of geometry that Leibniz wrote in about 1698 (published long ago by Gerhardt, but now all but forgotten), which Osvaldo Ottaviani and I have titled *Towards a Science of the Infinite*.[136] Here Leibniz explains how while "the absolute is the fount of reality and prior to the limited," "negation or limit is the origin of imperfection" (Gerhardt 1875, 600). Absolute predicates, like being extended per se, "consist in a purely positive reality and therefore indicate perfection, and only these are counted among the attributes of the supreme substance."

> Thus extension, or, if you prefer, continuous diffusion, is something absolute, which can be understood in the whole and in the part, in the great and in the small; but for a figure, like a circle or a square, besides extension per se itself, and thus taken absolutely, a limit or bound is required, and this is the negation of further extension. (600)

So created things participate in these absolute natures, although in a limited way: "things are extended through the benefit of the omnipresent substance, they receive active force by participation in the omnipotent, and the same applies in the other cases" (601). If we suppose a triangle to exist, Leibniz writes, but the kind of triangle is not defined by any reason, "by that very fact it is generated as a regular one" (602).[137]

> So, in just the same way, if one attributes to something a diffusion of presence, and supposes that no reason is assigned for it to have definite bounds, it will be immense or everywhere; if existence or duration is accorded to it, and nothing further of this kind is added in it, it will be eternal or necessary; finally, whenever simple and pure reality is understood, by that very fact is constituted the Maximum possible in things, ôr absolute infinite, in which duration, diffusion, power, cognition and anything at all that is in it, lacks limits, and in turn anything that can lack limits is in it; but the rest originate from it, and it is called God. (602)

What these elucidations show, I believe, is that the idea of extension as a continuous "diffusion of presence"—better, of compresence—is a constant throughout Leibniz's thinking about space, and that this continuous diffusion of compresence is the same as divine omnipresence or immensity. As Slowik has argued, Leibniz continues to uphold immensity as the basis of space in the

[136] This has been newly transcribed by Ottaviani and translated by myself in a projected collaborative volume of transcriptions and translations, *Leibniz on the Metaphysics of the Infinite*, forthcoming with Oxford University Press in the British Society for the History of Philosophy series.

[137] As we shall see in §3.5 below, this argument also occurs in the *Origo veritatum contingentium ex processu in infinitum* from the same period (A VI 4, 1664), and in Leibniz's discussion of the status of the principle of inertia. It is an instance of the curious argument form that, there being no reason to choose any one of a number of configurations over any other, the simplest must prevail.

Nouveaux essais of 1704 and in his correspondence with Clarke in the last months of his life (Slowik 2016, 69–70). In the former, for example, Leibniz writes that "these attributes are nothing other than the attributes of God... The idea of the absolute, in regard to space, is just the idea of the immensity of God and thus of other things" (Bk. II, ch. xvii, §3; A VI 6, 158); and a bit later he characterizes immensity as "an absolute or attribute without limits" (Bk II, ch. xvii, §18; A VI 6, 159; Slowik 2016, 70).

In these passages and in his remarks to Clarke, however, Slowik sees a change in Leibniz's position on space. He finds it significant that in these passages Leibniz no longer refers to "the extended per se" or to "divine magnitude" as in 1676 and 1689, respectively. Partly this is because he interprets the earlier passages as "spatializ[ing] God as a form of container of body/force properties," as evidenced by the reference to "adding" mass to the immensum in the first, or to endeavours being "in nothing other than God" in the second (Slowik 2016, 70). As we saw in §2.2, however, even in 1676 Leibniz tried to dissociate himself from such spatializing of God. On the one hand, immensity—like eternity, omnipotence, and omniscience—is an attribute of God, but it is not God himself. On the other, in agreement with Spinoza, he held that as a divine attribute, the extended per se is not divisible into parts, and is distinct from the extension of space, which is divided into different parts at different times. Thus in 1676 he writes: "Space is only a consequence of this [immensity]" (A VI 3, 391–2/LLC 55). What this means is that the extended per se should not be understood as something of actually infinite extension, of which finite existents are parts. An unbounded line, for instance, is not maximal in the sense of being an actually infinite line, greater than all finite lines that could be its parts, but in the sense of being *the pure attribute of extendedness to which no limits have yet been applied to deliver an actual extensum.* Points and boundaries can be conceived in it; but in itself it has no limits. This is, of course, a subtle distinction, and Leibniz himself was not always careful to distinguish divisible space from the indivisible attribute of immensity.[138] In his "New Characteristic for expressing Situation and Motion" of 1679, for instance, he equates extended space with the absolute, unbounded extensum: "it is clear that all points can be understood to be in one and the same extensum, which is the unbounded ôr absolute extensum, namely space; in which space, that is, there is nothing but extension" (CG xi 246; Appendix 3.2). Likewise, in the 1689 piece, Leibniz calls "real space" "something that is one, indivisible, immutable," thus blurring the distinction between space and the divine attribute of immensity, and talks of matter being created by the introduction of appetites

[138] Cf. De Risi, who remarks that in his youth "Leibniz himself had come close to regarding space as divine immensity" (565)—even while "refraining" from directly identifying space as a divine attribute in *De Origine rerum ex Formis* (De Risi 2007, 56)—describing it as a youthful folly that Leibniz would later "polemically and repeatedly" attack (56).

into this primary unity, as into God himself. Yet this real space "contains not only existences but also possibilities, since in itself, with appetite removed, it is indifferent to different ways of being dissected." Thus in this piece space is a "real relation" comprising the possibility of the compresence of bodies, that is, their possible relations of coexistence, or mutual situations; and what are added to it are the actual created things, whose boundaries are limitations or determinations of the absolute unbounded extensum.[139] Such an absolute, unbounded space, I contend, is not to be understood as an infinite whole, but as immensity, the "attribute without limits" he refers to in the *Nouveaux essais*.[140]

But Slowik sees further evidence for a change of position in Leibniz's late writings concerning the way in which God is supposed to be present. In the earlier writings, he maintains, God is "a form of container,"[141] a "holistic ontological platform for space" (2016, 79), so that his immensity would entail his substantial presence everywhere. In his later writings, however, God is present by his immediate operation. Thus in the *Nouveaux essais*, Theophilus/Leibniz responds to remarks by Philalethes/Locke about the infinitude of duration and expansion as follows:

> If God were extended he would have parts. But duration confers parts only on his operations. In regard to space, however, we must attribute immensity to God, which also gives parts and order to his immediate operations. He is the source of possibilities and of existents alike, the one by his essence and the other by his will. Thus space, like time, derives its reality only from him, and he can fill up the void whenever it seems good to him. So it is in this respect that he is everywhere.
>
> (*Nouveaux essais* II, xv, 2; A VI 6, 155)

Leibniz also urges this interpretation of God's presence against Clarke. In his rejoinder to Clarke's claim that "God perceives things, not indeed by his simple *presence* to them, nor yet by his *operation* on them, but by his being a *living* and *intelligent*, as well as an *omnipresent*, substance" (Second Reply §5; GP VII 360–1/LC 22), Leibniz replies in his Third Letter §12: "God is not present to things by situation but by essence; his presence manifests itself by his immediate operation"

[139] De Risi tells me he cannot follow me on this reading: "If space is an order of possible situations, and an ideal object insofar as it is a mere possibility, it cannot itself be grounded on God's immensity" (private communication). But on my reading, God's immensity for Leibniz consists precisely in the possibility of the compresence of bodies, since God is present by operation, and his (natural) operation consists in the continuous creation of substances, and therefore of bodies in mutual situation in every part of space, as I shall now proceed to argue.

[140] Contradicting Locke's assertion that we have no positive idea of immensity, Leibniz writes that we do have a positive idea of it, an idea "which will be true provided that it is not conceived as an infinite whole, but rather as an absolute or attribute without limits" (Bk II, ch. xvi, §18; A VI 6, 159).

[141] As we saw in §2.2 above, in his early letters to Thomasius Leibniz regards space as quasi-substantial and prior to the matter which fills it, something also noted by Parmentier: "Dans ses à lettres Thomasius, en 1699, Leibniz affirmait la nature substantielle de l'espace, objet de la géométrie" (CG 341). But this is not the same as regarding *God* as a container.

(GP VII 365/LC 28; Slowik 2016, 70). As Slowik remarks, the idea that God is present by his immediate operation "seems equivalent to the *extension of power* doctrine in Leibniz's late output, a conception that situates God's actions in space but not God's being" (Slowik 2019, 259).[142] He also remarks: "since God's immediate operation correlates with the continual production of the material world, the world's spatial order thereby situates that continual act of production" (Slowik 2016, 111).

I agree that God's being present through his continual production of the world is Leibniz's mature view. Indeed, as Slowik notes, Leibniz endorses it in the *Nouveaux essais*, in his discussion of the different ways God can conceived to be somewhere ("ubiety").[143] He tells Locke that a certain Scholastic doctrine does not deserve his scorn; according to this "repletive ubiety" conception, God "fills the entire universe in a more perfect way than minds fill bodies, for he operates immediately on all created things, continually producing them, whereas finite minds cannot immediately influence or operate upon them" (A VI 6, 221). The question, though, is whether this represents a radical change of view. Slowik interprets Leibniz's statement to Clarke that "God is not present to things by situation" as meaning that he has broken away from prior conceptions of immensity as omnipresence, and space in terms of the geometry of situation, in favour of a "more transcendent" conception of immensity, "leaving only the holistic, ideal conception of space that is posterior to the experience of the material world of discrete, bodily phenomena" (Slowik 2016, 79). For his part, De Risi also sees Leibniz as divorcing his conception of immensity from his mature treatment of space in terms of situations, although he argues from a diametrically opposite direction: while noting that "in his [youth] Leibniz himself had come close to regarding absolute space as divine immensity" (De Risi 2007, 565, n.86), he sees him as having come to oppose this in his dispute with the Newtonians in favour of a mature conception of space as ideal, the order of all possible situations.[144]

[142] Strictly speaking, it is not God's "being," but his *substance*, that is not situated. Slowik characterizes Leibniz's mature subscription to the *extension of power* doctrine as follows: "Leibniz seems willing to concede the general point that a finite entity, soul, angel, or monad, can be conceived as in place through its operations, but only on the condition that there is no influx or real causal interaction, since the influx has been replaced by the mediation of God's providence in establishing the harmony between the soul and its operations" (Slowik 2016, 111).

[143] The discussion occurs in Bk. II, ch. xxiii, §21 (A VI 6, 221); Slowik mistakenly references it as occurring in Bk. II, ch. xv, §2 (Slowik 2019, 251; 2016, 69, 111, 195, 258).

[144] According to De Risi, however, it is not space that is posterior to the experience of the material world, but *immensity as the basis of space*: "It seems more likely that Leibniz, referring to God's immensity as the foundation of space, had in mind that God actualized some of these possibles by creating spatial bodies. In this sense, the immensity of God should be the ground of extension, or extended matter, much more than the ground of space, even if Leibniz seems to equivocate a bit on words and to mention ('real') space explicitly in this connection" (De Risi, private communication). As we shall see in the next section, De Risi maintains (in direct opposition to Slowik) that it is the ideal space of geometry that "dominates and determines" the real space that he equates with extended matter.

But is there such a radical change in Leibniz's metaphysics of space? I have argued that, despite some minor inconsistencies in his phrasings, after 1676 Leibniz consistently denied a conception of space as a container. Neither God himself, nor his immensity, is an infinite whole containing finite things as parts. Thus it is no change of Leibniz's position when he takes Newton to task for his claim in the *General Scholium* of the second edition of his *Principia* that God "is omnipresent not only *virtually* but also *substantially*, for virtue cannot subsist without substance" (Newton 1713, 483/1999, 941). Since this follows Newton's claim that just as a person is one and the same person "in each and every organ of his senses," so "God is one and the same God always and everywhere" (483/941), Leibniz no doubt takes him to be advocating a similar position to More's early view, that "God, as far as the human mind can grasp, is whole everywhere, and his entire essence is present in all places ôr spaces and points of space," a doctrine known as *holenmerism*.[145] Leibniz takes Clarke to be advocating the same view when the latter declares in Newton's defence that God, "being *omnipresent*, perceives all things by his immediate presence to them," employing Newton's analogy of God's "being actually present to the *things themselves*" in the same way "as the mind of man is present to all the *pictures of things* formed in this brain" (First Reply §3; GP VII 353/LC 13)—only for Clarke to reveal in his Third Reply (§3) that he does not take space to be "an eternal and infinite *Being*," but rather "a *property*, or a *consequence* of the existence of a *Being* infinite and eternal" (GP VII 368/LC 31).[146]

At any rate the Newtonians are united in their rejection of the doctrine of nullibism, the doctrine that immaterial beings are not situated in space. Faced with Leibniz's description of God as an *intelligentia supramundana* in his Second Letter (§15; GP VII 358/LC 29), Clarke perhaps suspects him of such a view (Second Reply, §10; 361–2/23). But in his Third Reply (§15; 370/34), he declares himself satisfied that in using that term Leibniz was not thereby denying that God is in the world (§15; 366/29), and accepts Leibniz's explanation in his Third Paper that he viewed God's conservation as "an actual preservation and continuation of the beings, powers, orders, dispositions, and motions of all things" (§16; 366/29). In response to this, however, Clarke reiterates his and Newton's position that God is omnipresent in the sense that he "is really present to every thing, *essentially* and *substantially*" (§12; 370/33–4). He agrees with Leibniz that "His presence manifests itself indeed by its *operation*," but adds that "it could not *operate* if it *was not*

[145] Henry More, Letter to Descartes, March 5, 1649 (AT V 305), quoted in Slowik (2019, 235). See Slowik (2019) and (2016) for comprehensive discussions of this doctrine. Newton deletes this passage comparing the way souls are present in people's organs to the way God is present in space from the second edition of the General Scholium (1727). See Newton (1999, 941).

[146] Clarke very probably had Newton's assent on this point, for in the unpublished essay *De gravitatione* lying among his manuscripts Newton describes space as "an affection of being qua being," and as "an emanative effect of the first existing being" (Newton 2004, 25).

there" (370/34). So it is not nullibism that is at issue here, but substantial presence. All three thinkers agree that God is present everywhere he operates, and indeed that he operates everywhere, so that (in regard to divine presence) Leibniz is no less inclined to nullibism than Clarke or Newton. Where they disagree is on whether this entails God's *substantial presence as an extended subject.*

So while I fully agree with Slowik that it is a mistake to ascribe a Morian holenmerism to Leibniz,[147] I cannot concur with his contention that Leibniz's views evolved away from a conception of God as omnipresent in his early writings to "a thoroughly nullibist outlook at least by the time of the *New Essays* in 1704 (but more likely earlier)."[148] Leibniz conceives immensity as divine omnipresence throughout, but although "things are extended through the benefit of the omnipresent substance" (Gerhardt 1875, 601), "there is no subject which has nothing but extension in it" (GP VI 585); indeed, already in 1676 he was insisting that "immensity does not indicate extension or parts" (A VI 3, 484/DSR 43). By "a thoroughly nullibist outlook" Slowik means that Leibniz held all spirits to have no location in space, including monads. But Leibniz did not deny the location of finite spirits in space, at least mediately, through their bodies. Although he did not directly respond to Clarke's comment that a substance "could not operate if it was not there," Leibniz did respond to a precisely similar comment by Locke concerning finite spirits in his *Nouveaux essais*. To Philalethes' (i.e. Locke's) assertion that "Spirits cannot operate, any more than bodies can, but where they are, and at different times and places; thus I cannot but attribute change of place to all finite spirits," Theophilus (i.e. Leibniz) responds: "I think that is right, space being only an order of coexistents" (Bk. II, ch. 23, §19; A VI 6, 221); adding later that "one can always attribute a sort of motion to souls, at least by reference to the bodies to which they are united, or by reference to their mode of perception" (§21; 222). This is consistent with Leibniz's numerous claims that monads are situated where they act, that is, that even if they are not themselves directly located in space any more than a mind is, they must be where their body is, since the point of view they express is from where the organs of sense are situated.[149] In fact, in a manuscript

[147] See Robert Pasnau (2011) and Slowik (2019) for a refutation. Both Pasnau's book and Slowik's (2016) contain detailed discussions of holenmerism in the seventeenth century. I am indebted to Ed Slowik for our extended discussions on these points.

[148] "While his early writings favored a Neoplatonist-leaning conception of space in which God's being would seem to be present, Leibniz's views evolved over the next few decades, reaching a thoroughly nullibist outlook at least by the time of the *New Essays* in 1704 (but more likely earlier)" (Slowik 2019, 250).

[149] In the passages quoted by Slowik (2016, 258) and others to show that monads are not in space, Leibniz maintains that "monads in themselves do not even have situation with respect to each other, that is, a real one which extends beyond the order of phenomena" (to Des Bosses, May 26, 1712; LDB 255, L 604 AG 201), and denies that "they are crowded together in a point or disseminated in space" (to Des Bosses, June 16, 1712; LDB 241–3, L 602); he does not deny that they have a situation through the bodies they are united to. See my discussion of this issue in the book's Introduction above, and the additional quotations I give there.

that has recently been transcribed Leibniz is quite explicit in *combining* hole-nmerism with the "extension of power" doctrine to support the idea of the movability of souls:

> Hence, since the soul is whole in any part, i.e. since it is in a place only through its operation, it can easily be understood that the seat of the soul changes, that is, that its primary centre is in the place where it can most aptly continue its previous operations.[150]

In sum, concerning the "extension of power," as we saw in the quotation from the *Specimen Dynamicum II*, God's operation is manifested naturally through the force in each created substance, which is a limitation of divine omnipotence. It is only in miracles that God needs to act on matter by his immediate operation without doing so through the vehicle of the substances he has created. The limitation of the extended per se is the introduction of boundaries into the unbounded, so that figures with determinate magnitudes are formed out of what was otherwise a mere potential for the formation of figures, indeterminate magnitude. The figures in space are thus not *parts* of God's omnipresence, but rather actualizations of it through the bodies whose shapes they are. The reality of space lies in this potential for there to be figures everywhere, a reality residing in God's essence. But the actual filling of space, in the way that it is filled by the things God chooses to create, depends on his will.

This conclusion has further implications for the opposition that De Risi sees between extension, on the one hand, and situation, on the other. This was a problematic originally set out by Ernst Cassirer in his *Leibniz* of 1902. He writes of Leibniz's two "starting points" for the theory of space: from one "direction" space appears as a systematic whole—"not a whole in the sense of quantity, in which it would presuppose its individual parts as constitutive"—but "as the basis of the special elements," namely the figures marked out in extension. From the other direction, however, space "emerges from the point," in such a way as "to constitute it from the conditions of thought" (Cassirer 1902, 159). Marc Parmentier follows Cassirer's lead, speaking in his "postface" (CG 341–7) of extension and situation as the "two poles" of Leibniz's geometry in 1679, and of his "temptation to recover this duality by that of space and point" (CG 343). Similarly, De Risi writes of Leibniz being "unable to reduce the notion of space (or extension) to that of situation," and therefore having to "confine himself to opposing the two pairs: point and space, situation and extension" (De Risi 2007, 167). According to all

[150] This is from the manuscript *De incredibili rerum subtilitate*, LH 4, 2, 3 Bl. 5 [after 1698?], to be included in the compilation *Leibniz on the Metaphysics of the Infinite*, by myself and Osvaldo Ottaviani, forthcoming with OUP in the BSHP series New Texts in the History of Philosophy.

three thinkers, this creates a problem for Leibniz of connecting these two poles, the problem, in De Risi's words, "of demonstrating that space is extended" (167).

Now, this complementarity between space and point in Leibniz's theory of space is a theme he sounded again and again. It is stated particularly clearly in Fragment VII of the *Characteristica geometrica* written on August 11, 1679:

> *Space* is an absolute, pure extensum, ôr that which has all magnitude and no situation.
>
> A *point* is an absolute, pure situation, ôr that which has situation absolutely and per se, and no magnitude. (CG 138).

But does this constitute an unresolved problem in Leibniz's theory of space? It only does, I submit, if, like Cassirer and De Risi, one assumes that his goal is *to deduce space itself and its continuity* from its being an intellectual condition of possible experience. From the point of view of Leibniz's deep metaphysics of space, however, space is a manifestation of divine immensity and its diffusion throughout all of created reality.[151] What needs to be explained is how it is that we finite beings form our conceptions of situation and of space, and it is here that Leibniz's seminal arguments about the conditions for our perceiving things at the same time gain their effect, culminating in his justification for the definition of a straight line as the shortest path between two points in the *Initia Rerum*.

As we have seen, divine immensity may be understood in two different ways, as an attribute of God, and as how that attribute is manifested in created things. Understood as an attribute of God, it is the extended per se, indivisible, and unique, the basis of space; but considered in respect of things, it is the *immensum* or space itself, the unbounded extensum; it is "interminate" and continuous, possessing magnitude, but having no determinate magnitude if there are no limits in it. The limits are the surfaces, lines and points bounding the extensa that are limitations of it, and all congruent figures have the same situation (or at least, an equivalent one).

Looking at this from the standpoint of modern point-set topology, where no sense can be made of the distinction of continuity from contiguity, De Risi expresses his disappointment that Leibniz could not shake off the Aristotelian notion of contiguity, "that nasty concept" (2007, 188), so that it "eventually pollutes Leibniz's metaphysical argument" (2007, xiii). "Faulted at its very root, and hardly emendable as it is, such a notion can have no use in good reasoning" (xiii). I think, however, that this neglects a distinction between what pertains to existing things and what is said about them abstractly considered. If one has two bodies that are touching, for example, the contiguous bounds (endpoints, lines, or

[151] De Risi has joked with me (private communication) that he could agree with everything I have said in these two sentences, provided I replaced "deep" with "shallow"!

surfaces) of each are specific to the part they bound; but in representing this geometrically these bounds mark merely possible sections. Beginning in the 1680s, Leibniz comes to define a continuous whole in terms of "co-integrant parts," "parts which taken together coincide with the whole," and which are such that if they have something in common, this is at least a boundary.[152] But only possible parts can have a boundary in common; actual parts have their own boundaries.[153] The same goes, obviously, for situations. As Leibniz explained to Clarke, two bodies cannot have identically the same actual relation of situation to a given set of points, but merely an equivalent one. "But the mind, not content with agreement, looks for an identity, for something that should be truly the same, and conceives it as being extrinsic to the subjects; and this I what we here call place and space" (Fifth Letter, §47; GP VII 401). He had made a related point in *De primis Geometriae Elementis* of January 1680 with respect to the boundaries of two contiguous orbs of the Aristotelian model of the solar system:

> For, on the one hand, *situation* is nothing other than that state of a thing by which it happens that it is understood to exist in a certain way [*modo*] together with [*simul*] other extensa, that is, a mode of coexisting; whereas a *boundary* is that in an extensum which has the same situation as another thing has in another extensum; for instance, the convex surface of an inner orb and the concave one of an outer orb revolving on the inner one, for it is evident that even though these two surfaces are different, since when one is moved, the other remains at rest, they do not however differ in situation, but are congruent.
>
> (CG 276; Appendix 3.3)

The surfaces of the two spheres have their own individual boundaries, but they have the same situation with respect to other coexistents; or rather (to speak more precisely, as Leibniz does to Clarke), an equivalent situation, and the equivalence class of such situations is their place. Contiguity, therefore, does not enter into the mathematical theory. Places are equivalence classes of situations congruent to a given set of coexistents, and space is a system of such places.[154]

[152] This is quoted from the *Specimen Geometriae luciferae* (GM VII 284). But see also MT N.30 (Appendix 3.4) and MT N.106, both from the 1680s.

[153] Cf. Marshall (2011, 13–14), already quoted in my discussion of physical continuity in chapter 1: "the parts of the physically continuous quantity have their own separate boundaries which merely touch one another. This is the sense in which they are not really continuous, for if they were really continuous, two points which were next to each other would not have separate boundaries."

[154] Cf. also the *Specimen Geometriae luciferae* of 1695: "For in those things that are congruent, everything is the same besides position, so that they differ only in number. And in general, whatever can be done to or said of one congruent can also be done to or said of the other, with this one exception, that those things which are applied to one differ only in number ôr position from those things which are applied to the other" (GM VII 265). I will resume this discussion of contiguity versus continuity in the context of an alleged distinction between contiguous spaces and continuous space in the next section.

To summarize the argument of this section: On De Risi's phenomenological reading, the representation of the universe in each monad's perception is a mapping of the whole universe; it is a partial isomorphism (a homomorphism) from each monad onto the universe (which is identical to God's representation of it); this is, moreover, a spatial mapping, since phenomena are constituted by their representations in the individual monads as relations of situation. Substances themselves, moreover, are determined by their phenomenal representations, so that they are, in effect, reduced to relations. Space, on this view, is a representation. It is constituted by points in situational relations, but the possibility of being situated is a universal property of representation. This constitutes a transcendental deduction of space: the possibility of the production of figures in space is grounded on the uniformity of the underlying space, whose continuity (in the sense of completeness) Leibniz spelled out in the 1690s, but whose derivation as a condition of the possibility of sensible experience was not achieved until his phenomenological turn in his last years.

Against this, I argued that such an account neglects the role of the appetition of substances. Each substance has an inbuilt force or appetition, as well as perception. This relates crucially to space, in that it is the appetitions of these coexisting substances that give rise to the motions and resistances (antitypy) that account for matter's extension, and also the boundaries of the figures in it. The relations of situation among these figures, regarded as possible, constitute space. The way this connects with immensity is as follows. Divine immensity, the *immensum*, is an unbounded extensum. It is continuous in the sense that it can be arbitrarily divided into parts, but this requires the introduction of boundaries or limitations. The possible divisions issue in the various possible boundaries, which mark out the possible figures, and thus the situations. But there is no actual situation before the introduction of boundaries, which are the results of motions (more accurately, endeavours), and this requires the introduction of the appetitions of substances into what is otherwise an ideal order. The actual extension of matter is thus produced by the substances contained in it, each point of matter being differentiated by its differing motion (endeavour), resulting in a differing situation corresponding to its point of view or representation of the rest of the universe.

As we saw in the discussion of time in section 5 of chapter 1, the points of matter so constituted are such that there are always more between any two: they are dense. But the gaps between them are vanishingly small, so that the difference between this physical continuity and mathematical continuity is smaller than any measure that can be assigned. The mathematical continuum is an ideal order embracing all possible points of division, or situated points. Since (in the natural, i.e. not miraculous, course of things) God's presence is manifested through the actions of created substances, this presence is diffused everywhere in matter, constituting its extension as the actual relations of situation of existing things. The extension of space is this same extension regarded in terms of possibility: it is the order of all possible situations. Thus the diffusion of the relations of situation, relations of compresence manifested in terms of the possibility of co-perception, is the geometrical expression of the diffusion of divine presence (immensity). Thus, granting De Risi's points about Leibniz's improvements in his mathematical understanding of continuity, I see Leibniz as having maintained the doctrine of divine immensity as the basis of space throughout his mature philosophy (1676–1716); and, contra Slowik, not as instituting a change of his metaphysics of space from a container view in his middle period to a more transcendent conception in his maturity.[155]

At this point I want to turn to a further question that arises out of this discussion of the metaphysics of space. For, on either my interpretation or De Risi's, there is a distinction between the actual relations of situation among phenomenal bodies, and relations of situation conceived as possible. Does this constitute a distinction between two kinds of space—space as a given order of relations among bodies (actual space or phenomenal space), and space as the abstract order of relations that constitutes mathematical space? If so, might the two spaces have different structures? This topic is sufficiently involved to deserve treatment in a separate section.

2.5 Spaces: Real, Phenomenal, and Abstract

The question before us is whether Leibniz allowed *spaces*—whether real, physical, or phenomenal—in distinction from the space whose properties he has analysed with his *analysis situs*. As discussed in §2.2, in the spring of 1676 Leibniz argued that space is a mere aggregate of places, and that it is continually changing through time: "universal space is an entity by aggregation, and is continuously variable. It is a composite of spaces empty and full, like a net, and this net continually receives different forms" (A VI 3, 519/LLC 121).[156] He provides another visual image for this physical continuum[157] in the *Pacidius* later that year, that of the tunic:

> Accordingly the division of the continuum must not be considered to be like the division of sand into grains, but like that of a sheet of paper or tunic into folds; and so although there occur infinitely many folds, some smaller than others, a body is never thereby dissolved into points or minima. On the contrary, every

[155] "God's immensity remains as the foundation of space, of course, but in a more transcendent manner, so that only God's operations can be straightforwardly given a spatiotemporal location..." (Slowik 2016, 79).

[156] Leibniz's allowing here that some of the component spaces could be empty is noteworthy, given his commitment to a physical plenum in this period. That there *could* be vacua remains his position into his maturity. See Debuiche and Rabouin (2019) for discussion.

[157] We already treated the issue of the physical continuum in connection with time in §1.5; we will come back to it in the context of space below.

> liquid has some tenacity, so that although it is torn into parts, not all the parts of the parts are so torn in their turn; instead they merely take shape for some time, and are transformed; and yet in this way there is no dissolution all the way down into points, even though every point is distinguished from every other by its motion. It just as if we suppose a tunic to be scored with folds multiplied to infinity in such a way that there is no fold so small that it is not subdivided by a new fold; and yet in this way no point in the tunic will be assignable without its being moved differently from neighbouring ones, yet not torn apart from them. (A VI 3, 555/LLC 185)

Although Leibniz is here talking about a continuous body, in the dialogue (as elsewhere) he freely extends analogous considerations to space (as well as to time and change).

What this suggests, then, is the idea that each partition of space is a concrete instance of this continuously variable universal space. At each instant its parts are filled with matter, these parts being determined by the differing endeavours of the elastic matter contained in it. This tallies with what Leibniz told Clarke in his Fifth Letter:

> §27. The parts of time and place considered in themselves are ideal things, and therefore they resemble one another perfectly, like two abstract units. But it is not so with two concrete ones, or with two real times [*temps effectifs*], or two spaces filled up, that is to say, truly actual. (GP VII 395/L 700)
>
> §67. The parts of space are not determined and distinguished except by the things that are in it, and the diversity of the things in space determines God to act differently on different parts of space. But space taken without the things has nothing determining it, and is not even anything actual. (GP VII 407/L 708)
>
> §106.... If there were no created things, there would be neither time nor place, and consequently no actual space. The immensity of God is independent of space, as his eternity is independent of time. These attributes signify only, in respect of these two orders of things, that God would be present and coexistent with all the things that would exist. (GP VII 415/L 714)

Valèrie Debuiche and David Rabouin (in an interesting paper centrally concerned with the issue of the unity of space) note how these and similar texts have been interpreted by many scholars as supporting a distinction between an *abstract* space, a mathematical framework that is the same in all possible worlds, and an *actual* (or "real" or *concrete*) space that depends on the particular configuration of bodies in it for its properties, and so could have different structures in different possible worlds (2019, 172–5). Such concrete spaces could, it is alleged, be discrete, or perhaps even non-Euclidean, depending on the relations of the bodies in them.

On the use of the above texts to support such claims, Debuiche and Rabouin remark:

> Note, however, that these texts do not state that space "with things" in it is "real," and that this reality has to be related to material bodies. They rather affirm that there is no proper actuality or reality of space in and of itself (against Newton's reality of absolute space), and that space can only be said to be "actual" when it is considered as that in which actual things are. In this sense, talking about the actuality of space does not amount to claiming either that space is actual in the sense of being a material or physical entity or that space depends totally upon a certain disposition of actual bodies and has to change in some way with every change in the said disposition. (Debuiche and Rabouin 2019, 175, n.10)

This seems perfectly correct, especially as concerns his mature philosophy. But we still need to make sense of Leibniz's earlier views from 1676, where space is described in terms more nearly fitting for phenomena, that is, as an *ens per aggregationem*, and as a successive and changeable being, one that is differently partitioned at different times, and which undergoes changes from one partition to another.

One possible interpretation is this: in his earlier phase, Leibniz held that space is a well-founded phenomenon; but then in his mature work, he abandoned this construal for one in which he "has decided that space should join numbers, relations, lines, and points in the ideal realm." The latter quotation is from an influential article by Glenn Hartz and Jan Cover (1988, 497). They criticize the "two-level" view ascribed to Leibniz by the likes of Nicholas Rescher, Benson Mates, and John Earman, where there are monads or genuine substances, and then there is everything that results from them, the phenomena. To Hartz and Cover it is obvious that nothing like this two-level view "survives in the mature Leibniz" (495). Instead, there "Leibniz is operating with a *three-tiered* metaphysic," which they describe as follows:

> Monads, or *substantiae*, are at the ground floor metaphysical level. Immediately above it is the phenomenal level, where bodies, the *quasi-substantiae* (GP II 263/L 534) or *entia semimentalia* are stationed. Finally, all the way at the top, at the ideal level, are the *res mentales* (GP II 268) or *entia rationis* (GP II 189/A VI 6, 226–7). (Hartz and Cover 1988, 503–4)[158]

Leibniz does indeed describe bodies as "quasi-substances" (phenomena that are aggregates of substances), whose unity is merely mental, and space as ideal and a

[158] In quoting this passage I have taken the liberty of changing their citation notations to mine, and correcting their '*res mentalis*' to the plural form, *res mentales*.

being of reason. Whether this is appropriately described in terms of "ontic levels" is another question,[159] and one I tackled in the Introduction above. But leaving this issue aside for now, the main trouble with this analysis is that the "early period" where Leibniz is supposed to have included space and time as well-founded phenomena is designated by Hartz and Cover as from 1676 to 1688, and the mature period as from 1711 to 1716, with a "transition period" of 1696 to 1709 in between. Theirs is, of course, an old article, and now we have a much better selection of texts on which to base an assessment. They did not have the texts showing how Leibniz's casting of the continuum as ideal began already in 1676, not in 1695–96 as they suppose, and that the contrast between space and time as mathematical entities, and extension and duration as attributes of concrete things that are extended and endure, is also in place in 1676 (as we saw in section §2.2), and not an innovation first appearing in Leibniz's dialogical critique of Malebranche (*Conversation of Philarète and Ariste*), which they date as 1711. So their analysis is of no help in clarifying our problem, which is this: did Leibniz continue to entertain a conception of space as a continuously varying aggregate of its parts beyond the spring of 1676? And, relatedly, did he (ever) have a conception of a concrete or phenomenal space, in distinction from abstract space?

Two sets of considerations have contributed to widespread claims in the literature that Leibniz recognized such concrete or phenomenal spaces distinct from abstract, mathematical space. One of these derives from the *body-relativist* conception of Leibniz's relationism critiqued above in section §2.1. According to this view, if space consists in relations among actual bodies, then concrete or phenomenal space will just be any such system of relations among actual bodies, while abstract space will be such a system of relations conceived as possible. The other source for such a conception, and the main target for Debuiche and Rabouin's criticisms in their paper, is the notion that Leibniz advocated a *plurality of worlds*, each with its own space determined by the particular configuration of bodies contained in it. Some of the textual basis for this second line of interpretation lies in the papers Leibniz wrote in his "Paris Spring," so a consideration of these will be directly relevant to the question of whether there was any change in Leibniz's views on space after that period.

In the long fragment *De veritatibus, de mente, de Deo, de universo* from 15 April 1676, Leibniz engages in some deep reflections on the relationship between sense, existence and space, meditations which seem to be prompted by Descartes' own *Meditations*. He acknowledges that we can be immediately certain of our own existence. But when it comes to what exists apart from ourselves, we rely not only on sensation, but also on the congruence or coherence of our sensations. "And existence consists in this [congruence], in sensation conserving certain laws, for

[159] I will present some of my objections to ontic levels above in the course of my discussions of Leibniz on motion in §3.4. See also my discussion of concrete spaces and times in §1.4 above.

otherwise everything would be like dreams" (A VI 3, 511/DSR 63). Also, since different people sense the same thing consistently, Leibniz argues, what is sensed must have a common cause. Accordingly, we might say "that *space* is what brings it about that several perceptions cohere with one another at the same time" (511/65). He continues:

> These things being so, it does not follow from this that there is anything other than sense[160] and the cause of this sense and its consistency. Hence it follows, moreover, that there could be infinitely many other spaces and, in a word, other worlds, such that between these and ours there would be no distance—if, that is, there were certain minds to which other things appeared that were not at all congruent with ours.... Whoever asks whether there could be another world, another space, is simply asking whether there are other minds in no communication with ours. (A VI 3, 511–12/DSR 65)

Nicholas Rescher translates the second sentence here as "[T]here could exist an infinity of other spaces and worlds entirely different [from ours]. They would have no distance from us [nor other special relations to us] if the minds inhabiting them had sensations not related to ours" (Rescher [1996] 2003, 14, [1977] 2003, 95)[161]—thus converting the consequent of a hypothetical statement into a categorical assertion of the plurality of spaces and worlds. As Debuiche and Rabouin observe (2020, 176), this notion of plural spaces coheres with Rescher's interpretation of the mathematical space of Leibniz's *analysis situs* as a forerunner of topological space, a possible order of coexistences that could be instantiated by different actual (metrical) spaces in different worlds. Yvon Belaval, Ezio Vailati, and also Javier Echeverría, they note, have interpreted Leibniz's *analysis situs* similarly, enabling claims that the geometry of the world is, for Leibniz, contingent. In support of this Rescher writes,

> For while space is—everywhere—the "order of coexistence," it turns out that this order will be something different in different worlds—and necessarily so, since different worlds contain different and incompatible substances, and substances *internalize* such differences. (A difference in substances entails a difference in their relations which entails a difference in ordering relations).... Thus in no possible worlds can space be separated from the substances that 'fill' it.
> (Rescher [1996] 2003, 14–15, [1977] 2003, 95)

[160] The meaning of this is presumably that "it follows from this that there is nothing other than sense and the cause of this sense..."—rather as we say "I don't think that is accurate," intending it to mean "I think that is not accurate."

[161] Rescher's (2003) is a reproduction of papers published earlier with no attempt to eliminate redundancies, such as passages (often amounting to several pages) that are repeated verbatim from one paper to another. I quote the earlier papers with the pagination of (2003).

Rescher contends that this position he discerns Leibniz as taking during his Paris period is one "he continued to hold throughout his life" (2003, 14, 95). In early 1676, however, as Osvaldo Ottaviani has recently demonstrated, Leibniz did not yet have a conception of possible worlds. He had a conception of unactualized possibles, but he had not yet conceived of them as constituting a whole, a world. Prior to April 1676, "Leibniz recognizes one and only one 'series of things': the actual world, spatially, temporally, and causally unified" (Ottaviani 2021, 135). This is the actual world, every member of which "is connected to every other member by means of spatiotemporal and causal relations (what Leibniz sometimes calls 'relations of connection')" (135–6). The "worlds" that Leibniz refers to in this period are putative worlds existing at other times and places in the cosmos. But we can see the beginnings of a conception of possible worlds as systematic wholes in the continuation of his speculations in the *De veritatibus, de mente, de Deo, de universo* about the relationship between space and existence. For, he claims, there could be many spaces, each with its own law of nature, and asks whether "the immensity of God would be the same in all of them," and concludes that it would (A VI 3, 512–13/DSR 67). But here this line of thought breaks off. When he returns to it in Holland in December, he rejects the idea of such other worlds. One reason is that there would not thereby be more existents, since there are already infinitely many in this world and in any of its parts; but another reason is precisely that existence in space involves coexistence, things existing together at the same time:

> There is no need for the number of things to be increased by a plurality of worlds, for there is no number of things that is not in this one world, and indeed in any part of it.
>
> To introduce another genus of existents, and as it were another world which is also infinite, is to abuse the name of existence; for it cannot be said whether those things exist now or not. But existence, as it is conceived by us, involves some determinate time, that is to say, we say that a thing exists only when it can be said of it at some definite moment of time that "that thing now exists."
>
> (A VI 3, 581/DSR 103–5)

Leibniz repeats this in a second fragment bearing the same date, December 12, 1676, this time making clear that things existing at the same time have to bear a relation of distance:

> There is only one kind [*genus*] of world, that is to say, there are no beings besides bodies and minds, i.e. the kinds [*qualia*] of things we sense,[162] nor are there any

[162] It is possible that Leibniz has Spinoza in mind with this comment, since it is written at the time he was in conversation with the philosopher of The Hague, and he was aware of Spinoza's claim that "there are infinitely many other affirmative [divine] attributes besides thought and extension" (A VI 3, 385/LLC 43).

bodies but those that are at a certain distance from us. For if there were any such, it could not be said whether they are or are not now; which is contrary to our first principle [that anything either is or is not].

Two simultaneous existents, that is, must have a distance from one another, and therefore exist in the same space.[163] But in 1676 this space is the one unique space, and all possible existents must have a distance from all other existents in the one actual world in order to be possible.

It is not until Leibniz articulates his mature notion of compossibility in 1685–86 (as described in section §1.3 above) that he attains his mature notion of a possible world. Each such possible world would have its own law of general order, from which would follow the individual laws of each substance according to which its states would unfold, so that different contingent laws of physics could hold in other possible worlds. But the states of each substance would still come to be in an order determined by their individual laws so that equivalence classes of compatible and thus simultaneous states would be determinable, and at any time so determined, the body of each substance in that world would be situated relative to the others in that world, and therefore at a determinate distance from each other possible coexistent in its world. But this does not mean that these possible worlds have their own spaces and their own times, that there is a different structure of space or time in each. As I said in the case of time in §1.3, the order of possible successives is not an all-embracing container time: "we need only the idea that the time between any two states in the actual world could have been, or could be, instantiated by different histories." The same applies to space: it is the order of possible situations, and the particular situations can be instantiated by various arrangements of bodies without this implying that "the order of situations conceived as possible" is in any way different. Space, as I already argued in §2.1 above, is for Leibniz the order of possible situations, not an order or complex of situations of existing bodies.

With this in mind, let us return to Rescher's way of distinguishing Leibniz's "actual space" from the particular spaces that might instantiate it. He claims that

> It seems plausible to suppose that Leibniz's own project of *analysis situs* ('topology' as we nowadays call it) actually represents an attempt to devise a theory of spatial relationships that does not involve the whole range of specific commitments of a full-blown Euclidean geometry. Leibniz would surely have been neither surprised nor dismayed at the discovery of non-Euclidean geometries,

[163] Again, see Ottaviani (2021). This point is missed by Michael Futch in his otherwise perceptive discussion of Leibniz on the plurality of worlds. See "3.3.1. Multiple Worlds in Leibniz's Early Philosophy" in his (2008).

> and he would have no difficulty in assimilating such a diversity of spatial structures to his own theory of space.
>
> (Rescher [1996] 2003, 20, [1977] 2003, 102)

As we have seen above, Leibniz's *analysis situs* is *not* topology: it is a metric theory, and one designed precisely to ground the properties of the figures in Euclidean geometry in a theory of space. In fact, under De Risi's interpretation, in construing properties such as three-dimensionality and continuity as *properties of space*, it has a claim to be called the *first theory of Euclidean space*. Thus concerning Rescher's claims that space in Leibniz's metaphysics could be construed as contingent, that it could have been non-Euclidean "or who knows what else," De Risi declares:

> There is nothing more fallacious than such a supposition. Leibniz's definition of space as the order of all *possible* situations necessarily includes, in fact, all the situational configurations determined by the set of monads of the non-existing worlds. Absolute space is one and the same for all possible worlds. What changes is only the specific situational actualization of the order of possibilities. And even that, we have seen, merely consists in a different system of boundaries, and by no means in the determination of the curvature or dimensions of the ambient space, or anything else. (De Risi 2007, 566)

But if what changes in space is only the different systems of boundaries of the bodies it contains from one time to another, then, after all, we are not far away from the conception of space that Leibniz had already articulated in 1676. There, it is true, he had described space itself as changing, his "universal space" being a mere aggregate of places, one that it is continually changing through time.[164] Thus in previous publications I had called the space that is "continuously changing and becoming something different" "phenomenal space," as distinct from unchanging abstract space, and attributed it to Leibniz in his maturity.[165] Similarly, De Risi characterizes the difference between (1) space as the set of all possible situations, and (2) the specific actualizations of this from one time to another, as being the difference between (1) *ideal* or *absolute* space, and (2) *real* or *phenomenal* spaces.[166]

[164] Here I should note that Debuiche and Rabouin (2019, 177–8) ascribe to me what is actually *Leibniz's* (1676) characterization of universal space, as something continuously variable and taking different forms at different times.

[165] See Arthur (1987), (1994a), (2012/13).

[166] See his section "Real Space and Corporeal Substances" (D Risi 2007, 486 ff.), where he sets out "further to determine phenomenal space," that is, "to see what further determinations space must assume to become such a variegated world, and how such determinations might derive from the general concept of existence that we have analyzed" (486); he argues that *real space* (which he equates with matter) is "a space to which the determination of a system of boundaries has been added" (512), but that "the ideal dominates and determines the real. Thus the concept of possibility that determines the geometrical space logically precedes the notions of compossibility and existence necessary for constructing the real" (565–6).

Concerning Rescher et al., he writes "Hypnotized by the fact that for Leibniz *real* space depends on the actual monadic configuration, and oblivious that real space is not the ideal one, they have fantasized on how the spatial structure of Leibniz's metaphysics must have been contingent..." (2007, 566).

Debuiche and Rabouin have objected forcefully to this way of describing Leibniz's philosophy of space in terms of two different spaces, ideal and real, just because it lends itself to the idea that phenomenal (real, physical) space might have different properties from the ideal space, space proper. They write:

> Although abstract space corresponds to various (equivalent) concrete dispositions, and although Leibniz sometimes goes so far as to speak of an "actual" space to designate a particular instantiation of the system of places, it does not follow, as so many commentators have assumed, that there are two distinct spaces.
> (Debuiche and Rabouin 2019, 182)

So let's retrench, and see what evidence there is for ascribing a phenomenal (or real) space to Leibniz after 1676.

It will be useful to begin with an examination of the textual evidence. As we saw in §1.4, Jan Cover (1997, 303) dismisses the idea that there is any such evidence that the mature Leibniz advocated any view of time or space as phenomenal.[167] Actually, there are more instances of Leibniz's describing space and time as phenomena than Cover was aware of, extending throughout his career, but I do not think that they exhibit a *theory* of phenomenal space. Here are a number of relevant passages, with what is known about their dates of composition:

> But whether this space is some thing distinct from matter, or whether it is only a constant appearance or phenomenon, is not relevant here. (*Characteristica Geometrica*; CG 150–2; GM V 144)—[August 10, 1679]
>
> Similarly, mathematical things, such as space, time, a sphere, an hour, are only phenomena, which are conceived by us on the model of substances. (*Divisio terminorum*... AVI 4, 559–60/LLC 265)—[summer 1683–early 1685]
>
> It is the animate substance to which this matter belongs that is truly a being, and the matter taken as mass in itself is nothing but a pure phenomenon or well founded appearance, as also are space and time. (to Arnauld, GP II 118–19; A II 2, 249/LAV 253)—[October 19, 1687]

[167] This view has been surprisingly influential. With regard to time, for instance, in a review of a book by Adrian Niţă, Michael Futch writes: "As is well documented, through the middle period Leibniz sometimes characterizes time as a well-founded phenomenon, whereas in his later writings he usually identifies it with an ideal being of reason, or an imaginary being" (Futch 2011, 172–3). But Futch sees Leibniz's "characterization of time as an ideal continuum *and* as a set of relations that supervene on or result from changes" as "apparently conflicting" (173).

> Absolute space is no more a thing than time is, even though it is pleasing to the imagination; indeed, it can be demonstrated that such entities are not things, but merely relations of the mind trying to refer everything to intelligible hypotheses—that is, to uniform motions and immobile places, and to values deduced on this basis. (*Motum non esse absolutum*, A VI 4, 1638/LLC 333)—[1688–89?]

> Space, time, extension and motion are not things, but modes of consideration having a foundation. (*Primae veritates*...; A VI 4, 1648/L 270)—(deleted) [1689]

> Therefore it must be said that time is not composed of a definite number of instants. From these and similar considerations it is concluded that space, body and motion are not things but phenomena, although well-founded ones. (LH 4, 3, 5e Bl. 24)—[after 1695])

> Extension, like time and bulk, and the motion that consists of variations of these, disappears into the phenomena... (to De Volder, GP II 282/LDV 338–9)—(unsent) [January 19, 1706]

> And so the reality of bodies, of space, of motion, and of time, seems to consist in the fact that they are phenomena of God, that is, the object of his *scientia visionis*. (to Des Bosses, LDB 231–3)—[February 15, 1712]

I think it is evident that these various remarks made over the course of three decades, especially when viewed in the contexts from which I have abstracted them, do not constitute one well-defined theory of time and space as phenomena. If there is a common theme, it would appear to be that although we conceive or imagine such mathematical entities on the model of substances (*instar substantiarum*), they are not substances. Space and time are phenomena in the sense that they are pictured in the imagination, a picture that we then mistake as representing a substance (as, for instance, the Cartesians do when they take extension for a substance), but they are well founded in that they possess what reality they have through God.

Of course, none of this refutes or even addresses De Risi's reading, since the view he ascribes to Leibniz is supposed to be a production of his final years (or even months), when his "*Geometria situs* and metaphysics of space have finally merged" (2007, 383). So let us turn now to an examination of De Risi's account of phenomenal space, to which he devotes a substantial subsection of his book (2007, 382–428). He also equates this phenomenal or actual space with "real space," which he contrasts with "ideal space" in the concluding part of the book (486–577).

On De Risi's reading, phenomenal space is an "order of expressive relations" determined by "a mix of confusion and distinctness in the representation of things-in-themselves (an always residually non-expressive expressivity)," which is paralleled by "the determination of situation that Leibniz has provided in his geometrical writings" (382). That is, according to his analysis "two phenomena that differ in quality express a noumenal difference distinctly, while two phenomena that differ only in quantity express a difference between things-in-themselves only confusedly" (383). When two phenomena are identical in both quality and quantity, however, they cannot express any difference in the noumena, and so must necessarily be part of an equivalence class determined by equality and similarity together, i.e. congruence. "In other words," he argues,

> phenomenal space meant as the order of expressive relations must be regarded as a quotient space relative to the equivalence determined by equality and similarity together, that is, relative to congruence. Thus, situational relations, metaphysically assumed as relations that simply express deeper intermonadic relationships, must be geometrically interpreted as relations of equivalence relative to congruence. (2007, 383)

As we saw in the previous section, De Risi argues that it is because in 1679 the points of Leibniz's *analysis situs* "lack any relation of expressivity," that they "can only be distinguished through their situational relation, which is *a purely extrinsic* one" (412). In real or phenomenal space, by contrast, there is a more delimited set of relations expressed. Here, "in order for a real expressivity of space to be possible, the addition of the phenomenal distinction of each monad's organic body is necessary," and this involves the specification of a total system of spatial boundaries (492–3). These boundaries are those of each organic body and are produced, as I have argued above, by the activity of the monads that each organic body contains, and are represented more or less distinctly in the perceptions of any perceiving monad. According to De Risi's phenomenological interpretation, however, the attribution of an organic body to a substance is achieved through its relations of dominance and subordination with respect to all the other substances in the world.[168] Although he does not deny that corporeal substances are "independent of any perceptual act," nevertheless the only purpose they serve is that "of articulating the dominant monad's simplicity into a plurality of relations, which can be expressed by a perceptual act" (492).[169] Consequently, whereas in abstract or mathematical space what is being diffused is only the formal property of *situs* itself, the extension of real space depends on the particular set of

[168] "All the situational relations between the various parts of its organic body and the body itself as a whole express the relations between the dominant monad (to which we attribute the body) and the dominated monads" (De Risi 2007, 491).

[169] As noted in §2.4, De Risi's conception of corporeal substance as determined by representations to consciousness is indebted to Cassirer's: "Corporeal substance denotes the substance 'insofar as it is endowed with a determinate organic body, according to which it represents and desires' (E 678, n.5). This by no means records a new and alien moment in the monad, which seemingly treads close to the function of consciousness, but concerns only a determination in the content of consciousness itself" (Cassirer 1902, 408).

boundaries in existence, and thus on "the existence of some monads rather than other ones" (499). This means that "the extension of real space is a merely contingent property" (499).[170]

Debuiche and Rabouin, as already mentioned, object to such an attribution to Leibniz of two distinct spaces, "actual" or "real" versus "ideal" or "mathematical." The distinction should rather be understood in the same way as that between an abstract number and the thing numbered. As they observe, this is exactly what Leibniz says in the *Nouveaux essais*:

> Although it may be true that in conceiving body you conceive something more than space, it does not follow from this that there are two extensions [*etendues*], that of space and that of body. For it is like when, in conceiving several things at once, you conceive something more than number, namely the things numbered; and yet there are not two multiplicities, one abstract, that of number, the other concrete, that of enumerated things. In the same way, we can say, you should not imagine two extensions, the one abstract, of space, the other concrete, of body, since the concrete is such as it is only through the abstract. (II, iv 5; A VI 6, 127)[171]

Phenomenal extension, Leibniz insists, is the extension *of something*. He explains it in terms of a diffusion of το δυναμικὸν, the force of acting and being acted upon that constitutes substance.[172] Each substance resists a change in the boundaries of its organic body, this resistance being the *antitypia* which, Leibniz always believed, is what distinguishes matter from mere extension. Since there are monads in as small a portion of matter as you wish to consider, this resistance to compression is diffused everywhere in matter, and it constitutes what it means for each body to be extended. In Leibniz's mature philosophy—that is, once he has developed his dynamics in the 1680s—it is this force stemming from appetition that produces existing substances, namely when it is added to what would otherwise simply be the real condition for the possibility of the presence of created substances that is guaranteed by divine immensity.

[170] This may seem to contradict his objections to Rescher et al., presented above. But what is contingent, according to De Risi, is actual or phenomenal space as a system of boundaries, not the three-dimensional, continuous, metrical, abstract space defined by *analysis situs* that is presupposed by such a space. As he insists, "There is just one absolute space, and thus geometry is one as well" (De Risi, 2016, 49).

[171] I should note that De Risi quotes the same passage (2017, 562) to make a similar point about there being only a "*modal* difference" or "*ideal* distinction" between the two extensions; except that he conceives it as a distinction between ideal space and "the actualization of ideal absolute space into certain configurations of the world matter (called *real* space)," so that for him there is "a *real* identity between absolute space and matter" (561–2), and only an ideal difference between them.

[172] See, for instance, Leibniz's letter to De Volder of June 30, 1704 (LDV 305), and also "On Body and Force: Against the Cartesians" of May 1702 (GP IV 394/AG 251); see Arthur (2018, 229–37) for a non-phenomenalist reading of Leibniz's account of extension.

De Risi claims that matter "is not at all a new concept in our discussion, but just a different name for actual space, i.e. a space to which a system of boundaries has been added" (De Risi 2007, 512). According to his reading, "the only phenomenal difference" between this actual space and the ideal one is that "corporeal substances impede one another by generating some boundaries in space" (515). This impedance, manifested as a degree of impenetrability, "is what Leibniz calls the *materia prima* of each substance."[173] But as Leibniz tells Des Bosses, although it is true that "the places of monads may be designated through the modifications ôr boundaries of the parts of space," "mass and its diffusion result from monads, but space does not" (to Des Bosses, July 31, 1709/LDB 141).

In fact, we need to make a distinction here. For it is mass as *secondary matter* that is actually divided into parts by the varying degrees of impenetrability, resulting in the boundaries of the actual phenomenal bodies. Primary matter (or *materia nuda*), on the other hand, is, like space, something incomplete, abstract. But it is still distinguished from space by the fact that the property that is diffused is antitypy as such, and not situation as such, that is, an equivalence of magnitude and figure. In a passage from 1710 quoted by De Risi (516), Leibniz writes:

> Matter considered in itself, that is, bare matter, is constituted by antitypy and extension. I call antitypy that attribute through which matter is in space. Extension is continuation through space, ôr continuous diffusion through place. And so, as long as antitypy is continually diffused ôr extended through place and nothing else is supposed, there arises matter in itself, ôr bare matter. (GP VII 328)[174]

Modification or variety of antitypy, Leibniz explains, is variety of place, while variety of extension is variety of magnitude and figure. "From this it follows that matter is something merely passive, since its attributes and their varieties involve no action" (328). But "if we add in addition actual variation, that is, the principle of motion itself, there results something beyond bare matter" (328). Then, as Leibniz explained in his comments on Johann Georg Wachter in 1706, we obtain really existing "secondary matter or extended mass, something that is not at all a homogeneous body." "But that which we conceive as homogeneous and call primary matter is something incomplete, since it is merely potential" (Leibniz 2002, 6/AG 274). It consists in the indefinite repetition and continuous diffusion of antitypy, conceived as the homogeneous potential to be divided in the

[173] De Risi also writes of Leibniz being "at ease in speaking of a *real* identity between absolute space and matter," despite an ideal difference between them (2007, 562).

[174] This appears to be closely related in content and terminology to what Leibniz wrote to Rudolph Wagner (4 June, 1710; GP VII 528–9).

particular ways that it can be divided by the appetitions of the actual substances it could contain:

> Thus extension, or if you prefer, primary matter, is nothing but a certain indefinite repetition of things insofar as they are similar to one another, ôr indiscernible. But just as number presupposes things numbered, so extension presupposes things that are repeated, which contain accidents proper to them [*propria*] in addition to those they have in common. These proper accidents produce actual limits in size and shape, that beforehand were only possible.
>
> (Leibniz 2002, 6/AG 274)

Phenomenal extension, therefore, is the extension of matter. It is the result of the diffusion of antitypy through any space.[175] This means that the space is also extended, but this does not mean, as we have just seen, that there are two different extensions, the one concrete, the other abstract. In fact, both primary matter and space are abstract, incomplete, but their extension is the same. The system of boundaries created in what would otherwise be a merely abstract and homogeneous primary matter are accidents of the organic bodies existing at any time. Actual situations, that is, require actual phenomena taking place in space and time, and this involves the successive actualization of different systems of boundaries of bodies in space. But this is *phenomena in space*, not phenomenal space. This, I believe, resolves the issue of the plurality of spaces in agreement with Debuiche and Rabouin: "Space is not actualized in itself: it is a mathematical notion belonging to the realm of possible things. But it can be said to be 'actualized' when considered along with the actually co-existing things that fill it" (2019, 186).

There is, however, one important issue that still needs to be resolved. I had written previously (Arthur 2013c) that the difference between the phenomenal space that consists in a system of boundaries and the abstract one is precisely that the bodies whose boundaries delineate the parts of phenomenal space are merely contiguous to one another, as well as ephemeral, lasting only an instant; whereas there are no such boundaries in space itself, which is truly continuous. According to Leibniz's distinction between continuity and contiguity, then, it might seem that we have a difference of spaces: the ideal one is a continuum, and the phenomenal one is merely a contiguum.

This issue already came up in our discussion of the continuity of time in §1.5 above. For successive states or perceptions necessarily possess their own end-points, since they are contradictory to one another: the last moment of living

[175] De Risi writes of imagining each phenomenal body as expanding in space until it meets another body that resists its further diffusion, saying that this "may be described as the *diffusio* of each body's *antitypia*, according to the same model of the diffusion of situation that we saw Leibniz suggested in the transcendental constitution of ideal space" (515).

cannot be identical to the first moment of being dead. The issue is discussed by De Risi in one of his many extraordinary scholarly footnotes, this one n. 66 in chapter 2 (2007, 194–6). There, to show "how much damage the idea of contiguity can make," he cites or quotes several passages where Leibniz characterizes the parts as "indeterminate." Thus in the *De primis Geometriae Elementis* of January 1680, Leibniz writes:

> There are, in sum, two things to be considered in geometry, extension and situation. On the one hand, there is the *extensum*, a continuum in which parts that are simultaneous can be assigned; on the other, there is the *continuum*, that in which the parts are indefinite, ôr in which parts are merely designated by the mind. But when parts are designated in an extensum there already arise more extensa, possessing boundaries and situation.[176]

Likewise in a table of definitions from 1702–4 published by Louis Couturat, he writes:

> The *continuum* is a whole whose parts are outside one another [*partes extra partes*], and indeterminate: *outside one another*, that is, separately perceivable, so that they are distinguished from a gradual whole, whose parts interpenetrate, as when the <intension>[177] of qualities is estimated; whereas the *parts* of the continuum are *indeterminate*, since none are already assigned, but they can be assigned as you please, so that they are distinguished from the contiguous.
>
> (C 438–9)[178]

These passages, De Risi comments, "however vaguely phrased," show "how Leibniz availed himself of the most general concept of indetermination of parts to make a distinction between potential and actual partitions, which will prove absolutely inappropriate when you pass from the Aristotelian concept of extension to the concept of extension as a structured set of points" (2007, 195). The reason it is inappropriate, according to De Risi, is that two contiguous extensa do not share "something in common which is not itself a part" of the whole they comprise—that is, they do not have an extremum in common, since the Aristotelian definition requires that the contiguous extrema be "together, but not one." Yet their being one is a requirement of the definition of continuity that we saw him give in the *Specimen Geometriae luciferae* of 1695(?) in §2.3

[176] See my translation of this in Appendix 3.2.

[177] —reading *intensio* for Couturat's *intentio*.

[178] The subsection in which this passage occurs is titled "*Concreta Mathematica*." Leibniz explains: "A mathematical concretum is an extensum without resistance," an extensum being "a continuum with situation, ôr with an order of coexistents" (C 438); "Mathematical concreta are intelligible, and have no sensible qualities" (441).

above. In fact, it is one of two requirements for continuity, as stated in the *In Euclidis πρῶτα* of 1712:

> Moreover, two things are required for a continuum: first, that any two of its parts that together equal the whole have something in common which is precisely not a part; and second, that in the continuum there are parts outside one another [*partes extra partes*], as is commonly said, that is, that two of its parts can be assumed (but not ones that together equal the whole) which have nothing in common, not even a minimum. (GM V 184)

If the first requirement is not met, De Risi objects, the contiguous parts of extension cannot compose into a continuous extension, but must simply presuppose it; so Leibniz's earlier account depends on a "'substantialistic' concept of space" (De Risi 2007, 193), where the actual figures are formed by the introduction of a system of boundaries into a presupposed extension, unconnected with the mature theory of space as an order of situations.[179] The second requirement, De Risi claims, "essentially derives from the metaphysical concept of diffusion, which indeed calls for the reciprocal exteriority of the parts that are diffused" (191).[180] Yet this second condition, that of *partes extra partes*, is stated in the Table of Definitions (C 438) from ten or so years earlier than *In Euclidis πρῶτα*, as we can see above. Likewise, those definitions make clear that extension is a continuum taken with situation. So it is not evident that these are features only of the mature theory.

Regarding the first condition though, I refer back to what I said at the end of the last section. This is that what are contiguous are the boundaries of bodies conceived as actual existents, accidents proper to particular existents [*propria*]; but conceived as geometrical concreta, these boundaries do not belong to such particulars, and are abstracted from all sensible qualities, and are therefore indiscernible. Similarly, congruence, taken mathematically, involves an abstraction away from the particular existents. De Risi recognizes this indiscernibility of parts, but he sees it as an outcome of Leibniz's late construal of homogeneity.[181] On the contrary, it seems to me that the idea that the situated points constituting space (and the figures they determine) should be taken as indiscernible in themselves, is there in *analysis situs* from its inception—it is clearly stated, for instance, at the end of the *Characteristica geometrica* (CG 228–30), as we saw in §2.4.

[179] Cf. also: "What happens is that Leibniz must understand the various organic bodies as closed figures infinitely near one another—in short as contiguous spatial loci. Leibniz's entire metaphysics of real space rests therefore on the possibility of the concept of contiguity, which alone would allow matter to be determined and thus space to be expressive in the first place" (De Risi 2007, 495).

[180] "The process in which a homogeneous extension originates from the repetition *partes extra partes* of the identical will be called diffusion" (De Risi 2007, 413).

[181] Cf. "It is therefore homogeneity of real space that actually produces extension. Indeed, extension derives from the (infinitesimal) diffusion of *extensa* the parts of which are indiscernible" (2007, 509).

Thus two contiguous extended bodies will not share a boundary—each will have its own. But insofar as we represent the situations of these bodies, we abstract away from the individual properties of the bodies and conceive them purely as extensa filling that space. Even if the foundations of their situations (the degrees of expression in each body) are different in the two bodies, so that their situations as individual properties are numerically different, we are interested only in what could occupy *an equivalent situation*. This is stated explicitly by Leibniz in his correspondence with Clarke, where place is depicted as an equivalence class of situations (i.e. of things congruent to the same coexistents); but the indiscernibility of equivalently situated bodies is implicit in the very device of having abstract points stand for these situated coexistents, a device which, if I am right, Leibniz inherited from Hobbes. This means that in the mathematical representation of two contiguous bodies, they will be presented as sharing a boundary, since the contiguous extrema will be coincident. When we represent the space occupied by such bodies, then, we will represent the space as continuous: it will satisfy the definition from the *In Euclidis πρῶτα* quoted above (and the analogous one from the *Specimen Geometriae luciferae*).

This relates to the analogous case in time, discussed in §1.5. There we saw that although consecutive states of a monad are contiguous, this is because we have pictured them with precise instants marking their boundaries. In fact, however, enduring states are vague, and have no definite boundaries, as Leibniz argued in *Dans le corps il n'y a point de figure parfait* of 1686.[182] Similarly, as Leibniz also argued in the same manuscript, "there is no precise and fixed shape in bodies, because of the actual division of parts to infinity" (A VI 4, 1613/LLC 297). Actual bodies do not have precise shapes with settled boundaries, because they are being constantly bombarded by the jostling of the ever-smaller particles of matter around them. We *imagine* them, that is, picture them in our imagination, as having precise boundaries, and this allows us to represent them mathematically. As Leibniz explains in the *Specimen inventorum* of around 1688,

> Indeed, even though this may seem paradoxical, it must be recognized that the notion of extension is not as transparent as is commonly believed. For from the fact that no body is so very small that it is not actually divided into parts excited by different motions, it follows that no determinate figure can be assigned to any body, nor is a precisely straight line, or circle or any other assignable figure of any body, found in the nature of things, although certain rules are observed by nature even in its deviation from an infinite series. Thus figure involves something imaginary, and no other sword can sever the knots we tie for ourselves by misunderstanding the composition of the continuum. (A VI 4, 1622/LLC 315)

182 "All enduring states are vague, and have nothing precise about them" (A VI 4, 1613/LLC 297).

Now, if "figure involves something imaginary," it may be thought that this assertion completely undermines Leibniz's *analysis situs*, which depends crucially on the notion of figure and congruence among figures. But this is not so. As I shall argue in detail in §3.5, Leibniz follows Proclus in conceiving mathematics as the science of the imagination, while denying at the same time that there are any precise shapes (*figurae*) in the material world. The figures are produced by the imagination, but that does not detract from their correctly representing the forms to the intellect (Proclus), or the relations of situation possessed by corporeal substances (Leibniz).

Moreover, as we showed in the analogous case of the "vague states" in time (following the analysis of Hellman and Shapiro), sense can be made of the notion of congruence of intervals, and even of translation or instantiation of intervals anywhere in time, and also of bisections, without supposing that the intervals are either open or closed in the sense of set theory. Analogously, congruence, bisection, etc., may be defined on "vague" intervals of space—ones that are neither determinately open or closed—and points may be captured in terms of Cauchy sequences of such intervals.[183]

In this way, it is shown that Leibniz's assertion of the applicability of mathematics to the physical world is not an expression of wish-fulfilment, but has robust content. Because of the syncategorematically infinite division of matter into "physical points," each distinguished by a different endeavour, these points (arbitrarily small regions of matter) are dense, and not strictly continuous. That is matter, however, and not space. And even if these individual regions of matter have no precise spatial boundaries, nevertheless by continued bisection of space we can determine the region containing any given material point with arbitrary accuracy.[184] That is, the spatial interval between any two such points is fictionally infinitesimal: no matter how small the interval (difference) is between two such points, there is a still smaller one, as guaranteed by the Archimedean axiom. This is Leibniz's definition of an infinitesimally small difference, syncategorematically understood.[185] Consequently, we can take what he wrote to Pierre Varignon at face value: "Yet the actual phenomena of nature are arranged, and must be so, in

[183] A full discussion of this strange state of affairs would require an exploration of the whole issue of Leibniz's account of the role of imagination in mathematics, including the status of his notion of *imaginatio distincta*. See De Risi (2005) for some discussion, and translation of a relevant text, and the *habilitation* of David Rabouin (2020) for a profound analysis.

[184] De Risi (2007, 510, n.51) suggests something similar: "In fact, in order for quantity (i.e. a confused representation) to arise and permit a spatial representation, some indiscernible extensions without boundary must exist. In other words, around each point there must be a (small as one please [s]) extension without boundary"—although he suggests that such an extension "contradicts the actual subdivision of matter," which it will not, since the infinitesimal regions are of space, not matter, and they are arbitrarily small, but not actual infinitesimals.

[185] See Rabouin and Arthur (2020) for a treatment of Leibniz's construal of infinitesimals as syncategorematic.

such a way that nothing ever happens by which the law of continuity [. . .] or any of the other most exact rules of mathematics is violated" (GP IV 568/L 583).

In this connection, I should also say more about the charge (by Cassirer, Parmentier, and De Risi) that Leibniz did not manage to connect point and extension, at least until his last years. For it is not the case that the definition of continuity in terms of co-integrant parts was a development only of the 1690s. Now that more texts from that period have become available,[186] we can see that this notion, along with congeneity (homogony), were products of Leibniz's work in the 1680s.[187] In one of these texts, whose watermarks indicate a probable date of 1682–85, he writes:

> A *continuum* is a whole two co-integrant parts of which have something in common, which is called the bound of each of them. Now I call those parts *co-integrant* which taken together equal the whole, and have no part in common. Thus a bound is not a part, nor is it homogeneous with the things it bounds, otherwise it would be a part.[188]

In the same piece he recognizes that such a bound *is in* the whole it bounds because the whole can be understood to be generated from it by its increase:

> An *indivisible* is that which has no part and by the increasing of which something can be made that has parts, so that by increasing a point a line can be generated, by increasing a moment, an hour. Conversely, by decreasing an hour, just before it disappears into nothing, it ends in a moment.... (MT #30)

The notion of congeneity or homogony is nascent here, and described in terms of things "being of the same matter":

> Those things are *of the same kind ôr matter* whose simplest immediate requisites ôr minima are indiscernible from each other.
>
> A *transformation* is a change of a thing into another with the minima remaining the same and unchanged....

[186] These are the texts in MT, prepared by the ANR MATHESIS project in collaboration with the Leibniz Research Centre in Hanover (Leibniz 2021); some of the transcriptions were supplied by De Risi himself.

[187] One can also find the emergence of the concept of a "uniqueness" operation (to be discussed in §3.6) in the *Generalia* probably dating from 1685 or so (MT #106). See fn. [this one +2] below.

[188] This is "First Principles of Geometry (II)," MT #30 (39884), a translation of which is given in Appendix 3.4 below. Vincenzo De Risi referenced this text in his (2019, 140, n.62), while remaining agnostic on its date. As Siegmund Probst has since informed us, it is on paper with a watermark for 1678 that Leibniz often used in the period 1682–85.

> Point and body are of the same kind ôr matter, yet they are not homogeneous, since a body can be similar to another body, but a point cannot. (MT #30)

All this is made more explicit in a second piece, tentatively dated from 1685:

> A *continuum* is that of which two co-integrant parts have something in common, which is called a *bound. Co-integrant parts* I call those which together equal the whole and have no part in common. An *indivisible* here is that which has no part, but which *is in* the whole, ôr by the increase of which the whole can be made, ôr into which the quantum finally disappears by some continuous equable decrease. Being an immediate requisite [of another thing] I call *being in* or *constituting* [it]. There is no immediate requisite of a minimum. A *transformation* is a change of a thing with the same minima remaining.... *Congenia* are line and surface, even though they are not homogeneous. Intrinsic congenia are those that are in a thing.[189]

What these texts show is that Leibniz is already working on establishing a link between point and extensum, the link being the notion of a *transformation* between *congenia* (homogonous entities) that are heterogeneous. Evidently, this notion of a transformation between such entities of different dimensions has its origins in the Proclean idea of the generation of a line by the *flux* of a point, a surface by the flux of a line, etc. (which I will examine in detail in §3.5).[190] But Leibniz can be seen to be working all this into a serviceable mathematical notion of a transformation that is both bi-continuous and univocal and which applies to homogeneous and heterogeneous things alike. In these texts he is exploring the notion of a continuous transition from one thing to another, which involves a transition by means of a common element that is a minimum. In the *Specimen Geometriae luciferae* he describes this as a "quasi-transformation," a "a mental transformation" where "instead of minima we can employ quasi-minima, that is, indefinitely small parts" (GM VII 283). In this way a curvilinear thing can be continuously transformed into a rectilinear one by employing "something quasi-curvilinear, namely a rectilinear polygon having as great a number of sides as you wish," so that if this succeeds, "that is, if the error ôr difference between a quasi-transformation and a true one always comes out smaller and smaller so that it

[189] This is the text *Generalia*, LH 35 VII 30 Bl. 128, MT #106. As Siegmund Probst has pointed out to me, this piece is closely related to *De Calculus Geometrici Elementis* (A VI 4, 604–7); in fact, the wording in two of the paragraphs is nearly identical. That piece is securely dated as February–October 1685.

[190] For example, Leibniz writes in *De homogeneis et de homogonis* (MT #88), "Continuous change is a *flux*. As a line is made by the flux of a point, so homogeneous things are changed into each other by a flux, such as a point into a line, a circle into a square, even into a cone" (see Appendix 3.6).

finally becomes smaller than any given, then it can be inferred to be a true transformation" (283).

The idea that one can arrive at a point or instant by a continuous transformation of a line or a time is, of course, closely related to Leibniz's Law of Continuity. The latter involves a continuous transition to a limiting case, which may then be included in the series in question, at least in respect of one property that is conserved in the process. In discussing how magnitude can be obtained through the congruence of parts, Leibniz notes in a manuscript from 1715, that "when the law of continuity is added, those things that have common parts nevertheless vanish into those that do not."[191] One can see an argument of this kind in the *Scheda on situation and extension*:

> Let there be an extensum having something in common with another extensum; this is transformed in such a way that what is in common is continually diminished until nothing more can be subtracted from it, and when it reaches the point where it is necessary that if anything further is subtracted from it nothing in common will be left, then the common extremity will be a point. Thus it is demonstrated also that when an extensum is given, so is a point.
>
> (De Risi 2007, 589; Appendix 3.7)

This is close to the way of defining a point from a space in contemporary point-free topology—a significant improvement over simply assuming a transformation from a point to a line as effected by the movement of the point. Still, although one can see such improvements in sophistication in Leibniz's treatment of continuous transformations between extensa and their bounds, the connection between situated points and continuous extensum appears to be no late feature of his analysis situs, but there all the way through.

2.6 Conclusion

In this chapter I have tried to convey the richness of Leibniz's theory of space by expounding it in relation to his *analysis situs*, the geometry of situations. There has been a tendency, beginning with Russell but still in evidence today, to treat his theory of space in the context of his physics, his metaphysics, and even his theology, while ignoring his (admittedly incomplete and fragmentary) work in geometry. But Leibniz's *analysis situs* is not just relevant to his theory of space, it actually constitutes it. As has been shown by the meticulous studies of Echeverría and particularly De Risi, this novel foundation for the science of geometry was at

[191] Quoted from *Calculi situs fundamenta*, LH XXXV, 1, 13 18^r (De Risi 2007, 616); [1715].

the same time a highly original attempt to derive the basic properties of space. In fact, Leibniz was perhaps the first to have clearly conceived space itself as having the properties that tradition had ascribed to figures in space, as described by Euclidean geometry; and he was certainly the first to have set about systematically deriving them.[192]

This space, moreover, is not a merely mathematical space, but rather a mathematical construction of the very space in which bodies are situated, as I have argued in detail. Consequently, Leibniz's writings on *analysis situs* have a direct bearing on a number of issues concerning the adequacy of his theory of space as a foundation for his physics. Recognition that the subjects of his relational theory of space are equivalence classes of relations of situation, and not bodies together with ordering relations, is therefore of greatest consequence.

I have argued that according to Leibniz's earliest views, space has indeterminate magnitude and figure prior to its acquiring boundaries through the bodies in it. This is then married to a conception of situation, probably indebted to Leibniz's studies of Hobbes, according to which the situation of a group of bodies is determined by the figure in which the bodies are represented as points at its vertices. The situation of the bodies is then constituted by the boundary of the figure, consisting in the lines between pairs of vertices (including their mutual distances and angles), and the situation of any one body by the lines drawn from it to the other bodies with which it is compresent. What is then important is not the individual situations of any bodies, their relational accidents, but the relationships holding between equivalent situations: place becomes an equivalence class of congruent situations, and space a system of such places.

This exposition of Leibniz's early views not only helps to clarify the origins of Leibniz's *analysis situs* but also, I think, sheds light on some key features of his emerging metaphysics. It presents a clear picture, for instance, of what Leibniz means in the *New System* when he says of his *atoms of substance* that "mathematical points are their points of view for expressing the universe" (GP IV 480/ WFT 149). Conceiving all the organs of a corporeal substance as contracted into a physical point, its *point of view* consists pre-eminently in the situation of this point to all the other bodies in the universe. 'Expression' here is a technical term, involving at least a mapping (homomorphism) of the things expressed in the organs of sense of the perceiver. No doubt Leibniz's familiarity with the emerging projective geometry of Desargues, Pascal, and De la Hire also factors into his

[192] De Risi describes the attributing of properties like continuity to space rather than the figures in it as "almost unprecedented in the seventeenth century, when geometry was still largely regarded as the science of figures rather than a science of space. It is nonetheless perfectly in line with the whole Leibnizian project of *analysis situs*, that is to say, a new geometrical science which conceives space as a *structure* of situational relations (an *ordre de situations*), and takes it to be the proper object of geometrical investigations" (De Risi 2019, 161).

notion of point of view, but it is still striking how essentially this notion of perspective is embodied in Leibniz's geometry of situations.

Moreover, point of view as an aspect of *perception* is immediately suggested when the mathematical representation of the situation of a point is combined with the idea that the sense organs of an organic body can be considered as regarding other coexisting things as if from a point. A perception is a representation of all coexisting things in the universe, involving the situation of all of them to the point from which they are perceived. Leibniz connects this with the idea of quantity only being determinable through *compresence*, a necessary condition for congruence of situations. This is the basis of his distinction between quantity and quality. All the various qualities of a figure can be determined in it when it is taken by itself, but will then apply equally to any similar figure; only when the figures are brought into coincidence can their relative magnitudes be established. This means that space as a whole will only have a proportional metric: the actual magnitude of a given body requires the presence of a measure, such as a body representing a metre, by means of which it can be determined.

One feature of Leibniz's approach that has drawn some criticism (beginning with Cassirer) is the complementarity between extension and situation. It has been suggested that Leibniz's need to posit extension and its continuity *in addition to* relations of situation is an indication of the initial incompleteness of *analysis situs* as a theory of space. The latter is a structure of situational relations among points, a point being that which has situation but no extension; but extension, the extended per se, is an indeterminate whole: it has extension but no situation. It can be divided in arbitrarily different ways into bounded parts, and its continuity is expressed in terms of the possibility of such arbitrary division into parts. Continuity, that is, is construed in terms of whole and parts, i.e. mereologically, and this does not obviously connect with the idea of an *extensum* being constituted by points. In particular, the distinction between contiguity and continuity has seemed to many critics to present a major obstacle: the Aristotelian distinction between the endpoints of two adjacent things being together (contiguity), as against their comprising one point (continuity), is not reproducible in point-set topology.

Having raised this difficulty, De Risi sees Leibniz as having finally solved it only on his return to *analysis situs* in the last four years of his life. Putting to one side what is to my mind an over-Kantian interpretation in terms of a transcendental synthesis of perceptions, De Risi sees Leibniz as having forged the connection between the mereological and the point-set approaches with his notion of the *diffusion of situation*. Diffusion consists in the notion that the continuous extension of a line is produced by the successive congruence of situations of points, and justified by appeal to the possibility of their co-perception. More succinctly, "*extensio* is *diffusio* of *situs*" (De Risi 2007, 415). I have argued that this connection between extension and situation is there from the beginning. Even if Leibniz does not make the notion of diffusion (equivalently, repetition) explicit until 1693, the idea of extension as a continuous diffusion of compresence is a constant throughout Leibniz's thinking about space. Moreover, I argue, this diffusion of compresence is the same as God's omnipresence, the divine attribute of immensity. And the idea of immensity as the basis of space, I have argued, is a constant of Leibniz's thinking about space from 1676 till his death. The construal of extension as diffusion of situation is the formal expression of this idea, but the corresponding characterization of space as consisting in all points congruent with a given point is already prominent in the manuscripts of 1679.

There is, however, more to say about this notion of "all the points." When in the *Characteristica geometrica* of 1679 Leibniz defines space as "the locus of all the points (Y) congruent with Y, ôr Ȳ" (CG ix §89, 220), he is describing space as the totality of all points, and this suggests Dedekind-completeness. In fact, he is perfectly explicit that space so defined is the *collection* of all points: "Ȳ signifies all the points of the locus taken collectively, ôr the whole locus" (CG ix §95, 222). Yet for the last forty years of his life Leibniz had consistently maintained that there are no infinite collections: the notion of an infinite collection or set entails a contradiction (assuming the part–whole axiom that Leibniz takes to characterize quantity). Leibniz is committed to the actual infinite in the *syncategorematic* sense: no matter how many points are taken, there are more: "instead of 'infinite number' it should be said that there are more than can be expressed by any number" (to Des Bosses, March 2, 1706; GP II 305; LDB 30–3). What this entails is that space, as the locus of an infinity of points taken collectively, has the status of a fiction, just like the infinitely small. Thus Leibniz's absolute space, unlike Newton's, cannot be said to *exist*, and "has no place except in geometrical calculations."[193] This would be a problem if, as in a modern set-theoretical approach, an infinite domain had to exist in order for quantifiers to be well defined. Leibniz, however, recognizes a distributive sense of "all" that is applicable to actually infinitely many items, and his propositions concerning infinitely many mathematical objects must be taken in this sense. Thus, for instance, a continuous tract from one point to another is "that in which infinite points can be taken at will" (CG ix §17, 162), that is, "distributively, not collectively" (CG ix §19, 164). This distributive sense of universal quantification appropriate to the syncategorematic actual infinite is therefore adequate for Leibniz's constructions, and in particular for his construal of successive congruence that constitutes arcwise (or geodesic) connectedness.

[193] Cf. Leibniz's reply to Locke on space: "The idea of the absolute with respect to space is just the idea of the immensity of God and thus of other things. But it would be a mistake to want to imagine an absolute space that is an infinite whole composed of parts. There is no such thing: it is a notion which implies a contradiction; and these infinite wholes, like their opposites the infinitesimals, have no place except in geometrical calculations, just like the imaginary roots in algebra" (*Nouveaux essais*, Bk. II, ch. xvii; A VI 6, 158).

I have argued, moreover, that the contiguity–continuity distinction is not the stumbling block that De Risi supposed it to be. Contiguity applies to existing bodies, each of which has its own distinct boundary. In abstracting from the figures and situations of actual bodies, we are concerned only with equivalences of situation, so that in geometry two points that are "together" are treated as the same abstract point. Similarly, all coincident boundaries will be treated as identical, since they are situated in precisely the same way. This is where Leibniz's two-part definition of continuity in terms of co-integrant parts and the *partes extra partes* clause comes into play. As De Risi has argued, this definition of continuity by Leibniz "approaches Dedekind's notion of completeness given by the idea of a cut" (De Risi 2019, 143), enabling a correspondence between points of the continuum and the real numbers to be established. Thus for Leibniz, while any point of an actual body is at an instant distinguished from any other by its differing endeavour, so that such points are dense (they constitute a "physical continuum"), the points of an ideal continuum are complete as well as dense. They constitute all possible points of division of the continuum.

The link between extension and situation is provided on the one hand by the notion of diffusion of situation, the continuous diffusion of points arbitrarily situated with respect to any given point. On the other hand, it is supplied by the notion of homogony, where an extension is reducible to a point by a process of continual diminution, or where the extension in common between two overlapping extensa is diminished "until it reaches the point where it is necessary that if anything further is subtracted from it nothing in common will be left," rendering the overlap as a point (De Risi 2007, 589; Appendix 3.7). By this means an extensum may be regarded as "constituted" by an infinity of points (or more generally, objects of one dimension less) "taken in order" (i.e. parametrized), an anticipation of the idea of *fibration*.[194] An extensum is not thereby composed of points, but it can be regarded as composed of indefinitely small parts of itself having the same dimension. That is, the magnitude of one extensum can be compared to another by resolving both into a common measure that is infinitesimal and therefore fictional—"lines into indefinitely small linelets," and so forth:

> it can be shown that two extensa can be compared by resolving them into equal or congruent particles as small as we please, as though into a common measure, and the error will always be smaller than any one of these particles, or at least it will be in a finite increasing or decreasing ratio to any given measure; whence it is clear that the error in such a comparison is smaller than any given.
>
> (*Specimen Geometriae luciferae*; GM VII 260–95)

[194] Again, see De Risi's lucid discussion in his (2019, 131–4, 157–60).

Thus (suitably ordered) points may *constitute* a line, without composing it; while (under the appropriate conditions) an infinity of infinitesimal linelets may be regarded as composing a line, even though both the infinite number and the infinitely small magnitude are fictions.

All things considered, then, Leibniz's formal approach to the construction of space is as sophisticated as it is profound. Although Leibniz's definitions take geometry as foundational (in contrast to Dedekind's arithmetical approach), and leave something to be desired according to modern standards of rigour, his innovations concerning space and its continuity were far ahead of his time, and not improved upon until the work of mathematicians at the end of the nineteenth century. And as concerns the connection of Leibniz's formal theory with his metaphysical views, even though I cannot agree with De Risi's construal of this as a facet of an alleged move to an explicit phenomenalism in Leibniz's last years, I do agree with him about the profundity of the connection. It is no accident, in my estimation, that Leibniz developed his notion of perception as a substance's representation of the universe from its own point of view at exactly the same time that he was developing his theory of space as the order of situations among coexisting things or things that could be co-perceived. Space thus becomes a kind of transcendental condition for the possibility of things' coexisting or being perceived together. But the reality of space is not constituted solely by conditions on the possibility of sensible experience. Its reality derives rather from the diffusion of the divine attribute of immensity, the possible extensions and situations of bodies arising from limitations of the *immensum*. In this way the relations of situation of created bodies can be understood as "real relations," instantiations of relations existing in the divine understanding. The points on which these relations are defined, on the other hand, can be related to the extended lines representing the relations by continuous transformations. Absolute space is then the order constituted by all the points congruent to a given point, where the 'all' is taken distributively; if, on the other hand, it is taken as a whole consisting in the collection of all its situated points—as a "manifold"—it does not exist: it is a mathematical fiction.

3
De motu: Leibniz on the Relativity of Motion

> "Universally, when motion occurs, we find nothing in bodies by which it could be determined except change of situation, which always consists in relation (*respectu*) ... But this is to understand these things in mathematical rigour. Meanwhile, we attribute motion to bodies according to those hypotheses by which they are most aptly explained, and the truth of the hypothesis is nothing other than its aptness."
>
> (*Dynamica*, GM VI 508)

Introduction

Leibniz's statements about the relativity of motion have generally been found wanting.[1] Some of the main objections are as follows. First, there is his claim that the "equivalence of hypotheses" (abbreviated EH) about which of several moving bodies moving among themselves may be regarded as being at rest, applies universally: this is in conflict with the absoluteness of accelerated motion, as evinced especially in rotation.[2] Second, Leibniz's defence of Copernicanism on the basis of the EH is seen as problematic: if the various astronomical hypotheses (Ptolemaic, Tychonic, Copernican) are equivalent, how can the Copernican System be held to give the true motions? Third, it is hard to reconcile his upholding of the universal relativity of motion with his concession to Samuel Clarke that there is "a difference between an absolute true motion of a body, and a mere relative change of its situation with respect to another body."[3] Fourth, when Leibniz writes that one cannot determine the subject of motion if motion is taken as mere change of

[1] A good summary of the difficulties commentators have seen in Leibniz's views may be found in (Huggett and Hoefer 2019)—although I think the premise they take as foundational in their own analysis, that "in the metaphysics of monads that Leibniz was developing contemporaneously with his mechanics, *everything* is in the mind of the monads," is wholly mistaken. As explained in the Introduction above, I present an extended argument against this idealist reading in my (2018).

[2] See especially Stein (1977) and Earman (1989). At the other extreme, Anja Jauernig (2008) defends Leibniz's upholding of the EH. She claims, *inter alia*, that on one plausible reading, Leibniz's upholding of the universal relativity of motion in his dynamics is equivalent to General Relativity; see also Jauernig (2009).

[3] Leibniz to Clarke, 5th Letter, §53 [1716] (GP VII 404/L 706, AG 341).

Leibniz on Time, Space, and Relativity. Richard T. W. Arthur, Oxford University Press.
DOI: 10.1093/oso/9780192849076.003.0004

situation, whereas "force is something real and absolute," he has been taken to mean that the absoluteness of *vis viva* breaks the equivalence of hypotheses. It is hard to see how this would be so, since a body's *vis viva*, mv^2, is not invariant under change of hypothesis as to what is really moving.[4] Consequently some have suggested that Leibniz upheld a notion of "absolute speed," although this is not determinable;[5] while others have claimed that the equivalence of hypotheses applies not to phenomenal motions, but on a deeper dynamical level.[6]

Further difficulties have been found in making Leibniz's views on the physics of motion compatible with his metaphysics. Given his thesis of the ideality of relations, if motion is purely relative this would appear to imply that motion is only ideal. This, however is difficult to reconcile with Leibniz's many affirmations of the reality of the phenomenon of motion:[7] how could motion be only ideal if it has a reality grounded in force?[8] It has also been suggested that, given his assertion of the ideality of space, Leibniz would not have intended motion to be real anyway, and that it should instead be understood as a primitive monadic quantity.[9]

In this chapter I will attempt to throw some light on these issues by concentrating on the genesis of Leibniz's views in their own context, particularly in connection with his *Analysis Situs* and his philosophy of causation. In this way I hope to avoid some of the distortion that arises from forcing his views into modern conceptual frameworks, such as arises from reading the Equivalence of Hypotheses as a symmetry of spacetime, or from inquiring whether the relativity of motion he endorses is kinematic or dynamic, or whether he must endorse a Galilean or a full-blown Newtonian spacetime.[10] Although there is nothing wrong with trying to assess the contemporary relevance of seventeenth-century views, a

[4] John Earman interprets it this way in his (1989, 132); see also Stein (1977).

[5] Both Daniel Garber (1995, 308) and Roberts (2003, §4) suggest that there is a preferred frame of reference in which motions are real, even if this cannot be determined; in which frame velocity, and therefore mv^2, would be absolute. Huggett and Hoefer (2019) endorse this claim as "consistent" (§6.2.1).

[6] Ed Slowik maintained this in an early draft of his (2006), but rejected it in the published version. Anya Jauernig offers this as one of two ways of saving the consistency of Leibniz's views in her (2008).

[7] Thus in his *TMA* of 1672, Leibniz inverts Zeno's dichotomy argument, taking the negation of Zeno's conclusion that motion is unreal as a premise, using the dichotomy to prove that every extended subinterval must contain an unextended beginning (A VI 2, 264/LLC 339). In the *Pacidius Philalelthi* of 1676, the interlocutors agree that "it is not our place to call into question the reliability of the senses and to doubt the reality of motion" (A VI 3, 556/LLC 189).

[8] As we shall see below, although motion in itself "is not an entirely real thing" (§18 [1686], A VI 4, 1559/WFT 71), Leibniz believes that it is not wholly unreal, and receives its reality through force. See (GP IV, 369/L 393), and Paul Lodge's analysis in his (2003).

[9] Cf. Nick Huggett and Carl Hoefer, (2019, §6.4): "Given Leibniz's view that space is literally ideal (and indeed that even relative motion is not 'entirely real') perhaps the best answer is that he took *force* and hence *motion* in its real sense not to be determined by motion in a relative sense at all, but to be primitive monadic quantities." Jauernig also takes the question of whether motion is real to be whether it is a genuine attribute of a subject, but, in contrast, takes active forces to be the "intrinsic properties" of corporeal substances that ground real motion (2008, 6).

[10] Stein (1977, 33) considers and rejects an explanation of EH holding kinematically but not dynamically. Slowik avails himself of the distinction in his (2009), (2013), and (2016), as does

premature juxtaposition of Leibniz's statements with modern conceptions is apt to obscure rather than enlighten.[11] By treating his views in their own historical context, I hope to show that Leibniz's position on the relativity of motion turns out to be not only surprisingly coherent, but also still instructive in certain respects. This approach will not absolve his pronouncements of all shortcomings, but it will, I hope, make more understandable why he took the stances he did, even where they cannot now be defended.[12]

In §3.1 I describe the importance of Leibniz's distinction between motion as change of situation and motion with respect to cause. For *x* to be the cause of changes in *y* is for it to contain the reason for those changes, and in certain cases, I argue, this reason is identified according to the most intelligible hypothesis. Thus even though motion is entirely relative, so that the effects will be the same whichever of two bodies is regarded as being in absolute motion, the subject of motion—what is truly moving—can nevertheless be assigned by means of the most intelligible hypothesis for accounting for the changes in question. In §3.2 I show how this distinction is applied by Leibniz in defence of the Copernican Hypothesis as giving the true motions of the planets, and how this exonerates him from the charge of adopting an instrumentalist view of hypotheses.

In §3.3 I take on issues concerning Leibniz's dynamics, and the metaphysics of motion. Contrary to many commentators, I argue that for Leibniz force is *not* a sure criterion for absolute motion; rather it is what is real in what would otherwise be a pure phenomenon. Nor does the conservation of force require a preferred dynamical frame of reference, as has been alleged. On Leibniz's account it is the total absolute force that is conserved, and with respect to different frames this can be differently distributed between the directive force of an entire aggregate and the respective forces that are dissipated among the bodies constituting an aggregate. In the latter part of the section I critique the common assumption of recent commentators that Leibniz conceived various forces as operating on differing ontic levels, primitive forces on the monadic level, and derivative forces either on the phenomenal level, or between these levels. I argue that such talk of ontic levels

Jauernig (2009). I am inclined to be skeptical about applying this distinction to seventeenth-century natural philosophy generally, since most practitioners regarded a body that is not acted upon by anything external as having a *force of motion*. Newton himself thought of a body in motion as maintaining its state by virtue of its *vis inertiae*, an inherent force proportional to the quantity of motion. The use of the apparatus of spacetime theory to assess seventeenth-century natural philosophy has been critically examined (from differing perspectives) by Huggett (2006) and by Slowik (2013).

[11] Two very interesting studies of the ways in which Leibniz's relativism was misinterpreted as anticipating Einstein in the early twentieth century are Giovanelli (2011) and De Risi (2012).

[12] Since beginning work on the paper that became this chapter, I have published other related studies: Arthur (2013c), which is a translation and commentary of one of Leibniz's earliest writings on this topic, the "Mechanical Principles"; Arthur (2015b), which is a comparison of his views on relativity with those of Isaac Newton; and Arthur (2015a), which explores the significance of Leibniz's views on relativity for his adoption of substantial forms. See also Arthur (2018, ch. 5), for a related analysis of this issue in connection with his emerging theory of substance. Here I aim to engage more fully the accounts of modern commentators on Leibniz's views on relativity.

is nowhere to be found in Leibniz, and explain how I think the relationship between primitive and derivative forces should be understood.

In §3.4 I return to the issue of the relativity of motion, now in the context specifically of rotational motion, contrasting Leibniz's account with Newton's. I argue that there are some fundamental equivocations in Leibniz's account: (1) between phenomena in the sense of what appears to an eye at a certain location in matter, and phenomena in the wide sense of things that are known to occur, and need explaining; (2) between momentary motion (or *motio*) and motion diffused through a stretch of time (*motus*); and (3) between the composition of the former from component endeavours (for which the equivalence of hypotheses holds) and the composition of the latter from inflecting impulses that produce the curved trajectory over time, for which the EH is not valid.

This brings us to the issue of motion through time, and what kind of space is required to represent it, which I treat in §3.5. The question was already broached in §2.1 above, but not treated in the remaining sections of that chapter. We saw that the question of motion through space has been discussed by recent authors as though Leibniz were a modern relationalist, and the question is raised whether the kind of spacetime to which he would be entitled by his metaphysical assumptions would be adequate to ground his physics. But this discussion has been pursued without reference to his *analysis situs* which Leibniz explicitly devised in order to treat not just space, but also motion. Since abstract motions had traditionally been used to derive theorems in geometry since the beginning of the subject, it might be though that this has nothing to do with the treatment of the real motions of physical bodies. I treat this question first, before proceeding to Leibniz's famous treatment of motion in space in his Fifth Letter to Samuel Clarke.

3.1 Change of Situation and Motion with Respect to Cause

In order to understand Leibniz's position on relativity on its own terms, the first thing to appreciate is the firm distinction he draws between *motion considered geometrically* and *motion with respect to cause*. For the equivalence of hypotheses is taken by him to apply to *motion considered geometrically* or "formally," i.e. as mere change of situation, prior to any consideration of the cause of any of the motions. When causes are brought into play, as we shall see, he holds that it is possible to identify the subject of motion—which body is the agent, and which the patient, with respect to the phenomena to be explained—despite the relativity of motion as it is conceived geometrically.[13] A classic statement of this position is to be found in the *Discourse on Metaphysics* of 1686:

[13] The importance of the cause of motion in Leibniz's treatment of relativity is recognized by Paul Lodge in his careful treatment of the issue: "Leibniz holds that for there to be real motion, or motion

> For motion, if one considers only what it precisely and formally comprises, that is to say, change of place, is not an entirely real thing, and when several bodies change situation among themselves, it is not possible to determine, solely by a consideration of these changes, to which among them the motion should be attributed—as I could show geometrically if I wanted to stop to do so.
>
> (§18, A VI 4, 1558–9/AG 51, WFT 71)

Leibniz goes on to say that "the force or immediate cause behind those changes is something more real, and there is enough of a basis for attributing it to one body rather than another." But let us defer discussion of causes and forces until later, and concentrate now on motion treated geometrically.

Leibniz interprets treating motion "geometrically" to mean that "in itself" or "formally" it is *change of situation*. Once this is understood, the demonstration that he does not deign to give in the *Discourse* follows straightforwardly (at least for pairs of bodies in relative, rectilinear motion). For since the situation of any two such bodies is relational, a change of situation between them is also both relational and mutual. This is what Leibniz says in his unpublished *Dynamica* of 1689–90:

> Universally, when motion occurs, we find nothing in bodies by which it could be determined except change of situation, which always consists in relation (*in respectu*). Thus motion by its nature is respective. But this is to understand these things in mathematical rigour.
>
> (*Dynamica*, Section III, Proposition 19, GM VI 507–8)

Here situation is a technical term, the *situs* of Leibniz's *Analysis Situs*, a term whose conception (as we saw in the previous chapter) he probably owes to his reading of Hobbes's *De corpore*. Leibniz gives a particularly clear explanation of how the relativity of motion follows from this definition of it as change of situation in his early unpublished essay *Principia mechanica* ("Mechanical Principles"), probably dating from the summer of 1676.[14] He begins by defining a situation as "a mode according to which any body can be found, even though we recognize nothing in it specifically by which it can be distinguished from the others," adding that "this way of finding the body depends on knowing its distance from other bodies, and also on knowing the angle, that is, the figure which it makes with another body" (A VI 3, 103). As we have seen, what this means in practice is that

that may be attributed truly to a single body, there must be a body, some part of which is in relative motion with respect to some other body *and* which has the cause of that motion within it" (Lodge 2003, 294).

[14] It could have been written on Leibniz's return to Hanover in early 1677, although to me this seems less likely than the Summer 1676 date. For a translation and commentary on this piece, see Arthur (2013c). The title is supplied by the Akademie editors.

the situation of a point is determined by the distances from and angles it makes with several other point-bodies, taken as being at rest, so that together they form the vertices of a geometrical figure. Leibniz then proceeds to argue that this makes motion, as change of situation, entirely respective.

Leibniz's argument, briefly is this. He considers two bodies *A* and *B*. Things will appear exactly the same whether *B* is at rest and *A* is moving towards it with a uniform velocity $\underline{v}$, or *A* is at rest and *B* moving towards it with a uniform velocity $-\underline{v}$, or *A* and *B* are moving along a line towards one another with velocities ½$\underline{v}$ and -½$\underline{v}$, or *A* and *B* are moving uniformly in the same direction with a difference in velocities of $\underline{v}$. But suppose there are not only two bodies in the world? Leibniz appeals to an eye in a third body *C* (assumed immobile) observing the motions of *A* and *B*, to see whether "something certain can be determined concerning the absolute and proper speed of bodies" (A VI 3, 109). But again all the phenomena—all the mutual situations at each instant—will appear the same, and this is so even when *C* is allowed to move in the same direction as *B* but with half its velocity. Therefore, Leibniz concludes, not even an omniscient being will be able to determine which body is in absolute motion: "whatever speed or direction we attribute by assuming an absolute motion for one of the bodies, we will always find that anyone must then understand motion in the others in such a way that everything will appear as before" (109).

But how does that make merely geometric motion "not entirely real"? The idea is that things would *appear* exactly the same whichever of several bodies is considered to be at rest provided all their relative motions are the same. So if motion is understood as mere change of situation, Leibniz asserts, then the *phenomena of motion* will be the same whichever of them is assumed to be at rest. "From this it is therefore clear," he concludes in *Principia mechanica*, "that from the phenomena of the changed situation alone no certain knowledge can ever be had concerning absolute motion and rest" (A VI 3, 110).

Leibniz expresses the same position in the second (unpublished) part of his *Specimen dynamicum* of 1695 in a passage we quoted from in §2.1. If motion is conceived geometrically, then the same phenomena will be produced whichever body is taken to be really at rest:

> Now, it must be admitted that it is impossible for pure [*nuda*] extension, involving only geometric notions, to be capable of action and passion.... From this it follows that... motion considered apart from force—that is, insofar as only the geometric notions of size, shape and their variation are considered in it—is really nothing other than change of situation. Therefore *motion, as far as the phenomena are concerned, consists in a mere relation* [*respectu*]—as even Descartes acknowledged, since he defined it as translation from the neighbourhood of one body into that of another.... It must therefore be maintained that, if several bodies are in motion, it cannot be inferred from the phenomena which of

> them is in absolute, determinate motion or at rest; rather, rest can be attributed to any of them you choose and the same phenomena will still be produced.
> (*Specimen dynamicum II* [1695], GM VI 246–7)

One might wonder about this. Galileo, Kepler, and indeed the younger Descartes, all thought of motion geometrically, but tended to conceive all the motions—of the Sun and the planets, say—as occurring in some kind of space—what Descartes called "generic space." If one assumes such a space as a really existing container, then real motion would be a change of location in this absolute space. But Leibniz was alive to this. In one of the autobiographical passages with which he typically introduced his views, this one in a dialogue on his new doctrine of force written in Rome, the *Phoranomus*, he admits that he himself only came to doubt the existence of absolute motion after he had begun doubting the existence of such a space, under the influence of Descartes' arguments in his *Principia* as well as Aristotle's, that place or situation is defined with respect to neighbouring bodies:

> Moreover I began to have misgivings about the nature of motion. For when I once conceived space as a real immobile place, endowed only with extension, I could define absolute motion as a change of this real space. But gradually I began to doubt whether there existed in nature such a being [*ens*] as we call space, and from this it followed that there could also be doubts about absolute motion. Certainly Aristotle had said that place is nothing but the surface of the surrounding matter, and Descartes followed him, defining motion (i.e. change of place) as change of neighbourhood. Hence it seemed to follow that what is real and absolute in motion does not consist in what is purely mathematical, such as change of neighbourhood or situation, but in motive power itself; and if there were no such thing, this would mean that absolute and real motion would be abolished. (*Phoranomus II*, [1689])

Scholars have tended to be suspicious of the accuracy of Leibniz's autobiographical vignettes. Here one might doubt his account of what prompted him to reject absolute motion, and surmise that a more likely cause was his reading of Newton's *Principia* in 1688. But we may lay such misgivings to rest in this case. For in manuscripts Leibniz wrote soon after his arrival in Hanover in 1677, we see that he did indeed begin to doubt the reality of absolute motion in connection with the relational nature of space:

> If *space* is a certain thing supposed in pure extension, whilst the nature of *matter* is to fill space, and *motion* is change of space, then motion will be something absolute; and so when two bodies are approaching one another, it will be possible to tell which of them is in motion and which at rest; or, if both are moving, with what speed they are moving. And from this will follow those conclusions which

> I once showed in the *Theory of motion abstractly considered*. But in reality space is not such a thing, and motion is not something absolute, but consists in relation. ("Space and motion are really relations," A VI 4, 1968/LLC 225)

The reference here to his *Theoria motus abstracti* of 1671–72 illustrates an important feature of Leibniz's thinking about the relativity of motion. For the rules of collision he had given there, like Descartes' own rules in the Third Law of his *Principles*, take for granted an immobile space in which the trajectories of the bodies are traced, and with respect to which they have their velocities—what we would call a privileged reference frame. That is, they are not invariant under a change of hypothesis (as to which body is to be taken as at rest). If a privileged space (call this 'P') is thus identifiable as that in which the rules of collision hold, then the equivalence of hypotheses (call this 'R' for relativity) must be false: ($P \rightarrow \neg R$). By contraposition, therefore, if one accepts the relativity of motions (even of only inertial motions), then the laws of collision do not permit the identification of a privileged or absolute space ($R \rightarrow \neg P$).[15]

In fact, one can discern two distinct criticisms in Leibniz's thought at this stage. First there is the a posteriori argument that the laws of collision do not allow the identification of a privileged space, while the phenomena of collisions are consistent with the relative nature of motion. Second, Leibniz mounts the a priori argument that if bodies and their motions are conceived *entirely* geometrically, then the laws of collision cannot be derived. For the power a body has to resist a change in its state of motion is not derivable from the Cartesian understanding of body in terms of "the geometric notions of size, shape and their variation," as Leibniz took his own failed efforts to have sufficiently demonstrated. He was of course well aware that his contemporaries (Huygens, Wren, Borelli, Pardies, etc.) took for granted this power of bodies, enshrined in what we call the law of inertia.[16] But since this power of bodies to resist change of their motion could not be derived from purely geometric principles, he inferred that it is a property of bodies that must be added to a purely geometric understanding of motion in order to deliver the laws of mechanics.[17] As his views developed, Leibniz would insist

[15] Notice that the inference in the conclusion is from the relativity of motion to the non-absoluteness of space, not the other way around. Nevertheless, Leibniz does infer the relativity of motion from his analysis of space in terms of situations in 1676, as we shall see below.

[16] As has been observed by many others, the law of inertia in this sense, that of a body's power for resisting a change in its state of motion, is not necessarily the same as the tendency of a body to continue in uniform, rectilinear motion. More on the latter is §3.5.

[17] Cf. what Leibniz wrote in an unpublished piece from December 1675 concerning the law of conservation of quantity of motion: "This fact has been derived from the phenomena, but no one has shown its origin in nature itself. We have assumed by a kind of prejudice that a greater body is harder to move..." (A VI 3, 466/LLC 31). Newton took a similar position about the passive nature of the mechanical principles, including what we call inertia, arguing that the laws of motion should be understood as passive principles that need supplementing with active principles in order to explain the phenomena. See Garber (2012) and Arthur (2015b) for further discussion.

that this "primitive passive power" is an inherent, non-geometrical principle constitutive of matter, explaining not only bodies' resistance to change of motion (their inertial mass), but also their resistance to penetration, the diffusion of such resistance constituting their extension. This would not actually make any difference to the acceptance of the law of inertia or the conservation of (directed) quantity of motion, for example, but it would provide a metaphysical justification for concepts on which they are based. Precisely this issue arose in Leibniz's later controversy with Clarke and Newton, in that the latter claimed his laws were "mathematical" and did not require metaphysical grounding.

Moreover, the passage quoted above from Part II of the *Specimen dynamicum* begins by stating a third, independent shortcoming of a purely geometric physics that had troubled Leibniz since the 1670s: its inability to support a distinction between action and passion. If all there is to motion is change of situation, then it will be impossible to identify the subject of motion, and therefore which is the agent and which the patient in any collision. That this was a concern for Leibniz is indicated by the marginal note he made at the head of his dialogue on motion and its continuity, *Pacidius Philalethi*, soon after its completion in late November 1676:

> Here are considered the nature of change and the continuum, insofar as they are involved in motion. Still to be treated are, first, the subject of motion, to make it clear which of two things changing their mutual situation motion should be ascribed to; and second, the cause of motion, i.e. motive force.
>
> (A VI 3, 529/LLC 129)

For if no body can be said to "have" the motion rather than any other (since any of them may equally be regarded as at rest, as long as the relative motions are all preserved) then it appears that the system of relative motions is more nearly a property of the world as a whole. Leibniz suggested just this in February 1677,

> A remarkable fact: motion is something relative, and one cannot distinguish exactly which of the bodies is moving. Thus if motion is an affection, its subject will not be any one individual body, but the whole world. Hence all its effects must also necessarily be relative. The absolute motion we imagine to ourselves, however, is nothing but an affection of our soul while we consider ourselves or other things as immobile, since we are able to understand everything more easily when these things are considered as immobile.
>
> ("Motion is something relative"; A VI 4, 1970/LLC 229)[18]

[18] The implications of this passage for the development of Leibniz's views on relativity are discussed by both Lodge (2003) and Garber (2009).

This discussion of absolute motion is not directed at Newton, whose views on space did not become known until his *Principia* was published ten years later. What would have been on Leibniz's mind at this time, rather, were more likely the views of Spinoza, which he had discussed extensively with Tschirnhaus in Paris the previous year. If only the whole world is the subject of motion, and therefore of the actions and passions of the individuals in it, Leibniz must have realized, then this was tantamount to Spinoza's one-substance view.[19] Not only motions, but actions and passions too, would be modes of one substance. The appearances of absolute motions in the things about us would then be mere phenomena or appearances produced in us according to our point of view. Spinoza held bodies to be modes of extension (of *natura naturata*), moving within this extension according to (some variant of) the rules of collision outlined by Descartes. The incompatibility of these rules with the relational nature of motion therefore explains Leibniz's motivation for raising the issue with Spinoza when they met in late 1676 (and also why Spinoza might have been moved to express concern).[20] Either motions are absolute motions in a real extension, as the laws presuppose—but this is at odds with the relativity of motion; or they are merely relative modes that vary with the point of view, consistently with the relativity of motion—but this would remove the foundation for distinguishing action from passion. Leibniz was persuaded by Spinoza's account of action, according to which that body would be identified as the cause in which the reason for the action could be more clearly and distinctly understood: he would later adopt it in the form "that whose expression is more distinct is deemed to act, and that whose expression is more confused to be acted upon... And that thing from whose state a reason for the changes is most readily provided I adjudged to be the cause."[21] But he was not persuaded by the compatibility of such an account of action with the relativity implicit in the geometric conception of motion: the asymmetry of action and passion would be undermined if motion could as well be attributed to one body (the cause) as another (the effect). What Leibniz desired was an account according to which individuals could correctly be conceived as the real subjects of their own actions.

[19] For further discussion of the relation of Leibniz's views on action and the relativity of motion to those of Spinoza, see Arthur (2014, chapter 5), and on how they influenced him to adopt substantial forms, see Arthur (2015a).

[20] Foucher de Careil gives a quotation from Leibniz to this effect, but the source now appears to be lost. On this matter, see Garber's discussion in his (2009, 105). Even if Spinoza had begun to doubt Descartes' laws late in life, this argument would still have been of concern to him because of the conflict between his explanation of agency and the relativity of motion.

[21] Leibniz, *Specimen inventorum* [c.1688], (A VI 4, 1620/LLC 311); cf. Spinoza: "I say that we act when something happens in us or outside us, of which we are the adequate cause, that is (by D1), when something in us or outside us follows from our nature, which can be clearly and distinctly understood through it alone. On the other hand, I say that we are acted upon when something happens in us, or something follows from our nature, of which we are only a partial cause" (*Ethics* II139/D2). See Martha Kneale (1972) for an illuminating comparison between Leibniz and Spinoza on activity.

This brings us back to the distinction outlined above, between motion as change of situation and motion with respect to cause. For all the discussion so far has concerned motion only insofar as it is considered formally, not insofar as it can be attributed to subjects if one knows something about the causes involved. Thus the above quoted passage from "Motion is something relative" of 1677 continues:

> It should be noted, however, that when we consider motion not formally as it is in itself, but with respect to cause, it can be attributed to the body of that thing by whose contact change is brought about. (A VI 4, 1970/LLC 229)

Leibniz had given a much fuller discussion in his prior "Mechanical Principles." "Change of situation," he writes there, "is not yet sufficient for us to judge which of two things that have changed situation with each other we should ascribe the motion to" (A VI 3, 104). For when people sail away from port, for instance, they do not hesitate to "ascribe motion to the ship rather than the earth, because they see the sails billowing with the wind, and the sea foaming under the impacts of the oars." Similarly, when they go for a walk, "they believe they are approaching the town, rather than the town approaching them, because they feel some fatigue and exertion in themselves, and besides they are accustomed to attributing motion to small things rather than big ones" (104). From such considerations

> it is clear that in the case of two bodies, motion should be attributed to that one which contains the cause of their mutual situation having changed, because we have seen it receive a blow, or because it is dislocated and deformed, or shows signs of having received some other blows and of having had a change wrought in it as a result of this. If such signs are absent, we judge from what could happen more easily, or from what has usually happened up till now. (A VI 3, 104–5)

Thus even though a stagecoach is moving relative to trees, and these are attached to the Earth's surface by their roots, "no one doubts that the coach moves over the ground rather than the ground under the coach" (A VI 3, 105). Still, applying such considerations to the motion of the stars and planets, people commonly judge that it is those bodies that are moving and the Earth that stands still. Philosophers, however, have "weighty reasons for the contrary opinion," since it is far more reasonable to attribute a relatively modest motion to the Earth than to believe "that the observable universe should be moved with some insane rotation" (105).

This tallies with Leibniz's mature conception of how the relativity of motion is connected with his philosophy of cause, as can be seen by comparing it with the following remarks in an essay—incomplete, but perhaps intended for publication—probably written in Vienna some twelve years later:

> And that thing from whose state a reason for the changes is most readily provided is adjudged to be the cause. Thus if one person supposes that a solid moving in a fluid stirs up various waves, another can understand the same things to occur if, with the solid at rest in the middle of the fluid, one supposes certain equivalent motions of the fluid <in various waves>; indeed, the same phenomena can be explained in infinitely many ways. And granted that motion is really a relative thing, nonetheless that hypothesis which attributes motion to the solid, and from this deduces the waves in the liquid, is infinitely simpler than the others, and for this reason the solid is adjudged to be the cause of the motion. Causes are not derived from a real influence, but from the providing of a reason.
> (*Specimen inventorum*, [*c.*1688], A VI 4, 1620/LLC 311)

Similarly, the passage quoted above from the later *Dynamica* continues:

> ...But this is to understand these things in mathematical rigour. Meanwhile, we attribute motion to bodies according to those hypotheses by which they are most aptly explained, and the truth of the hypothesis is nothing other than its aptness. (*Dynamica*, Section III, Proposition 19, GM VI 508)

That is, we can say that a body is truly in motion (with respect to cause) when we have identified the most intelligible (or most apt) hypothesis that accounts for the production of the relative motions. This echoes the account Leibniz had already laid out in his "Mechanical Principles." Moreover, in that essay Leibniz applies such considerations to the Copernican controversy. He acknowledges that "the beauty and simplicity of the Copernican System easily attracted every talented person onto its side." "For... with the Earth transferred into the place previously ascribed to the Sun, many imaginary circles, many eccentric circles, and many anomalies vanish"; but also the hypothesis of the annual motion of the Earth—that is, that it "changes its situation to the fixed stars"—"would certainly be sufficiently corroborated" by such phenomena as changes in the apparent diameter of the fixed stars and parallax. Likewise, its diurnal motion is established by the alleged facts that "hanging lamps constantly swing from east to west, or that waves impinge only on eastern and western shores" (A VI 3, 105–6).

This, I think, is very interesting. Given his commitment to the universal relativity of motion, it might be thought that Leibniz should deny the truth of the Copernican hypothesis. But his distinction between "motion as it is in itself" and "motion with respect to cause" allows him to identify, in a given scenario, which motions are merely apparent. The coach is moving and the ground is still, because we know the horses are pulling the coach. Of course, the ground itself could be moving, but that is not relevant to explaining which of the two relative motions is real and which apparent in this scenario. In other words, *the distinction between real and merely apparent motion is to be crafted with respect to the explanation given*—in a word, the distinction is *aetiological*. Likewise, with respect to the fixed stars—that is, taking them to be at rest—it is clear that the simplest hypothesis to explain the phenomena of planetary motions (and to distinguish real from apparent ones) is that according to which the Sun is taken to be at rest and the Earth in motion, both around its own axis and around the Sun.[22] Again, the supposedly fixed stars might themselves be in motion, but that is not relevant for the explaining of the motions in the Solar System. This is, however a distinct question from that of determining an *absolute motion*, that is, a motion that is not determined in relation to bodies, as was discussed above in §2.1. As we saw there, the distinction between true and apparent motion was common property to both those who thought places in space are absolute, and those who held that situations are always relative.

This account of a true motion being distinguished from a merely apparent one by an appeal to cause, then, is sufficient to account for Leibniz's agreement with Newton and Samuel Clarke in the last year of his life on that matter, without foisting on him any change of position or backsliding.[23] In his Fifth Letter, he writes.

> I grant that there is a difference between an absolute true motion of a body and a mere relative change of its situation with respect to another body. For when the immediate cause of the change is in the body, that body is truly in motion, and then the situation of other bodies with respect to it will be changed consequently, though the cause of that change be not in them. (§53; GP VII 404/LC 74)

Leibniz should not, however, have called this an "*absolute* true motion," since that suggests an agreement with Newton and Clarke that such motions are not relative. The position I think he is quite entitled to is that there is a difference between a true motion—namely, that identified by the most intelligible hypothesis—and an apparent motion. But this does not entail that the true motion is absolute in Newton's sense.[24] Leibniz does not allow that any motion is absolute in this sense, since an addition of a velocity in a certain direction to everything in the universe would leave all relative motions unchanged. (As we

[22] In fact, and according to Newton's exposition, the Sun is not absolutely at rest, but revolves around a point not too far from it as a result of the attractions of all its planets. Of course, the Sun is also rotating around its own centre, with a period of approximately once every twenty-five days at the equator. Galileo inferred such rotation from his observations of the sunspots, but I am not aware that Leibniz discussed it.

[23] This point has been lucidly made by Vincenzo De Risi in his (2012, 12–13): "his above-mentioned definitions of true [as opposed to merely apparent] motion recurred throughout his scientific production, and certainly through the many years preceding not only the correspondence with Clarke, but even the publication (1687) of Newton's *Principia*."

[24] Leibniz had written about "absolute motion" before encountering Newton's views, of course, but he should have recognized that his usage was not the same as Newton's. These days also it is normal

would now say, the phenomena would be invariant under a transformation of all velocities by a finite velocity boost in a given direction, a Galilean transformation.) In fact, Leibniz had expressed himself more accurately in his *Specimen inventorum* of around 1688: "... like place, motions too consist only in relation, as Descartes correctly recognized. Nor is there any way of determining precisely how much absolute motion should be assigned to each subject" (A VI 4, 1622/ LLC 315).

So for Leibniz, the attribution of causes does not establish an absolute motion in Newton's sense. He agrees that in certain cases knowledge of causes can help us determine which of several bodies in relative motion is the cause of the relative motions, and then the one containing the cause can be said to be moving truly as opposed to merely apparently. But this is an *aetiological* criterion—it gives us the cause in the sense of the (fallibly) true explanation—and not an ontological one. It does not give us a "sure criterion" for whether that body is moving in the absolute sense required by Newton, since the phenomena might all look the same if that body were at rest relative to some other bodies taken as fixed.[25] Thus I think we have a satisfactory answer to the second objection to Leibniz's account mentioned in the introduction. Although any changes of situation of one body relative to another are mutual, a knowledge of the immediate causes of these changes of situation allows a proper distinction between those motions that have been caused—the true motions—and those that are merely apparent, the motions that are simply effects of the mutuality of changes of situation. Leibniz is perfectly well able to account for a difference between true motions and merely apparent ones by reference to the causes of motions, even though he does not agree with Newton's interpretation of true motions as motions in absolute space.

According to this account of true motion, then, one can say that despite the fact that motion, considered geometrically, is merely relative, the Copernican System gives the true motions of the bodies in the solar system, since it gives the most apt

practice for commentators to identify 'true' with 'absolute', as is done by Huggett and Hoefer in their (2019). Two notable exceptions are Robert Rynasiewicz (1995) and Ori Belkind (2013), who both point out that the distinction between "true" and "apparent" motion was common property in the seventeenth century, and that this leaves undecided how "true" motion should be determined: whether with respect to a container space, or relative to bodies. See also Slowik's valuable analysis (2016, 23 ff.) of "true" versus "apparent" motions in the seventeenth century.

[25] Compare De Risi's lyrical comment: "What we are left with is only partial and local theories that, here and there, from time to time, give us, *per speculum et in ænigmate*, across and beyond phenomena, an intimation or a glimpse of true motion" (2011, 22). My reservation about this way of expressing it is that it suggests that there is a universally true motion (of which we obtain glimpses), whereas I think Leibniz's true motions are only local, relative to the system of phenomena to be explained. For example, the motion of the Moon across the rooftops is merely an apparent motion caused by my walking down the street, and the Moon is properly taken to be at rest in order to explain these phenomena; likewise its motion through the night sky is an effect of the Earth's rotation, which is taken as a true motion in this explanation.

or most intelligible explanation of them, and the truth of the hypothesis, as we have seen, "is nothing other than its aptness."[26]

Now, Leibniz did not just propose this account in his private papers, but advocated it publicly on several occasions as a way of finessing the problem of realism in cosmology. But in advocating an "Equipollence of Hypotheses" in astronomy, he has been accused of adopting an instrumentalist position, either as a compromise in order to have his own Keplerian account of planetary motions accepted in Catholic circles, or as his official philosophy on the role of hypotheses, thus undermining his claim in the *Tentamen* to "have come close to the true causes of celestial motions" (GM VI 149). This issue is deserving of an extended discussion.

3.2 Copernicanism and Instrumentalism

In discussions with leading intellectuals during his six months in Rome in 1689 Leibniz learned that the possibility of lifting the censure of Copernicanism was being actively discussed in some quarters close to the Holy See. Since he had just published a mathematical account of the planetary motions in accord with the Copernican Theory and Kepler's realist physical interpretation of it (in his *Tentamen de Motuum Coelestium Causis* written in Vienna in 1688), he certainly had ample motivation to lend his support to the cause.

We have four manuscripts relating to this episode, the first three of which were written when Leibniz was in Rome in 1689:[27] a preliminary note, the *De motu in rigore mathematico accepto* ("On Motion taken in Mathematical Rigour") (A VI 4, 2065–7) (hereafter *De Motu*), and two further drafts, published in the Akademie edition as *De praestantia systematis Copernicani* ("On the Superiority of the Copernican System") (A VI 4, 2068–9, 2070–5) (hereafter *De praestantia I* and *De praestantia II*). I have translated these in Appendix 4, together with an account of the circumstances of their composition. The fourth is the memorandum included by Couturat in his selection of previously unpublished manuscripts (1903), and titled by Ariew and Garber "On Copernicanism and the Relativity of Motion" (C 590–3/AG 91–4); I shall refer to it by its *incipit*, *Cum geometricis*. It was also written on paper with a Rome watermark, and Couturat mistakenly took it to be a preface for the *Phoranomus*, Leibniz's essay on dynamics in the form of

[26] Formerly I had described Leibniz's attribution of real motion on the basis of the most intelligible hypothesis as "only honorific" (Arthur 1994, 232). I would no longer describe it that way; I now believe Leibniz thought that one can identify what is really in motion by identifying its cause, this being that which is truly acting; but that one can still give equivalent descriptions of what is going on physically according to other hypotheses, as I explain below.

[27] See Bertoloni Meli's path-breaking paper (1988) for a detailed historical analysis, and Adomaitis (2019) for an up-to-date account of the manuscripts and their relations taking into account the publication of the first three in the Akademie edition.

two dialogues also written at that time. But if it is a preface, it is more likely one for a revised version of the *Tentamen*.[28] At any rate, among all four documents there is considerable agreement in aim, philosophical content, phrasing, and parts of the argumentative structure.

The last mentioned of these, *Cum geometricis*, has been most discussed, partly because it has been available in Couturat's edition since 1903 and in English since 1989 when Ariew and Garber published it in their translation volume (AG 90–4). In it, Leibniz argues that most of his contemporaries believed the Copernican hypothesis to be true, but distinguished astronomers among the Jesuits were prevented by the Censure from promoting it.[29] But if only it were recognized "that the truth of a hypothesis should be taken to be nothing but its greater intelligibility, all difference would vanish between those who presented the hypothesis as more agreeable to the intellect, and those who championed it as the truth" (C 592/AG 92). For such is the nature of this matter, he continues,

> that these are both the same thing, nor should one look for a greater or different truth. And since it is permitted to prefer the simpler hypothesis, it will also be permitted to teach it as the truth in that very sense. In this way the authority of the censors will also be preserved in such a way that they would never need to issue a retraction in the future no matter what new things were eventually discovered in the heavens or on Earth, and yet no violence will be done to the distinguished discoveries of our age (on the pretext of the Censure).[30]
>
> (*Cum geometricis*; C 592/AG 92–3)

This proposal has been variously interpreted. Because the argument begins with a strong statement of the Equipollence of Hypotheses, it would seem to undermine any claim that Copernicanism can be interpreted as literally true. From this point of view, the proposal to interpret "true" as "having greater intelligibility" appears as a compromise: if the Copernicans would accept this, they could teach the Copernican system openly, and in return would not insist that the words of the

[28] *Cum geometricis* does not address Leibniz's innovations in dynamics. I agree with Meli's suggestion that it was more likely a preface for the second redaction of the *Tentamen* (the "*Zweite Bearbeiten*," GM VI 161–87), tailored to be acceptable to the Censors in Italy and elsewhere (Bertoloni Meli 1988, 25, 36–40).

[29] As Bertoloni Meli discusses (1988), Leibniz had spelled this out in detail in the postscript of a letter to Landgrave Ernst that he wrote from Vienna on his way to Italy, recording how very able Jesuits like Claude Deschales were only prevented from expounding the Copernican hypothesis by censorship, and how the Jesuit Honoré Fabry and the Minim Friar Marin Mersenne regarded it as provisional (A I 5, 185–6). Compare with *De praestantia II*, Appendix 4.3.

[30] This sentiment echoes those expressed in *De motu* and *De praestantia II*: "nor in this way will there be a fear that the Censors would be forced into a retraction at some time by new discoveries" (A VI 4c, 2067), "And so with this explanation of the original censure the scruple which many people wrongly had would be destroyed, and [these considerations] would prevent the Censors from having to revoke a necessary decree at some point" (2074).

Bible be reinterpreted.[31] This interpretation appears to be in conflict, however, with the Keplerian interpretation of hypotheses given by Leibniz in his just-published *Tentamen*, in which, he hoped, he had "come close to the true causes of celestial motions" (GM VI 149).[32] Consequently, Leibniz's proposal has been interpreted either as an adventitious fix for the purpose of avoiding censorship; or, as an instance of his enduring commitment to an instrumentalist interpretation of theories or hypotheses, undermining the Keplerian pretensions of the *Tentamen*.

Thus Alexandre Koyré argued that *Cum geometricis* was "intended to show the authorities of the Catholic Church that, if they recognized, as they should, that motion was a relative concept, they would also recognize that the systems of Copernicus, Tycho Brahe, and Ptolemy were equivalent, and that therefore the condemnation of Copernicus was meaningless and should be lifted" (Koyré 1965, 125). But as Laurynas Adomaitis has convincingly argued, if the rival systems were equivalent with respect to the truth in planetary astronomy, then that would hardly be an argument for lifting the censorship, since arguments from the scripture would then carry the day.[33] Moreover, if they were equivalent in astronomy, the Church could choose to censor those that conflicted with scripture without detriment to science—contrary to Leibniz's expressed motivations for his intervention.[34] "The equivalence of hypotheses," therefore, "seems to ground Copernican censure rather than argue against it" (Adomaitis 2019, 69).

So how does the equivalence of hypotheses figure into the argument? Leibniz begins *Cum geometricis* with a strong statement of the relativity of motion:

> Since I have already derived through geometrical demonstrations the equipollence of all hypotheses about the motions of any bodies whatsoever, however numerous, that are moved solely by corporeal impacts, it follows that not even an

[31] Michael Friedman, for example, writes: "Leibniz was concerned with both furthering his goal of the reunification of Protestant and Catholic churches and, at the same time, using Clement XI's interest in astronomy to promote a grand compromise between the Church and Copernicanism in which the Church would allow Copernicans to hold that their opinion is the simplest and most intelligible—and in this sense 'truest'—hypothesis, and the Copernicans would concede that there is no need to reinterpret Scripture from a heliocentric point of view. Of course Leibniz failed in both of these grandiose schemes" (2010, 508).

[32] As Bertoloni Meli points out, Kepler claimed in the *Astronomia Nova* to be practising astronomy by means of physical causes, not fictitious hypotheses (1988, 38).

[33] "But the answer only plays into the hands of the opponent because since they are equivalent, the Church is free to censure one of them without any harm to knowledge, and, of course, the reason to do it is the word of Scripture in Joshua 10:13: 'So the Sun stood still'" (Adomaitis 2019, 69).

[34] As Meli observes, Leibniz's main motivation for wanting the Copernican censure lifted was his claim that it "hinders the development of science" (1988, 27), given that for him "the advance of human knowledge is a fundamental point which has to be put in the first place" (1988, 21). Meli fortifies this point by means of a detailed analysis of Leibniz's correspondence with Landgrave Ernst of Hessen-Rheinfels.

> angel could discern, in mathematical rigour, which of many bodies of this kind is at rest, and is the centre of motion of the others.[35] . . . To summarize, (since space without matter is an imaginary thing) motion in mathematical rigour is nothing but the change of situation of bodies to each other, nor is it something absolute, but consists in relation. (C 590/AG 91)

Here Leibniz reiterates that "motion in mathematical rigour" is mere change of situation, and is entirely respective. And, as he had argued in the *Principia mechanica*, "from the phenomena of the changed situation alone no certain knowledge can ever be had concerning absolute motion and rest" (A VI 3, 110). So this does not yet address motion with respect to cause. The tract *De motu* also begins with a strong statement of the relativity of motion, this time explicitly extending it to the phenomena of impact, where a body possesses some motive force and can exert a force of percussion:[36]

> I have demonstrated that when motion is taken in mathematical rigour, there is no principle for determining which of two bodies is absolutely at rest or is moving with such an absolute motion.
>
> Thus it can happen that a body can exert a kind of percussion or other motive force [*vim movendi*] and yet not be moving absolutely in mathematical rigour, but remain at rest—as when a ball thrown by hand in a ship runs in a horizontal straight line from the bow towards the stern, and meanwhile the ship is moving with equal speed in the opposite direction, from the stern towards the bow; there will be no absolute motion of the ball, even if the ball were also to receive an impulse from the hand and made an impact on another ball at rest on the ship.[37]
>
> And, specifically, I have demonstrated this to be a wonderful law of nature, that all phenomena, not only of bodies moving freely, but also those colliding with one another, take place in the same way, wherever rest or absolute determinate motion finally exists. (A VI 4c, 2065–6)

[35] The figure of the angel occurs also in *De praestantia II*: "so that not even an angel could determine something absolutely with mathematical certainty" (A VI 4c, 2072). See Appendix 4.3. What an angel could or could not do is, of course, a common trope of Leibniz's methodology. See Butts (1985) for a discussion.

[36] This corresponds to the difference in Leibniz's *Dynamica* between Proposition 16 of section II and Proposition 15 of section III (GM VI 501). Concerning the former he writes: "That is, we understand what is and can only be recognized in purely mathematical motion, namely situations and their changes. But the physical part, namely, the subject of the cause of acting and forces, and of apt explanations giving reasons, is another consideration" (GM VI 484–5).

[37] Leibniz gives the same example of the ball being thrown on the ship in *De praestantia II* to support the same conclusion, namely that "it is therefore true that not even a force of acting [*vis agendi*] is a certain indicator of absolute motion" (A VI 4c, 2072).

'Absolute determinate motion' here connotes motion that is not merely relative to other bodies, but a "change of real space" i.e. of position in an assumed background space—the "space without matter" derided as "an imaginary thing" in *Cum geometricis*. In both documents Leibniz then turns his attention to the distinction between true and apparent motion:

> Since, nevertheless, people do assign motion and rest to bodies, . . . it must be seen in what sense they can be judged not to have spoken falsely. And it must be replied that one should choose that hypothesis which is more intelligible; and that the truth of a hypothesis is nothing but its intelligibility.
>
> (*Cum geometricis*, C 590–1/AG 91)

Likewise, in *De motu* Leibniz advances the criterion of greater intelligibility to distinguish between rival hypotheses: "it follows that whenever one system is preferred to another this can only be accepted in the sense that one is more intelligible than the other and is better accommodated to the things it is proposed to explain" (A VI 4c, 2066). In that tract he continues: "therefore when one hypothesis is said to be true and the other false, this is to be understood of it in its own respect." Consequently, "the Spherical doctrine of Ptolemy is just as true as the Theoretical planetary system of Copernicus, because the true way of explaining, which is the right and appropriate and best one, is the same for both" (2066). This relativization to a particular respect is repeated in *De praestantia II* (A VI 4c, 2073), and also in *Cum geometricis* in a passage that commentators have taken for an explicit statement of instrumentalism:

> And since one hypothesis might be more intelligible than another and more fitting for a given purpose in a different respect—not so much in respect of people and their opinions, as rather of the very matters that need to be treated—so also one will be true in a different respect, the other false. For a hypothesis to be true, therefore, is for it to be properly employed. So, although a painter can present the same palace through different scenographic projections, we would judge that he made the wrong choice if he preferred one which covers or obscures parts that are important to know in the circumstances. In just the same way, an astronomer makes no greater mistake by explaining the theory of the planets in accordance with the Tychonic hypothesis than if he used the Copernican hypothesis for teaching spherical astronomy and explaining day and night, thereby burdening the student with a difficulty that has no place here. And the observational astronomer [*Historicus*] who insists that the Earth moves, rather than the Sun, or that the Earth rather than the Sun is in the sign of Aries, would speak improperly, even though he followed the Copernican system; nor would Joshua have spoken less falsely (that is, less absurdly) had he said "be still, Earth!"
>
> (C 590–1/AG 91–2)

Commenting on these passages from *Cum geometricis*, Ed Slowik writes "one would be hard pressed to find a more straightforward elaboration of an instrumentalist or pragmatist theory of truth, since the truth of a theory's claims about the individual states of motion are linked to that theory's being 'more appropriate for a given purpose'" (2013, [7] 56–77). Thus the upshot of the Equivalence of Hypotheses, according to Slowik, "is that the truths of individual bodily motions are relative to the theory used to explain those motions." But is this what Leibniz says? What he seems to be alluding to with the example of the Tychonic hypothesis is that it would be a mistake for an astronomer to use it for explaining the appearance of the planets, since that would obscure things that are important to know in the circumstances. In other words, the aptness of a hypothesis is its fitness to explain the phenomena concerned. But this is not the same as saying that truth is dependent on the *theory* (i.e. hypothesis) in question. If that were so, Leibniz would just be contradicting himself. He would be saying that the equivalence of hypotheses means that "the same body may be assigned different states of motion, and hence truths, from the perspective of these different hypotheses" (Slowik 2013, [6]), and yet that the Copernican hypothesis should be preferred as the true one, since it is more intelligible and more apt than any of its rivals. It would also fall foul of Adomaitis' criticism of Koyré: if truth were relative to the hypothesis adopted, then this would give no reason for the censure to be lifted, nor justify Leibniz's claims that the progress of science is being jeopardized by it.

But Adomaitis himself has also proposed an instrumentalist interpretation. Like Slowik, he stresses that the equating of the truth of a hypothesis with its aptness or fit is not an opportunistic move, but Leibniz's consistent position through the years.[38] And like Slowik, he interprets Leibniz as denying that true motions can be assigned by hypotheses. But the instrumentalism he ascribes to Leibniz is somewhat different: "The interpretation that we are developing here denies the ability to 'actually assign motion and rest' (Garber 2009, 115) and, therefore, might be called *instrumentalist*, with a proviso that this should not be taken in a contemporary sense of the term" (Adomaitis 2020, 131).[39] It is also not to be confused with the kind of instrumentalism advocated by Andreas Osiander and Cardinal Roberto Bellarmino, and opposed by Kepler.[40] On Adomaitis' reading, "causal relations are an important factor in calculating the simplicity of a hypothesis but they do not make hypotheses certain" (131). Thus on his reading,

[38] See Slowik (2013, [7] & [9]) and Adomaitis (2019, 66).

[39] In his (2019), Adomaitis writes: "The modern sense [of instrumentalism] would be that of John Dewey or Charles S. Peirce that includes many premises which Leibniz does not accept, e.g. Leibniz does not see science as a 'doubt-inquiry process' and applies the instrumentalist criteria only to equivalent hypotheses, not the laws or axioms" (2019, 80, n.52).

[40] "The sense in which they [Osiander and Bellarmino] think of equivalent hypotheses," he writes, "is that they are incompatible states of affairs but there is no way of deciding between them. Leibniz seems to be endorsing the view that since they are indistinguishable (sc. relative), they cannot be truly contradictory because the very matter that they describe is relative" (Adomaitis 2019, 68, n.14).

no assignment of true motions can be made, since all hypotheses are equivalent; yet the Copernican hypothesis can still be preferred as the simplest. Although simplicity is a criterion for choosing the best hypothesis, this does not render the hypothesis true, in the sense of demonstrable. That is, taking the term 'true' as a synonym for demonstrable or absolutely certain, Adomaitis insists that hypotheses cannot be true, and so he uses "*true" to signify true in the sense of being the most intelligible hypothesis, as opposed to demonstrated:

> Although we can demonstrate physical laws and prove that they are true, we can only accept hypotheses as instrumentally true or, in our notation, *true because if judged by the absolute standard of truth: "since [hypotheses] cannot be consistent, they are all false" (A6.3.110, tr. Arthur). The criterion for the truth of physical laws is their demonstration; whereas the criterion for the *truth of hypotheses is simplicity. (Adomaitis 2020, 128)

It must of course be granted that hypotheses are not demonstrated, since once a hypothesis is demonstrated it is no longer regarded as just a hypothesis. And in his PhD thesis Adomaitis (2020) has made a strong argument for regarding Leibniz as having demonstrated certain of the axioms of physics using the principle of the Equipollence of Cause and Effect and the principle of substitution *salva vi*. But I do not see why we should simply identify truth with being demonstrated, with its consequence that Leibniz is prohibited from calling some hypotheses true or certain. According to his philosophy of science, axioms, as necessary truths, must be demonstrable a priori; but even contingent truths may be demonstrable a posteriori. They will then be demonstrated, but not absolutely (i.e. *necessarily*) true, for they will depend upon empirical premises, making them fallible. But if a hypothesis satisfies a wide range of phenomena it may be regarded as (at least morally) certain, and as giving the true causes. On Adomaitis' reading, astronomical hypotheses for Leibniz are "frames of motion," ascribing "different configurations of moving objects in which motion and rest cannot be ascribed to any of the bodies with certainty."[41] But Leibniz, on the contrary, is explicit that some hypotheses *can* be held for certain. As he wrote in the preface to his planned Elements of Physics of 1680,

> *Some hypotheses can satisfy so many phenomena and so easily that they can be held for certain. Among other hypotheses, those are to be chosen which are simpler, and in the interim they can be employed instead of true causes.* The conjectural

[41] "Astronomical hypotheses and other relative configurations (which we call frames of motion) described above have essentially become Leibniz's new vision of hypotheses. Hypotheses are different configurations of moving objects in which motion and rest cannot be ascribed to any of the bodies with certainty" (Adomaitis 2020, 127).

> method *a priori* proceeds by hypotheses, assuming certain causes, albeit without proof, and showing that what would happen now according to those assumptions does indeed take place.... Hence from the success of a hypothesis no firm demonstration can be deduced. Yet I shall not deny that the number of phenomena which can be happily explained by some hypothesis may be so great that it can be taken as morally certain. And hypotheses of this kind may even suffice for general use.... Phenomena are virtually contained in the hypotheses from which they can be deduced, and so those who remember the hypothesis will easily be able to recall these phenomena when they want, even if they know that it is false and certain other phenomena may be found with which it conflicts. Thus the Ptolemaic hypothesis may suffice for novices in astronomy, who wish to remain content with certain notions of the common people about celestial things.
> (*Praefatio Ad Libellum elementorum physicae*, A VI 4c, 1999, 2000)

What seems clear from this discussion is that, even though hypotheses cannot be regarded as firmly (i.e. a priori) demonstrated, their whole purpose is to assign true causes, which they do with varying degrees of certainty or verisimilitude; and that those that most intelligibly explain the relevant phenomena may be taken as (at least morally) certain or true;[42] and that even those that are known to be false, like the Ptolemaic hypothesis, may still be the most apt to explain certain things in certain circumstances. This, I contend, is what Leibniz means by saying in *Cum geometricis* that "for a hypothesis to be true is just for it to be properly used," and that as a result, when explaining certain phenomena relative to the Earth taken as fixed reference point, an Earth-centred system (whether Ptolemaic or Tychonic) may be the proper one to use in that context.

Adomaitis makes much of the fact that in *Principia mechanica* Leibniz claims that no hypothesis can be refuted: "For since no hypothesis can be refuted rather than others through certain demonstration, not even by someone omniscient, it follows that none is false rather than others; that is, (since they cannot be consistent) they are all false" (A VI 3, 110/Arthur 2013c, 115). But the hypotheses Leibniz is referring to here are ones attributing an absolute motion or rest to the bodies in question—one that is not relative to particular bodies. But as I have observed, this is a separate question from that of determining true rather than apparent motions in a given scenario, which is achieved by reference to the most likely cause. Thus immediately after the above quotation, Leibniz adds: "In this case, however, we will be permitted to choose the simpler mode of explaining, which involves reference to a cause from which the remaining changes may be

[42] Cf. the discussion in Leibniz's *De modo distinguendi phaenomena realia ab imaginariis* (A VI 4, 1502/L 364), and also in "Metaphysical Definitions and Reflections" of 1678–79 (A VI 4, 1398/LLC 243–5), where Leibniz says that it suffices for things that appear to be called *true* if they "agree with each other and obey certain laws, and accordingly leave a place for human prudence and predictions."

derived more easily.[43] Hence we will not hesitate to attribute motion to the solid body from which we can deduce the undulations of the surrounding liquid, rather than thinking of these undulations as originative..." (110/115).

Regarding astronomical hypotheses, it is Copernicanism, of course, that gives the most intelligible explanation of the astronomical phenomena. The retrograde motions of planets, for example, are depicted in the Copernican system as merely apparent: the true cause is the motion of the Earth around the Sun. In keeping with this, all four texts present ringing endorsements of Copernicanism. Thus in *De motu* Leibniz writes "And indeed so superior is the Copernican system for explaining the phenomena of the planets, that it must be admitted that an astronomer who did not understand this would be groping around in the greatest darkness" (A VI 4c, 2067)—a sentence repeated almost word for word in *De praestantia II* (A VI 4c, 2067), with the same theme of those not adopting it being condemned to darkness recurring in *Cum geometricis* (C 592/AG 93). And in the latter Leibniz specifies some of the phenomena brought to light more intelligibly in the Copernican system: its dispatching at a stroke the "labyrinths of stations and retrograde motions of the planets," and its exhibiting the "magnetic law" evident in the binding of the moons of Jupiter and Saturn.[44]

But in all four texts Leibniz indicates that it is Kepler who has done most towards identifying the true cause of the motions of the planets. As he says in *Cum geometricis*, Kepler has outdone Copernicus, "observing that all the phenomena are produced if it is supposed that the Earth and all of the primary planets are borne in an ellipse whose focus is the Sun, and that it is a law of motion in an orbiting planet that the areas swept out from the Sun are proportional to the times" (C 593/AG 93). In each of Leibniz's texts this serves as an introduction to a brief summary of the main argument of his *Tentamen*.[45] This is stated most succinctly in *De motu*:

> And we ourselves, having taken the universal laws of planetary motion discovered by Kepler in his principles, namely, the harmonic circulation of celestial matter as deferent orbits of planetary fluid around the Sun, and the rectilinear approach of the planet as something heavy and towards the Sun as if that were a magnet, and having geometrically resolved them, have, if I am not mistaken, contributed something towards obtaining the best explanation of the phenomena, that is, the truest hypothesis. (A VI 4c, 2067)

[43] Cf. "It is impossible to judge whether and to what extent a given body is at rest or moves, unless one has a reason for its greater explicability" (*De praestantia II*; A VI 4c, 2072).

[44] Characteristically, Leibniz does not enter into the details that have persuaded astronomers to prefer the Copernican hypothesis over the Tychonic, which latter can also explain the differences between the true and apparent motions of the planets in the same way.

[45] Leibniz gives the fullest description of "certain thoughts which seem to us to reveal in good measure the secrets of the system and physical causes of celestial motions" (A VI 4c, 2068) in *De praestantia I*. See Appendix 4.2.

Here Leibniz is adhering quite closely to Kepler's own account of planetary motion, not only in adopting the inverse proportionality between the velocity of circulation and the distance of the planet from the Sun (which he called a *harmonic motion*), but also in conceiving the motion accounting for the variation in the distance of the planet as having a different cause.[46] Moreover, as Meli has argued, it is important to see that Leibniz is not just presenting himself as an heir to Kepler's version of Copernicanism and his three laws, but as offering, in keeping with Kepler's philosophy of science, a *physical hypothesis* about the causes of the celestial motions. This harks back to Kepler's defence of Tycho Brahe against the criticisms of Raimarus Baer (Ursus). Baer had argued that "accurate prediction and retrodiction of apparent celestial coordinates does not guarantee the truth of astronomical hypotheses," and that "all hypotheses involve blatant absurdities" (Bertoloni Meli 1993, 21). Kepler countered that whereas geometrical hypotheses, like Ptolemy's hypothesis of an equant point, can perhaps accurately account for some observations, many other geometrical hypotheses might do so too. True hypotheses, on the other hand, also give a causal explanation, and this has to be justified in a wider context, on both physical and philosophical grounds, thus reducing the chances that the agreement is merely accidental.

It is quite likely, Meli suggests, that Leibniz was moved to stress his agreement with Kepler in the *Tentamen* in response to the appearance of Newton's *Mathematical Principles of Natural Philosophy*. For Newton was proposing that natural philosophy could be erected on "mathematical principles" alone, and then compared with the phenomena, independently of any commitment to their causes. In contrast, Kepler had written in his *Epitome Astronomiae Copernicanae*, "you should give *causae probabiles* for your hypotheses which you propose as the true cause of appearances, and thus establish in advance the principles of your astronomy in a higher science, namely physics or metaphysics."[47] For Leibniz and his contemporaries, this meant proposing a causal hypothesis in agreement with the mechanical philosophy. This explains what motivated Leibniz to compose the *Tentamen*. As a mere deduction of the inverse square law it would have been largely superfluous, given Newton's magnificent achievement—even if Leibniz had, in his derivation, ingeniously pioneered the use of differential equations in physics. The idea was to show how Newton's mathematical results could be achieved on the basis of a sound mechanical hypothesis, and Leibniz could see

[46] See Eric Aiton's (1972) for a very readable and accurate analysis of Kepler's theory. In the *Tentamen*, Leibniz credits Kepler with "opening the way to the investigation of causes [*rationes*]"; "For to him we owe the first indication of the true cause of gravity and of the law of nature on which gravity depends, which is that rotating bodies endeavour to recede from the centre along the tangent" (GM VI 148/Meli 1993, 127). For a penetrating analysis of how Leibniz used Kepler and adapted him to his own purposes, see Bertoloni Meli's (1993), esp. 27–33.

[47] From Kepler's collected works (KGW VII, 25), quoted from Bertoloni Meli (1993, 22), who himself is quoting the translation given by Jardine, *Birth of History and Philosophy of Science* (Cambridge 1984), 250.

no more rhetorically effective way of achieving this than by showing how Kepler's own hypotheses could be interpreted in terms of the mechanical philosophy. So he attempts to deduce the inverse square law starting from the hypothesis of harmonic circulation, where the planets are carried by a circulating vortical fluid, together with a second hypothesis accounting for the "magnetic attraction" along the radius toward or away from the Sun.[48]

Leibniz was by no means alone, of course, in holding that the Copernican system needed accounting for by an appeal to hypotheses that are in conformity with the mechanical philosophy. This was also the case for Huygens, who had proposed his own vortex theory, and the two of them had discussed Leibniz's ideas in letters exchanged in 1692–94. In his letter to Huygens of June 22, 1694 Leibniz reiterates his view that although "the phenomena could not provide us (or even angels) with an infallible basis for determining the subject or degree of motion," nevertheless "if motion… Is something real, it is necessary that it have a subject" (GM II 184/L 418). He recalls how when they were in Paris together twenty years earlier, Huygens had at that time been of the opinion that circular motion could serve as a criterion for absolute motion, as Newton now maintained (185/418). Resuming that topic in his letter of September 14, 1694, Leibniz mentions that Huygens had mentioned this to him one day in Paris, when he himself "believed he could already see that circular motion has no advantage in this respect":[49]

> And I see you are now of the same opinion as me. I maintain, therefore, that all hypotheses are equivalent, and when I assign certain motions to certain bodies I do not have, nor could I have, any other reason than the simplicity of the hypothesis, believing that one may hold the simplest hypothesis (all things considered) to be the true one. Thus, having no other criterion, I believe that the difference between us is only in the manner of speaking, which I try to accommodate to common usage as much as I can, *salva veritate*. I am not even very far from your own view, and in a little paper that I communicated to Mr. Viviani and which seemed to me suited to persuade Messrs. of Rome to allow the opinion of Copernicus, I accommodated myself to it.
>
> (GM II 199/L 419)

Thus Leibniz acknowledges that in the tract he sent to Viviani he agrees with Huygens' position, namely that all motion is essentially relative, so that all

[48] As Meli notes, however, "it should not be overlooked that for Kepler the outward motion of a planet was due to magnetic repulsion, whereas for Leibniz it was an effect of the planet's own orbital motion: no repulsive force was needed in his account" (Bertoloni Meli 1993, 28).

[49] In the light of these remarks—and the facts themselves!—it is quite remarkable that Leibniz should have written to Thomas Burnett three years later (May 18, 1697) that he and Huygens had *both* been of the view that centrifugal force was a criterion for absolute motion "when Newton's book appeared," but had each independently changed their opinions since (GP III, 205). Gianfranco Mormino gives a nice analysis of this strange assertion in Mormino (2011).

hypotheses are equivalent. He believes that the only difference of opinion between the two of them is that he is trying to save what truth there is in common expressions—as when we talk of the Sun rising in the East—by "accommodating his own opinion to common usage" as far as possible, while preserving the truth. He believes that the true hypothesis is the one that is most probable, explaining the phenomena in the simplest way. Thus, he believes, he can say that if one is simply explaining the phenomena of the relative motions of the planets from the point of view of spherical astronomy (concerning the phenomena as seen from Earth), then the hypothesis "most apt" to explain them is the Tychonic, where the Earth is taken to be at rest. This is analogous to the way we still talk of the Sun rising in the East, even though we know that the apparent motion of the Sun is due to the Earth's spinning on its axis. Leibniz gives this very example in a letter he wrote to Landgrave Ernst von Hessen-Rheinfels in Vienna in Summer 1688 on his way to Italy:[50]

> If Joshua had been a disciple of Aristarchus or Copernicus, he would not have changed the way he expressed himself, otherwise he would have shocked the people present as well as common sense. All Copernicans, in their ordinary speech and even among themselves, when the issue is not scientific, will always say that the sun has risen or set, and will never say the same of the earth. These terms pertain to phenomena, not to their causes. (A I 5, 186/Bertoloni Meli 1988, 23)

We can find Leibniz giving a fuller account of this concern for accommodating accounts of motion and cause to common usage in the *Système nouveau*. There he writes:

> Ordinary ways of speaking can still be preserved. For one may say that when the disposition of a certain substance provides a reason for the change in an intelligible way, so that we can judge that the other substances have been adapted to it in this regard from the beginning according to the order of God's decrees, then that substance should be thought of as *acting* upon the others in this sense. Moreover, the action of one substance on another is not an emission or transplanting of some entity, as is commonly supposed.... It is true that in matter we can easily conceive of both emissions and receptions of its parts, through which one can rightly explain all the phenomena of physics mechanically; but as material mass is not a substance, it is clear that action with respect to substance itself can only be what I have just described. (GP IV 486/L 459)

It may be thought that such considerations are only pertinent in metaphysics, where it is crucial (for morality and much else) to be able to identify the agent

[50] Cf. also *De praestantia* II, (A VI 4c, 2071) in Appendix 4c.

responsible for a given action. We can identify (perhaps fallibly) the reason we have for an action in ourselves, and by extension can guess the reasons others have for theirs. But as this is not possible for bodies, it may be thought that in physics, as opposed to metaphysics, such "agent causation" is inapplicable.[51] Were that the case, then Leibniz's position would reduce to a mere instrumentalism: we can assign motions according to the hypothesis that is most useful to us, but there is no fact of the matter in physics as to which body causes which to move. But in the continuation of the above passage from the *Système nouveau*, Leibniz makes clear that the "providing of a reason for a change in the most intelligible way" is also of use in physics, both for establishing the correct laws of motion and for identifying the "true motions" of bodies:

> These considerations, however metaphysical they may appear, are still of remarkable use in physics for establishing the laws of motion, as our *Dynamics* will be able to elucidate. For one can say that in the collisions of bodies each suffers only from its own elasticity, caused by the motion that is already within it. And as for absolute motion, nothing can determine it mathematically, since everything ends in relations [*rapports*]. As a result, there is always a perfect equivalence of Hypotheses, as in Astronomy, so that however many bodies one takes, it is arbitrary to assign rest or even any degree of speed to any of them you choose, without the possibility of being refuted by the phenomena of motion, whether rectilinear, circular or composite. However, it is reasonable to attribute true motions (*véritables mouvemens*) to bodies following the supposition which provides the reason for the phenomena in the most intelligible way, this denomination being in conformity with the notion of Action that we have just established. (GP IV 486–7/L 459)

There are two outstanding problems, however. One is the matter of whether the equivalence of hypotheses does indeed apply to all motions, or whether there is a criterion for absolute motion in circular motion, as Huygens had once believed, and Newton had argued in his *Principia*. Before coming to that, however, we need to treat the question of where force fits into this picture. For Leibniz claimed that "motive force, or the power of acting, is something real, and can be discerned in bodies" (A VI 4, 1622/LLC 315). And, as he wrote to Huygens, after stating the equivalence of hypotheses as above, "you will not deny, I think, that each body does have a certain degree of motion, or if you wish, force, in spite of the equivalence of these hypotheses about their motion" (GM II 185/L 418). Does this not break the equivalence of hypotheses?

[51] Laurynas Adomaitis has sought to persuade me of such a position in private conversations, and this is the line he takes in his PhD thesis for the Scuola Normale in Pisa (2020). The distinction between "agent causation" and "event causation" (only the latter of which Adomaitis thinks applicable to Leibniz's physics) is due to Jonathan Lowe.

3.3 Force and the Metaphysics of Motion

In the discussion so far I have concentrated on those aspects of Leibniz's views that remained constant over the last four decades of his life. But, of course, his natural philosophy and metaphysics underwent a profound change in the late 1670s with the rehabilitation of substantial forms, and the subsequent fashioning of his dynamics in the 1680s and 1690s. Now, clearly Leibniz's notion of force is of the greatest relevance to his understanding of the causes of motion, and thus to the problem of the relativity of motion. How then does force come into the picture?

In the *Discourse of Metaphysics* Leibniz refers to "the force or immediate cause behind those changes" being "something more real," and as providing "enough of a basis for attributing [motion] to one body rather than another" (§18, A VI 4, I558–9/AG 51). The context makes clear that Leibniz is referring to his new concept of force announced in his *Brevis demonstratio*, whose measure is mv^2. This suggests that an appeal to force can override the equivalence of hypotheses that applies when motion is understood simply as change of situation. Similarly, Leibniz writes to Arnauld in January 1688:

> Motion in itself, separate from force, is only a relative thing, and its subject cannot be determined. But force is something real and absolute, and since its calculation is different from that of motion, as I demonstrate clearly, it should not be surprising that nature retains the same quantity of force and not the same quantity of motion. (A II 2, 274/LAV 279, 339)

Commentators have struggled to make sense of these claims. If, as John Roberts asserts, "what Leibniz is referring to here by force is clearly the quantity mv^2 that he claims is conserved in all interactions" (Roberts 2003, 558), then it is hard to see how this would allow him to escape the relativity of motion. For, as Bertrand Russell had already objected, if force is measured by mv^2 as its effect, "the effect can only be measured by means of motion, and thus the pretended escape from endless relativity breaks down" (Russell 1900, 86). Roberts, however, does not see that as an insurmountable objection, and has suggested that Leibniz's meaning is that there is a frame of reference in which the body has an absolute speed corresponding to this measure of force, although it is not physically determinable.[52] This is similar to the analysis previously given by Daniel Garber, according to whom

[52] Roberts argues that "mv^2 is a conserved quantity in any inertial frame, but it leaves open the possibility that there is no way of empirically determining the true *vis viva* of any body or of establishing a reference frame that attributes to each body its own absolute speed" (Roberts 2003, 567). Cf. Huggett and Hoefer (2019): "At the very least, Leibniz does say that there is a real difference between possession and non-possession of *vis viva* (e.g., in Section 18 of the *Discourse*) and it is a small step from there to true, privileged speed. Indeed, for Leibniz, mere change of relative position is not 'entirely real'... and only when it has *vis viva* as its immediate cause is there some reality to it." Slowik

> There is, in this sense, a correct frame for determining motion, the frame in which the motions observed are the effects of real underlying forces which are their causes. *But such a frame could never be identified.* Motion has a foundation, but one that makes no real (or apparent) difference in the world of physics; in this way the theory of motion would seem to float free of its foundations in the notion of force. (Garber 1995, 308)[53]

The idea here is that Leibniz's force, insofar as it is the foundation of motion, plays no part in the determination of mechanical motion.[54] This is true, I would suggest, only if one interprets force as *primitive force*. For one line of argument Leibniz has for the reality of force is a metaphysical one: active force consists in the inherent activity of each substance, the tendency of any of its states toward later states. As Leibniz tells De Volder, "when I speak of the enduring primitive force, I do not mean the conservation of total motive power about which we were once concerned, but the entelechy that always expresses that total force, as well as other things" (June 20, 1703; LDV 262/263). Since Leibniz insists that such metaphysical concepts should not be appealed to in physical explanations, it is true that insofar as force provides a real foundation for motion in this sense, it cannot be appealed to in explaining the phenomena of motion.

As noted above, however, it is not the primitive active force that Leibniz is appealing to in referring to "the calculation of force" in the passages quoted above, but rather the derivative active forces of his physics.[55] In that case, though, it would be unreasonable for Leibniz to argue that one can use such a calculation to demonstrate that it is the quantity of force, not of motion, that is conserved (and in this sense absolute) if, as Roberts and Garber would have it, the forces underlying real motions can never be determined or identified. Rather, it would

also endorses the privileged reference frame hypothesis, but finds that it "would necessitate a potentially infinite set of separate reference frames for the potentially infinite set of particle collisions in Leibniz's plenum" (2006, 631).

[53] H. G. Alexander had made a similar suggestion in his edition of the Leibniz–Clarke correspondence: Leibniz "may therefore have held that the distinction between absolute and relative motion is metaphysical, not physical: that is, the absolute motion of a body can never be experimentally determined; and so the concept of absolute motion is of no use in physics" (LC, xxvi–xxvii).

[54] Cf. Huggett's claim (in an as yet unpublished ms.) that "force, being form, while necessary for the laws, cannot play a role in particular mechanical explanations. Moreover, since force belongs to the dynamical realm, which ontologically precedes the phenomenological world to which spatial concepts belong, force and hence motion in the sense of LM are at root non-spatial" (Huggett 2006, 14). Here "LM" denotes "Leibnizian motion," involving "possession of non-zero force, and more specifically the particular speed implied by the value of the force" (Huggett 2006, 13).

[55] As Anya Jauernig rightly observes, although some of Leibniz's references to force as a foundation of motion are to primitive force, "Leibniz states as least as often that the real in extension is force, where the context makes it clear that he is talking about physical forces, in particular the passive forces of impenetrability and resistance" (Jauernig 2008, 9). But she thinks that these "expressions clearly do not refer to the active forces of simple substances, but to the physical forces of corporeal substances," erecting a distinction where I think none exists; see Arthur (2018, ch. 6) for an analysis of this topic. That Leibniz pursued two different lines of reasoning for the reality of force was pointed out earlier by Paul Lodge (2003, 283).

seem that the force that can be calculated must be the derivative active force, which would therefore be determinable. The way such a calculation of forces is supposed to provide "enough of a basis for attributing [motion] to one body rather than another," however, needs further discussion.

Leibniz's insistence that force is an essential component in the reality of motion is directed primarily at the Cartesians, as Paul Lodge has stressed.[56] He quotes Leibniz from his critique of the Cartesian philosophy, the *Animadversiones in partem generalem Principiorum Cartesianorum* from 1692:

> If motion is nothing but change of contact or of immediate vicinity, it follows that we can never define which thing is moved. For just as the same phenomena may be interpreted by different hypotheses in astronomy, so it will always be possible to attribute real motion to one or the other of the two bodies which change their mutual vicinity or position. Hence, since one of them is arbitrarily chosen to be at rest or moving at a given rate in a given line, we may define geometrically what motion or rest is to be ascribed to the other, so as to produce the given phenomena. Hence if there is nothing more in motion than this reciprocal change, it follows that there is no reason in nature to ascribe motion to one thing rather than to others. The consequence of this will be that there is no real motion. (GP IV 369/L 393)

As Lodge observes, this is a hypothetical argument, whose initial premise and conclusion Leibniz does not accept (Lodge 2003, 278–9). In fact, Leibniz's overall conclusion of this discussion is: "Thus in order to say that something is moving, we will require not only that it change its position with respect to other things but also that there be within itself a cause of change, a force, an action" (GP IV 369; L 393).

To throw some light on the connection of force with a cause of change, Lodge makes use of an interesting sequence of unpublished drafts written by Leibniz some time between summer 1678 and winter 1680–81, of which the second, *De motu corporis* (A VI 4, 2011–12), seems to be a first draft of the fifth, *De causa motus* (A VI 4, 2017–19), a treatment of the problem of the cause of motion. The first piece begins by defining what it means for a body to be in motion:

> *That body is in motion* which is the proximate efficient cause why a certain part of it changes its situation with respect to other bodies; otherwise it is said to be *at rest*. (A VI 4, 2011)[57]

[56] "I have suggested that Leibniz's main response to the argument from relativity relies on the introduction of a conception of body that is richer than the Cartesian one" (Lodge 2003, 297).

[57] As he explains, he refers to 'parts' in the definition because "if a body is moved around its own axis, it cannot be said to have changed its situation with respect to others, even though its part changes situation with respect to others" (A VI 4, 2012)

If two bodies, like the Earth and Sun, are changing situation with one another, Leibniz explains, we could not on that basis say which is really moving. "But if we knew what the proximate cause of the changed situation is," he argues, "we would attribute the motion to that";

> For example, if there were in the Earth a certain intelligence which moved it around its orbit; or, if we suppose it had been moved shortly before, since once a motion has begun it does not stop without a reason; or, if certain things were to occur in the fluid surrounding the Earth which result from a certain force of action of the Earth's being in motion, but which could not otherwise be explained by a certain cause—to wit, if the surrounding aether receded from the Earth in such a way that it would have to be repelled by it if the Earth moved—certainly this would establish that the Earth moved, for otherwise the various motions in the ambient fluid could not be reduced to a single cause.

The first of these cases refers to Kepler's idea of intelligences moving the planets, which he used to buttress his realist interpretation of Copernican orbits. In the second, as Lodge notes, it looks as though Leibniz means to identify the proximate cause of the Earth's motion with the initial cause by an appeal to a principle of inertia (an appeal to its inertia would explain why the body was moving, but not in what direction and velocity as a result of collisions). The third is consistent with the examples we have seen earlier of a solid body's motion in a fluid where, although the convection of the fluid around the body *could* be explained by positing numerous causes for such motions with the body at rest, it is vastly more intelligible to attribute it to the single cause of the body's being really in motion. Lodge claims that these examples show "that Leibniz thinks bodies whose relative motion is caused by a soul or active principle have real motion" (Lodge 2003, 289), but I do not see any reason to attribute that view to him on the strength of this passage. Leibniz seems to have firmly rejected the conception of causation as involving an action of created souls on bodies from 1676 onwards, as I shall argue further below.

Still, given the Akademie editors' dating of these texts (summer 1678–winter 1680/81), they were written by Leibniz in the immediate aftermath of his articulation of his new conception of force in February 1678, at a time when he would have been trying to work out its implications. Lodge draws attention to a passage in the *De causa motus*, where he appears to present "the communication of motion" from one body to another as a criterion for the first being really in motion. Daniel Garber comes to the same conclusion in analysing this passage in his (2009, 112–15). In the passage Leibniz first of all gives his (now familiar) distinction between motion mathematically considered, and motion with respect to physical causes:

> Moreover, for motion I require not only a change of situation from other things, but also that the cause be in that to which motion is ascribed. For if we consider only change of situation, or what is merely mathematical in motion, it cannot be said to which of several bodies changing situation among themselves motion should be ascribed. And so it happens that in Astronomy too, so many hypotheses can be imagined, of which kind the one that places the Moon at the centre of the world is by no means probable.[58] But if we come to physical causes, then we can more easily establish that to which motion should be ascribed.
>
> (A VI 4, 2017–18)

Then he presents a scenario where, when a body *c* is assumed to collide with a body *d* it communicates motion to it, and is therefore really moving: "for then we pass from the simple mathematical consideration of the change of situation to physics, namely to some action, for then some communication of motion will follow, from which it will be evident that some action is to be ascribed to *c*" (A VI 4, 2018). I confess that it is not so obvious to me that Leibniz intends this as a stand-alone criterion for whether a body is really in motion. For he immediately proceeds to argue that a body that is at rest relative to surrounding points taken as immobile could still have a force of acting. He gives the example of a man who is walking along the deck of a ship from the prow to the stern with an equal and opposite speed to that of the ship with respect to the bank. The man is not moving relatively to the bank, and yet

> it should nevertheless be said that he is in motion; for that man certainly feels himself getting tired by walking, and can push or pull something by his motion; and therefore he acts. Since, then, we attribute motion to that in which there is a cause of change of situation, ôr that which acts, we certainly say that the body *e* moves, even though the place that it has in the world and which is taken with respect to fixed points, does not change. (A VI 4, 2019)

The criterion of the man feeling fatigue already occurred in the *Principia mechanica* (A VI 3, 104), as we saw above. But this is no more a purely mental criterion of real motion than it was there:[59] he feels tired because he has done work to cause

[58] Adomaitis (2020, 127) has plausibly suggested that here "Leibniz is probably referring to Kepler's *Somnium, sive Astronomia Lunae* (1634). In it, Kepler imagines the Moon as the island of Levania and Levanians believe that the Moon is the center of the universe."

[59] Cf. Lodge (2003, 289): "In each of the examples that I have mentioned we find Leibniz appealing to a notion of force that might be regarded as mental rather than physical in character" (but I am not suggesting that he was intending this as a strict dichotomy). Adomaitis (2020, 132) interprets Leibniz as relying on a "feeling," and suggests that feelings cannot count any more than "seeings" as criteria for true motions: "Feeling is not a privileged way of perceiving. Yet, for example, Leibniz would not agree that since a person sees that the Sun moves, it must move. The same goes for feeling, namely, phenomenal experience can be misleading." As I shall argue below, however, being aware that one has expended force (energy) is a perfectly good empirical criterion for being an agent and not a patient.

his own motion, and likewise, through his work he could cause new motions in other things. We will come back to the ability to do work as a criterion in a moment. But the point of the example is made clear in the paragraph following, where Leibniz raises objections to "those who would define the motion of a body as translation from the vicinity of immediately contiguous bodies, which are regarded as though at rest in the vicinity of other bodies" (A VI 4, 2019)—namely, the Cartesians. This definition, Leibniz says, was concocted to enable them to deny that the Earth moves, "since it is carried off with the heaven or its vortex, and does not therefore recede from the surrounding bodies." Putting aside the fact that this is false, "since, even if it is granted that it is moved with its vortex, it will, however, be moved more slowly than the subtle matter of its vortex," nevertheless, even if we suppose it to be true, it will not tell against the motion of the Earth:

> For someone might deny that logs move when they are carried off by the river, yet they are sometimes moved with so much force that they even knock down bridges. And so we attribute motion to that thing which has a force of acting.
>
> (A VI 4, 2019)

As noted, both Lodge and Garber see Leibniz in this text as advocating the communication of motion as a criterion for what is really in motion. Perhaps that is so; this is, after all, a transitional draft, and we cannot assume that he has already attained his mature conceptions. But it is possible to see him as instead drawing attention to the inadequacy of a purely geometrical foundation: there has to be something else in body to explain how the logs have a force to destroy the bridge when they are relatively at rest in the water. As Leibniz develops his dynamics, this becomes the passive force necessary to explain both the mass of the logs and their resistance to penetration; the logs are less massive than the water (or they wouldn't float!), but the water (observably) has less far resistance to penetration (which explains why the logs, and not the water, would destroy the bridge). But force in this sense is not a criterion for what is really in motion (as both Lodge and Garber rightly object), since whether it is the really moving logs that hit the bridge or, in a hypothetical scenario where the same bridge is launched with equal and opposite speed against the logs sitting at rest in the water, the same impact and destruction of the bridge would result.

As we shall see, Leibniz himself makes this point perfectly clearly in other writings: force is *not* a certain criterion for absolute motion.[60] In a collision, the phenomena of motion will be the same whatever hypothesis is made about which of two bodies is in absolute motion. That is not the same, however, as saying that

[60] As we saw, Leibniz writes this explicitly in two of the manuscripts on Copernicanism written in Rome: in *De motu in rigore mathematico accepto* (A VI 4, 2065–6), and in *De praestantia II*: "And it is therefore true that not even a force of acting is a certain indicator of absolute motion" (A VI 4, 2072).

there is no criterion for which of two bodies is the agent, which the patient, that is, which body may truly be said to be in motion. Here I recall the distinction (discussed in §2.1) that was common property to all disputants in the seventeenth century, whatever their position on whether place is relative or absolute, between true and apparent motion. Leibniz is advocating as a criterion for true motion that a body should contain the cause (i.e. reason) for the changes under discussion. As we saw, he has defined "*That body [to be] in motion* which is the proximate efficient cause why a certain part of it changes its situation with respect to other bodies; otherwise it is said to be *at rest*" (A VI 4, 2011).

Thus, as in the case of explaining the waves generated by a ship moving through the water by a conspiracy of motions in the fluid against a resting ship, the latter explanation of the collision of logs and bridge would be "remarkably convoluted [*perplexa*]" (A VI 3, 110), and the explanation in terms of a single cause is far more intelligible. So, Leibniz concludes, the Cartesians are quite wrong to believe that "what is real and positive in motion belongs equally to each of two contiguous bodies receding from one another. For the force of acting could be in one alone, and so be the cause of the changed situation" (A VI 4, 2019).

Leibniz expands on this point in *Specimen dynamicum II*. He chastises Descartes for not having noticed that "*the equivalence of hypotheses is not changed by the impact of bodies upon one another*," so that his Rule of Collision "which claims that a body at rest can in no way be driven from its place by another smaller body, hardly squares with this" (GM VI 248/AG 131, L 445–6). On the contrary, he insists,

> it follows from the respective nature of motion that *the action of bodies on one another ôr their percussion is the same, provided they approach each other with the same [respective] velocity*. That is, if the appearance of the given phenomena remains the same whatever finally may be the true hypothesis, that is to say, to whichever of them we may finally truly ascribe motion or rest, the same outcome will be produced in the phenomena sought ôr resulting phenomena, even with respect to the action of the bodies on one another.[61] And this is just what we experience, for we would feel the same pain whether our hand ran into a stone at rest, suspended, if you like, from a string, or the stone hit our resting hand with the same speed. Meanwhile, we speak as the matter requires, according to the more apt or simpler explanation of the phenomena. It is in just this sense that we appeal to the motion of the *primum mobile* in Spherical Astronomy, while in the

[61] The contrast between the given [*data*] and the sought [*quaesita*] phenomena occurs also in Leibniz's Law of General Order of 1687: "When the cases (or givens [*data*]) continually approach one another in such a way that one finally disappears into the other, it is necessary that the same thing happens in the corresponding consequents or outcomes (or things sought [*quaesita*])—which depends on a still more general principle: *as the given things [data] are ordered, so also are the things sought [quaesita]*" (GM VI 129); in French at (GP III 51).

> theory of the planets we should use the Copernican hypothesis, so that those disputes conducted with such fever (in which even the theologians were involved) already completely disappear. For even though force is something real and absolute, motion belongs to the class of respective phenomena, and truth is to be seen not so much in phenomena as in their causes.
>
> (GM VI 248/AG 131, L 445–6)

Let us now return us to the other criterion for real motion that Leibniz gave in the *De causa motus*, that of the man's ability to do work. This criterion points to a different notion of force in relation to cause. It is connected with the principle to which Leibniz gives pride of place in his explanations of the connection between causes and forces, what he will later call his Principle of Equipollence, namely that "the full cause is equivalent to the entire effect." He had already formulated this principle in 1676, using the same reasoning about the necessity for there to be a conservation of force in an isolated system. In the 1676 tract he argues

> Any full effect, if the opportunity presents itself, can perfectly reproduce its cause, that is, it has forces enough to bring itself back into the same state it was in previously, or into an equivalent state. In order to be able to measure equivalent things, it is therefore useful that a measure be assumed, such as the force necessary to raise some heavy thing to some height.... Hence it happens that a stone that is constrained by a pendulum and that falls from some height, can, if nothing interferes and it acts perfectly, climb back to the same height but no higher, and if nothing of the forces is removed, no lower either.
>
> (A VIII 2, 136)

At this point in the development of his thought (in 1676), however, Leibniz was still operating with a notion of "force" as directed quantity of motion, in keeping with his contemporaries. As late as January 1678, when he begins his concerted effort to develop this insight about equipollence, the "force" that is conserved in every motion "is the quantity of effect, ôr, what follows from this, the product of the quantity of the body and the quantity of velocity" (Fichant 1994, 71).[62] As has become clear with the recent publication of further writings on physics from this period, Leibniz was operating in a neo-Galilean conceptual scheme where the motive force (*vis viva* or "live force"[63]) of a moving body was conceived as proportional to the quantity of motion, infinitely greater than the dead force

[62] Cf. *Axioma de Potentia et Effectu* (A VIII 2, 235), dated February–September 1676.

[63] In translating *vis viva* as "live force" rather then "living force" I am following the lead of Jennifer Coopersmith (2017, 16), and bucking a long tradition. A body is not rendered a living being by having such a force, on Leibniz's conception, any more than a "live wire" is thought to be literally alive on ours.

(*vis mortua*) involved in statics, which is proportional to a moment or endeavour (*conatus*) to move.[64]

This conception of *vis viva* changed in February 1678. In the course of his investigations summarized in *De corporum concorsu*, Leibniz came to the realization that if causes are to be estimated in terms of their ability to bring about a certain effect, then the correct measure of the force of a body's motion could not after all be proportional to its quantity of motion. For a pendulum bob at rest at the top of its swing has zero quantity of motion, yet possesses a force capable of producing the motion of the bob at the bottom of its swing; by the Principle of Equipollence, taking the raising of the bob through this height as the entire effect (neglecting the effects of friction), these must be equivalent. And the motive force the bob possesses at the bottom of its swing must in turn be convertible into an equivalent dead force at the top. As Huygens had shown, since by Galileo's Law of Fall the velocity of the bob after it has fallen through a height h is proportional to $\sqrt{h}$, the force necessary to raise it through this height must be proportional to the body's mass m and the square of its velocity. Thus live force or *vis viva* must be proportional to mv^2. Extrapolating from this case, Leibniz was led to propose the universal conservation of force: each isolated body or system of bodies has its own invariant measure of this force, whether this is manifested as live force or as an equivalent dead force, namely the capacity a body has for producing motion when it is not yet in motion. As he writes in the tract *A New Principle of the Mechanical Universe*, written some time after 1680 and first published by Michel Fichant in 1994,

> In every entire machine, ôr in every total aggregate of bodies being subjected to some action, the power remains the same before and after the action. Hence when the sensible power of the agents declines little by little, this happens not because the impetus perishes, but because it is redistributed in the insensible parts of the surrounding bodies (which I count as being in the entire machine). And since the whole universe is a most perfect entire machine, for no body can be assumed outside it that would take away part of the impetus, it follows that the same power ôr force always endures in the world.[65]

The significance of these considerations for the relationship between cause and force, for Leibniz, is that causes should not be understood in terms of impulsive forces, but in terms of the capacity of a body to do work.[66] Force, understood in

[64] See, for instance (A VIII 2, 119, 134 and 162). Evidence of this Galilean origin lasts into Leibniz's mature dynamics, as in the *Specimen Dynamicum* and the *Tentamen*. There Leibniz presents dead force as an infinitely small element of live force. More on this below.

[65] *Principium Mechanicae Universae Novum (A New Principle of the Mechanical Universe)*, LH XXXV, 10, 5, fo 1–2, 3–4; date uncertain, probably after 1680; quoted from (Fichant 1994, 290).

[66] In this respect, there is great irony in Sam Westfall's charge that "Even in Leibniz, 'force' refers to the activity of a body, and not an action on a body. The development of a conception of force as action

this way, has the dimensions of mv^2, whether this is manifested in motion (live force), or remains a potential for producing an equivalent motion in a body not yet moving (dead force).[67] This allows Leibniz to say that each corporeal substance (or aggregate of corporeal substances) has the capacity to act even when (as a whole) it has vanishingly small total mv^2—its live force is simply converted into an equivalent dead force, which is the sum of the live forces in its constituent parts. This, he claims, is precisely what happens in collision. The live force of the impinging body is gradually converted into elastic force, conceived as carried by intestine motions within it on ever smaller scales; as the body rebounds, some of this elastic force is then converted back into live force. In an inelastic collision some of the live force of the body as a whole will be redistributed among these constituents without restitution. In each case the *total absolute force* remains the same, being the sum of the *directive force*, by which the body as a whole can act on other bodies, and the *respective force* "by which the bodies contained in an aggregate can act on each other" (*Specimen dynamicum*, GM VI 239; cf. *Dynamica* Section III, Proposition 11, GM VI 495). Thus a body or system of bodies neither gains nor loses force, provided the system is isolated. Leibniz insists on this point against Clarke and Newton:

> The objection is that two soft or inelastic bodies meeting together lose some of their force. I answer that this is not so. It is true that the wholes lose it in relation to their total motion, but the parts receive it, being internally agitated by the force of the meeting or shock. Thus this loss occurs in appearance only. The forces are not destroyed, but dissipated among the small parts. There is here no loss of forces, but the same thing happens as takes place when big money is turned into small change. I agree, however, that the quantity of motion does not remain the same....[68]

This completely disposes of Russell's objection. The effect can indeed "only be measured by means of motion" (Russell 1900, 86).[69] But when a body loses some

on a body to change its state of motion, a conception that contributed greatly to the further elaboration of mathematical mechanics, was inhibited by the mechanical philosophy during the [seventeenth] century" (Westfall 1977, 138).

[67] Strictly speaking, according to Leibniz, no body is absolutely at rest. In his early work this followed from interpreting endeavour as an infinitely small motion; in his mature dynamics, rest can be considered as limiting case of motion, by the law of continuity. Such a conception motivates Leibniz to consider dead force as an infinitely small live force, but I shall not go into this matter further here.

[68] Fifth Paper, §99, (GP VII 414/LC 87–8). See also *Dynamica*, Part II, Prop. 7: "In any system whatever of bodies not communicating with others, the power is the same," and Prop. 8: "There is always the same power in the universe" (GM VI 440); and Part III, Prop. 10: "The respective active force of bodies through their action on one another does not change in quantity, nor does the respective speed therefore change, provided the bodies only act or are acted on by each other, and no portion of the forces is retained by their parts" (GM VI 494).

[69] Bertoloni Meli has made a similar response to Russell's criticism in terms of the distinctions between directive, respective and absolute force, although he wants this to rest on the scalar nature of mv^2 (Bertoloni Meli 1988, 25–6).

of its directive force, this deficit remains in the body as respective forces, dissipated among its small parts and carried by these internal motions.[70] This also makes Roberts' suggestion of a preferred dynamical frame otiose: although there will be a preferred frame for the purposes of explanation, it is the total absolute force, not the live force mv^2, that is absolute.

Quantity of motion, on the other hand, will be lost in inelastic collisions. In the standard account, shared for instance by Huygens and Newton, the action of one body on another is understood in terms of its production of a change in quantity of (directed) motion in the second body, $\Delta m\underline{v}$.[71] Against this, Leibniz argued that if action and reaction are understood in terms of the exchange of quantity of motion, then in a collision either body may equally well be regarded as being the cause, since such exchanges depend only on the relative velocity of the colliding bodies: "it follows from the respective nature of motion that *the action or impact of bodies on one another will be the same provided the speed with which they approach each other is the same*" (*Specimen dynamicum II*, GM VI 248).[72]

But if force is regarded instead as a body's capacity to do work, force that may be converted into motion, then the subject of motion can be determined by reference to the most intelligible hypothesis for explaining the phenomena in question. For example, when a set of skittles is hit by a ball, it is more intelligible to see the cause of the falling of the skittles in the motion of the ball than to see the ball as at rest and colliding with several skittles moving in such a way as to hit it precisely at the point of impact. No force is actually gained or lost by any body in a collision, according to Leibniz, whereas on the Newtonian view, by contrast, action requires the transfer of force (quantity of motion) from agent to patient. For Leibniz that body acts in which the reason for the changes is more intelligibly located. It is that body which is the "proximate cause of the changes," and this is the way in which we are to determine "to which body the motion belongs" (*Discours* §18, A VI 4, I 558–9/AG 51).

Once the proximate cause of the changes has been identified, then, we will be able to determine not only which body is truly in motion, but also its degree of speed. This tallies with Leibniz's answer to Clarke, discussed in §3.1 above: "when the immediate cause of the change is in the body, that body is truly in

[70] On this conception of elastic force, as carried by motions of fluids within bodies on ever smaller scales, the total force in a body is live force, whether directive or respective. But Leibniz also characterizes elastic force as a dead force, an element of live force, which is aggregated from it over time. I will not attempt to resolve or explain this discrepancy here.

[71] At least, this is so for discrete impact forces. In the case of a continuously acting forces, Newton takes each change of motion over a moment of time Δt, and then proceeds to the limit as Δt is shrunk to 0, producing a force that is proportional to an instantaneous acceleration. So by our lights, his force concept is dimensionally ambiguous. For a discussion of the rival Newtonian and Leibnizian notions of force, see Garber (2009, 172 ff.) and Arthur (2015b).

[72] Cf. Corollary 3 of Newton's Law 3: "*The quantity of motion, which is determined by adding the motion made in one direction and subtracting the motion made in the opposite direction, is not changed by the action of bodies on one another*" (Newton 1999, 420).

motion, and then the situation of other bodies with respect to it will be changed consequently, though the cause of that change be not in them" (Fifth Paper, §53; GP VII 404/LC 74). Motion as something merely relative is an *extrinsic denomination*, to use the scholastic parlance: it is something attributed to a subject by reference to other things extrinsic to the subject. If motion were merely relative, as Descartes had defined it, then a body could be said to move just because a second body with respect to which its motions is judged is moved, without any basis for this change of situation in the first body. According to Leibniz, however, "there are no purely extrinsic denominations" (see the discussion in Appendix 2): there must be a way of determining the true subject of motion, that body which contains the immediate cause of the change of their relative situation.

Interestingly, Newton makes the same point in criticism of Descartes in the Scholium to the Definitions in his *Principia*, and it will be instructive to compare his criticism with Leibniz's. In fact, Newton had directed the criticism explicitly at Descartes in his earlier unpublished essay, *De gravitatione*. There he notes the discrepancy between Descartes' definitions of "motion in the common sense" (according to which the planets are moving in a vortex) and "motion in the philosophical sense," namely, "the translation of one part of matter or one body from the neighbourhood of those bodies immediately contiguous to it, and which are regarded as at rest, to the neighbourhood of others" (*Principles*, Part II, §24–5, and III, §28; Descartes 1985, 233, 252). For the rotating of bodies around the Sun in the philosophical sense "does not cause a tendency to recede from the centre, which a rotation in the common sense can do" (Newton 1962, 93). So the Earth "endeavours to recede from the centre of the Sun on account of a motion relative to the fixed stars," but "relative to its own orbit it has no endeavour, because it does not move in it" (95). "What should rather be said," Newton argues, "is that only the absolute motion of the Earth, thanks to which it endeavours to recede from the Sun, coincides with its unique natural motion, and that its translations with respect to external bodies are but external denominations" (94). From this he concludes that "the definitions of places, and thus of local motion, [must] be referred to some immobile being such as extension alone, or space insofar as it is regarded as something truly distinct from bodies" (98).

Leibniz, of course, does not agree that this conclusion follows: positions in a supposed absolute space could not be determined independently of bodies. The Earth's true motion, both diurnal and annual, can be identified in accordance with the fact that the Copernican hypothesis gives the most intelligible explanation of the system of planetary motions. But this is still a motion relative to the fixed stars, assumed as stationary—we would be assuming a relative space, in Newton's terms, but Leibniz held (as we shall see in the next section) that all motions of bodies among themselves would remain the same in such a space however it is assumed to be moving. Thus for him, even though true motions are determinable according to the most intelligible hypothesis, there exists no criterion for an absolute motion in Newton's sense.

In sum: Bodies understood "geometrically" as mere vehicles of motion have no capacity to act. Insofar as actual bodies do have a capacity to resist change of state (inertia), they must have within them a power for this resistance; and insofar as they have the capacity to do work, they must have within them a power to act. The power to resist changing state is their passive power, and the power to change their own states is their active power. Bodies only have these powers or forces insofar as they act, and since a substance is a thing that acts, bodies contain substances within them whose nature is force. The capacity to act is what Leibniz calls the primitive active force, although this is not a mere potentiality, but always involves an endeavour to act; the latter is manifested in the derivative forces of bodies, which are modifications of the primitive active forces existing at each instant; and this is what is real in motion.

At this juncture, I need to say more about Leibniz's metaphysics of force. For in the preceding analysis I have assumed that the derivative forces manifested in bodies are instantaneous modifications of something enduring and substantial—namely primitive force—and that, consequently, such enduring and substantial primitive forces must be presupposed everywhere in bodies, grounding the derivative forces and what is real in the bodies' motions. In so doing I have made no appeal to any distinction of different levels of reality on which force might be operating, or on which motion might be taken to be relative or absolute. This puts me at odds with the preponderance of modern commentators, who have interpreted the relationship between types of force and motion in terms of tiered *ontic levels*: primitive force is thought to exist *at the metaphysical or monadic level*, while motions occur *at the phenomenal level.*[73] Accepting this, however, then creates a difficulty for the ontic status of the derivative forces of Leibniz's dynamics: if derivative force is what is real in motion, while motion would otherwise be phenomenal because purely relative, then derivative force presumably should not be located on the phenomenal level. Yet it cannot be assigned to the purely metaphysical level either, where primitive force or form resides. So what is its ontic status? In order to resolve this impasse, it has been suggested that there is a further ontic level intermediate between the level of bodies and mechanical motions, on the one hand, and monads and changes of perception on the other, namely the *dynamical level.* Thus Daniel Garber suggests

[73] An early example is Sam Westfall, who contends that for Leibniz "nature is mechanical only on the level of phenomena, and that ultimate reality consists of centers of activity, a conception utterly opposed to the complete passivity of matter in the mechanical philosophy" (Westfall 1977, 138). More recently, Paolo Bussotti (following Garber and Jauernig) has written: "it is well known that Leibniz' s ontology is stratified. There are many levels of reality, and although their laws and properties are connected, they are different" (Bussotti 2015, 32; cf. also 112–13).

> that there are at least two levels in Leibniz's natural philosophy. At the surface, as it were, is the mechanical philosophy, in which everything is explained in terms of the notions of size, shape, and motion, assuming that motion satisfies certain laws. This, I think, is what Leibniz often thought of as physics proper. But below physics proper stands the science that treats force and the metaphysical entities, the corporeal substances to which force, properly speaking, pertains and from which motion and its laws derive. This science is what Leibniz called dynamics. (Garber 1995, 283–4)[74]

And Jauernig, while disagreeing with the way Garber characterizes these levels, criticizes rival accounts as deficient in not taking the different ontic levels into account.[75] She suggests that there is no difficulty reconciling Leibniz's apparently conflicting statements about the reality of motion "because he is a relativist about motion at the merely phenomenal level and an absolutist about motion at the dynamical level" (Jauernig 2008, 12). She claims that "the spatiotemporal structure of the dynamical level of Leibniz's ontological scheme must be rich enough for absolute accelerations and absolute speeds to be well defined," and offers two alternatives for "the structure of the merely phenomenal level": on the "strong relativist reading," the structure is what Earman has dubbed "Leibnizian spacetime, in which no absolute motions of any kind are well defined," and a "weak relativist reading," on which the structure is "the so-called Galilean spacetime, in which absolute accelerations are well defined" (2008, 13).[76] Before getting embroiled in the details of any of these interpretations, I should first like to express severe reservations about their founding assumption, the supposition of these ontic levels in Leibniz's scheme.

My main charge is that, although talk of ontic levels is rife among commentators, it is nowhere to be found in Leibniz. I believe this applies across the board,

[74] Garber allows, however, that "the two levels are difficult to separate completely and treat entirely independently" (1995, 284). Slowik (2013) also appeals to different levels, writing of a "true ontology at the foundational micro-level of monads" to which "none of the geometrical hypotheses that treat macro-level bodily phenomena actually correspond." Only one level pertains to dynamics, however, the "bodily level," or "level of well-founded phenomena." For Slowik, Leibniz's primitive force exists at "the level of simple substances," and manifests itself as derivative force at the bodily level. One could, I believe, express the gist of this without invoking levels, by saying simply that the primitive force in bodies manifests itself phenomenally in the derivative forces they exhibit, which are modifications of this primitive force.

[75] Jauernig finds fault with the treatments of Leibniz on relativity by Roberts (2003), as well as the accounts by Parkinson (1969) and (Arthur 1994), for failing to take these different levels into account. Concerning Garber's account, which allows that there could be absolute motion at the phenomenal level even though it is not determinable, she insists that "at this level there *is* no such correct frame and there *are* no absolute motions" (Jauernig 2008, 14).

[76] As if this were not already a sufficiently rococo framework, Jauernig suggests further finer structure, distinguishing corporeal substances of a first, second and possibly third kind (2008, 10), and three phenomenal levels that "differ with respect to the criteria of degrees of 'reality' or 'well-foundedness' that their phenomena conform to" (11). For some insightful criticism of Jauernig's analysis, see Tamar Levanon's (2010).

but I shall restrict the discussion here to the idea of ontic levels in Leibniz's natural philosophy. To be sure, he insists that forms are metaphysical, and should not be appealed to in physical explanations. But this injunction pertains to *kinds of explanation*, not to ontic levels. By it, Leibniz means to rule out a whole slew of kinds of explanation given by his predecessors and contemporaries: Kepler's explanation of intelligences as the causes of planetary motions, Severinus's "elaborator spirit," the revival of the idea of non-mechanical plastic powers to explain the formation of organic bodies, and so forth. What all these explanations have in common is the proposal that forms can act directly on matter. Leibniz himself was still entertaining such explanations as recently as 1676, when he speculated that the mind could activate its own vortex.[77] Against all such notions, by 1678–79 Leibniz would insist that physical explanations be either in terms of an efficient cause construed entirely mechanically (i.e. in terms of bodies operating by the laws of mechanics), or in terms of a final cause, namely in terms of an appeal to principles of optimization, such as light taking the most determinate path. These are the two "kingdoms" that Leibniz identifies.[78] They are two different domains in which one may offer explanations of events, or, better, two different (but wholly compatible) ways of explicating what is going on in the one actual world. But to the best of my knowledge, Leibniz nowhere characterizes them as two different ontic levels. Even in demarcating the two kingdoms, he is at pains to point out that they "interpenetrate," and that both kinds of explanation can be given "even in corporeal nature":[79]

> The smallest parts of the universe are ruled in accordance with the order of greatest perfection; otherwise the whole world would not be so ruled... It is for this reason that I usually say that there are, so to speak, two kingdoms even in corporeal nature, which interpenetrate without confusing or interfering with one another—the realm of power, according to which everything can be explained *mechanically* by efficient causes, when we have sufficiently penetrated into its interior, and the realm of wisdom, according to which everything can be explained architectonically, so to speak, or by final causes, when we understand its ways sufficiently. (GP VII 272–3; cf. Garber 2009, 258–9)

More specifically, I do not know of any occasion on which Leibniz claims that force pertains only to "metaphysical entities" and not to "physics proper" (Garber

77 What seems to have been decisive for Leibniz in his turn against the action of souls on matter was not so much the arguments of the Cartesians against mind–body interaction (with which he would have been familiar throughout his time in Paris), as his reading (and translation) of Plato's *Phaedo*, which he later quotes in the *Discourse*. See Arthur (2018, 148, and the discussion in ch. 4).

78 See for example Leibniz's clear statement of this in the first part of his *Specimen dynamicum* (1695).

79 For an insightful discussion of Leibniz's "Two Realms," see Jeffrey McDonough's (2008).

1985, 284),[80] or operates only at the "level" of corporeal substances and not on the level of bodies (Jauernig 2008, 6).[81] This is to effect a kind of schism between force on the one hand, and matter and motion on the other.[82] An analogous worry was expressed by Leibniz's correspondent Paul Pellisson-Fontanier, who feared that Leibniz's descriptions of substance as "a species of force" could easily be misconstrued as making substance have only a virtual presence in body, as it did for the Calvinists.[83] Given the analogy, it will be worth exploring Leibniz's response to Pellisson.

He begins by distinguishing two ways of talking about "substance": "the subject itself, as when one says that the body or the bread is a substance; and the essence of the subject, as when one says the substance of the body or the substance of the bread, and then it is something abstract."

> Thus when one says that the primitive force makes the substance of bodies, one means their nature or essence—also Aristotle says that their nature is the principle of movement or rest—and the primitive force is nothing but this principle in each body, from which arise all its actions and passions. I consider matter as the first internal principle of passion or resistance, and it is through this that bodies are naturally impenetrable; and the substantial form is nothing but the first internal principle of action and rest, ἐντελέχεια ἡ πρώτη. [first entelechy].... Thus no one could formalize [*se pourra formaliser*] who does not take substance *in abstracto* as primitive force, which always remains the same in the same body,

[80] Although Garber refers to corporeal substances as "metaphysical entities" (1995, 284), he may mean only to allude to Leibniz's claim that whether a given body is that of a corporeal substance, or simply an aggregate, is a metaphysical issue of no concern to the physicist. My point is that the forces that apply to corporeal substances apply to aggregates of them, and indeed, the forces in bodies are aggregated from the forces of the substances they contain.

[81] Jauernig writes: "the real motions in the world are the motions of corpor[e]al substances that possess active forces, which, given that active forces are intrinsic properties, means that these motions can be attributed to particular corporeal substances without reference to other entities" (2008, 6). She appears to interpret corporeal substances as minute particles that move only rectilinearly. Ed Slowik also appeals to dynamical levels in his (2009), where he maintains that "if only physical forces are implicated, then there may be a more basic level of force (of a more teleological sort) that brings about mv^2 and its conservation, but which lies both beyond our understanding and at a more fundamental level than the mere measure of mv^2" (Slowik 2009, 135)

[82] Prior to Garber and Jauernig, Bertoloni Meli had also attributed a distinction in Leibniz between motion understood dynamically and motion as it could be discerned from the phenomena: "Leibniz has a twofold concept of motion: from the point of view of dynamics, motion is intrinsic to one body, is absolute and cannot be known from phenomena; from the point of view of geometry, or of kinematics, motion is relative, that is, it is a change of position with respect to other bodies, and has the status of a relation[37], that is, it is a purely ideal entity" (1988, 26).

[83] Pellisson-Fontanier was a key ally of Leibniz's in his attempts at Church unification. Leibniz had proposed that his metaphysics was not susceptible to the same shortcomings as the Cartesian philosophy in accounting for the real presence of substance in the Eucharist. In his letter of December 31, 1691, however, Pellisson advised Leibniz that when he (Leibniz) "explained substance as a species of force that can be applied in diverse places," he could be taken as denying the "veritable real presence" of substance in the Eucharist in favour of "a presence of force and of virtue," in line with the Calvinists and contrary to the Augsburg Confession (A II 2, 473).

and gives rise successively to the accidental forces and particular actions, which are all only a consequence of their nature, or primitive and subsisting force, applied to other things. (January 8/18, 1692, A II 2, 486–8)

From this it is clear that Leibniz conceives primitive force as *being in bodies*, and indeed as constituting their natures. The bodies themselves, as Leibniz explains elsewhere, are either the organic bodies of the substances whose form they are, or aggregates of such. "It is true, however," Leibniz continues in his reply to Pellisson, "that substance *in concreto* is something other than force, for it is the subject taken together with this force." Although this was said to assuage Pellisson's worries that Leibniz's appeal to primitive force as substance could be taken as undermining the "real presence" of the substance in the body, it applies equally to Jauernig's interpretation. For having declared that for Leibniz "the question of whether there are any true and real motions, should, in the first place, be understood as the question of whether there are any motions that can be univocally attributed to specific subjects," she takes this to mean that "the real motions in the world are motions of corpor[e]al substances" (Jauernig 2008, 6)—interpreting the latter as "multiple minute substances" on the lower, dynamical level. On the contrary, I maintain, the specific subjects to which true motions may be ascribed are substances *in concreto*, whether these are mere aggregates, like a loaf of bread or a ship, or corporeal substances like you or me or a microscopic animal. The "accidental forces" that Leibniz refers to in the letter to Pellisson are the derivative forces of his physics, such as the elastic force of bodies subject to compression or stretching, the centrifugal force of a body revolving around a centre, and the active forces of bodies in motion. All these derivative forces apply to bodies, and are most definitely part of Leibniz's "physics proper." They are not, however, intrinsic to bodies, being "only a consequence of the... primitive and subsisting force, applied to other things."

Leibniz gives the clearest accounts of the nature and status of derivative force in his correspondence with De Volder. The Dutch philosopher was especially attracted to Leibniz's claim that substance is intrinsically active, and he wanted Leibniz to provide a demonstration of this.[84] But as the correspondence proceeds De Volder expresses increasing frustration with Leibniz's appeal to primitive force. If derivative forces are all that operate in the realm of physics, he asks in a letter of 7 October, 1702, what is the point of appealing to primitive force? Why

[84] See in particular De Volder's letter of 12 November 1699, where he claims to have "jumped for joy" when he read in the *Acta eruditorum* Leibniz's claim that "there is active force in all substances" (LDV 143). But we should not forget Leibniz's later correspondence with Rudolph Christian Wagner in the 1710s, where he relates the same views about primitive and derivative forces in response to Wagner's misconstrual of him as attributing activity directly to (bare) matter, similarly to how De Volder had misconstrued him (GP VII 528 ff.). (I mistakenly call R. C. Wagner "Gabriel Wagner" in my 2018, 235).

shouldn't derivative forces alone suffice? "Surely everything that will happen in the universe will follow from this alone? So why is it necessary to postulate primitive forces and those indivisible things?" (LDV 250–3). Leibniz replies:

> I regard the substance itself, endowed with primitive active and passive power, like the Ego or something similar, as the indivisible or perfect monad, not those derivative forces that are continually found to be one thing after another.... And indeed derivative forces are nothing but modifications and results of primitive forces.... Every modification presupposes something lasting. Therefore, when you say, "Let us suppose there is nothing to be found in bodies except derivative forces," I reply that the hypothesis is not possible, and again that it gives rise to error, insofar as we are assuming incomplete notions as the complete concepts of things. (to De Volder, June 20, 1703, LDV 262–3)

Responding to further objections from De Volder the following January, Leibniz explains that in regarding derivative forces as modifications he is far from denying that they are active; "rather, from the fact that they are active and yet modifications I infer that there are some active primitive things of which they are modifications" (to De Volder, January 21, 1704, LDV 286–7). Amplifying, he adds

> Derivative force is the present state itself insofar as it tends towards a following state, ôr pre-involves a following state, just as the present is pregnant with the future. But the persisting thing, insofar as it involves all the cases, has primitive force, so that primitive force is like the law of the series, and derivative force is like a determination that designates some term in the series. (LDV 286–7)

This echoes what he had told De Volder earlier: "I recognize primitive entelechy in active force exercising itself in various ways through motions, and, in a word, something analogous to the soul, whose nature consists in a certain perpetual law of the same series of its changes" (to De Volder, April 3, 1699, LDV 74–5). Primitive force is "an attribute from which change follows, whose subject is substance itself" (LDV 73), and the changes are the derivative forces, which are its instantaneous modifications.

In sum, Leibniz nowhere talks of forces as operating on differing ontic levels.[85] Derivative forces, being modifications of primitive forces, cannot exist on a different plane of reality from them; and primitive forces, even if they manifest

[85] Regarding the relation between the phenomena of motion and force, I fully agree with Adams when he writes that the derivative forces of Leibniz's physics are "identical with the intrasubstantial derivative forces that are modifications of primitive forces. Thus one and the same derivative force would have both an intrasubstantial effect, the passage from current to future perceptions, and a phenomenal effect, the physical motions" (Adams 1994, 385–6).

themselves phenomenally in the created world only through the derivative forces that are their limitations, are nonetheless the entelechies, natures or forms of their organic bodies, in which these forces are located. Primitive force itself, of course, does not do any impelling of bodies, and is not appealed to in explanations of the phenomena.[86] Derivative forces, on the other hand, urge bodies in different ways at each instant, and motions over time result from sequences of the changes they produce. Just as bodies result from the aggregation of monads, so their motions result from the aggregation of derivative forces. Consequently,

> In the phenomena, i.e. in the resulting aggregate, everything is now explained mechanically, and masses are understood to impel one another. And in these phenomena there is no need for a consideration of anything but derivative forces, once it is accepted whence they result, namely, the phenomena of aggregates from the reality of monads. (to De Volder, June 20, 1703; LDV 260–1)

Moreover, these considerations are relevant to another way in which Leibniz conceives force as providing a foundation for motion. As he explains on numerous occasions, as in an earlier letter to Pellisson-Fontanier quoted by Garber, "motion is a successive thing, which consequently never exists, any more than time does, since all of its parts never exist together" (July 1691; A II 2, 434; cf. Garber 2009, 154–5). In contrast with that, he tells Pellisson, "force or effort exists completely at each moment, and must be something true and real." Derivative force exists completely at each moment as the momentaneous modification (or accident) of the permanent attribute of primitive force. But, as I noted in the Introduction above, this no more delivers it onto a distinct ontic level from primitive force than shape's being an accident of an extensum puts those things on different levels. Motion as a whole consisting of successive parts never exists together (*simul*); you could say the same of derivative forces if you took them (as De Volder wanted to) for independently existing things. But they are successive accidents of the permanent attributes of primitive force. Leibniz repeats this claim that motion never truly exists because it has no coexisting parts in the *Specimen Dynamicum* (GM VI 235/L 436), adding in the unpublished second part: "Also, since only force and the nisus arising from it exist at any moment (for motion never exists, as discussed above), and since every nisus tends in a straight line, it follows that *all motion is either rectilinear or composed of rectilinear motions*" (1695, GM VI 252/L 449). The importance of this last assertion will become clear in the next section.

[86] "Properly and rigorously speaking perhaps it will not be said that a primitive entelechy impels the mass of its body, but only that it is joined with a primitive passive power that it completes, that is, together with which it constitutes a monad" (to De Volder, June 20, 1703; LDV 260–1).

3.4 Rotational Motion: Leibniz vs. Newton

Leibniz, we have seen, upholds Copernicanism as the true hypothesis in astronomy, citing (in *Principia mechanica*) such phenomena as changes in the apparent diameter of the fixed stars and parallax as evidence for Earth's annual motion, and alleged facts about the constant swinging of hanging lamps from east to west, and waves impinging only on eastern and western shores in favour of the diurnal motion (A VI 3, 105–6). And yet, almost in the same breath, he claims that "however many observations are made, it is still never possible to demonstrate which motion is the absolute and proper motion in bodies" (110), so that it is "clear that from the phenomena of the changed situation alone no certain knowledge can ever be had concerning absolute motion and rest" (110). How can this be? How can one appeal to the phenomena to vouchsafe the truth of the Copernican hypothesis, yet claim that there is no basis in the phenomena for choosing one hypothesis over the other?

Part of the explanation, I have argued, is that Leibniz insists that the equivalence of hypotheses applies to motion *more geometrico* (or "in mathematical rigour"), as change of situation; whereas true motions are assigned on the basis of causes, most intelligible hypotheses. (This explains the qualification "from the phenomena of the changed situation alone" in the above quotation.) Thus in *Cum geometricis* he claims that because of the equivalence of hypotheses, "no eye, in whatever point of matter it is placed, could discern by certain indications from the phenomena where there is motion, and how much and of what kind it is—'whether God moves everything around it, or turns it around itself'" (C 591/AG 91).[87] This allusion to the eye placed in matter clearly evokes the discussion in the *Principia mechanica* where Leibniz imagined an eye placed in different locations still seeing the same relative motions between pairs of bodies undergoing uniform rectilinear motion. But those were uniform rectilinear motions, and the hypotheses in astronomy concern motions in curved orbits.

To illustrate the difficulty we need only to return to the dispute in astronomy. For example, the diurnal motion of the Sun and all the heavenly bodies will appear the same to an eye on the Earth's surface whether it is the Earth spinning (as in the Copernican hypothesis), or everything spinning around it (as in the Ptolemaic).

[87] The italicized phrase is a rough quotation of Seneca's allusion to Aristarchus of Samos in his *Natural Questions*: "*circa nos omnia Deus an nos agat* (whether God turns everything around us, or turns us around)" which Leibniz had earlier quoted in a letter to Claude Perrault (see my note in LLC 381–2), and also obliquely alluded to in his own notes on Descartes' *Principia* at the end of 1675: "Ejection along the tangent does not argue the real motion of the rotating thing, since it would be the same if everything moved around it" (A VI 3, 215/LLC 25). Again, his claim in *Cum geometricis* that "not even an angel could discern, in mathematical rigour, which of several bodies of that kind is at rest, and is at the centre of the others" evokes the similar claim in the *Principia mechanica* that "if absolute motion cannot be distinguished from other phenomena, not even by Him to whom all the phenomena are revealed, it follows that motion and rest taken absolutely are empty names" (A VI 3, 110).

But these are just visual appearances; the resulting phenomena will be different, as Leibniz himself reports, with his examples of the tides and swinging lamps (perhaps an anticipation of Foucault's pendulum of 1851). So there seems to be an equivocation here between phenomena in the sense of what appears to an eye at a certain location in matter, and phenomena in the wide sense of things that are known to occur, and need explaining.

Moreover, there is a second difficulty, this one concerning situations. On the Ptolemaic hypothesis the annual motions of Venus will not account for the differences in the apparent size of the planet at apogee and perigee, whereas this fact is explicable on the Copernican hypothesis. This difference in explanation is due to the different mutual situations (angles and distances) of the three bodies (Earth, Sun, and Venus) to one another at apogee and perigee of Venus's orbit according to the two different hypotheses. So it is not the case that there is no difference of phenomena between the two hypotheses; moreover, this difference in the phenomena depends on a difference in situation.

Of course, this difference in the apparent size of Venus can be accommodated on the Tychonic hypothesis, where the planets revolve around the Sun, but the Sun revolves around the Earth. Indeed, the Tychonic system is "precisely equivalent mathematically to Copernicus' system," to which it may be transformed "simply by holding the sun fixed instead of the earth."[88] This is true because the angles and distances of any planet seen from Earth, as well as their rates of change, will be precisely the same whether the Sun goes around the Earth or the Earth around the Sun, provided the other planets circle the Sun. There will however be a difference in terms of the Earth's situation with respect to the fixed stars, a difference that would in principle show up in their different angular sizes at different times of the year and in the phenomenon of stellar parallax. Leibniz notes both of these facts in his *Principia mechanica*:

> But if also the apparent diameters of the fixed stars change, or different parts of the heavens are vertically above the same point of the earth at different times, and this agrees with the hypothesis of the annual motion of the earth, according to which the earth changes its situation to the fixed stars, this hypothesis [*sc.* Copernicus's] would certainly seem to be sufficiently corroborated.
>
> (A VI 3, 105/Arthur 2013c, 101)

Even though the stellar parallax was not confirmed until 1838 by Friedrich Bessel, many Copernicans (including Leibniz) were convinced that its not yet having been detected was simply because of the very great distance of the stars.[89] Thus

[88] This is argued by Thomas Kuhn in his classic *The Copernican Revolution* (1957, 202, and 204).

[89] There were several attempts to detect this parallax, notably by Robert Hooke in 1674, Jean Picard in 1680, and later John Flamsteed (recorded in a letter to John Wallis, published in 1600). Flamsteed

Leibniz notes that philosophers, "having understood the immense distance of the stars, believed it was more in keeping with reason to attribute a modest motion to the earth, like that found in most everyday bodies in our era, than that the observable universe should be moved with some insane rotation" (105/101). On the Tychonic system, moreover, the orbit of Mars intersects the orbit of the Sun, so that in principle their distance apart could be zero. Even though this does not actually happen because the mutual motions of these bodies are the same as on the Copernican system, this intersection of their orbits was fatal to the Ancients' crystalline spheres; and once the planets were cut loose from their orbs, it was crucial to explain how they remained in orbit. Here, as Descartes had already suggested, the hypothesis of their being carried by a vortical fluid gave promise of a mechanical explanation, one not feasible for Tycho's system.

But why was Leibniz so convinced that the equivalence of hypotheses applies to *all* motions? One reason, as we have seen, is that he saw it as applying not only to motion geometrically conceived, but also to physical motions. Leibniz maintained that all the phenomena in collisions of bodies—and thus all the phenomena in the mechanical philosophy, strictly speaking, since they are all produced by collisions—will be the same no matter which of the impinging bodies is taken to be in motion and which at rest. Even taking into account the fact that bodies have a moving force, so that a body moving faster has greater force of impact, this will not alter the relative motion of two colliding bodies; for given the conservation of quantity of motion in a given direction established by Huygens, Wren, and Wallis—what Newton called 'momentum', and Leibniz 'quantity of progress'—the force of impact in a collision will be the same for the same relative motion, and will therefore not license any assignment of absolute motion. Still, as before, these are rectilinear, uniform motions; and Leibniz ascribes the Equivalence of Hypotheses to objects in any kind of relative motion, so that, as he expresses it in the *Système nouveau*, "however many bodies one takes, it is arbitrary to assign rest or even any degree of speed to any of them you choose, without the possibility of being refuted by the phenomena of motion, whether rectilinear, circular or composite" (GP IV 486–7/L 459).

What is his argument for this? It is, in a nutshell, that all circular or composite motions are composed from rectilinear motions, so that "it follows that if the equivalence of hypotheses holds in any rectilinear motions whatsoever, then it will also hold in curvilinear ones" (*Specimen Dynamicum*, 1695, GM VI 253).

did detect notable annual differences which he imputed to stellar parallax, but he was forced to retract this after Jacques Cassini's criticisms. What he had not realized—as was demonstrated by James Bradley in 1728—was that the discrepancy in the Pole Star's positions that he had measured were the result of stellar aberration, due to the finite speed of light (a result established by Ole Rømer in 1676, but rejected by Cartesians like Cassini and his father, Giovanni Domenico Cassini). See Williams (1979) for a discussion.

This does *not* in fact follow, although the reasons why are subtle, and the argument requires some unpacking.

In his unpublished *Dynamica* Leibniz attempts to lay out the argument in systematic fashion.[90] In Part II, Section II, Chapter 2 (On the Direction and Shape of Motion), he states the Equivalence of Hypotheses for uniform rectilinear motions:

> **Proposition 16.** The same phenomena of bodies (i.e. their mutual situations at given times) can be obtained through various hypotheses (i.e. assignments of motions and states of rest). And if the phenomena can be produced in one way though uniform rectilinear motions, it will be possible to render such motions in innumerable other ways still, and in the end to assign rest or a given uniform rectilinear motion to any movable body whatever. (GM VI 484)

This is the equivalent, in Leibniz's terms, of Newton's Corollary 5 to his Laws of Motion, and he would have approved of Newton's justification:

> **Corollary V.** *The motions of bodies enclosed within a given space are the same with respect to one another, whether the space is at rest, or moving uniformly in the same straight line without circular motion.*
>
> For the differences of the motions tending in the same direction, and the sums of those tending in opposite directions, are the same at the beginning in either case (by hypothesis), and from these sums and differences arise the collisions and impetuses by which bodies strike one another. Therefore by Law 2 the effects of the collisions will be equal in each case, and thus the motions of the bodies with respect to one another in the one case will remain equal to the motions with respect to one another in the other. The same thing is clearly corroborated by experiment. All the motions in a ship happen in the same way whether it is at rest or moving uniformly in a straight line. (Newton 1687, 20–1)

Note that the spaces Newton is referring to here are movable and therefore relative spaces, and relative space is "determined by our senses from the situation of the space with respect to bodies, and is commonly used instead of immovable space" (Newton 1687, 11). These spaces with respect to bodies considered as immobile are therefore equivalent to the spaces in which Leibniz claimed we represent what we imagine as absolute motion (except that he considers them not as portions of a pre-existing absolute space, but as mere representations).[91]

[90] See Bouquiaux (2017) for an illuminating analysis of Leibniz's writings on relativity in his *Dynamica*.

[91] Cf. what Leibniz wrote in February 1677: "The absolute motion we imagine to ourselves, however, is nothing but an affection of our soul while we consider ourselves or other things as immobile, since we are able to understand everything more easily when these things are considered as immobile" ("Motion is something relative," A VI 4, 1970/LLC 229).

Both authors then take the same law or corollary to apply to momentaneous motion (endeavour, *conatus*), which (following Descartes and Hobbes) they take to be essentially rectilinear. Accordingly, the Equivalence of Hypotheses applies to "nascent" motions, uniform rectilinear motions within each fictional moment of the duration of a moving body. This is clear from the way both Newton and Leibniz (like their contemporaries) apply the parallelogram of motion to the elements of motion existing in each moment of a body's trajectory. Thus in his *Philosophiæ Naturalis Principia Mathematica* of 1687, Newton analyses the motion of a body in a circular orbit around a centre in any nascent moment of the trajectory as composed of an inertial motion in a straight line tangent to the curve, and an increment of motion in a straight line directed towards the centre (Proposition 1, Book 1; Newton 1687, 32). These were compounded into a resultant motion represented by the diagonal of the parallelogram (Corollary 1, p. 17).[92] Leibniz adopted the same representation in his *Tentamen*, representing increments of motions in equal elements of time by straight lines.[93]

In the same proposition of his *Principia* Newton showed how a curvilinear motion could be treated as resulting from a succession of such vanishingly small rectilinear motions, each successive motion being compounded from the previous rectilinear motion and a new impulse towards the centre. When the size of the moments is reduced indefinitely as the number of them is correspondingly increased, the result is that the polygon is effectively a continuous curve, and Newton was able by this means to give an exquisite demonstration of Kepler's Area Law. Leibniz accepted this construction, and for him it was the very model of how all curvilinear motion should be treated. (He did not, however, accept Newton's construction for his Proposition 6 of Book 1, which, in assuming that Galileo's t^2 law held within each moment, assumed a momentaneous acceleration.[94] This Leibniz took to be incompatible with a founding principle of the mechanical philosophy, that all changes of motion are produced by local impacts. I will come back to this.)

As we have seen, from the fact that curvilinear motions can be compounded from rectilinear ones in this way Leibniz inferred that the equivalence of hypotheses holds for these motions too. Thus in Proposition 19 of section III of Part II of the *Dynamica* (On the Collisions of Bodies) Leibniz generalizes Proposition 16 of section II as follows:

[92] The mathematical practice of composing motion under gravity from a downward impulse due to gravity and a uniform motion in a straight line at each moment was clearly laid out by Huygens in his *Horologium oscillatorium* (Huygens 1673; 1986). The Parallelogram Law was also common property for mathematical physicists of that time.

[93] By this means Leibniz proved, for example, that "in equal elements of time the increments of the angles of harmonic circulation are inversely as the squares of the radii" (Leibniz 1689, 90). For expositions of Newton's method of working within nascent moments and the parallel with Leibniz's use of conservation of *conatus*, see Bertoloni Meli (1993) and Arthur (2008).

[94] See Bertoloni Meli's excellent analysis in his (1993, 78 ff.), and my (Arthur 2013a).

> **Proposition 19.** The Law of Nature that we have established concerning the equipollence of hypotheses[95] is true not only in rectilinear motions (as we have already shown), but in general: that is, that a hypothesis once answering to the present phenomena will then always answer to the consequent phenomena, in whatever way the bodies should act among themselves, provided, of course, that the system of bodies is not communicating with others, or that no external agent supervenes. (*Dynamica*, GM VI 507)

Again we may compare this with Newton, this time to his Corollary 6 of the Laws of Motion, and the justification he gives of it:

> **Corollary VI.** *If bodies are moving in any way whatsoever with respect to one another, and are urged by equal accelerative forces along parallel lines, they will all continue to move in the same way with respect to one another, as if they were not incited by those forces.*
>
> For those forces, by acting equally (in proportion to the quantities of the moving bodies) and along parallel lines, will (by Law 2) move all the bodies equally (with respect to velocity), and so will never change their mutual positions and motions. (Newton 1687, 21)

Leibniz gives a similar justification for Proposition 17 of section III, which he appeals to in support of Proposition 19:

> **Proposition 17.**[96] All motions are composed of uniform, rectilinear ones.
> For all motion is in itself uniform and rectilinear, but all action on bodies consists in motion. Therefore rectilinear motion cannot be inflected except through the impression of another motion supervening upon it, a motion that is also rectilinear in itself (leaving the prior one intact), and hence no origin of curvilinear or non-uniform motion is intelligible except through the composition of uniform rectilinear ones. (GM VI 502)

The difference is, of course, that Newton's accelerated motions are in parallel lines. Leibniz has inferred from the fact that curvilinear motions can be constructed

[95] As this wording shows, Leibniz treated the terms "equivalence of hypotheses" and "equipollence of hypotheses" as synonymous. Jauernig (2008, 15) insists that they should be distinguished, since Proposition 19 of Section III concerns the continuation of motion according to a given hypothesis, whereas Proposition 16 of Section II concerns the equivalence of competing hypotheses. But as the comparison with Newton shows, both concern the motions of bodies among themselves remaining the same through time whatever hypothesis is made about the motion of that space.

[96] There is a clerical error in Leibniz's enumeration here, since this "Proposition 17" comes immediately after Proposition 15. So his propositions 17 and 19 in the third section should be 16 and 18.

from rectilinear ones by successive compositions of rectilinear motions in each (ultimately vanishing) moment that the resulting curvilinear trajectory is essentially still a polygon. Where we would say that in the limit it has become a continuous curve, on Leibniz's understanding the trajectory is a result of a continuing bombardment of smaller and smaller impulses to infinity, but with each such impulse occurring in a straight line over a finite moment of time. So there is no moment for him in which a composition of rectilinear motions is not occurring. But even if we grant this, it does not follow from the fact that the equivalence of hypotheses holds in each moment of the construction, that it must hold for the resulting motion over time. The composition of rectilinear motions to give a resultant motion *within each fictional moment*—to which momentaneous motions the equivalence of hypotheses applies, as noted above—is not the same as the composition of the curved orbit out of rectilinear motions *over time*. Indeed, if it were valid to infer from what is the case at each moment to what is the case over a stretch of time, then one could argue along with Zeno of Elea that because there is no motion in each moment while an arrow is in flight, there is no motion over time; or that, because there is no centrifugal force in a curved trajectory in any of the moments, just a uniform rectilinear motion, then there can be no centrifugal force in motion around a centre.[97]

Jauernig erects an elaborate defence of Leibniz in terms of her two-level scheme. She notes, correctly, that Leibniz believes that a truly curvilinear trajectory, such as the curved orbit of a planet, only appears as such to an observer. In reality, according to him, all motions are rectilinear. The latter is enshrined in Proposition 17 of Part II, Section 3 of his *Dynamica*, which we already quoted:

> **Proposition 17**. All motions are composed of uniform, rectilinear ones.
> (GM VI 502)

This suggests to her that Leibniz means *either* that all phenomenal motions are, despite appearances, really only rectilinear—which she calls the "weak relativist reading"; *or* that there are "relative curvilinear and relative non-uniform motions at the phenomenal level, even though at the dynamical level all motions are uniform and rectilinear" (Jauernig 2008, 25)—the "strong relativist reading." On the latter reading, "all *corporeal substances* naturally move uniformly and rectilinearly: the nisus or conatus of corporeal substances, i.e. the smallest increments of their motions, which (supposedly) implies that the motions of corporeal substances are naturally rectilinear" (24).

[97] As I have observed elsewhere, Wolfgang Rindler argued just this, in a mistaken effort to show the necessity of the clock hypothesis in relativity theory. But as I wrote in rebuttal, the young Newton gave an ingenious (and sound!) derivation of the formula for centrifugal force with exactly such a construction. See Arthur (2010).

The latter reading is objectionable on several grounds, some of which I have articulated already. First, there is no "bottom level" of corporeal substances; second, Leibniz nowhere ascribes a type of motion to corporeal substances that he does not ascribe also to bodies; and third, there are, according to Leibniz, no smallest elements of motion in reality: each stretch of motion is actually further divided into subsidiary motions without limit. Leibniz does, of course, represent elements of motion, his endeavours (*conatus* or *nisus*), as differentials. As such, these are idealized limit-motions, taking place within moments under the understanding that the moment is to be reduced to zero duration. It is on this understanding that they can be taken to be naturally rectilinear and uniform.[98] But they are not infinitely small actual motions which compose to form a continuous motion—that would be to invite the contradictions attendant on the composition of the continuum.

The last point is not just a difficulty for this reading of Leibniz on relativity, but points to a problem with his own justification for the rectilinearity and uniformity of the motions in each moment. One may accept that the *nisus* or *conatus* are directed in straight lines tangent to any curve, but these are not motions: they are tendencies to move. Moreover, insofar as the motions within each moment are *mathematical constructions*, there is no reason why they must be uniform: Newton's construction for his Proposition VI in book 1 of the *Principia*, which depends on composing a non-uniform motion with a uniform one in every given moment (prior to taking the limit), is perfectly sound. Leibniz rejects it because he is considering the motions as *actual* motions caused by impacts.[99] But then since these collisions occur on all scales without limit, there are in fact no uniform rectilinear motions that are not inflected by such impacts.

Although Jauernig defends the plausibility of her "strong relativist reading," she notes some difficulties, and expresses a preference for the "weak relativist reading." According to the latter reading "the only physically possible motions for physical bodies are uniform and rectilinear" (2008, 23), so that a curved trajectory is in reality an infinite polygon. This interpretation has the curious feature that it involves a distinction between how phenomenal motions really *are*—they are rectilinear and uniform—and how they *appear*, namely sometimes as curvilinear. But the phenomenal motions are supposed to be how moving bodies appear! Nevertheless, as we already saw in the last chapter, Leibniz himself stresses that there is something "paradoxical" about the geometrical representation of motion.

[98] See, for instance, the passage from Leibniz's letter to Claude Perrault in 1676 quoted by Bertoloni Meli (1993, 75): "And firstly I take it as certain that everything moving along a curved line endeavours to escape along the tangent to this curve; the true cause of this is that curves are polygons with infinitely many sides and these sides are portions of the tangents" (A II 1, 264). I return to a discussion of this point in §3.5 below.

[99] On all these matters the reader should consult Bertoloni Meli's excellent discussion in chapter 4 of his (1993).

On the one hand, possible situations and motions are portrayed geometrically, that is, as they are construed by the imagination, where they are represented as straight lines. Yet, because of the actual division to infinity of any body, "no determinate figure can be assigned to any body, nor is a precisely straight line or circle, or any other assignable figure of any body, found in the nature of things" (A VI 4, 1622/LLC 325). But this leads Jauernig to posit further levels, a "more fundamental phenomenal level" and a "less fundamental" one (2008, 35–6, n.37). In the same footnote, though, she also argues that in Leibniz's thinking, phenomena incorporate abstract elements to varying degrees,[100] and I think this is right. At a certain scale, the repeated impacts of tiny bodies will produce an irregular polygon of sufficiently many sufficiently small sides that we will not be able to discriminate it in sensation from a truly continuous curve. It will appear to us as a curved trajectory, and this is how we represent it in the imagination. But then the representation in the imagination of the small sides as straight lines is itself an abstraction, since there is no scale on which the motion of the body will not be interrupted by further jostling from the particles around it. So in this sense, even the straight-line motions are phenomenal—they involve something imaginary.[101]

Jauernig prefers the weak relativist reading because, she maintains, if there are really only uniform rectilinear motions, then "it meets the difficulty raised by the mere Galilean invariance of Leibniz's conservation law," which the strong reading does not without further "fleshing out" (2008, 26).[102] She reasons as follows. If at any moment *all* motions are in fact rectilinear, then all motions at any moment are invariant under a Galilean transformation: that is, as Leibniz stated in his enunciation of EH in Proposition 16 of Part II, Section II, of the *Dynamica*, the same phenomena will be produced even if all motions are boosted by a certain velocity in a given direction. From this Jauernig infers that Galilean relativity will hold for all motions. This is true, but it does not mean that the EH holds for all motions. For that is to commit the very fallacy exposed above: from the fact that the EH holds in each moment, it does not follow that it holds for a motion through time. Leibniz's justification of Proposition 19 is unsound.[103]

[100] "In calling something ideal or imaginary, Leibniz does not always mean to contrast it with the phenomenal, but he sometimes merely wants to indicate that the phenomen[on] in question is of a special kind, namely, one that contains ideal features, i.e. features due to abstraction or confusion" (Jauernig 2008, 35–6).

[101] Cf. Adams' remark: "We might put this by saying that the other sensible qualities are appearances *of* sizes, shapes, and motions—and as such appearances of appearances" (Adams 1994, 229). (I owe this reference to one of OUP's readers of my manuscript.)

[102] See Slowik (2009) for a critique of this "fleshing out," and Jauernig (2009) for her response.

[103] Thus I wholly agree with Howard Stein when he says that the exact meaning of "composition" in Leibniz's proposition 17 of section III is "not immediately clear—not even from its own proof; but it *is* quite plain that, *whatever* it means, it cannot justify an inference from Galilean to 'general' relativity" (1977, 4–5); and I also agree with Nick Huggett and Carl Hoefer (2019) when they observe that "even in a pure collision dynamics the phenomena distinguish a body in uniform rectilinear motion over time, from one that undergoes collisions changing its uniform rectilinear motion over time: the laws will hold in the frame of the former, but not in the frame of the latter."

We can express this point using terminology Leibniz himself introduced in criticizing the Cartesian definition of quantity of motion: he says they failed to distinguish whether the motion involved is momentaneous, for which he coined the term *motio*, or rather "motion diffused through a stretch of time," *motus* (GM VI 237). The composition that produces new motions and inflections in a bent trajectory occurs at the beginning of each moment, and the resulting *motio* within each moment does indeed satisfy the equivalence of hypotheses. But this is not the same as the composition that includes all the inflecting impulses and produces the curved trajectory over time, the *motus*, for which the EH is not valid.

But what of the alleged difficulty for the strong relativist reading, which the weak reading is supposed to avoid? According to Jauernig this is "the old problem stemming from the mere Galilean invariance of Leibniz's conservation law," the conservation of mv^2—namely, the fact that "merely Galilean invariant laws do not hold in all frames of reference, and thus allow the phenomena to distinguish between some of these frames, which constitutes a violation of General Relativity" (2008, 25).[104] This difficulty reduces to the fact that, granting Galilean invariance, certain phenomena, such as the centrifugal effects cited by Newton, will nonetheless serve to distinguish a body that is really rotating, even if it has been supposed to be at rest or moving with a uniform velocity in a straight line, thus violating the universal relativity that Leibniz held to be essential to motion. But since the Galilean relativity of the uniform rectilinear motions (and therefore of *motio*) of Proposition 16, Part II, Section II, does not validly generalize to the Galilean relativity of all motion (*motus*), Jauernig's "weak relativist reading" does not avoid this difficulty either.

Of course, Leibniz recognizes that he must account for the phenomenon of centrifugal endeavour. Indeed he has to, given that he has analysed the gravitational tendency to the centre in terms of a difference between the centrifugal endeavour outwards of the fluid matter of the circulating vortex and that of the (solid) orbiting body. Thus, referring to Newton's (and previously Huygens') argument that the centrifugal force of a body whirled on a cord would give a criterion for circular motion,[105] he writes in his *Dynamica*,

> I admit that these things would happen in this way, if this were the nature of a cord or firmness and thus of circular motion, as these things are wont to be commonly conceived. But in truth, when everything is considered precisely, it is

[104] Here by "General Relativity" Jauernig presumably means a universal relativity of motion to whichever bodies are taken to be at rest, inviting a potential confusion with Einsteinian General Relativity—on which more below.

[105] As I argue in (Arthur 1995, 332), Newton was not the first to suggest that rotation constitutes an absolute motion; apart from the fact that it was conjectured long before by Huygens (but not published by him), it is (as I argue below) already suggested by Isaac Barrow in his *Mathematical Lectures*, which Newton would have heard when Barrow delivered them in 1664 (see Feingold 1993).

> found that circular motions are nothing but compositions of rectilinear ones, and that there are no other cords in nature but the laws of motion themselves.
>
> (GM VI 508/Stein 1977, 42)

The cohesion of the cord, that is, cannot be assumed as given. As Leibniz had long argued, if cohesion is explained by the existence of hooks and handles, or "other tangled textures," this only pushes the onus of explanation further down: for "what connects the parts of the fibres or hooklets?" (GM VI 510/Stein 1977, 44). This is not a superficial difficulty, but a product of Leibniz's commitment to a strictly mechanical philosophy which abjures Newton's unexplained "tractions" or actions at a distance. Cohesion, Leibniz insisted, must be explained in terms of motion in common. But this involves him in intractable difficulties: not only does he have to explain gravitational traction in terms of differential centrifugal force, he has to explain the origin of this force in terms of the pressure of the surrounding fluid, and at the same time the cohesion of the solid bodies involved, both in terms of their motion in common. These difficulties have been well documented and explored by others, and I shall not pursue them further here.[106]

Of course, one of Leibniz's reasons for holding fast to the equivalence of hypotheses for all motions was his conviction that all motion is essentially relative, that is, relative to whatever set of bodies has been identified as being at rest—the hypothetical "fixed existents" that are necessary for describing motion. In this regard it is interesting to consider what response he could give to a particular argument of Isaac Barrow's for the need to posit an absolute space. Barrow had argued in his *Mathematical Lectures* that, assuming the world as a whole is spherical and rotates, all the mutual situations of its parts will remain the same through time. So, if absolute space is rejected, and motion is mere change of relative situation, then "this sphere could not be transferred or rotated around its own axis, even by God"; such a rotation of the whole sphere can only be conceived, rather, as "a successive change of space, so that one part enters into the space vacated by the previous one" (Barrow [1683] 1860, 156). Clarke raised the same objection in his Fifth Reply (§§26–32). Although Leibniz's death precluded a response, he would presumably have replied just as Huygens had earlier replied to Neile in 1669: if we are considering the motion of the whole world, then there is no possible set of coexistents apart from it with respect to which it could be said to move. So, indeed, God could not make it rotate; not even God could make the world rotate with respect to something that does not exist.[107] As we have seen,

[106] See, for example, Stein (1977), Earman (1989), Slowik (2006, 2013), and Jauernig (2008).

[107] This argument would seem to have particular force against Barrow, since he characterized absolute space as a "mere capacity," and denied that it is an actual existent. As De Risi has observed (2012, fn.19), Reichenbach saw Leibniz's point of view here as anticipating General Relativity: "[Clarke] concludes from Leibniz's point of view that the parts of the rotating sun would have to lose their centrifugal force if all external matter around them were destroyed. This conclusion is indeed correct,

however, the same argument will not work for the rotation of a body *within* the world, and with respect to the other bodies in it.

Of course, having demonstrated that only relative linear velocities are involved in impact, and assuming that all changes of motion are by impact, it was natural for Leibniz to assume that this relativity would generalize to all motions produced by such impacts. In partial mitigation, Einstein had a similar blind spot about rotational motion, and this was due to a similar conviction that all motion is relative to whichever bodies are taken to be at rest.[108] Interpreting this as a relativity to reference frames conceived as co-ordinate systems defined by material rods and clocks, Einstein was led to seek a General Relativity where the same laws of physics would apply to bodies in co-ordinate systems in any state of motion. As has been explained in detail elsewhere, however, his attempts to deliver such a universal relativity, first by means of the Equivalence Principle, then through the requirement of the general covariance of physical laws, then through Mach's Principle, were all unsuccessful.[109] Einstein nevertheless produced a magnificent theory of gravitation, that merged together inertia and gravity into one unified field. But although in the resulting theory of General Relativity there are mathematical transformations that can make rotational motion apparently disappear, this does not eradicate the physical difference between a body rotating relative to this inertio-gravitational field and the field rotating with respect to the body. The mass of the Sun is the source of the inertio-gravitational field in which the Earth traces a geodesic, not the other way around. Copernicanism is just as true in a general relativistic context as it is in a Newtonian one.

Thus Leibniz was on the right track when he said that the Copernican system should be preferred to the Tychonic because it is "more in keeping with reason to attribute a modest motion to the Earth... than [to believe] that the observable universe should be moved with some insane rotation" (105/101). But here, I contend, he should have appealed to the criterion of expenditure of force (capacity for work done) that we saw him appealing to elsewhere: *it would have required immensely more force to set the observable universe in motion around the*

and modern physics admits it" (Reichenbach 1924, 429; my translation of De Risi's excerpt from the German). De Risi contests this, observing that according to GR (given reasonable boundary conditions and a Schwartzschild spacetime) a centrifugal force would be produced by the Sun even if all other matter were annihilated. On the other hand, though, that the world as a whole would have zero angular momentum is a prediction of the version of General Relativity proposed by Barbour and Bertotti (1977) on Machian principles.

[108] "It has, of course, been known since the days of the ancient Greeks that in order to describe the movement of a body, a second body is needed to which the movement of the first is referred. In physics the body to which events are spatially referred is called the coordinate system" (Einstein 1954, 229).

[109] For this reason, it is very misleading to speak of the universal relativity supposed by Leibniz as "General Relativity." For an illuminating discussion of Einstein's attempts, see especially Janssen (2012). He begins: "It is so obvious today that the general theory of relativity does not extend the relativity principle from uniform to arbitrary motion that it has become something of a puzzle how Einstein could ever have claimed it did" (2012, 159). For (what I hope is) a clear introductory exposition of General Relativity, see Arthur (2019c, ch. 7).

Earth than to set the Earth spinning. Had Leibniz done so, he could have agreed that the Copernican system, as refined by Kepler, was the true hypothesis in a stronger sense than the one he admitted: the Earth really has a rotational motion with respect to the rest of the universe, and not the other way around—this is easily the most intelligible hypothesis, and is therefore (fallibly) true. This would then have broken the equivalence of hypotheses for this and analogous cases of circular motion, without committing him to this motion being an absolute motion in Newton's sense, one with respect to an assumed absolute space. (But, of course, this would have ruined his argument for removing the Copernican censure!)

That he did not is attributable to various factors. One of these, I suggest, stems from his understanding of the conservation of force. In every elastic collision, on Leibniz's conception, active force is not only conserved overall, but also in each of the colliding bodies, where a body's live force is converted into elastic force and then back into live force. This entails that, even though force is the measure of a capacity to do mechanical work, in a collision no new force is introduced into a system (although some might be dissipated as friction, sound or ejected particles in an inelastic collision).[110] Thus in thinking always of an equivalence of force it would not have occurred to Leibniz to argue that more force would be needed for one body to be in relative motion rather than the other. Secondly, we have seen from all the discussion so far that a fundamental weakness in his philosophy has been in his treatment of motion through time. Having subscribed to the traditional argument for the unreality of time, Leibniz was committed to the view that what is real in time is momentaneous change, and what is real in motion is derivative force, that is, momentaneous tendencies or endeavours towards future states. Motion, as a successive thing, does not really exist now, only the derivative forces from which it results are real. This, at any rate, is one explanation of why he was so convinced that motion through time is nothing more than an aggregate of its linear compositions at each instant.

We have not yet discussed, however, how Leibniz himself sought to treat motion through time mathematically, and to this I now turn.

3.5 Motion through Space and Leibnizian Spacetime

So we turn now to the question of motion through space and time. This brings us back to the issues we were examining in §2.1, where modern critics have found

[110] Cf. Leibniz's correction of Johann Christoph Sturm's claim in his *Elective Physics* that "when motion is transferred from one globule to another through several intermediary ones, that last globule is moved with the *same force* as the first": "It seems to me rather that it is moved by an equivalent force, but not by the same force, since each one (as surprising as this may seem) is set in motion by *its own proper force*, namely elasticity, when it is repelled by the pushing of the one next to it" (*On Nature Itself*, §14; GP IV 514–15/WFT 220).

fault with Leibniz for not having sufficient resources in his relational view of space and time to provide a proper foundation for motion. This criticism is well summarized by John Earman: "to the modern reader the most glaring inconsistency in Leibniz's analysis is that it assumes an absolute or invariant notion of straight-line motion, a notion that is simply unavailable in the relationally acceptable Machian and Leibnizian space-times" (Earman 1989, 72). But this verdict depends on a conception of relationalism that is not Leibniz's, one in which, in contrast to Newton's theory of the physical world, "bodies alone exhaust the domains of the intended models of the relationist's world" (114). In chapter 2 we described in detail Leibniz's characterization of space as an "order of situations," where places are equivalence classes of relations of situation of a body to other bodies, indeed all possible such relations of situation, and space a system of such places. Space is thus something abstracted from the bodies which might be in it (that is, it is a system of equivalence classes of relations abstracted from bodies which might instantiate those relations): it is not body-relative in the way that Earman and others have supposed. This opens up distinct possibilities for a relationalist account of motion in space that is immune to the standard objections.

From the outset Leibniz explicitly conceived his *analysis situs* as a theory of space *and motion*. As we saw, his *Principia mechanica* of 1676 deals with motion as a change of situation.[111] And an essay on his *Characteristica* from September 1679, *New Characteristic for Expressing Situation and Motion* (reproduced in Appendix 3.2), begins:

> In order to perfect the new Characteristic that I have conceived for expressing Situation and Motion, certain things must be explained more distinctly, and some demonstrations must also be given. (CG xi 246)

Indeed, appeals to motion are prevalent in the manuscripts on *analysis situs*. To give just two examples from the various drafts Leibniz wrote in the 1670s:

> The generation of the straight line and the circle can in fact be obtained without already presupposing a straight line or circle, namely, if any body is moved while two of its points remain unmoved. Each [moving] point of the body describes a circle by its motion; while all the points remaining at rest fall on a straight line. (CG iii 66 [1676?])

[111] "Whenever we observe something to move," he writes there, "we notice first of all some change or state of things different from what we remember sensing shortly before. In fact this difference consists in the situation of the bodies... Situation is a mode according to which any body can be found, even though we recognize nothing in it specifically by which it can be distinguished from the others" (A VI 3, 103/Arthur 2013c, 108).

> If a point *A* is moved from *B* to *C* and by this very fact all its places are simply determined, all those places are said to be in a *straight line*.
>
> (CG iv 76 [February–August 1679?])

This warrants an extended discussion, though, since it cannot automatically be assumed that the use of motion in geometry is the same as motion of material bodies through space and time. In fact, there was a long tradition of the use of motion in geometry stemming from Euclid's *Elements*, and according to many commentators through the centuries, such motions should precisely not be understood as spatio-temporal ones. Thus there are appeals to motion in Book XI of the *Elements*, in the definitions of the sphere, the cone and the cylinder (which are described, respectively, by the revolution "of a semicircle about its diameter," "of a right-angled triangle about one of the sides containing the right angle, which side remains fixed," and "of a right-angled parallelogram about one of the sides, which remains fixed");[112] and also in the demonstrations of the congruence of figures by the transporting of one figure so as to lie on top of another in Propositions I.4 and 1.8 (where one triangle is "applied" to another), and III.24 (where one segment of a circle is "applied" to another) (Euclid 1933, 10, 14, 96). Given that Leibniz was aiming to give a superior foundation for the *Elements* with his *analysis situs*, it is not surprising, then, that the same kinds of motions are found in his writings on the subject.

The status of this use of motion in the *Elements* was a vexed issue, and a topic much discussed by geometers in their commentaries. The interpretation of congruence in terms of a transportation of figures was rejected, as we shall see, by the influential sixteenth-century commentator on the *Elements*, Jacques Peletier du Mans (1517–82), but he had no objection to understanding the definition of a circle in terms of motion. Here, for example is the gloss he gives on Euclid's definition of a circle:

> This definition will be made evident from the given abstract description of the circle. When indeed the given straight line *ab* is completely rotated around the point *a*, by its motion the point *b* causes the circumference, and the fixed point *a* remains in the centre of the circle. (Peletier 1557, Def. I.15)

This may be compared with the following definition given by Leibniz in *Scheda* 2 of his *Characteristica geometrica* of 1679:

> A *circle* is the path of a bounded straight line in a plane while one of its points remains unmoved, continued until the line returns to its former place.
>
> (Leibniz, CG vi 106)

[112] Quotations are from the Toddhunter edition (Euclid 1933, 221–2).

Such characterizations of figures in terms of rotations were traditionally described in terms of *phora*, the "carrying" of one figure to create another. In fact, *phoronomy* is Leibniz's preferred term for geometry involving motions and time.

This kind of motion was assimilated to another tradition of distinctly Pythagorean origin, according to which the basic geometrical objects are generated by the motion of an object of the next lower dimension—a line by the motion of a point, a surface by the motion of a line, etc. In his *De Anima*, Aristotle indicates the metaphysical origin of this idea among the Pythagorean "*mathematici*," for whom it was a motion in the soul, an instance of the emanation of the Many from the divine One.[113] He gives it short shrift: "How are we to imagine a unit being moved? By what agency? What sort of movement can be attributed to what is without parts or internal differences?" (*De Anima*, 409a, 2–3).

Here Aristotle's allusion to a point as "that which is without parts" is consistent with the Euclidean definition.[114] For him such an entity was an abstraction incapable of change, a subject of mathematics, but not of physics, and the same would apply to Euclid's definitions of a line as "a breadthless length" and of a surface as "having length and breadth only."[115] In keeping with this, as Douglas Jesseph remarks, commentators accepted the Euclidean definitions of line, surface and volume, but "typically interpreted them as expressing the fundamentally immaterial nature of geometrical objects, since no physical bodies can satisfy the definitions" (Jesseph 1999, 78). Thus geometers such as Omar Khayyam and al-Tusi maintained that while it is admissible to use phoronomic motions as didactic aids, they should not be used for the essential definitions of geometrical figures. Others, such as Hero of Alexandria (*c*.10 AD–*c*.75AD), defended the use of motion in geometry on the grounds that what is appealed to is a purely intellectual form of motion, not a physical one.[116]

[113] This Pythagorean origin is also testified by the influential sixteenth-century commentator Christopher Clavius (1538–1612), who reports that "the mathematicians [*mathematici*] also, in order to teach us the true understanding of the line, imagine a point, as described in the previous definition, to be moved from one place into another. Now since the point is absolutely indivisible, there will be left behind by this imaginary motion a certain path lacking in all breadth" (Clavius 1589, I, 13; quoted from Jesseph 1999, 80, and Axworthy 2014, slide 18a).

[114] Although Euclid of Alexandria lived after Aristotle, this characterization of a point was probably part of the Pythagorean tradition Euclid inherited. Erhard Weigel, like others of his time, confused the author of the *Elements* with Euclid of Megara, a Socratic philosopher who lived before the Stagirite, and interpreted Aristotle as being inspired by the *Elements*. See Brancato (2015, 129).

[115] See Angela Axworthy's forthcoming book (2021) for a full account of the use of kinetic definitions in geometry. I am much indebted to Axworthy for providing me with a transcript of a presentation she gave on this topic in (2014), and for subsequently providing me with copies of her articles (2017), (2018), and book manuscript (2021).

[116] Cf. Hero (1976, 14–15), Df. 1: "just as the unit is the origin of number, the point is the origin of geometrical objects, and it is indeed the beginning of a progression (ἔκθεσιν), but not as if it were a part of the line as the unit is a part of number, but as intellectually presupposing it; then the line is conceived from the motion (κινηθέντος, from κινέω) of a point, or rather from the intellection of the point in the state of flowing (ῥύσει, from ῥύσις) . . ." quoted from Axworthy (2021, 4–5).

In the sixteenth century such a defence of the use of motion was much fortified by the rediscovery of the *Commentary on the first book of Euclid's Elements* of Proclus of Lycia (412–85), published in 1533 (Proclus 1533). Proclus argued that the imagination (φαντασία) is a faculty intermediate between the understanding (διάνοια, also translated as 'discursive knowledge') and sense perception. It provides "a kind of 'screen' or 'mirror' onto which discursive thinking projects mathematical forms, or λόγοι, and makes them 'visualizable' to an inner gaze of the mind" (Nikulin 2008, 153):[117]

> What projects the images is the understanding [διάνοια]; the source of what is projected is the form in the understanding; and what they are projected in is this "passive intellect [νοῦς]" that unfolds in revolution about the partlessness of the genuine intellect [νοῦς]... (*In Eucl.* 56/Proclus 1970, 45)

For Proclus geometrical figures therefore have a kind of existence that is intermediate between the ideal forms, on the one hand, and bodies as they appear to sense, on the other: they are unchangeable like the forms, but extended, like the objects of sense.[118] Thus motion in geometry does not connote a motion in the material world, but is the projection of a figure in the imagination, one that emulates or "follows" the progress of thought. The generation of figures does not give their essential definitions or true causes, since the forms (such as the sphere in itself) are unchangeable, but it enables the unfolding of their properties in the mind's eye.[119] Accordingly, Proclus distinguishes the metaphysical generation of forms—number from the One, a line from a point—for which he uses the term ῥύσις or flow, from the motion of figures as they are produced in the imagination, an inferior kind of *kinesis*. Thus in his *Commentary on the first book of Euclid's Elements*, on the first postulate ("That a straight line may be drawn from any one point to any other"), he writes

[117] David Rabouin places Descartes in this same Proclean tradition: "Indeed imagination or *phantasia* is described in the *Regulae* at the same time as a faculty of the mind and as a kind of proxy (a *subjectum* in Descartes' parlance) on which human reasoning projects and manipulates basic mathematical relations" (Rabouin 2018, 4753). Rabouin identifies Alessandro Piccolomini as relying on Proclus's theory of the imagination in claiming "that the unity of mathematics was given by a universal theory expressed on the background of a kind of 'matter', or 'subject', called by him 'imagined quantity'...These specific features of Proclean epistemology are also key elements of Descartes' epistemology in the *Regulae*" (4757). And, according to Rabouin, Leibniz explicitly recognized these Proclean elements in Descartes' philosophy of mathematics.

[118] Thus in the opening sentence of the Prologue of Part I of his commentary, Proclus writes: "Mathematical being necessarily belongs neither among the first nor among the last and least simple of the kinds of being, but occupies the middle ground between the partless realities—simple, incomposite, and indivisible—and divisible things characterized by every variety of composition and differentiation" (Proclus 1873, 3 ll. 1–7/1970, 3).

[119] Cf. Axworthy (2014, 2): "In this context, the generations of geometrical objects were conceived as complementary to their definitions, which exhibit the essential properties, or the features which universally belong to the object."

> The drawing of a line from any point to any point follows from the conception of the line as the flowing (*ῥύσιν*) of a point and of the straight line as its uniform and undeviating flowing (*ῥύσιν*). For if we think of the point as moving (*σημεῖον κινούμενον*) uniformly over the shortest path (*κίνησιν*), we shall come to the other point and so shall have got the first postulate without any complicated process of thought. (Proclus 1970, 145; quoted from Axworthy 2021, 15, n.64)

The influence of Proclus's commentary on geometers of the sixteenth century was profound. It can be seen for example in the commentary of Oronce Fine (1494–1555) on the same postulate in 1536:

> For any given point can describe a straight line by *abstractively flowing* [*abstractivè fluendo*] through the shortest path to any other point, even one that is imagined anywhere. (Fine 1536, Post. 1; Axworthy 2021, 33, n.133)

and in the glosses on the definitions of Jacques Peletier in his 1557 commentary:

> From the *perpetual flow of a point in length* [*Ex Puncti fluxu perpetuo in longum*], we imagine the generation of the line.
> (Peletier 1557, def. 1.1, 2; Axworthy 2017, 459, n.1)
>
> And just as the line originates from the *continuous flow of the point* [*ex Puncti fluxu in continuum*], so the surface originates from the *transversal drawing of the line* [*in transversum ductu*].
> (Peletier 1557, def. 1.4, 3; Axworthy 2017, 459, n.1)

As we shall see, there is perhaps a hint of Proclus in Leibniz's own attempts to derive the line as the shortest path between two points from a metaphysical consideration of the motion of a point—even if, like others in the period, he ignores Proclus's distinction between *ῥύσις* or *fluxio* (flow) and *κίνησις* or *motus* (locomotion). In fact, one can find more than a few echoes of Proclus in Leibniz's writings, particularly in his conception of mathematics as "the science of the imagination."[120] But before getting into that there is more to say on the status of motion in geometry in the period.

This notion of the generation of continuous geometric quantities by motion was fortified by (and mixed up with) an equally old tradition, the kinematic definition of curves. Thus, for example, an ellipse could be generated by the motion of a point in such a way that the sum of its distances to two given points (the foci of the ellipse) is held constant—that constitutes a kinematic definition of the ellipse. Similarly, an Archimedean spiral is generated by the uniform motion of a point

[120] For an illuminating treatment of Descartes' and Leibniz's conception of mathematics as a science of the imagination, see Rabouin (2020).

along a line which is itself rotating in a circle at constant angular velocity. This use of moving points to describe figures in geometry goes all the way back to its origins in ancient Greece, where Archytas of Tarentum (428–347 BCE) was instrumental in promoting the use of such "mechanical motion," and the pseudo-Aristotelian *Mechanical Questions* ensconced a tradition of using such methods that was to last through the centuries.[121] In the seventeenth century, Descartes and Fermat construed curves as generated by continuous motion of the ordinate, and the "kinematic conception of curves" was featured prominently by Schooten. Here, however, the issue of mechanically produced curves became embroiled with controversies over what could be counted as genuinely geometrical, issuing in Descartes' division of all curves into either geometrical or mechanical.

Part of the appeal of this conception of the generation of magnitudes by motion is that (assuming the continuity of motion) it circumvents the problem of the composition of the continuum. Rather than conceiving a continuous line as composed of indivisible points it could be conceived as generated by a flowing point, and similarly with surfaces and volumes, each being generated by the flow of a magnitude one dimension lower. Thus we find Jacques Peletier writing in 1557:

> And just as the line arises from the *continuous flow of the point*, the surface originates from the *transversal drawing of the line*... So unlike number, which comes from the accumulation of units, the line does not come from the *addition of points*, but rather from their *continuous flow*. And in this the continuous differs from the discrete, because the continuous can be infinitely divided, without us ever reaching the point in the way the unit is reached in discrete quantity. (Peletier 1557, def. 1.4, def. I.2)

Similarly, Isaac Barrow helped himself to the notion of continuous flow in geometry, despite his overall nominalist position. For him not only were lines, straight and curved, generated by the flow of a point, but also "Time may be conceived as the trace of a continually sliding instant" (LG 161, 165). Unlike Peletier, however, he declared himself agnostic about composition, saying that it did not matter whether a line was regarded as formed by the motion of a point or by the accumulation of elements.[122]

[121] According to Diogenes Laertius' *Lives of the Eminent Philosophers*, Archytas "was the first to bring mechanics to a system by applying mathematical principles; he also first employed mechanical motion in a geometrical construction, namely, when he tried, by means of a section of a half-cylinder, to find two mean proportionals in order to duplicate the cube" (Diogenes Laertius 1925, Book VIII, §83). In my account of "Aristotle's Wheel" in (Arthur 2012) I speculate that Archytas might well have been the originator of that thought experiment, which was a live issue in Leibniz's time as a result of Galilei's use of it in his *Discorsi*.

[122] "To every instant ôr indefinitely small particle of time (I say instant or indefinite particle, for it makes no difference whether we understand a line to be composed of points or of indefinitely small linelets, nor, in the same way, whether we suppose time to be made up of instants or of innumerable minute timelets..." (Barrow [1670] 1860, 167–8/Barrow 1916, 38).

Not so Barrow's distinguished successor in the Lucasian Chair of Mathematics at Cambridge, Isaac Newton, who followed him in adopting this kinematic conception of curves and generation of quantities by motion. For Newton was adamant that, despite his use of infinitely small quantities in the *Principia*, he regarded continuous quantities as generated by motions and not as composed of infinitely small elements:

> I don't here consider Mathematical Quantities as consisting of *indivisibles,* whether least possible parts or infinitely small ones, but as described by a continual motion. Lines are described, and by describing are generated, not by any apposition of Parts, but by the continuous motion of Points, Surfaces by the motions of Lines, Solids by the motion of Surfaces, Angles by the Rotation of their Legs, Time by a continual Flux, and so on in all the rest. These *Geneses* are founded upon Nature, and are every Day seen in the motions of Bodies. (Newton, *De quadratura curvarum* [1693??])[123]

What is striking here is that Newton has not only collapsed any distinction between metaphysical flow (*rhusis*) and locomotion, but also any distinction between motion in geometry and motion through space and time, assimilating the generation of geometrical figures to the motion of physical bodies in nature. He has thus trampled roughshod over the traditional distinction between pure and mixed mathematics; and quite consciously so, as evidenced by the Preface to the Reader in the first edition of his *Principia*:

> For the description of both straight lines and circles, upon which *geometry* is founded, belongs to *mechanics*.... To describe straight lines and circles are problems, but not geometrical problems. Their solution is postulated from *mechanics*, and *geometry* teaches the use of problems thus solved. And it is the glory of geometry that from so few principles sought from elsewhere it is able to produce so much. Therefore *geometry* is founded in mechanical practice, and is nothing but that part of *universal mechanics* which renders the art of measuring into accurate propositions and demonstrations. But since the manual arts are employed especially in making bodies move, it happens that *geometry* is commonly thought to pertain only to magnitude, and *mechanics* to motion.
>
> (Newton 1687, i)[124]

[123] (Newton 1962, 141); translation amended by comparison with Newton's revision in Newton (2008 [MPN VIII], I, 2 §2, 106–7).

[124] Newton certainly seems inspired by Neoplatonists such as Proclus, however. Compare, for instance, his claim in *De gravitatione* that "there are everywhere all kinds of figures, everywhere spheres, cubes, triangles, straight lines, everywhere circular, elliptical, parabolical, and all other kinds of figure, and those of all shapes and sizes, even though they are not disclosed to sight" (Newton 1962,

This is a radical ontologizing of geometry that goes well beyond the kind of mathematization of the mechanical philosophy envisaged by Descartes and Galileo. But it has a precedent in Hobbes, who also took his inspiration from Galileo's analysis of motion in his *Discorsi*. Hobbes's philosophical outlook, of course, is fundamentally different from Newton's, having the explicitly stated goal of giving a thorough-going materialistic account of mathematics—something that was anathema to the pious Cambridge don.[125] Nevertheless, in reducing geometry to a branch of universal mechanics whose postulates are taken from mechanics, and in which motions of geometrical objects are treated as motions in space and time, Newton's view of geometry has this much in common with that of the despised philosopher from Malmesbury. This is all very much to the point here, since we are trying to understand how Leibniz conceived motions in geometry, and we know that he derived much of his initial understanding of geometry and its foundations from his close reading of Hobbes, prior to studying it in earnest under Huygens' guidance in Paris. It will therefore be worth entering into a short digression on Hobbes's programme.

In his *De corpore* Hobbes gives an extended treatment of the foundations of mathematics, taking up all of chapters 12 to 19, leading up to their intended culmination in chapter 20, a triumphant solution to the ancient problem of squaring the circle—the first in a series of increasingly embarrassing such attempts that winnowed away Hobbes's initial reputation as a mathematician of note. The point of these efforts, as Douglas Jesseph has explained, is to show that if geometry itself—the paragon of sound reasoning—could be established on materialist foundations, and thereby reformed and improved, the same and more could be expected of all knowledge based on the same principles, involving only bodies and their motions.[126] This was not lost on John Wallis, who took up the challenge (on behalf of mathematicians, universities and "the Christian world") by attacking the mathematical foundations of Hobbes's system.[127]

100/2004, 22) with Proclus's "Consequently, all the limits are everywhere, and each comes to light in its proper place, their appearances varying according to the power that prevails in them" (Proclus 1873, 92/Proclus 1970, 75).

[125] Thus, in his *De gravitatione*, Newton gives a hypothetical account of bodies as parts of space that God has endowed with mobility, impenetrability and perceptibility, and then writes that "we cannot posit bodies of this kind without at the same time positing that God exists, and has created bodies in empty space out of nothing, and that they are beings distinct from created minds, but able to be united with minds" (Newton 1962, 106–9/2004, 28–31).

[126] "Because the fundamental nature of a body is clearly apprehended, this program (so Hobbes thinks) will make short work of the most difficult geometric problems, and the truth and power of the Hobbesian philosophy will be made manifest by its success in the solution of outstanding problems" (Jesseph 1999, 70).

[127] See Wallis's letter to Christian Huygens of January 11, 1659, quoted by Jesseph: "... And so it seemed necessary that now some mathematician, proceeding in the opposite direction, should show how little he understands his mathematics (from which he takes his courage)" (HOC 2: 296–7; Jesseph 1999, 70).

Thus Hobbes set himself the task of showing how all geometry can be properly founded on the notions of body and motion. He took issue with the Euclidean definition of a point as "that which is without parts" (as we explained in §2.3 above), replacing it with his own definition as "that whose quantity is not considered." Similarly, "if the magnitude of a body that moves be not considered (even though there always is some magnitude), the path along which it passes is called a *line*, or *one single dimension*, the space through which it passes, a *length*, and the body itself a *point*; this is the sense in which the Earth itself is usually called a point, and its annual path, the line of the ecliptic" (*De Corpore*, II, ch. 8, §12; LLC 359). His motivation for these definitions, as Jesseph has explained (1999, 73 ff.), is to ground all mathematical quantities in the nature of body and its motions. This is in keeping with his insistence that philosophy should proceed by deriving the properties of things from a knowledge of their generation. Thus in his description of philosophical method in Part I Hobbes notes the standard definitions of place as "the space which is adequately filled by a body" (*De Corpore*, I, ch. 6, §6; Hobbes 1839b, 62), and motion as "the privation of one place and acquisition of another" (62–3), and that these are then followed by "their generations ôr descriptions, so that a line, for example, is made by the motion of a point, a surface by the motion of a line, one motion by another motion, etc." (63). But, he insists, "the names of things which are understood as being able to have a cause ought to have in their definition the very cause or mode of their generation, as when we define a circle to be the figure brought about by the rotation of a straight line in a plane, etc." (72).

Here a contrast with Proclus's philosophy of mathematics will be most instructive. For Proclus had insisted against his predecessors in the Academy that mathematical objects are not the subject of the understanding directly, nor are they simply abstractions from sensible things. Rather, although the understanding (διάνοια) contains forms (λόγοι), it is unable to see them without the aid of the imagination: it "unfolds and exposes them and presents them to the imagination sitting in the vestibule" (*In Eucl.* 54–5/Proclus 1970, 44). But for Proclus the "screen" on which they are projected in the imagination is an "intelligible matter," and the understanding unfolds or explicates its knowledge of the forms, "happy in their separation from sensible things, and finding in the matter of imagination a medium apt for receiving its forms" (*In Eucl.* 55/1970, 44).[128] Thus, in contrast to the unit of arithmetic,

[128] I should add that there is a certain ambiguity to the status of the imagination (φαντασία) in Proclus. It is conceived as the "screen" or surface onto which the figures are projected, but also as an active component in the mind doing the imagining, a fourth cognitive faculty (below νοῦς) and intermediate between διάνοια and αἴσθησις. See Nikulin (2008 159–61); see also Rabouin (2020), who sees a strongly analogous dual role for the imagination in Descartes' conception of it.

> the point is projected in imagination and comes to be, as it were, in place and embodied in intelligible matter. Hence the unit is without position, since it is immaterial and outside all extension and place; but the point has position because it occurs in the bosom of imagination and is therefore enmattered.
> (*In Eucl.* 96/Proclus 1970, 79)

To such a conception of motion occurring in "intelligible matter" separate from sensible things, Hobbes retorts:

> But everyone knows that nothing except body can be moved, nor can motion be conceived of in anything except body. And every body in motion traces a path with not only length, but also breadth. Therefore the definition of a line should be as follows: *a line is the path traced by a moving body, whose quantity is not considered in the demonstration.*[129]

Thus where Proclus insisted that there has to be a material substratum on which figures are traced, "a particular matter that is capable of accommodating extended figures (cf. *In Eucl.*15.5–6; 51.9–20)" (Nikulin 2008, 161), for Hobbes, space itself is the "screen" on which figures are represented in the imagination: "an imaginary space indeed, because a mere phantasm, yet that very thing which everybody calls space" (*De corpore*, ch. 7, §2; Hobbes 1839b, 82). Thus "*space is the phantasm of an existing thing insofar as it is existing*; that is, with no other accident of the thing considered apart from its appearing outside the person imagining" (§2, 83). Figures are phantasms of shapes of actual bodies, whose magnitude Hobbes identifies with "real space."[130]

Likewise, motion also produces a phantasm, Hobbes argues:

> Just as a body leaves a phantasm of its magnitude in the mind, so also the motion of a body leaves a phantasm of its motion in the mind, namely, an idea of the body passing by a continuous succession, first through this space, then through that. (§2, 83)

Proclus had rejected the definition of a line in terms of the motion of a material point because it "appears to explain [the line] in terms of its generative cause and sets before us not line in general, but the material line" (*In Eucl.* 79; quoted from Jesseph 1999, 80). Hobbes begs to differ:

> The reason why I say that those things that have a cause and generation should be defined in terms of their cause and generation is this: the aim of demonstration is

[129] *De Principiis et Ratiocinatione Geometarum*, 2; quoted from (Jesseph 1999, 81).
[130] Cf. Piccolomini's "imagined quantity"—Rabouin (2018, 4753) (see fn. 117 above).

> the knowledge of the causes and generation of things, which, if they are not contained in the definitions, cannot be contained in the conclusion of even the first syllogism that is made from these definitions...
>
> (*De Corpore*, I, ch. 6, §13; Hobbes 1839b, 73).

Hobbes's requirement that definitions of geometrical objects include their causes is not in itself novel, as Aristotle's theory of demonstration in the *Posterior Analytics* also required that proofs proceed from causes.[131] The difference is that Hobbes is construing cause solely in terms of efficient cause, eschewing the material, formal and final causes identified by Aristotle.[132] Consequently, for him a real or causal definition must exhibit the essential properties of a thing by showing how it is generated. This is his interpretation of the traditional distinction between *real* and *nominal definitions*: whereas a *nominal definition* can be sufficient to identify a thing through certain of its properties, a *real definition* should exhibit the possibility and nature of the thing defined by giving its generation. He therefore criticizes Euclid's definition of a circle—as "a plane figure contained by one line, called the circumference, such that all the straight lines drawn from a certain point within the figure to the circumference are equal to one another" (*Elements* I, Def. 15)—in that it fails to give a causal account of the circle's generation. He proposes instead a definition corresponding to Peletier's gloss on Euclid's Definition 15 that we quoted above, where the circle is generated by the motion of the endpoint of a line whose other endpoint remains fixed.[133]

As we have seen, Leibniz gave a similar definition of the circle in *Scheda* 2 of his *Characteristica geometrica* of 1679. This is consistent with his adoption[134] of Hobbes's insistence on the importance of giving *real* as opposed to *nominal definitions*, as well as Hobbes's way of characterizing the former in

[131] See the discussion of Jesseph (1999, 201), and the references contained there.

[132] Cf. what Leibniz wrote in the Latin fragment from 12 February 1676, which I quote and translate in footnote 135 below.

[133] Hobbes's definition is as follows: "If a straight line existing in a plane is so moved that, with one endpoint remaining fixed, the whole line is carried around together until it returns to the place from which the motion began, it will have described a plane surface bounded everywhere by a curved line, that line, namely, which the endpoint carried around will have described. This surface is called a *circle*..." (II, ch. 14, §4; Hobbes 1839b, I, 157). Hobbes's definition of the circle as the surface rather than the circumference may appear curious, but (as David Rabouin has pointed out to me) it is already to be found in Antiquity.

[134] I would say "endorsement"; but curiously Leibniz, who on other occasions readily admitted debts to Hobbes, does not acknowledge him as his teacher in this connection. In fact, he uses this distinction "to satisfy a difficulty raised by Hobbes," a difficulty Leibniz foists on Hobbes by misinterpreting him as holding that "all definitions are arbitrary and nominal" (A VI 4, 542), which manifestly was not Hobbes's view. See Jesseph (1999, 198 ff.) for a clear discussion of Hobbes's nominalism and Leibniz's misinterpretation of it.

terms of generation.[135] Explaining this distinction in his essay "Of Universal Synthesis and Analysis" of around 1683 (A VI 4, 541), Leibniz gives the example of the phoronomic definition of the circle, misremembering it as Euclid's own definition of a circle, rather than Peletier's gloss or Hero's or Hobbes's definition. He maintains that a circle, for example, could be defined as that curve such that "any point on each segment of it makes the same angle with the two endpoints of the segment." But—even though this is indeed a property of the circle—that would not suffice to establish that such a curve is possible. In contrast,

> the definition of the circle proposed by Euclid, that it is the figure described by the motion of a straight line in a plane around one extremum, does provide a real definition, since it is clear that such a figure is possible. It is useful, therefore, to have definitions that involve the generation of the thing, or, failing that, its constitution, that is, a way in which it is shown to be either producible or at least possible. (A VI 4, 541/MP 12–13)

Thus far, then, Leibniz is following quite closely in Hobbes's footsteps. Where he goes beyond Hobbes, however, is in his treatment of *congruence*. Of all the uses of motion in the *Elements*, this had been the most controversial. As De Risi has explained (2017, 279), debates centred on its use in establishing congruence by superposition, involving the movement of one figure to coincide with another. This stemmed from its use in the fourth of the Common Notions of Euclid's *Elements*, the one dealing with congruence as superposition "Things which coincide with one another are equal to one another." Peletier, for example, notwithstanding his endorsement of Proclus's defence of motion in geometry, had rejected the appeal to motions in the "applying" of one figure to another in Propositions I.4 and 1.8 and III.24. This for him was to introduce something essentially mechanical into geometry:

> To superpose figures on figures is something mechanical; but to perceive by the intellect [*intelligere*] is purely mathematical. For if we admit the superposition of lines and figures in proofs, nearly all of geometry will be full of applications of this kind...Although application would be somewhat more tolerable than superposition, it is, however, rejected in geometry. Indeed, it is surely not

[135] Cf. this remark made by Leibniz in February1676: "Every property which contains the efficient cause of a thing, ôr its generation, suffices for discovering all its attributes. That is to say, whatever contains the generation of the thing, as Hobbes says. [*Proprietas omnis quae continet efficientem rei causam, seu generationem, sufficit ad omnia ejus attributa invenienda. Scilicet quicquid generationem rei continet, ut loquitur Hobbes.*" I am indebted to Osvaldo Ottaviani for supplying me with this fragment from a mathematical manuscript, dated by Leibniz 12 February 1676 (LH 35, 3B 15, Bl. 6 recto).

> permitted to transport a line in order to describe a circle according to its magnitude, if there has not first been drawn an equal line.
>
> (Peletier, 1557, Prop. I.4, 16)

Clavius, however, was highly critical of Peletier's rejection of motion in superposition on these grounds. In his opinion, the motion involved in superposition is no more mechanical than that involved in the generation of a line:

> Peletier seems not to have understood sufficiently how geometers use this superposition. Indeed, they do not want the superposition of figures to be made in fact (for this would certainly be something mechanical), but only in thought and in the mind, which is the work of reason and the intellect.
>
> (Clavius 1589, Prop. III.16, 368; translated from the quotation in Axworthy 2018, 27, n.156)

Clavius therefore assimilates the motion involved in superposition to the kinematic motions defended by Proclus in terms of flow. Thus, commenting on Postulate 1, "that a straight line may be drawn from any one point to any other," he writes:

> For since the line is the imaginary flow of a certain point, and since therefore a straight line is a flow progressing along an absolutely straight path, it turns out that, if we have conceived the point as something moving straight towards another point, a straight line will be correctly drawn from one point to another.
>
> (Clavius 1589, Post. 1, 57–8; translated from the quotation in Axworthy 2018, 32, n.178)

Hobbes presumably thought that he had finessed such controversy about motion in geometry by conceiving all motions as taking place in the same imaginary space, rejecting the idea that geometric ones take place in some "intellectual matter." Noting how some had interpreted *place* as an immovable, real space, he countered that if one takes into account that "nothing is general or universal besides names or signs," one should realize that place in general "is nothing but a phantasm in the mind or memory of a body of such a magnitude and such a figure."[136] Conceiving the motions of geometric figures is thus nothing different from conceiving motion generally, where "the motion of a body leaves a phantasm of its motion in the mind, namely, the idea of a body passing by a continuous succession, first through this space, then through that" (*De corpore* II, ch. 7, §2; Hobbes 1839b, 83).

[136] (II, ch. 8, §5, Hobbes 1839b, 94). Cf. also "That space, by which word I here understand imaginary space, which is coincident with the magnitude of any body, is called the *place* of that body" (II, ch. 7, §2, 82).

Leibniz, however, did not agree that the possibility of superposition could be established by an appeal to motion, and saw things the other way around. Agreeing with Hobbes on the need for establishing the possibility of geometric operations through genetic definitions, he found the definitions through motion advanced by Hobbes and Clavius to be themselves in need of demonstration. Thus at the start of the concluding four paragraphs of the *Characteristica geometrica* (see Appendix 3.1), he writes:

> §104. Moreover, all the common definitions of the straight line are not yet sufficiently perfect, in that, since there is still always a need for a demonstration, it can be doubted that such a straight line is possible. It seems that this, however, should be supposed to be among the simplest things, and so there is a need for a definition that makes it immediately evident that a straight line is possible.
>
> (CG ix 226)

His basic line of attack is apparent in the earliest version of the 1679 *Characteristica* (Scheda 1). Rather than base congruence on motion (whether mechanical or only in the intellect), he sought to found motions on congruence:

> *Any straight line whatever can be moved upon another*, i.e., is successively congruent with it. For two straight lines can be so located that they are parts of one straight line. This being so, if one is at rest, the other moving towards it in space along the same straight line indefinitely, it will pass through its place—assuming, of course, that they can penetrate one another; but it can move indefinitely in space, since any straight line can be applied to any greater straight line.
>
> (Leibniz, CG v 90 [August 11, 1679])

The same considerations prompt Leibniz in his *Characteristica* (CG ix) to refashion the definitions of straight line and circle so that the possibility of these figures is demonstrated in terms of successive congruence. To this end he articulates a notion of a rigid displacement in terms of a "trace" [*tractus*].[137]

> (76) Let us review some things: From any point to any other one can understand a line to be drawn, a rigid one.
>
> (77) *A.B.* signifies the situation of *A* and *B* to one another, that is, a rigid trace through *A* and *B*, which trace suffices for us to be a line. Thus *A.B.C.* signifies a rigid trace through *A.*, *B.*, and *C.*

[137] 'Trace' is my translation of Leibniz's term *tractus*; as noted by Parmentier (who translates it into French as *tracé*), it means the figure traced, the *vestigium*; when it is moved, we get a path [*via*] or trajectory (CG fn.47, 167). It seems to correspond to what Leibniz called a trace (French: *trace*) in his Fifth Letter to Clarke, on which, more below.

> (78) Whatever is posited in space can be moved, whether it be a point or a line, or some other trace, that is to say, to anything in an extensum something can be assigned congruent to it. Hence *A.* ≅ *X.*, *A.B.* ≅ *X.Y.*, *A.B.C.* ≅ *X.Y.Z.*, or *A.B.C.* ≅ *ωX.Y.Z.*[138]

After framing various other properties of motion in terms of equivalences of congruence, such as that "(82) Every trace can be moved with one of its points remaining fixed: *A.B.C.* ≅ *A.X.Y.*" (CG ix 216), Leibniz refashions his definitions of straight line and circle accordingly:

> (83) A *straight line* is a trace which cannot move when two points in it are at rest; that is, if a certain trace is moved while two points remain immobile, then, supposing further points in it remain at rest, all these points are said to be situated in a straight line, or to fall on a trace that is called a straight line. Equivalently, if *A. B.C.* ≅ *A.B.Y.*, it is necessary that *C* ∞ *Y*. That is, if one finds some point *C* situated in a straight line with the points *A* and *B*, the trace *A.B.C.* (or *A.C.B.*) cannot be moved with *A* and *B* immobile in such a way that *C.* is transferred into *Y.*, and so the trace *A.B.C.* will be congruent with *A.B.C.*... Namely, if *A.B.Y.* ≅ *A.B.X.* and therefore *Y* ∞ *X*, then ϖ*X* (∞ ϖ*Y*) will be a straight line.[139]

As De Risi has observed (2007, 220 ff.), Leibniz improves this definition in later manuscripts by means of the operator of *uniqueness*, denoted "*unic.*" Thus "'*A.B unic*' means that *A.B* ≅ *A.C* implies that *B* coincides with *C*" (DeRisi 2007, 220): that is, *A.B unic* $=_{\text{def}}$ *A.B* ≅ *A.C* → *B*∞*C*.[140] Then the straight line defined according to §83 above is the locus of all the points that are unique for the situation to two given points:[141] for a given *A*, *B*, it will be ϖ*Y* such that *A.B.Y unic*.

Similarly, Leibniz's revised definition of a circle is again a phoronomic one, but with the possibility of motion demonstrated in terms of the possibility of congruence:

[138] Again, I have systematically used the symbol '≅' for congruence in place of Leibniz's 'γ'. The notation '*ωX.Y.Z.*' means 'any of those in the situation *X.Y.Z.*', and 'ϖ*X*' signifies 'all the *X*'.

[139] Leibniz notes that it still needs to be demonstrated rather than assumed "whether there are points of this kind situated in a straight line, and whether they compose a trace, and whether the trace is that line." He writes: "Now the path of a point so moved will certainly be a *straight line* if it passes through all the points of this kind; certainly the locus of all points taken in a straight line with two points will be no other trace than a line."

[140] See *Rectam definio*... in Appendix 3. The definition of the uniqueness operation can now be traced back at least to the mid-1680s. It is found among the texts transcribed by the Mathesis Project in collaboration with the Leibniz-Archiv in Hannover, for instance, in MT #46 , *Notiones in Geometria explicanda* [LH 35 I 5 Bl. 49, 4 °] (Appendix 3.5 below), and in MT #106, *Generalia*, [LH 35 VII 30, Bl. 128, 2°].

[141] Cf. the text *Magnitudinis nomine*... from January 1712 transcribed by De Risi: "A *straight line* is the locus of points unique to two points.," "If YAB is unic. then $\bar{Y}$ will be a straight line" (De Risi 2007, 598, 599). Also, from *Geometria determinatoria* of 1702: "A *straight line* is the locus of all points uniquely situated to two given points" (627).

> (84) If, when two points *A.B.* of the trace *A.C.B.* remain at rest, the trace itself is moved, the line which its point *C* describes when it is moved is called *circular*. But whether it is possible for some trace to be moved with two points remaining at rest must not be assumed, but should be defined by a demonstration. If *A.C.B.* ≅ *A.Y. B.*, then ⓐ*Y* will be a *circular* line, and if there are however many points *C.D.E.F.* and *A.B.C.* ≅ *A.B.D.* ≅ *A.B.E.* ≅ *A.B.F.*, they are said to be in one and the same circular line. This definition of a circular line does not presuppose that a straight line and plane are given, as does Euclid's definition.

What characterizes these moves by Leibniz, then, is that he is *demonstrating*, not assuming, the possibility of the motions necessary for establishing the theorems of geometry. The definitions of the various figures in geometry in terms of motion are given in such a way that they provide formal demonstrations of the possibility of the motions involved. The possibility of rigid motions in space is equated with the possibility of congruence, that is, of a second extensum's being in a situation congruent to the first, all such congruent situations being equivalent—this comprises the uniformity of space. That is, the possibility of motion amounts to the substitutability of situations that are equivalent through congruence. What is crucial here is that the equivalence is justified by the *indiscernibility* of congruent situations: whereas similar objects are those that cannot be distinguished from each another when they are taken one by one, *singulatim*, congruent objects are those that cannot be distinguished from each another when they are brought into coincidence and compared together at the same time (*simul*).

De Risi aptly summarizes this approach of Leibniz's as follows:

> First, Leibniz gives a definition of congruence independently of any consideration of motion, Then, he characterizes space through uniformity (i.e. the possibility of bringing an object from any place to any place through congruence.) In this way, he founds the possibility of motion on the spatial property of uniformity. Finally, he affirms that the motion that has been made possible in this way fully characterizes the group of isometries, in the sense that the abstract definition of congruence must perfectly coincide with the one obtained through rigid motion and superposition. (De Risi 2007, 278)

Now let us compare this approach with the others we have been considering.

Unlike Proclus and the traditional geometers, Leibniz is no more inclined to distinguish motions in geometry from mechanical motions than are Hobbes and Newton, with the proviso that a full consideration of mechanical motions involves the causes of motion, and thus dynamics.[142] For him, as for Proclus and Hobbes

[142] Cf. De Risi's remark that in this approach, "geometrical motion is certainly legitimate, but it transforms geometry into kinematics, thus failing to distinguish the two sciences. Probably however

(and indeed, Descartes), figures are representations in the imagination. Like Proclus, Leibniz argues that there are no precise shapes (*figurae*) in the material world: "There is no precise and fixed shape in bodies, because of the actual division of parts to infinity" (A VI 4, 1613/LLC 297).[143] Similarly, as we saw in chapter 1, there are no precisely enduring states, even though we represent them as such in the imagination. But Leibniz goes beyond Proclus, Hobbes and Descartes in seeing that the relations between these figures in the imagination can be represented analytically—not analytically as in the algebraic geometry of Viète and Descartes, but in a more direct way where the symbols stand for the situations as depicted in the imagination. As a result, he envisages the mechanical application of the results of geometry even more directly than Newton or Hobbes. Where for Newton geometry has to defer to mechanics for its postulates, and where Hobbes too had relied on given motions being presupposed in geometry, Leibniz has the idea of a formal science by which results can be derived directly not only in geometry, but also in mixed mathematics:

> (7) But if we are once able to represent figures and bodies by letters, not only will we be able to make marvellous progress in geometry, but also we will treat optics, and phoronomy and mechanics, and in general whatever is subject to the imagination, by a definite method and, as it were, an analysis, and by this marvellous art we will bring it about that the invention of machines will be no more difficult than constructing problems in geometry. In this way, without pain and expense, very complex machines, the natural realities themselves, can be described and transmitted to posterity without the help of figures. (CG ix 148)

That is, as he explained in the letter to Huygens accompanying a sketch of his *analysis situs*, he was attempting "a new characteristic, completely different from Algebra and which will have great advantages for representing exactly and naturally to the mind, although without figures, everything that depends on the imagination" (GM II 20; quoted from Rabouin 2020, 1).

Thus even though Leibniz agrees with Proclus and Hobbes that figures are constructed in the imagination, he does not have to presuppose an imaginary

this is a point Leibniz would graciously concede to his adversary, for a commixture of geometry and kinematics does not bother him at all. What matters to him in fact is to preserve the difference between geometry and dynamics" (2007, 281).

[143] Compare with what Nikulin (1998, 161) says of Proclus, "there are no geometrical figures either in the physical world (where they are all imprecise) or in discursive thinking as pure λόγοι" (*In Eucl.* 49.25). Compare also "the λόγος of a circle is not itself circular, i.e. is not itself a geometrical figure.... This perfect yet extended circle that is conceived or 'seen' in and by imagination facilitates discursive logical reasoning about the circle and its properties" with what Leibniz wrote in *Infinite Numbers* of April 1676: "entities of this kind, i.e. polygons whose sides do not appear distinctly, are made apparent to us by the imagination, whence there arises in us afterwards the suspicion of an entity having no sides. But what if that image does not represent any polygons at all? Then the image presented to the mind is a perfect circle" (A VI 3, 499/LLC 91).

space onto which they are projected, the imaginary space "which everybody calls space" (*De corpore*, ch. 7, §2, 82). Rather, the production of the relations of situation that constitute the figures and traces is itself the production of space.

For Hobbes, as we have seen, "*space is the phantasm of an existing thing insofar as it is existing*" (*De corpore*, ch. 7, §2, 83), and places are parts of such a space that are "coincident with the magnitude of any body."[144] This idea that space is an immediate concomitant of *existence* is also expressed by Newton in his important essay *De gravitatione*, unpublished until the 1960s: "And hence it follows that space is an emanative effect of the first existing being, since if any being whatsoever is posited, space is posited" (Newton 2004, 25). Of course, the contrast here is that Newton, in conceiving space as an immediate concomitant of *every* existing being, conceives it as an effect emanating from the "first existing being," namely God, so that space can and does exist in the absence of bodies, and indeed, of any people imagining. Geometrical figures are conceived by him as being latent in this absolute space, and are made perceptible by the direct action of God, who can endow movable portions of space with the property of being accessible to the senses.[145]

Leibniz, on the other hand, makes space an effect of *coexistence*, not existence. It is because any two things can be simultaneously co-perceived that we perceive them as existing in space: "When we conceive two points as simultaneously existing, and ask why we say they are simultaneously existing, we will think the reason is that they are simultaneously perceived, or at least that they can be simultaneously perceived" (§108; CG ix 228). And, as we have seen, he conceives a path connecting any two situated things to be a concomitant of their being perceived together:

> When two things are simultaneously perceived to be in space, by that very fact a path is perceived from one to the other. And since they are congruent, by that very fact is conceived the path of one into the place of the other. But two points are congruent to one another. So what is perceived when two points are simultaneously perceived is thus a Line, that is, the path of a point. (CG ix 228)

[144] In this respect, Hobbes is in agreement with Barrow, for whom space is "a pure, unmitigated potentiality, the mere capacity, placeability or (if I may be allowed the term) interposability of any magnitude" (*Lectiones Mathematicae X*; Barrow 1860, 158). But see Strong (1970, 161–2) for an account of Barrow's fierce objections there (1860, 179–80) to Hobbes's characterization of space as a phantasm, and Arthur (1995) for a comparison of Barrow's philosophy of space and time with Newton's.

[145] As I argue in my (1995, 332), whereas "Barrow denies that space is an actual existent, or that it has any actual figures, dimensions or parts distinct from those of magnitudes placed in it," Newton, in his *De gravitatione*, "unambiguously maintains the actuality not only of space but even the figures in it: 'There are everywhere spaces contiguous to spaces, and everywhere all kinds of figures, everywhere spheres, cubes, triangles, straight lines…even though they are not disclosed to sight' [Newton 1962, 100, 133]."

A path, however, like all the traces or extensa, is something perceived as existing at an instant. It is a possible path that could be traced in time, but in fact, because it is not traced in time, it signifies a continuity of simultaneous coexistence:

> When we perceive something as existing, by that very fact we perceive it to be in space, that is, we perceive that there can exist indefinitely many other things that are in no way discernible from it. Or, what is the same thing, we perceive it to be movable, or that it can just as well be in one place as in another, and because it cannot be in many places simultaneously, or be moved in an instant, we therefore perceive that place as continuous.
>
> (CG ix 228–30 [modified after De Risi 2007, 411])

Leibniz's claim here that "whenever we perceive something as existing, by that very fact we perceive it to be in space" may sound reminiscent of Newton's view. But the crucial difference is that for Leibniz it is actually *coexistence* (the possibility of being co-perceived) that produces the idea of space, and this requires simultaneity, compresence. This is why Leibniz characterizes space not only as the order of situations, but also as the order of *simultaneous* existents (or perceptions). As he explains to the Newtonians in his Third Letter to Clarke,

> For myself, I have indicated more than once that I hold space to be something purely relative, as is time; I take it to be an order of co-existences, as time is an order of successions. For space indicates, in terms of possibility, an order of things which exist at the same time, inasmuch as they exist together, without entering into their particular manners of existing. And when we see many things together, we are aware of [*s'apperçoit de*] this order of things with each other.
>
> (Leibniz, Third Paper to Clarke, §4 [February 25, 1716], GP VII 363)

Now it is precisely at this point that the criticisms of modern commentators might be thought to have traction. For if this is all there is to space, an order of things that are *simultaneous*, then it would seem that Newton, Stein and Earman are correct in claiming that it is insufficient for representing the motions of real bodies through real time.

I have already given one rejoinder to this above (in §2.1). This is that the objection assumes, with Newton, that motion is only possible if space is a structure that continuously exists through time, within which structure places and changes of place may be assigned. What Leibniz has done, by contrast, is to show how motion can be conceived in terms of change of situation, where situation is conceived *relationally*, and not in terms of places in a presupposed space. I will say more on this in a moment.

But first let us examine the charge that the "notion of straight-line motion... is simply unavailable" to Leibniz as a relationalist (Earman 1989, 72). As we have seen, Leibniz has demonstrated (at least by his own standards) the possibility of straight-line motion with his notion of successive congruence. Hobbes, as we have seen, presages something like successive congruence with his notion of "the idea of a body passing by a continuous succession, first through this space, then through that" (*De corpore*, ch. 7, §2, 83)—his characterization of the "phantasm of motion." But he identifies this "phantasm" with time itself. Leibniz, by contrast, founds the possibility of motion or rigid transport on the uniformity of space, characterized by the equivalence of all congruent situations, and a straight line on the uniqueness of the situation of the points on it to two given points. So if a body has moved from *A* to *B* in a straight line, then a prolongation of this motion is uniquely determined to be in a straight line.

Now Leibniz follows Descartes in holding that in a moment there is no possibility of variation of motion, so that *conatus*, the endeavour to move in a moment, must be rectilinear.[146] That is not quite accurately stated, however. As we have seen, the idea of a state involving no variation is a fiction: it is how we represent a state of motion in the imagination. In fact there is always further variation within a given state, it is just that it may not be discernible at a certain level of analysis or, indeed, of perceptual acuity. Leibniz addresses this in his *Dynamica* with his idiosyncratic notion of a "simply simple motion" (GM VI 341 ff.). This occurs "when neither one state of a point is discernible from another before or after it, nor one point from another, insofar as only their motion is considered" (341). That is, there is "no principle for discerning them, even if the points may perhaps be discerned in respect of bodies, namely through the situation they have in bodies, which is inevitable, not to speak of the qualities by which the parts of bodies are varied, which does not concern us here" (341). As De Risi observes, such simply simple motions "define uniform motion through self-similarity, and therefore they determine it as straight according to Leibniz's well-known characterization of a straight line" (De Risi 2007, 544, n.73). Thus the uniformity and rectilinearity of motion is established for an incipient motion, an endeavour. This therefore constitutes Leibniz's justification for his long-standing agreement with Descartes that motion in itself is rectilinear, and that at any instant a body's endeavour is in a straight line.[147]

[146] Leibniz is consistent throughout his corpus in holding that a body in curvilinear motion tends at any instant to move along the tangent, at least from 1675 onwards. De Risi (2007, 544) notes a counterinstance in the *Hypothesis Physica Nova*, where Leibniz follows Galileo, but that appears to be constrained motion.

[147] De Risi objects to Descartes' argument for the rectilinearity of motion in a straight line (in *Le Monde* (AT XI 43–7) and in the *Principia* (AT VIII 63–4)) that "of course, since an instant is unextended, in it there can be no motion, thus not the rectilinear one either" (2007, 541, n.72). But in extending this criticism to Leibniz (as he does in fn.73, 544), De Risi fails to take into account that for Leibniz a moment is, fictionally, extended. As he writes in his *Specimen Geometriae luciferae* of 1695, "a line is composed of indefinitely small lines, a surface of indefinitely small surfaces, and a body of indefinitely small corpuscles; that is, it is shown that two extensa can be compared by resolving them into equal or mutually congruent particles, however small, as if into a common measure; and the error

This is not all there is to it, however. For although as we have seen "all motions are in themselves rectilinear," there are nevertheless motions in curved trajectories. Leibniz wishes to account for the fact that in real motions, there are impacts within even a motion that is represented as momentaneous. Thus what is rectilinear on one level of analysis can prove to be inflected when the analysis is pursued at a deeper level: a given difference in velocity *dv* can be comprised of an infinity of solicitations *ddv*. Consequently, in the *Dynamica* Leibniz calls the instantaneous tendency to move in a straight line a *striving* [*nisus*], having the dimensions of velocity times mass, but distinguishes it from endeavour [*conatus*], explaining that while *nisus* is always rectilinear, "endeavours may be rectilinear, as well as also circular or otherwise curvilinear."[148] As we have seen, Leibniz argues correctly that two endeavours can be composed in any moment—say, an existing endeavour in a straight line and a centripetal impulse to move towards a centre; and also that a curved trajectory of a body moving through space and time can be conceived as the limit of a polygon consisting in infinitely many such small lines resulting from such momentaneous compositions. But from these correct premises he fallaciously concludes that "All motions [through time] are composed of rectilinear uniform motions" (Part III, Proposition 17; GM VI 502). The mistaken (and wholly unnecessary) notion of a curvilinear endeavour seems to be another indication of his confusion of thought on this point.

Leibniz does explore other definitions of the straight line, though, including the Archimedean definition of it as the shortest path between two points, and this seems to open up a different approach to justifying inertial motion. In the *analysis situs* manuscripts of his last years that De Risi transcribes in his book, Leibniz tries to derive a straight line as the shortest path from his definition of it as "the unique path from *A* to *B* that is determined solely by *A* and *B*." Thus, he argues in *Magnitudinis nomine*..., of January 1712, "If the shortest path is unique, it will be a straight line," and it will also fall between the endpoints *A* and *B* (De Risi 2007, 600). He gives a more extended argument to this effect in *Spatium absolutum*... of August–September 1714 (609–10), which he then summarizes as "A *straight line* is a line through points of it that are uniquely situated to two proposed points. A bounded straight line is the shortest line between the endpoints" (611).

Here, De Risi suggests, "an opportunity arises for Leibniz to suggest that the direction of inertial motion could be architectonically determined according to the

will always be smaller than one particle of this kind, or at least of a finite, either constant or decreasing, ratio to it" (GM VII 273). This is Leibniz's syncategorematic interpretation of infinitesimals (on which, see Rabouin and Arthur (2020)).

[148] "Direction together with velocity multiplied by quantity of matter, yet with flexibility at the same time, we designate by the term striving [*nisus*], but more generally by the term endeavour [*conatus*]. Whence strivings are all rectilinear, whereas endeavours may be rectilinear as well as also circular or otherwise curvilinear" (GM VI 471). This notion of a curvilinear endeavour is an unnecessary mistake, indicating a residual confusion in Leibniz's conception of curvilinear motion, since he also argues that two rectilinear endeavours compose to form a further rectilinear endeavour.

principle of minimum space" (2007, 542). That is, given his teleological notion of *appetition*, Leibniz might well have defined an inertial motion to be the motion of a monad's body that is locally determined in such a way that "by tending to its own realization it minimizes space" (542). He might thereby have anticipated the modern notion of inertial motion foundational to relativity theory, where an inertial path is the minimum spatial path in spacetime between two spacetime points. What is intriguing about this is that in some papers from the 1690s Leibniz tries to provide "a general method of determining the minimum line on any surface whatsoever," thus implicitly recognizing that the shortest line is not necessarily unique; and he attempts this by means of differential methods "in much the same way as we do now, i.e. [by] referring to tangent space" (2007, 236–7, n.99).[149] Nonetheless, Leibniz declines to define a straight line a priori in terms of inertial motion.

That such an a priori deduction of the law of inertia was a live option for Leibniz, De Risi argues, is also illustrated by the case of his contemporary, the Cartesian Johannes Clauberg. In his *Disputatio physica*, xxi, §3, Clauberg had written: "And at every moment a moving body is disposed to continuing its motion in a straight line, not a curved one... Only, that is, in a straight line, which is defined as the shortest extension from one point to another, through all of them, and is congruent to itself and any other straight line..." (Clauberg 1691)—"which of course," De Risi adds, "reminds us of Leibniz's definition through self-similarity" (542, fn.72).

De Risi suggests that the reason Leibniz declined to define inertial motion as the shortest path in space is that the kind of architectonic characterization of it given by Clauberg would make it a necessary property of motion, whereas Leibniz wished to maintain the contingency of the laws of motion,[150] including the principle of inertia, although "he is hesitant to attribute too much contingency to inertia" (2007, 543). He gives as an example Leibniz's claim (in a subsequently expunged passage of a letter to Arnauld in April 1687) that if God had decreed that a stone given a circular motion by a slingshot "would continue to go freely in a circular line when it was released from the slingshot, without being pushed or restrained by I know not what, this would be a miracle, since according to the laws

[149] As De Risi observes, Leibniz's motivation for these attempts was probably to validate Archimedes' determination of the straight line as the shortest path between two points, rather than to establish the possibility of a space with non-zero Riemannian curvature (2007, 236). Moreover, by integrating on a field of tangent vectors representing attraction to a centre "he will not get to the minimization of length—he gets to the minimization of action instead" (237, n.99), which is certainly of great interest concerning priority for the Principle of Least Action, and for the history of the origins of the calculus of variations.

[150] As Laurence Bouquiaux observes, for Leibniz the inertial property of matter, its resistance to change of motion, goes hand in hand with the principle of the equality of action and reaction, both of which he regards as being among the contingent laws of nature (Bouquiaux 2017, 56).

of nature it ought to continue in a straight line along the tangent".[151] De Risi argues that the very fact that Leibniz acknowledges that God could contravene the principle of inertia entails that it is contingent. To move non-inertially is not self-contradictory, but as a miracle, it is contrary to the principle of reason: there would be no reason why the stone should go in a circular orbit rather than any other.[152] Now this argument form, involving the principle of sufficient reason, also occurs in Leibniz's mature definition of distance in the (second essay of the) *Initia Rerum*, and the consequent characterization of a straight line as the shortest distance between two points:

> There is, moreover, a certain order in the passage of perception[153] when it passes from one to another through others between them. And this order may be called a *path*. But since this order can be varied in infinite ways, it is necessary for there to be one simplest one, which is, namely, by proceeding according to the very nature of the thing through determinate intermediaries, that is, through those which are related as simply as possible to both the extremes. For if this were not the case, then there would be no order, no reason for distinguishing among coexisting things, since it could go from one given thing to another given one through any path whatever. And this is the shortest path from one to another, whose magnitude is called *distance*.... (GM VII 25/L 671)[154]
>
> From one given point nothing further is determined. But from two given points is determined the simplest path from one to the other, which we call a *straight line*.

[151] A note in passing: De Risi (2007, 545, n.73) quotes the passage from a French edition, claiming that it is "not included in the Gerhardt edition," but in fact it is, at (GP II 93); in LA it is included as one of the passages Leibniz added later when considering the correspondence for publication (LA 358, 364).

[152] De Risi also maintains that Leibniz could not agree with the Cartesians that the principle of inertia could be derived from first principles concerning extension and its modes—as we have seen in the case of Clauberg—but must involve dynamical considerations: "Leibniz is right when he forcefully insists against the Cartesians that just knowing extension and its modes is not enough to solve a mechanical problem in general, and that we also need to take something else into consideration... namely, living force as an expression of the entelechy of a composite substance, and inertial mass" (2007, 530).

[153] The Latin is "*in percipiendi transitu*." Loemker (L 671) translates this as "in the transition of our perceptions," lending itself to a distinctly phenomenological interpretation. But the context shows rather that what are being perceived are coexisting things (and also those that we can infer to have continued to exist during the time of the perception), *situation* being the order among these. Likewise, the "intermediaries" are the things perceived, not "our perceptions" or "intermediate states," as Loemker translates *intermedia*.

[154] Similarly in the first essay in the *Initia rerum* Leibniz argues: "*In each of the two orders* (time and space) *things are judged to be* ***nearer to*** *or* ***farther from*** *one another from according as more or fewer things are required for understanding the order between them.* Hence two points are nearer for which the maximally determined interposed things yield a simpler configuration [*aliquid simplicius*]. Such a maximally determined interposed [configuration] is the simplest path from one to the other, at the same time minimal and maximally self-similar [*aequabilis*], namely the straight line, which is shorter [*minus interjecta*] between nearer points" (GM VII 18/L 667).

> Hence it follows, *first*, that a straight line is the shortest path from one point to another, ôr that its magnitude is the distance between the points. (GM VII 26)

The point is that this derivation of the straight line as the shortest path rides on the principle of sufficient reason: there could be no reason for distinguishing among coexisting things unless there were one maximally determinate, and thus simplest, path from one to the other. In fact, "there would be no order," and thus no space, if that were not so.[155] But if the definition of the straight line (and the existence of space itself) rest on the principle of reason, it seems that geometry is not derivable from the principle of contradiction alone. From the parallel with the case of the circular motion of the stone, De Risi infers that the principle of inertia is likewise "a hybrid between truths of fact and truths of reason, which [Leibniz's] epistemological system was not fully ready to receive" (2007, 545–6).[156]

This does seem to constitute a mismatch with Leibniz's stated epistemology, where mathematical truths, as truths of reasoning or necessary truths, are supposed to be derivable from the principle of contradiction[157] alone: "Truths of reasoning are necessary, and their opposite is impossible" (*Monadologie* §33, GP VI 612). But the argument form is curious: Leibniz is not arguing that anything but the simplest form would be contrary to sufficient reason, but that, there being no reason to choose any of the forms over any other, the simplest must prevail. As Debuiche and Rabouin note (2019, 185), the same argument form occurs elsewhere in the Leibnizian corpus.[158] It is used in the *Origo veritatum contingentium ex processu in infinitum* from around 1689, in the case of God's choosing an equilateral triangle, everything else being equal. Moreover, he explicitly compares this case with his derivation of the straight line:

> God's first decree or resolution is to act in all cases with supreme reason. So if we imagine [*fingeremus*] a case in which it is agreed that there should exist a triangle

[155] As pointed out by Debuiche and Rabouin (2019, 193), this formulation entails that Leibniz "posited the existence of a *via simplicissima* as a condition for the existence of space itself."

[156] Concerning this contingency of the inertial principle and Leibniz's declining to derive it from the consideration of minimal distance, De Risi contends that "the determination of real space (matter) versus ideal space entirely rests on some architectonic principle," namely "the one condition of minimum on distance" (549); and yet Leibniz does not deploy this argument. The "intrinsic reason" for the lack of such a demonstration, he concludes, is "that the phenomenal nature of real space prevents physical laws from ever being *a priori* deducible even *ex hypothesi* of the architectonic minimum principle" (549). But it is not a question of two different kinds of space, *real space* and *ideal space*, where the latter is supposed to have contingent properties that could in principle differ from those deduced *a priori* in geometry (as we have already argued in §2.4 above); the issue is about the contingency of *inertial motion* in the one, unique space characterized in *analysis situs*. See also Debuiche and Rabouin (2019) for an incisive investigation of this whole issue.

[157] By the *principle of contradiction* Leibniz means the principle "by virtue of which we we judge that which involves a contradiction to be false, and that which is opposed or contradictory to the false to be true" (*Monadologie* §31, GP VI 612).

[158] A similar instance of the argument form occurs, for example. in the *Theodicy*, §196 (GP VI 232–3/H 249).

> with a given perimeter, but without anything in the data from which one could infer the kind of triangle, one would have to say that God would produce an equilateral triangle, freely indeed, but without a doubt. For there is nothing in the data that would prevent any other triangle from existing, and so an equilateral triangle is not necessary. Nevertheless, in order for no other triangle to be chosen, it suffices that in no other triangle apart from this one is there a reason for it to be preferred. And it is the same if, given that one has to draw a line from one given point to another without any data to determine the species or the magnitude of the line, a straight line will certainly be produced—but freely, for just as there is nothing to prevent one drawing a curve, there is nothing to persuade one to draw a curve either. (A VI 4, 1664)

By analogy, one may say the same in the case of the motion of the stone released from the slingshot: without any data to determine either the velocity or direction of its subsequent motion, a uniform rectilinear motion will certainly be produced; but freely, since there is nothing to prevent God moving it in some other curve, even though there's no reason for him to do so either. So it is not self-contradictory for God to make the stone freely move in a circle, but also there is nothing to persuade him to do so either.[159]

This clearly contrasts with some of Leibniz's famous arguments in his dispute with Clarke, particularly the "Leibniz shift" arguments: there, if God has no reason to choose between two alternatives, such as the whole world's being shifted forwards with a certain velocity and its remaining in its present state, then God *cannot* freely choose one or the other, as they are the same state. But what is the difference between the two cases? It seems to me to devolve around *indiscernibility*. For it is the abstract or ideal status of the parts of space and time, the fact that they denote indiscernibles, that is the engine for the Leibniz-shift arguments.[160] God cannot create the world a year sooner because times are only identifiable with respect to the things that occur at them; without this concrete reference, the supposedly different times that we can imagine are in fact indiscernible from one another. By the identity of indiscernibles, they are therefore identical. So it is impossible for God to choose one over the other. In the case of the stone continuing in a circular orbit with respect to surrounding bodies,

[159] Cf. Michel Fichant: "Leibniz understands an *a priori* proof to mean a proof instituted on the basis of causes or reasons, and a proposition will not cease to be a contingent truth by having been demonstrated a priori, since the antecedents from which it is deduced are themselves both irreducible to the principle of contradiction and yet non-deducible from experience. Now this is the status of the propositions that Leibniz calls metaphysical" (Fichant 1995, 64; quoted from Bouquiaux 2017, 55–6, my translation).

[160] Cf. Rabouin (2020, 306): "In fact, ideal objects, precisely because they are classes of indiscernibles and are entirely resorbed into relations of essence, can be rediscovered by means of the mere reduction to the identical and the principle of non-contradiction (which, as I indicated in the chapter, constitutes for Leibniz the true guarantor of objectivity, since it exposes the thing to the ultimate risk of impossibility)" (my translation).

however, this is not indiscernible from its having continued in a straight line with respect to those bodies. So it is not in itself impossible, and God could choose to make it so.

But in the latter case we are talking about an actual, concrete motion; and in order for there to be such a concrete motion, there must be some reason for it to occur in the determinate way that it does, a reason either in the body itself, or in the bodies with respect to which it is moving. As Leibniz wrote in a supplementary study to some notes he made on Des Bosses' letter of December 12, 1712:

> A thing having situation is said *to move* when it contains the cause of the changed situation, that is, when the reason for its change of situation with another is provided from it. But if a sufficient reason is provided from it, this one moves while the rest remain at rest; whereas if it is less than sufficient, then several things move simultaneously. (LDB 312/313)[161]

Thus for God to make the stone continue in a circle, without any reason in the body or its surrounds, would be a miracle. And this is, of course, Leibniz's objection to Newton's positing of an action at a distance. He insists, as we have seen, that a body's moving in a circle (or other curved orbit), contrary to its inclination to continue in the same direction, requires a supervening cause.[162]

The contrast, that is, involves the distinction we have already explored between motion conceived geometrically (as mere change of situation), and motion conceived with respect to cause. The tendency of any body to continue in uniform rectilinear motion in the absence of any cause for it to do otherwise is motion conceived geometrically, without respect to its cause; it is "simply simple motion." It is uniform because it is self-similar, and therefore rectilinear. And when we consider motion simply as a change of situation, it is intrinsically relative. If one body *A* changes its situation relative to another *B*, then by that very fact *B* has changed its situation with respect to *A*: they are in relative motion. The two cases are thus indiscernible from one another. This means that the laws, and also the phenomena (if not their description), will be invariant under a change of velocity, i.e. of uniform, rectilinear motion. The indiscernibility leads to an invariance under a transformation.

As we have seen, however, Leibniz extends this invariance under a change of velocity also to rotational motion (i.e. to angular velocity): *all* motion is relative, so

[161] Cf. the definition in the first essay of the *Initia rerum*: "*Motion* is change of situation. That thing *is moved* in which there is a change of situation and at the same time a reason for the change" (GM VII 20/L 668).

[162] Compare Leibniz's definition of *direction* in the *Dynamica*: "*Direction* is the straight line in which a mobile is moved from an initial point on a line towards another point on the same line, *unless impeded by a supervening cause*" (Section 2, Ch. 2; GM VI 469; my italics).

that, mathematically, all hypotheses are indiscernible.[163] This would entail invariance under a broader group of transformations. Leibniz claims that this invariance is broken in actuality by appeal to the cause of motions. In complex cases, as we have seen, this cause is identified by a hypothesis, the simplest or most apt hypothesis being taken as (defeasibly) true. Thus, for example, whether the Earth is stationary—or, equivalently, moving uniformly in a straight line—or moving in an orbit around the Sun, will depend on which hypothesis is simpler or more apt. This gives Leibniz a serviceable criterion for distinguishing a true motion from a merely apparent one. But it does not alter the fact that, according to him, competing hypotheses about which bodies are rotating will be indiscernible from one another. So, if motion through space is encapsulated in terms of invariances, it entails too broad a group of transformations.

Now, to come back to the objection that space, as an instantaneous order of coexistents, is not such that motion can be represented in it, two points are in order. First, space is that in which situations and their changes may be represented, rather than a pre-existing canvas on which they are traced. Provided that changes of situation may be represented through the same abstract structure, space does not need to be presupposed as an existing entity—even though, as an order of possible situations, it has its own reality in the divine mind. Second, the changes of situation can be represented in terms of successive congruence within such an order. One body may be transposed into the place of another congruent to it. Moreover, this may be effected without presupposing places as parts of space, but only as equivalence classes of the situations involved in the successive congruences.

To see this we should return to the famous paragraph §47 of Leibniz's last letter to Clarke, which begins:

> §47. Here is how people come to form the notion of space. They consider that many things exist at once, and they find there a certain order of co-existence, according to which the relation of one thing to another is more or less simple. This is their situation or distance. When it happens that one of those coexistent things changes its relation to a multitude of others which do not change their relations among themselves, and that another thing, newly come, acquires a relation such as the former had to the others, we say it has come into the *place* of the former; and this change we call a *motion* in that body which contains the immediate cause of the change. (GP VII 400/L 703)

[163] Cf. *Dynamica*, Scholium to Proposition 19: "In general, when motion occurs, nothing can be found in bodies by which it can be determined except change of situation, which always consists in relation [*respectu*]" (GM VI 507); and "in Proposition 19 it was shown that, on account of the respective nature of motion, hypotheses are indiscernible" (GM VI 509).

Note here how Leibniz is careful to stress that change of situation alone is not sufficient for motion: there needs to be a reason, drawn from aetiological considerations, for assigning motion to one of two bodies in relative motion rather than the other. He continues:

> And even when many, or even all, the coexistent things change according to certain known rules of direction and speed, one can still always determine the relation of situation which each one acquires to every other, and even that relation which each one would have, or which it would have to any other, if it had not changed at all, or if it had changed in any other way. (400/703)

This is evocative of the example Leibniz had given in the *Nouveaux essais* of the pieces on a chess board (discussed above in §2.1). Just as the Arab horsemen are able to picture the relative positions of the pieces in their imagination, so we are able to form a representation of the motions of bodies in ours, whether the positions are committed to paper or not. Knowing the initial situation of any body to a set of existents taken as fixed, then given the "certain known rules of direction and speed" according to which changes of situation might occur, we can calculate the subsequent situations of that body. But this does not require a fixed background space on which the motions are traced, any more than horsemen need to carry a board on which their chess moves are played.

But the situation of something must be with respect to other coexistents; if some subset of coexistents are hypothetically taken to be at rest, then the situation of other coexistents can be taken with respect to the same set of hypothetically immobile individuals. To this end Leibniz introduces the notion of "fixed existents," these being "those in which there has been no cause of any change in the order of their coexistence or (which is the same thing) in which there has been no motion" (400/703). This is a *fiction*, we *feign* that there are individuals whose mutual situation does not change:

> And supposing or feigning that among those co-existents there were a sufficient number of them which had had no change among themselves, we may say that those that have such a relation to the fixed existents as others had to them before, now have the same *place* which those others had. And that which comprises all those places is called *space*. Which shows that in order to have an idea of place, and consequently of space, it suffices to consider these relations and the rules of their changes, without needing to fancy any absolute reality apart from the things whose situation we consider. (400/703)

Thus we may define *place* to be "that which is the same for *A* and *B* when the relation of coexistence of *B* with *C*, *E*, *F*, *G*, etc. agrees perfectly with the relation of coexistence which *A* had with the same *C*, *E*, *F*, *G*, etc." Here Leibniz stresses that his use of *agreement* rather than *identity* is because relations of situation "agree only," since the situation of *A* is an individual accident of *A*, not of *B*. "But the mind, not contented with an agreement, looks for an identity, for something that should be truly the same, and conceives it as being extrinsic to the subject; and this is what we call *place* and *space*." Thus the hypothesis of fixed existents allows us to define place in terms of an equivalence class: it is the class of all things that bear an equivalent situation to our (fictitious) fixed existents. Places are thus defined relationally by their place in an ideal or abstract order, like the places in a genealogical tree.

As for the objection that space, as the order of *simultaneous* existents, cannot be such as to represent motions through time, Leibniz denies that coexistence needs to be restricted to those things that can be simultaneously perceived. At the beginning of the second essay of the *Initia rerum mathematicarum metaphysica* he writes:[164]

> But we recognize as coexisting not only those things which are simultaneously perceived, but also those which we perceive successively, provided it is supposed that during the transition from the perception of one thing to the perception of the other, the earlier thing has not been destroyed nor the later one brought into existence. From the latter hypothesis it follows that both coexist now, when we attain the perception of the later one; from the former it follows that both coexisted when we examined the earlier one. (GM VII 25/L 671)

It is on this assumption of the continued existence of a given thing through time that we can compare ratios of the spaces traversed by it with ratios of the times taken, assuming a uniform motion, and thus apply quantity to time. The possibility of motion requires successive congruence, the substitutability for one another of equal intervals of space traversed by a uniform motion.

Of course, at this point we are confronted with the unfortunate fact of the incompleteness of Leibniz's theory of space. On the one hand he has laid the groundwork for a great many of the properties of Euclidean space, many of which are now expressed in terms of groups of transformations. For example, as De Risi notes, the straight line and circle that Leibniz had defined in terms of motion are now commonly defined instead as "the figures that are invariant when acted upon by the subgroups of translations and rotations of the group of plane isometries" (2007, 234); his definition of homogeneity as essentially a local homeomorphism, yielding the topological invariance of dimension (201), and of homogony in terms of a continuous change (170); and so forth. On the other hand, however, the

[164] Cf. also *De homogeneis et de homogonis*, (MT N. 88 (40938)) "Situation can hold for future things as well as for those coexisting now, for future things are determined from present ones, and conversely."

"known rules of direction and speed" appealed to by Leibniz are expressed, not in terms of situations, but algebraically. Still, we may assume that the foundations that Leibniz attempted (for Euclidean space, together with absolute time) have been completed by the work of later scholars (in ignorance of Leibniz's own far-sighted attempts).[165] Then we may enlist as the main bridge between his foundations and the modern formulations the idea of *transformations*. From this point of view, just as space can be expressed in terms of groups of transformations of equivalent situations, so motions through space can be expressed in terms of groups of transformations of equivalent changes of situation.

In conclusion, we may say that it is not the case that Leibniz could not account for motions through space and time. He could (at least, assuming the success of his foundations) account for uniform rectilinear motions, and this and the other transformations that he envisaged, such as invariance under a change of instantaneous velocity in a given direction, would indeed determine the allowable motions. And such transformations, evidently, will determine a structure; and we may even, with a whiff of anachronism, call this structure a "spacetime." But rather than possessing too weak a structure, as in the mooted "Leibnizian spacetime," the spacetime Leibniz implicitly describes is too strong. The transformations he envisages are too broad, including as they do invariance under a change of angular velocity: a transformation of all rotational motions with respect to the same set of fixed existents will not lead to the same phenomena (as evidenced by the case of stellar parallax, for example, which would distinguish a Tychonic system from a Copernican one).

Also, finally, we are not obliged to regard this structure as an existing thing. Although situations in space are conceived as lines and other geometrical figures traced in the imagination, this does not require some species of Proclean "intelligible matter" on which they must be traced, since the relations of situation can be represented symbolically. Similarly, we can conceive motion along trajectories through time resulting in those traces, without there having to be an immobile canvas on which they must be traced, since the changes of the relations of situation can likewise be expressed symbolically. As Leibniz explains at the end of §47 in his Fifth Letter to Clarke,

> I remark, finally, that the traces that moving bodies sometimes leave on the immobile ones on which they move, have given people the occasion to form in their imagination such an idea, as if there still remained some trace when there is no immobile thing. But this is only something ideal, and signifies only that if there were some immobile body there, the trace might be marked upon it. And it

[165] I refer the reader to De Risi's book, which is replete with mini-studies of the relationships between Leibniz's situational geometry and the geometrical studies of later authors such as Graßmann, Pasch, Dehn, Pieri, and Veronese.

> is this analogy that makes one imagine places, traces and spaces, although these things consist only in the truth of relations, and not at all in some absolute reality.
> (to Clarke V, §47: GP VII 402)

These "places, traces and spaces" moreover, are determined relative to a given system of bodies taken as reference. So they are not something pre-existing in which the motions occur, but are calculable relative to the situation determining the initial conditions. This is relevant to the interpretation of General Relativity, as I will discuss further below.

3.6 Conclusion

In this chapter I have defended Leibniz against various objections that have been made to his stated views on the relativity of motion. In summarizing them here, I shall also reflect on what I take to be their implications for our understanding of related issues in contemporary philosophy of physics.

The key point missed by critics of Leibniz's views on the relativity of motion is his distinction between motion conceived geometrically and motion with respect to cause. The former is mere change of situation, which (between any two bodies) is intrinsically relative and mutual; the latter involves identifying the subject of motion by reference to the best explanation of the change. This distinction enables Leibniz to differentiate between true and merely apparent motions within the compass of a relational theory, without being committed to motion as a change of position in a presupposed absolute space.

This distinction was necessary to Leibniz as a convinced Copernican. He wished to allow for the fact that, although motion treated purely geometrically is entirely relative, there is a fact of the matter about which of the world systems is adequate to the phenomena, and therefore gives the true motions. This is secured by appeal to the most intelligible hypothesis that accounts for the motions to be explained. This criterion is aetiological: it gives the true motions relative to the context of the phenomena to be explained. Leibniz thought that by this means he could negotiate a means for allowing the Copernican hypothesis to be regarded as true in the context of planetary astronomy, while leaving intact the censure of Galileo for contradicting the Scriptures.

Part of the reason why this strategy was unconvincing, I argue, is that it rests on an equivocation about satisfying the phenomena. If there were a complete equivalence of hypotheses, as Leibniz maintained, then there would not even be the possibility of phenomena such as stellar parallax distinguishing the Copernican hypothesis as more apt to explain them. Conversely, if such phenomena are sufficient to determine the true motions among the planets (at least defeasibly), then it can be expected that this will make a real physical difference, with other different phenomena resulting—as in fact the Earth's orbiting the Sun (rather than the other way around) makes for different phenomena, such as those that are exhibited by Foucault's pendulum or produced by the Coriolis force. So whilst Leibniz's position is not an instrumentalist one à la Osiander, since he maintains that the Copernican hypothesis is the one delivering the true motions in planetary astronomy, it is not quite a realist position either, since he does not sufficiently recognize that the true hypothesis will result in a discernible difference in the resulting phenomena.

Leibniz held that to enlist the causes of motion is to leave the realm of geometry for that of dynamics. Consequently interpreters have sought to identify force as Leibniz's criterion for true motions. This it could not be; for if force is understood as impulsive force (change of directed quantity of motion), in any collision the phenomena will be the same no matter which body is held to be at rest or in uniform rectilinear motion. On the other hand, the total live force, *vis viva*, is conserved in any isolated system of bodies, and in a collision of two bodies (treated as an isolated system), although the directive force will change under a change of hypothesis about which bodies are at rest, the total force will not: it will be differently distributed between the directive and the respective forces making up the total force. Force makes motion real in a different sense, I contend. Derivative force is what is real in motion at any instant, and the derivative forces of bodies are the instantaneous manifestations of the primitive forces of the monads they contain.

Here I have argued against a prevalent understanding of Leibniz's dynamics as involving "ontic levels," where the primitive forces are located "on the monadic level," and derivative forces "on the phenomenal level," or even on some supposed "dynamic levels" between these. Rather, the primitive forces are permanent dispositions to act or be acted upon, which continually manifest themselves in bodies as the derivative forces of action and reaction at each instant. The endeavours to act manifest themselves psychically as appetites and aversions, and physically as derivative active forces. These correspond to the two realms of causation, final and efficient; but these are compatible (and intertwined) realms of explanation, not ontic levels. The derivative forces are modifications of the primitive ones, and so cannot be located on a different plane of reality.

Returning to the issue of the Equivalence of Hypotheses for *all motion*, I examined Leibniz's arguments for this in detail. The explanation he gives for extending it from uniform rectilinear motion to curvilinear motion too rests on the composition of motions. But here, I agree with his critics, he is in error. He has conflated two different sorts of composition, one a composition of the endeavours that combine into a resultant endeavour *in any moment*, the other where a succession of infinitesimal straight-line motions compose into a motion through a curved path in space *over time*. That the endeavours at any moment satisfy the equivalence of hypotheses does not entail that a multiply inflected motion through time will do so too. Leibniz seems to have persuaded himself that the fact that the motion of the Earth around the Sun and that of the Sun around the Earth are mathematically indiscernible entails a general relativity of curvilinear motion. But with respect to a fixed reference system, like the fixed stars, the changes of situation in the two cases are not physically equivalent, as evidenced by the possibility of stellar parallax. Since the situation of each of two bodies to one another is mutual and therefore relative, their changes of situation will be too; but this does not entail that the situation of the two bodies to a third will be the same at various times under different hypotheses about which is orbiting which.

It is important to see, however, that this shortcoming of Leibniz's defence of the relativity of motion does not undermine his general relationist position. The objection just outlined was in fact made from a relationist perspective: the absoluteness of rotational motion consists in its discernibility (through the phenomena consequent on it) *relative to a set of bodies* taken as immobile, the "fixed" stars in the case in question. This point is generalizable, I believe, and applies even in the case of contemporary astrophysics, notwithstanding the radical changes in our understanding of space and time introduced by Einstein. General Relativity is not a universal relativity of all motion with respect to whichever bodies are taken to be at rest—if it were, it would make no difference whether the Earth, for instance, were taken to be rotating on its own axis, or at rest with the whole universe rotating around it. The equivalence between gravitation and acceleration means that the gravitational field of the Sun, for example, is represented in the theory by distortions of spacetime, so that the Earth traces an inertial path (a straight path in spacetime) in orbiting the Sun. There is no physically equivalent scenario where a stationary Earth has a gravitational field represented by a spacetime in which the Sun and all the other bodies in the universe are moving inertially around it. In any application of the theory, the sources of gravity are given data, and the question of which bodies are rotating relative to the others is determinable, and not arbitrary. The absoluteness of rotational motion is relative to the other bodies in the universe that determine the structure of the spacetime in question.[166]

Minkowski's innovation of replacing consideration of space and time separately by that of *spacetime*, was of course unanticipated by seventeenth-century thinkers. His interpretation of spacetime points as world-events paved the way for Einstein's General Relativity, and changed the rules of the game. For now the structure of spacetime was no longer considered a backdrop against which processes take place, but the fabric of physical reality itself. This has had the unfortunate consequence of completely reifying the mathematics, the very thing that Leibniz had opposed in the Cartesians. Here there is perhaps something to be learned from Leibniz: if spacetime is conceived not as a physical object interacting

[166] It may be objected that rotation is representable even in an empty spacetime; this is true, but it presupposes an otherwise empty spacetime as having a given structure, one that is not determined by the distribution of matter.

with the bodies in it (an impossibility, since it is a four-dimensional object in which change is already supposed to be represented), and is instead conceived as representing the possibilities for the motions and evolutions of processes subject to the nature of their environment, then it has a definite reality, but is not a self-existing thing. For in order to apply the theory to a given situation, the environment itself needs to be modelled and factored in.[167]

As we have seen, one of the chief criticisms that has been laid at Leibniz's door in recent times is the charge that his relational theories of space and time are inadequate to support the kind of structure necessary for his physics. A simple Cartesian product of the three-dimensional Euclidean space he conceived with a one-dimensional continuum for time yields what has been called a "Leibnizian spacetime," whereas more structure is required—specifically, an affine connection of points of space through different times—in order to ground the law of inertia to which he subscribed. As I hope my discussion has made clear, there is a good deal of irony in these criticisms. In the first place, it is assumed that space was already known to be Euclidean. But, as De Risi has persuasively argued, Leibniz seems to have been the first to have conceived space itself as Euclidean, rather than the figures in it, and to have attempted to derive its properties, such as continuity and isotropy, as we saw in chapter 2. Critics of Leibniz like Howard Stein have readily attributed a spacetime to Newton on the grounds that in his criticisms of Descartes he presupposed the self-identity of points of space through time, such a "kinematic connection" being a requisite for defining motion along a straight line at a constant speed through space and time. But Newton nowhere attempts to derive properties of his absolute container space that would support its having either a Euclidean spatial structure or an affine structure through time. It is therefore highly ironic to take Leibniz to task for not seeing that space itself needs to have certain properties in order to support the possibility of such motion. As I have tried to show in this chapter, Leibniz conceived his *analysis situs* from its inception as a foundation for representing both situation and change of situation—that is, motion conceived geometrically. Even if his attempts at deriving all the properties of space (and, allowing some anachronism, spacetime) fall short, and required the considerable later labours of mathematicians to establish its topological, affine, differential and metrical structure, it cannot be said that he failed to recognize the need to establish the structure required to ground the inertial motion of bodies. And indeed, he made some admirable strides in this direction, anticipating not only arcwise connectedness with his conception of successive congruence, but even partially foreshadowing modern notions of connectedness, completeness, and fibration.

[167] For example, in producing his solution of Einstein's field equations to predict the motion of Mercury around the Sun, Schwartzschild ignored the influence of all other masses in the solar system, modelled Mercury by a nonrotating object of negligible mass, and assumed that the gravitational field is static in time, spatially symmetric in space, and converges to a flat Minkowski metric at infinity. See Arthur (2019c, ch. 7, esp. p. 197) for discussion and references.

APPENDIX 1

A Formal Exposition of Leibniz's Theory of Time

A1.1 The Relational Core: Version 1

Suppose M is the aggregate of all monads m, and let us denote the states of a given monad m by the indexed set S_m.

In what follows I am going to leave existence assumptions implicit: we shall simply assume that there are monads, each one having a series of states (which we will prove to be actually infinite in the syncategorematic sense), that there are other series compossible with any given series, and so forth.[1] I am not assuming that a monad simply *is* such a series; formally, a monad has the structure of a *fibre bundle*, more specifically, a *tangent bundle*: it has a series of states, each with a tendency to go into future states in the same series according to a given law. Second, in what follows the universal quantifier is to be understood *distributively*, in keeping with Leibniz's denial of the categorematic infinite: that is, although I am using set theoretic symbols, $(\forall x \in S_m)$ should be understood as "For any x that is a state in the series S_m," and not as connoting that there is an infinite set or collection of such states of which it is a member.

Now, according to Leibniz, "My earlier state involves the reason for the existence of my later state": that is, each of the states of any series of states *involves the reason for*, or *contains the ground for*, any subsequent state in the series, so that the states in each series are ordered by this relation of ground containment G. The fact that it is a *series* suggests that G is a serial ordering of states, and a serial ordering is one that is transitive, asymmetric and connective. This gives us the following preliminary axioms for G as a relation defined on the states of a given monad, S_m:

Axiom G1 (Transitivity of G on S_m):
$(\forall x, y, z \in S_m)\ \{(x\mathrm{G}y\ \&\ y\mathrm{G}z) \rightarrow x\mathrm{G}z\}$
Axiom G2 (Asymmetry of G on S_m):
$(\forall x, y \in S_m)\ (x\mathrm{G}y \rightarrow \neg y\mathrm{G}x)$
Axiom G3 (Simple Connectivity of G on S_m):
$(\forall x, y \in S_m)\ \{x \neq y \rightarrow (x\mathrm{G}y \vee y\mathrm{G}x)\}$

These axioms apply only to states which are ordered within each individual monadic series by G as the result of the law of that series. In other words, the only states we know to contain reasons for one another are states of individual series. But as we shall see, these are not the only states that Leibniz supposes to involve one another's reasons. So let us now look at the more general relation of reason-inclusion, R, which we can consider as applying to any pair of states of monads in every series in the same world, W_m. Now, just as a monad is not simply an

[1] This account is much indebted to John Winnie's (1977), although his account of the causal basis of spacetime depends on a symmetrical relation of causal connectedness, later relaxed to causal connectibility; as well as on standard Cantorian set theory.

aggregate of states in certain relations, so a world is not simply an aggregate of states in certain relations, but an aggregate of monads whose series of states are compossible. In what follows, $(\forall x \in W_m)$ should be understood as "For any x that is a state in any monadic series of states S_n that is *compossible* with S_m, where S_m denotes the states in a given monadic series." (We will come back to the definition of compossibility in A I.3 below.)

Here we may note that if it is only pairs of states in the same series that are related by R, then the first two axioms will automatically apply to all the states in a given world, for in that case if xRy, then x and y will be in the same series. Leibniz, as we shall see, extends the notion of reason-inclusion so that it may also relate states x and y in different monadic series in the same world; but it still makes sense to assume that the first two axioms will hold for all states in such a world:

Axiom 1 (Transitivity of R on W_m):
$(\forall x, y, z \in W_m)\ \{(x\mathrm{R}y \ \&\ y\mathrm{R}z) \rightarrow x\mathrm{R}z\}$
Axiom 2 (Asymmetry of R on W_m):
$(\forall x, y \in W_m)\ (x\mathrm{R}y \rightarrow \neg y\mathrm{R}x)$

From the second of these axioms it immediately follows that no state involves its own reason:

Theorem 1 (Irreflexivity of R on W_m):
$(\forall x \in W_m)\ \neg x\mathrm{R}x$

Proof: Suppose there is a state a which involves its own reason, giving aRa. By asymmetry we have $a\mathrm{R}a \rightarrow \neg a\mathrm{R}a$, so $\neg a\mathrm{R}a$ by *modus ponens*. But this gives a contradiction; so there is no state which involves its own reason, $\neg(\exists x)x\mathrm{R}x$, or $(\forall x)\neg x\mathrm{R}x$.

We may now capture Leibniz's definition of simultaneity in terms of states of things "none of which involves its opposite." This in turn depends on the Aristotelian intuition that two states of the same thing existing at different times are "opposites," i.e. incompatible.[2] But obviously if we define simultaneity as non-opposition, and opposition in terms of different times, we will be presupposing times in order to define time, just as Russell had objected. The trick, then, is to define simultaneity as non-opposition, but to define opposition directly in terms of reason-inclusion. Since different states of the same monad will be opposite, and two states of the same monad will be different if either involves the reason for the other, we generalize this for all states in the same world: opposite states are those either of which involves the reason for the other. Thus we have:

Axiom 3 (Opposition of States):
$(\forall x, y \in W_m)\ (x\mathrm{O}y =_{\mathrm{def}} x\mathrm{R}y \vee y\mathrm{R}x)$

From this axiom and Theorem 1, it is easy to prove the *symmetry* of O on W_m:

Theorem 2 (Symmetry of O on W_m):
$(\forall x, y \in W_m)\ (x\mathrm{O}y \rightarrow y\mathrm{O}x)$

Proof: Suppose there are two states a and b such that a$\mathrm{O}b$. By Axiom 3 we have $a\mathrm{R}b \vee b\mathrm{R}a$. Now suppose $\neg b\mathrm{O}a$. Then $\neg(b\mathrm{R}a \vee a\mathrm{R}b)$ by Axiom 3. But taken together with $a\mathrm{R}b \vee b\mathrm{R}a$ this yields a contradiction; so $b\mathrm{O}a$ by *reductio ad absurdum* and Double Negation. By Conditional Proof, $a\mathrm{O}b \rightarrow b\mathrm{O}a$ and the theorem follows by Universal Generalization.

[2] Cf. Van Fraassen's discussion in his *An Introduction to Time and Space* (1970, 35–7), where he situates Leibniz's account firmly in the Aristotelian tradition.

Definition 1 (Simultaneity as non-Opposition):
$(\forall x, y \in W_m)(x\mathrm{S}y =_{\mathrm{def}} \neg x\mathrm{O}y)$

Trivially, Axiom 3 and Definition 1 yield

Theorem 3 (Simultaneity as mutual lack of reason-inclusion):
$(\forall x, y \in W_m)[x\mathrm{S}y \leftrightarrow (\neg x\mathrm{R}y \;\&\; \neg y\mathrm{R}x)]$

If simultaneity, so defined, can be shown to be an equivalence relation, then this theorem will encompass the weak connectivity of R on W_m:

Theorem 4 (Weak connectivity of R on W_m)
$(\forall x, y \in W_m)\{(\neg x\mathrm{R}y \;\&\; \neg y\mathrm{R}x) \rightarrow x\mathrm{S}y\}$

Now the *symmetry* of S follows from Definition 1 and Theorem 2:

Theorem 5 (Symmetry of S on W_m):
$(\forall x, y \in W_m)(x\mathrm{S}y \rightarrow y\mathrm{S}x)$

Proof: Suppose aSb. Then $\neg a\mathrm{O}b$ by Definition 1. But $b\mathrm{O}a \rightarrow a\mathrm{O}b$ by the symmetry of O (Theorem 2). Therefore $\neg b\mathrm{O}a$ by *modus tollens*. Therefore $b\mathrm{S}a$ by Definition 1. Thus $a\mathrm{S}b \rightarrow b\mathrm{S}a$ by conditional proof, and the theorem follows by universal generalization.

The *reflexivity* of S is also provable:

Theorem 6 (Reflexivity of S on W_m):
$(\forall x)\; x\mathrm{S}x$

Proof: Suppose $\neg a\mathrm{S}a$. Then $\neg(\neg a\mathrm{R}a \;\&\; \neg a\mathrm{R}a)$ by Theorem 5. Suppose $\neg a\mathrm{R}a$. Then $\neg\neg a\mathrm{R}a$ by Conjunctive Syllogism, giving a contradiction. So $a\mathrm{R}a$ by *reductio ad absurdum*. Therefore $\neg a\mathrm{R}a$ by asymmetry of R (Axiom 2), again giving a contradiction. So $a\mathrm{S}a$ by *reductio ad absurdum*. The theorem follows by universal generalization.

Now, for simultaneity to be an equivalence relation it needs to be symmetric, reflexive and transitive. But as things stand it is impossible to prove the *transitivity* of simultaneity across states of different monads, since there is no connection across different monadic series. Leibniz realized the necessity for such a connection, for he explicitly supplies it with a proposition deriving from the "connection of all things": "and since, because of the connection of all things, my earlier state involves the earlier state of the other things as well, it also involves the reason for the later state of these other things, so that my earlier state is in fact earlier than their later state as well." Here I interpret "*x* involves y" as "*x* is an immediate requisite for *y*," so that *x* and *y* must be compatible. The axiom of connection may thus be parsed as follows:

Axiom 4 (Axiom of Connection):[3]
$(\forall x, y, z \in W_m)\; \{(\neg x\mathrm{O}y \;\&\; y\mathrm{R}z) \rightarrow x\mathrm{R}z\}$

from which Leibniz's formulation follows trivially by Definition 1:

Theorem 7 (Simultaneity form of Axiom of Connection):
$(\forall x, y, z \in W_m)\; \{(x\mathrm{S}y \;\&\; y\mathrm{R}z) \rightarrow x\mathrm{R}z\}$

[3] Winnie calls this the "Leibniz Postulate," which in his exposition is: "[For any events e_1, e_2 and e_3,] if e_1 and e_2 are simultaneous and e_1 is causally connected with e_3, then is e_2 also causally connected with e_3" (Winnie 1977, 138).

In conjunction with Theorem 5, we may now derive the transitivity of S:

Theorem 8 (Transitivity of S on W_m):
$(\forall x, y, z \varepsilon W_m)$ $\{(xSy \,\&\, ySz) \rightarrow xSz\}$

Proof: Suppose *aSb* and *bSc*, but ¬*aSc*. Now ¬*aSc* entails *a*Oc (Def. 1), and thus *a*R*c* v *c*R*a* (Axiom 3). Suppose *a*R*c*. From *aSb* we have *bSa* by symmetry (Theorem 5). Thus we have (*bSa* & *a*R*c*), and thus *b*R*c* (by Theorem 7). But *bSc* entails ¬*b*R*c* by Theorem 3 and simplification, and we have a contradiction, *b*R*c* & ¬*b*R*c*. Therefore ¬*a*R*c* by *reductio ad absurdum*. Therefore *c*R*a* (by Axiom 2), and by Conjunction we have (*bSc* & *c*R*a*). An application of Theorem 7 now yields *b*R*a*. But *bSa* entails ¬*b*R*a* by Theorem 3 and simplification, and again we have a contradiction. Therefore *aSc* by *reductio ad absurdum*, and the conclusion follows by conditional proof and universal generalization.

Simultaneity, therefore, being symmetric, transitive and reflexive, is an *equivalence relation* on W_m. It will therefore define an *equivalence class*, the class of all those states in W_m that are simultaneous with one another.

This is all we need to construct a theory of temporal ordering of states. We will say that one state occurs at the same time as another if and only it is simultaneous with it; and it is before another if and only if it contains the reason for it. For this purpose it will be convenient to define a *moment* as M_n is the class of states simultaneous with a given state *n*:

Definition 2 (a moment, M_n):
$M_n =_{\text{def}} \{x|xSn\}$

It follows that that a moment is the quotient of all the states in a given world by S:

Theorem 9 (simultaneity partitions a world into moments)
$M = W_{m/}S$

To say that a state *x* happens at a moment M_y is to say that *x* is simultaneous with *y*:

Definition 3 (to be at a moment):
$(\forall x, y \,\varepsilon\, W_m)$ $[x \,\varepsilon\, M_y =_{\text{def}} xSy]$

One state will be earlier than another if and only if the first includes the reason for the second:

Definition 4 (Temporal Ordering of States, B –'is before'):
$(\forall x, y \,\varepsilon\, W_m)$ $\{xBy =_{\text{def}} xRy\}$

From these definitions it is easy to prove that this ordering is a serial ordering:

Theorem 10 ('Earlier than' is a Serial Relation)
10a (Transitivity of B on W_m):
$(\forall x, y, z \,\varepsilon\, W_m)$ $\{(xBy \,\&\, yBz) \rightarrow xBz\}$
10b (Asymmetry of B on W_m):
$(\forall x, y \,\varepsilon\, W_m)$ $(xBy \rightarrow \neg yBx)$
10c (Weak Connectivity of B on W_m):
$(\forall x, y \,\varepsilon\, W_m)$ $\{(\neg xBy \,\&\, \neg yBx) \rightarrow xSy\}$

This is a theory of the temporal ordering of states. But it is a long way short of a complete theory of time. In particular, we have not yet defined change, and nothing has been said about the relative sizes of states, or their continuity. We will treat these matters in A 1.5 below.

A1.2 The Relational Core: Version 2

In this version, following the suggestion of Jan Cover (1997), we stick with the relation of ground containment G defined only on the individual monadic series S_m, for which we postulate the same three axioms *G1-G3* as above. Now Leibniz proposes that each monadic state expresses or represents, more or less confusedly, and from its own point of view, "the same state of the universe." This is an objective phenomenal *state of affairs* which is *maximal* in the sense that it is that which is the object of all possible representations of the same thing. Following Cover, we shall call it a "world state" (1977, 312); the idea is that the monadic states express these world-states partially and confusedly from their own particular point of view, and that this is what accounts for the difference among the diverse but harmonious monadic states. So we posit the existence of a set Ω_m of world-states ω for each world W_m of monads, each of which is expressed by a monadic state in this world:

Axiom E1 (Expression of world-states):
$(\forall\omega \in \Omega_m)(\exists x \in W_m)\ x\text{E}\omega$

such that every monadic state expresses a unique world-state:

Axiom E2 (Maximality of world-states):
$(\forall x \in W_m)(\forall \upsilon, \omega \in \Omega_m)\ [(x\text{E}\upsilon\ \&\ x\text{E}\omega) \leftrightarrow \upsilon = \omega]$

(Cf. Cover's proposition that two world-states "are identical iff neither includes a state of affairs the other does not" (312).) Now we lay down the axiom that two monadic states are *simultaneous* iff they express the same world-state:

Axiom E3 (Simultaneity as expression of identical world-states):
$(\forall x, y \in W_m)\ (\forall \upsilon, \omega \in \Omega_m)\ [x\text{S}y =_{\text{def}} (x\text{E}\upsilon\ \&\ y\text{E}\omega) \leftrightarrow \upsilon = \omega]$

This captures Cover's proposal that "two states are simultaneous iff the world-state contained by one is identical with the world-state contained by the other" (312). Since identity is symmetric, transitive and reflexive, it follows that simultaneity will be too, establishing at a stroke that simultaneity is an equivalence relation.

A1.3 Compossibility

Let us begin by assuming that Axioms 1 and 2 above can be extended across all monadic series, including incompossible ones:

*Axiom 1** (Transitivity of R across states in any 3 possible monadic series):
$(\forall x \in S_m)(\forall y \in S_n)(z \in S_o)\{(x\text{R}y\ \&\ y\text{R}z) \rightarrow x\text{R}z\}$
*Axiom 2** (Asymmetry of R across states in any 2 possible monadic series):
$(\forall x \in S_m)(\forall y \in S_n)(x\text{R}y \rightarrow \neg y\text{R}x)$

This is defensible even though only states in the same possible world will contain the reasons for one another; for in that case they will not fulfil the antecedents of the conditionals. An analogue of Theorem 1 will follow as before:

*Theorem 1** (Irreflexivity of R on monadic states):
$(\forall x)\ \neg x\text{R}x$

We will also assume that Axiom 3 can be extended across all monadic series:

*Axiom 3** (Opposition of states in any 2 possible monadic series):
$(\forall x \in S_m)(\forall y \in S_n)\ [x\text{O}y =_{\text{def}} x\text{R}y \vee y\text{R}x]$

From Axiom 3* and Theorem 1*, the *symmetry* of O will follow as before:

*Theorem 2** (Symmetry of O in any 2 possible monadic series):
$(\forall x \in S_m)(\forall y \in S_n)\ (x\text{O}y \rightarrow y\text{O}x)$

We also need to make explicit the fact that a possible world is a maximal set of compossible series:

Definition 5 (Possible world as maximal set of compossible series):
$W_m =_{\text{def}} \{S_n | S_n\ comp\ S_m\},\ (\forall S_m)(\forall S_n)[(S_m \in W_m)\ \&\ (S_n\ comp\ S_m) \rightarrow (S_n \in W_m)]$

(Again, here there is an implicit existence assumption, that there exist infinitely many such series compatible with any one.) The success of these definitions depends on our being able to prove that *comp* is an equivalence relation. If we import our axiom of connection unchanged,

Axiom 4 (Axiom of Connection):
$(\forall x, y, z \in W_m)\ \{(\neg x\text{O}y\ \&\ y\text{R}z) \rightarrow x\text{R}z\}$

then together with Definition 5 this entails that if two distinct series are compossible, the Axiom of Connection applies:

Theorem 11 (Connection applies between compossible series):
$(\forall x \in S_m)\ (\forall y, z \in S_n)\ \{m \neq n \rightarrow (S_m\ comp\ S_n \rightarrow [(\neg x\text{O}y\ \&\ \text{R}yz) \rightarrow x\text{R}z])\}$

Now we need to prove that it applies *only* between compossible series. According to Leibniz there is a Law of General Order which determines the notion of each possible world, and therefore everything that will be consequent on a given state of any individual substance in that world. This, then, will be the foundation for the fact that every state in this world involves the reason for every subsequent state. In other words, if x and y are states belonging to two distinct monadic series, and xRy, then the two series are *compossible*. This then allows us to lay down the following condition on compossibility:

Axiom 5 (Axiom of General Order):
$(\forall x \in S_m)(\forall y \in S_n)\ \{m \neq n \rightarrow (x\text{R}y \rightarrow S_m\ comp\ S_n)\}$

Suppose, then, that x and y are states from distinct incompossible monadic series, so that $m \neq n$ and $\neg S_m\ comp\ S_n$. It follows from Axiom 5 that $\neg x\text{R}y$ and that $\neg y\text{R}x$, i.e. $\neg(x\text{R}y \vee y\text{R}x)$, and therefore that $\neg x\text{O}y$ by Definition 1*. We have therefore shown that states from distinct incompossible monadic series do not contain one another's opposite:

Theorem 12 (Compatibility of states from incompossible monadic series):
$(\forall x \in S_m)(\forall y \in S_n)\ \{m \neq n \rightarrow (\neg S_m\ comp\ S_n \rightarrow \neg x\text{O}y)\}$

To reiterate: since no state involves the opposite of any state of an incompossible series, non-opposition does not pick out any unique corresponding state in the incompossible series. Suppose now that a is a state from one monadic series and b and c are states from a distinct incompossible monadic series such that bRc. By Axiom 5, $\neg a\text{R}c$. But since by Theorem 12 $\neg a\text{O}b$, we have $(\neg a\text{O}b\ \&\ b\text{R}c)$. If the Axiom of Connection held, we would have aRc, giving a contradiction. So that axiom fails across such distinct incompossible series: for such states, a from one series, and b and c from another incompossible series, $\neg[(\neg a\mathrm{O}b \ \&\ b\mathrm{R}c) \rightarrow \neg a\mathrm{R}c]$.

Theorem 13 (Connection applies only between compossible series):
$(\forall x \in S_m)\ (\forall y, z \in S_n)\ \{m \neq n \rightarrow (\neg S_m\ comp\ S_n \rightarrow \neg[(\neg x\mathrm{O}y \ \&\ \mathrm{R}yz) \rightarrow x\mathrm{R}z])\}$

Theorems 11 and 13 together entail

Theorem 14 (Connection applies if and only if the series are compossible):
$(\forall x \in S_m)\ (\forall y, z \in S_n)\ \{m \neq n \rightarrow [(\neg x\mathrm{O}y \ \&\ \mathrm{R}yz) \rightarrow x\mathrm{R}z] \leftrightarrow (S_m\ comp\ S_n)\}$

Now if we import our definition of simultaneity unchanged,

Definition 1 (Simultaneity as non-Opposition):
$(\forall x, y \in \mathrm{W}_m)[x\mathrm{S}y =_{\mathrm{def}} \neg x\mathrm{O}y]$

then, since by Definition 7 compossible series belong to the same world, non-opposition entails simultaneity if two distinct series are compossible. Thus the axiom of connection holds if and only if S_n and S_m are distinct *compossible* series, and for such series simultaneity is an equivalence relation; whereas if we take two distinct series that are *incompossible*, the axiom of connection fails to apply, and to any state in the first there is no unique state corresponding to it in the second; and simultaneity is not defined. It follows that simultaneity is an equivalence relation among states of those and only those monadic series which are mutually compossible. Thus two monadic series S_m and S_n are *compossible* iff for any state of S_m there is a unique compatible (i.e. simultaneous) state of S_n:

Theorem 15 (Compossibility of monadic series):
$S_m\ comp\ S_n \leftrightarrow (\forall x \in S_m)\ [(\exists y \in S_n)\ y\mathrm{S}x \ \&\ (\forall z \in S_n)\ (z\mathrm{S}x \rightarrow z = y)]$

From this theorem the transitivity, symmetry and reflexivity of the compossibility relation are readily provable: *comp* is an equivalence relation.

Now if we define the *universe* U as the union of all monadic series:

$$\mathrm{U} =_{\mathrm{def}} \cup \{S_n\}$$

the aggregate of series that are *compossible* with a given series S_m, the world W_m, will be the quotient of U by the equivalence relation *comp*, as desired:

Theorem 16 (Compossibility partitions the universe of possibles):
$W_m =_{\mathrm{def}} \mathrm{U}/S_m = \{S_n | S_n\ comp\ S_m\}$

A1.4 Temporal Counterparts

The aim here is to make sense of claims such as that Adam, in taking the apple, could have done otherwise. Adam—identically the same Adam—could not. But we can make sense of such claims in terms of a "counterpart-Adam," Adam*, who can have as many predicates in common with Adam as we wish. We assume, however, that an ever finer discrimination showing the difference between Adam and Adam* will always be possible by way of more minute perceptions. So let us introduce the notion of two states being indistinguishable below a certain threshold Δ. We will suppose that two individuals m and n share a finite collection of predicates during a given time if and only if their corresponding states are indiscernible with respect to some threshold Δ during that time. Implicit in this proposal is the idea that the finer the threshold of discrimination under which the states are indiscernible, the more properties the two counterparts will share.

We can make this more precise by introducing the idea of a *stage* of a monadic series S_m, which is some truncated sequence of states of that series running from state *s* to state *t*, which we denote $S_m|_s^t$.

If for every state of $S_m|_s^t$ there is a unique corresponding state of another individual *n* that is indiscernible from it relative to Δ, *n* is a *counterpart* of *m* during that stage. The unique corresponding state that is indiscernible from some state *x*, with $s \leq x \leq t$, is *c-simultaneous* with *x* relative to Δ.

Thus if we consider "Adam" as characterized only by a certain set of properties, such as "the first man, whom God puts in a garden of pleasure, and from whose side God draws a woman," various counterpart-Adams are definable as above, who can be considered as having states indiscernible from his during a certain stage of his existence, but to whom other predicates not applicable to Adam may apply, such as "refuses to take the apple." A counterfactual such as "Adam might not have sinned" is then explicable in terms of there existing in God's consideration another possible world in which a counterpart-Adam—an individual sufficiently similar to Adam during some segment of his life as to be indiscernible from him within threshold Δ—does not sin. There are, I conclude, sufficient resources within Leibniz's theory of time to account for the same individual (identical within a certain threshold of discrimination Δ) acting otherwise than she or he in fact does at a given time.

A1.5 Change and Continuity

In A 1.1 we presented a theory of order for monadic states. Nothing was said about the relative sizes of states, nothing about change, and nothing about contiguity versus continuity. The states as defined so far could even be punctual, or they could be atomic but of finite size. In order to treat these matters, we will need to introduce mereological notions such as that one state "is a part of" another. Using this notion, we will be able to define change, and what it means for a change to be in a state.

First, let us define being a part of a given state:

Definition 6 ("part of," ≤):
x is *a part of* state *y*, written $x \leq y$.

We will assume that a part of a state is itself a state at a higher level of resolution, and lay down the following axioms on the relation:

Axiom 6: the relation $x \leq y$ ("*x* is a part of *y*") is reflexive, antisymmetric and transitive:
6a (Reflexivity of ≤): $\forall x\, x \leq x$
6b (Antisymmetry of ≤): $\forall x\, \forall y\, \{(x \leq y\ \&\ y \leq x) \rightarrow x = y\}$
6c (Transitivity of ≤): $\forall x\, \forall y\, \forall z\, \{(x \leq y\ \&\ y \leq z) \rightarrow x \leq z\}$

We now define two derivative relations that will be convenient in what follows, that of one part *overlapping* another, and that of two parts being *separate*:[4]

[4] Here, and in what follows, I am leaning heavily on the treatment of Geoffrey Hellman and Stewart Shapiro in their *Varieties of Continua* (Hellman and Shapiro 2018).

Definition 7: (x *overlaps* y, $x \circ y$):
$x \circ y =_{\text{def}} \exists z\ [z \leq x \ \&\ z \leq y]$

Definition 8: (x is *separate* from y, $x|y$):
$x|y =_{\text{def}} \forall z\ \neg[z \leq x \ \&\ z \leq y] =_{\text{def}} \neg x \circ y$

We now lay down the following axiom:

Axiom 7: $x \leq y \leftrightarrow \forall z\ [z \circ x \rightarrow z \circ y]$, or equivalently,
$x \leq y \leftrightarrow \forall z\ [\neg z|x \rightarrow \neg z|y]$

From this axiom one can easily derive

Theorem 17 (the Extensionality Principle):
$x = y \leftrightarrow \forall z\ [z \circ x \leftrightarrow z \circ y]$

So far this is equivalent to a standard mereological system.
Now we define pairwise distinct parts of a state:

Definition 9: (x_1 and x_2 are *pairwise distinct parts* of x, $x_1 \ll x \ \&\ x_2 \ll x$):
$x_1 \ll x \ \&\ x_2 \ll x =_{\text{def}} [(x_1 \leq x \ \&\ x_2 \leq x) \ \&\ x_1|x_2]$

Moreover, if either of these distinct parts involves the reason for the other, then they themselves are states (substates of x) that could be discerned at a higher resolution. So let us define pairs of substates of x as follows, writing "x_1 is a substate of x" as $x_1$sub$x$:

Definition 10 (*substates* of x):
$x_1$sub$x$ & $x_2$sub$x =_{\text{def}} [(x_1 \ll x \ \&\ x_2 \ll x) \ \&\ x_1 \mathrm{R} x_2]$[5]

With these definitions we can define when *change* occurs, compatibly with Leibniz's thesis that "change is an aggregate of two opposite states in one stretch of time" (A VI 4, 556).[6] There is a *change* in x, which we'll symbolize $\exists c\ c \subset x$, iff x has pairwise distinct parts x_1 and x_2 one of which contains the reason for the other:

Definition 11 (occurrence of *change* in state x):
$\exists c\ c \subset x =_{\text{def}} \exists x_1, x_2\ [(x_1 \ll x \ \&\ x_2 \ll x) \ \&\ x_1 \mathrm{R} x_2]$

Notably, this way of defining an occurrence of change is compatible even with the idea that "all enduring states are vague" (A VI 4, 1613-14), so that *they do not have definite boundaries*. That is, they are not assumed to be either bounded or unbounded. Now we take into account Leibniz's thesis that "during any state whatsoever some other things are changing" (A VI 3, 559/LLC 197) with the following crucial axiom:

Axiom 8 (the Axiom of Change):
$\forall x\ \exists x_1\ \exists x_2\ (\exists y_1, y_2\ \varepsilon\ S_y)\{((x_1 \ll x \ \&\ x_2 \ll x) \ \&\ [(y_1 \mathrm{S} x_1 \ \&\ y_2 \mathrm{S} x_2) \ \&\ y_1 \mathrm{R} y_2]\}$

From Axiom 8 and Definitions 10 and 11 it follows that

Theorem 18 (Every state is divided into a pair of substates):
$\forall x\ \exists x_1\ \exists x_2 [(x_1 \ll x \ \&\ x_2 \ll x) \ \&\ x_1 \mathrm{R} x_2]$

[5] One might expect "& $(x_1 \mathrm{R} x_2 \vee x_2 \mathrm{R} x_1)$]" in this definition, but the second term is already covered in the definition on interchanging x_1 and x_2.

[6] I do not attempt to symbolize the "in one stretch of time"; what this connotes is that the change occurs within an extended duration or series of states; but Leibniz is explicit that "there is no moment of change common to both states, and thus no state of change either" (A VI 3, 566/LLC 211)—as there is for Hegel, for example.

Proof: According to Axiom 8, for any given state n we have two pairwise distinct states $a \ll n$ & $b \ll n$ and two states c, d of monad y such that cSa and dSb, and one involves the reason for the other, cRd. Since a is simultaneous with c and cRd, then aRd by Theorem 7. The theorem follows by Conjunction and Existential Generalization.

Thus there is no state which is not divided into pairs of substates, by Definition 10. The states in any monadic series are therefore *dense*.

And by the definition of occurrence of change (Def. 11), it follows as an immediate corollary that there is no state whatsoever in which there is no change:

Theorem 19 (Every state contains a change):
$\forall x \; \exists c \; c \subset x$

Thus the changes in any state are also *dense*, in keeping with the analysis of §1.4.

Next, we proceed to the issue of temporal continuity. First we need to define the location of change in a given state x. We should note that by definition 4, $xBy =_{\text{def}} xRy$, so that Definition 10 may be rewritten

*Definition 10** (*substates* of x):
$x_1\text{sub}x \;\&\; x_2\text{sub}x =_{\text{def}} [(x_1 \ll x \;\&\; x_2 \ll x) \;\&\; x_1 B x_2]$

Now, regarding the change c: it *is in* x (although it is not a part of it). It is not, however, in either of the two substates x_1 or x_2 (since change is defined in terms of the aggregation of the two opposed substates). So if it is in x but not in either of the substates x_1 or x_2 into which it divides x, then it must be temporally between them (here I am tacitly allowing that the relation B orders changes as well as states):

Axiom 9 (Location of Change):
$\exists c \; c \subset x$ (with x_1 and x_2 the resulting substates such that $x_1 B x_2$) $\rightarrow (x_1 B c \;\&\; c B x_2)$

We are now in a position to tackle the issue of continuity. According to Leibniz's mature definitions, two things are required for continuity.[7] One is the traditional requirement that there be what he calls "parts outside one another (*partes extra partes*)," that is, parts that do not overlap. This is satisfied by our Definition 9 above: *pairwise distinct parts* are precisely those that do not overlap, and we have seen that changes induce the existence of such parts.

As things stand, however, there could be a gap on either side of a change. This is where the second of Leibniz's requirements comes into play. It depends on what he calls "co-integrant parts": these are parts which together equal the whole, and which are such that if they have something in common, this is not itself a part. If pairwise distinct parts have something in common that is not a part, it would have to be a shared boundary. But in keeping with the idea of states being vague, we may leave it undetermined whether they have such a boundary in common. So let us first define the notion of a *sum* of the substates of a given state, with substates defined by Definition 10* above:[8]

Definition 12 (x's substates x_1 and x_2 *sum to* x, written $x_1 + x_2 = x$):
$\forall x \; \forall x_1 \forall x_2 \; \{(x_1 + x_2 = x) =_{\text{def}} \forall w[w \leq x \rightarrow (w \leq x_1 \vee w \leq x_2)]\}$

[7] Cf. Leibniz's definition in *In Euclidis* πρῶτα of 1712: "Moreover, two things are required for a continuum: first, that any two of its parts that together equal the whole have something in common which is precisely not a part; and second, that in the continuum there are parts outside one another [*partes extra partes*], as is commonly said, that is, that two of its parts can be assumed (but not ones that together equal the whole) that have nothing in common, not even a minimum" (GM V 184).

[8] I am indebted to Filippo Costantini here for his suggestion that I define a sum of parts before proceeding to co-integrant parts.

Now we may formulate Leibniz's *Axiom of Continuity*. This is that if a change divides a state x into substates x_1 and x_2, then these substates sum to x:

Axiom 10 (Axiom of Continuity):
$\forall x \{\exists c\ c \subset x$ (with substates x_1 & x_2 such that $x_1 Bc$ & $cBx_2) \rightarrow x_1 + x_2 = x\}$

As we have seen, x_1 and x_2 are non-overlapping parts, and now they also satisfy Leibniz's definition of being co-integrant parts of x. For although they are separated by the change c, still they sum to the whole interval containing only themselves and the change, which is not itself a part of the interval. So even though x_1 and x_2 do not share a boundary (since c is not in either substate), they seem to fulfil the intent if not the letter of Leibniz's second requirement for continuity, being co-integrant parts of x that are bounded by something that is not part of the interval, c.

In keeping with this notion of continuity and with Leibniz's conception of a monadic series as a continuous series of states, let us also define the *duration* of a monadic series between any two changes c_1 and c_2, written $D_{c1}{}^{c2}(S_m)$, as the sum of all its states between those changes, where every pair of successive states x_k and x_{k+1} are such that $x_k B x_{k+1}$ and the states sum to $x_1 + x_2$ and are thus co-integrant parts of the whole duration:

Definition 13 (Duration as a Continuous Series of States):
$D_{c1}{}^{c2}(S_m) =_{def} \sum \{x_k | x_k B x_{k+1}$ & $(c_1 B x_k$ & $x_k B c_2)\}$

So defined, *duration is a property of the series*, and not to be equated with time, which is a general concept by means of which absolute magnitudes may be assigned to durations.[9] Nevertheless, since a (proper) part is smaller than any whole of which it is a part, durations have relative magnitudes.

Now according to Theorem 18, every change c divides a state x into substates x_1 and x_2, and according to Axiom 9, these are such that $x_1 Bc$ & cBx_2. So by Axiom 10 these states are co-integrant parts of x. Now by the Axiom of Change each of these states will in turn be divided by a further change: x_1 will be divided by some change c_1 into x_{11} simultaneous with some further state v_1 and x_{12} simultaneous with a state v_2 (of some monad v); and x_2 will be divided by some change c_2 into x_{21} simultaneous with some further state w_1 and x_{22} simultaneous with some further state w_2 (of some monad w), with $v_1 R v_2$ and $w_1 R w_2$, and therefore $v_1 B v_2$ and $w_1 B w_2$. So $x_{11} B x_{12}$ and $x_{21} B x_{22}$, and by axiom 10, $x_{11} + x_{12} = x_1$ and $x_{21} + x_{22} = x_2$. So now the change c is flanked by two smaller states, $x_{12} < x_1$ and $x_{21} < x_2$, with $x_{12} Bc$ & cBx_{21} (by axiom 9), with x_{12} being the interval between c_1 and c, and x_{21} that between c and c_2. This process, of course, can be reiterated *ad libitum*.

From this it follows that the interval between any two changes dividing the duration of a monad is *infinitesimal*, in the strict, syncategorematic sense employed by Leibniz. For suppose we have a monadic series of states in which the smallest state between two of the changes is of magnitude ε. Then there will always be changes in other monads that split this smallest state into substates of still smaller magnitude $\delta < \varepsilon$, since each of them is a pairwise distinct, co-integrant part of the original state. Thus there is no smallest interval; rather the intervals between the changes in any given monadic series are arbitrarily small, *infinitesimal*: for any given minimum magnitude ε that might be preassigned, they are of still smaller magnitude, $\delta < \varepsilon$. But these are not actual infinitesimals, elements of a non-Archimedean continuum. For Leibniz rejects such infinitely small actuals as contradictory. Such intervals

[9] Cf. *On Magnitude*: "Duration is a continuity of existing. Time is not duration, any more than space is collocation.... Time is a certain continuum according to which something is said to endure" (A VI 3, 484/DSR 41).

are on the point of vanishing, rather than being absolute nothings. Nevertheless, differences that can be made smaller than any given difference in this way, have a magnitude that is null.[10]

This means in turn that a monadic duration, as defined by Definition 13, will consist in an actually infinite sum of infinitesimal states. This corresponds to Leibniz's conception of an integral, as argued in §1.5. Moreover, the infinitude of these states is an actual infinite understood syncategorematically. This can be demonstrated as follows. Suppose that there is only a finite number of states between any two changes. No matter how large this finite number is taken to be, it follows (by the above argument) that the number of states is greater. Therefore the number of states between any two changes is (syncategorematically) actually infinite.[11]

It therefore also follows that *moments* as defined by Definition 2 above, equivalence classes of all the states that are simultaneous with the original state containing the change c, are also infinitesimal, and have zero magnitude (in comparison with any finite duration).[12] They thus have a punctual character, and are properly speaking *instants*.

It is beyond the scope of this work to extend these considerations any further. But the interested reader should consult Hellman and Shapiro (2018), where the authors give a detailed formal construction of a continuum without assuming any boundaries or points. In chapter 2 of the book they construct a model of such a "gunky" continuum for the 1-dimensional case, in which they are able to model the *congruence* of intervals in the continuum, again without assuming that these intervals are either bounded or unbounded. In chapter 3 (co-written with Øystein Linnebo) they extend this treatment to what they call an "Aristotelian" continuum, eschewing the actual infinite involved in their previous chapter in favour of an Aristotelian potential infinite. This work raises the prospect that a similar treatment should be possible for Leibniz's conception of time as the order of all possible successions, so that the actually infinitely many partitions entailed by the axiom of change are each instantiations of the possible partitions of any continuous time.

[10] For detailed studies of Leibniz's syncategorematic conception of the infinitesimal, see Arthur (2013c) and Rabouin and Arthur (2020).

[11] For an account of Leibniz's syncategorematic actual infinite, see Arthur (2018c).

[12] Cf. the first essay of the *Initia Rerum*: "*Duration is the magnitude of time.* If the magnitude of time. is diminished uniformly and continuously, time disappears into a *moment*, whose magnitude is zero" (GM VII 17/L 666).

APPENDIX 2

On Leibniz's Treatment of Relations

"I hold that all accidental forms are only modifications of substantial ones, just as figures are modes of space, and so qualities no more subsist per se without substances than figures without a thing figured, or relations without those things which are related to one another."[1]

"There is no term which is so absolute or so detached that it does not involve relations and is not such that a complete analysis of it would lead to other things and indeed to all other things". (A VI 6, 228)

Introduction

It has so long been held that Leibniz 'denied relations'—that is, sought to eliminate them by reducing relational statements to ones involving only intrinsic properties of the relata—that this has become one of the standard dogmas concerning Leibnizian metaphysics. It was Bertrand Russell who first made the claim (in his (1900)) that Leibniz wanted to reduce relations to attributes of the relata—the Reduction of Relations thesis—in association with his thesis (shared with Louis Couturat) that Leibniz derived his metaphysics from a commitment to subject-predicate logic. Since that time the idea that Leibniz wished to eliminate relations from his fundamental ontology has become the received view. It has been endorsed in one form or another by C. D. Broad (1975, 36 ff.), G. W. R. Parkinson (1969), Nicholas Rescher (1967), Benson Mates (1986), Donald Rutherford (1995), and Jan Cover and John O'Leary-Hawthorne (1999). Although these commentators reject Russell's too-easy derivation of the reduction of relations thesis from Leibniz's commitment to a subject-predicate logic, they take it to be implicit in the nominalistic rephrasings Leibniz gives of certain relational statements, as well as in his denial of purely extrinsic denominations, and in his claims that relations are ideal. The following assessment by Donald Rutherford is typical:

> Thus, it is not only the case that the relations of things are necessarily grounded in intrinsic properties of the things related; but, in terms of ontological commitment, those relations imply nothing more about the created world than that certain individuals possess intrinsic properties between which some similarity or connection can be apprehended. In light of this, it is undoubtedly right to attribute to Leibniz a thesis about the *reducibility* of intersubstantial relations.[2] (Rutherford 1995, 184)

[1] «*Statuo omnes formas accidentales esse tantum substantialium modificationes, ut figurae sunt modii spatii, adeoque per se subsistere non magis posse qualitates sine substantia, quam figuras sine figurato aut relationes sine iis quae ad se invicem referuntur.*» This is from the manuscript LH I, 20, Bl. 216 (dating from some time after 1695), transcribed and sent to me by Oswaldo Ottaviani, for which I am very grateful.

[2] Here Rutherford follows Rescher in restricting the application of the relation-reducibility thesis to relations obtaining among distinct individual substances, so that is only "intermonadic relations" that are reduced (Rescher 1967, 74); cf. also Mugnai (1992).

According to this consensus, since Leibniz regards time and space as intersubstantial relations, they are at best mental entities, results of the perceptions of substances, but with no reality independent of their being perceived.

In Leibniz's actual scientific practice, however, relations take pride of place. In his metaphysics the principle of pre-established harmony involves the notion that each state of a substance *represents* or *reflects* the whole of the rest of the universe. In his theory of time, as recounted in these pages, Leibniz takes the relations of *involving the reason for* and *being a* (mediate or immediate) *requisite of* as foundational; in his theory of space, his innovative new approach to geometry, prosecuted over the last forty years of his life, is based on *relations of situation*; and motion, according to his accounts over the same forty years, is essentially relative. In broad brush strokes, in fact, his methodology may fairly be described as consisting in a reduction to equivalence relations: instants or moments are determined by states that are simultaneous with one another, and places by those things sharing equivalent relations of situation, conservation of force is based on an equivalence in the capacity to perform work, and so forth. If this does not establish that Leibniz was a "realist" about relations, it certainly casts serious doubt on the idea that he meant to "deny relations."

In what follows I shall try to explain how Leibniz's position should be understood. Certainly, his stance on relations is very complex, and I make no claim give a comprehensive evaluation. Instead I will present a brief sketch of various interpretations that have been given of Leibniz's philosophy of relations, aiming to distil the main points for and against each of the rival interpretations with primary reference to the relations of time and space. First I shall sketch the chief features of Russell's eliminativist interpretation, attempting a sympathetic understanding of its basis as well as pointing out where his assumptions led him seriously astray. I proceed from there to a brief evaluation of the rival interpretation of Leibniz on relations proposed by Hidé Ishiguro, Jaakko Hintikka, and others, according to which Leibniz upheld the reality of relations as properties, and that the elimination of relations in all senses was never his intention. This relational properties interpretation was subsequently subjected to detailed critique by Mates and Massimo Mugnai, among others. They pointed out that it would trivialize some of Leibniz's principal doctrines, namely, the Identity of Indiscernibles and his denial of purely extrinsic denominations, as well as undermining any motivation for his nominalistic parsings. In order to get around such objections, a variant interpretation was proposed by Lawrence McCullough (1978) and at book length by Dennis Plaisted (2002), according to which it was not relational properties that Leibniz wished to spare from the need for reduction, but the individual accidents of substances regarded as tropes, each specific to the substance of which it is the accident. Against this interpretation, Mugnai, in particular, has urged that such an interpretation is in conflict with the textual evidence, and indeed to the whole nominalist-conceptualist tradition in which Leibniz was working. Nevertheless, there are some important divergences Leibniz makes from that tradition that are consequent on his own "correction" of the traditional metaphysics of substance (Leibniz 1694), leading him to a novel construal of the reality of the spatial and temporal orders, as we shall see.

A2.1 Russell's Thesis

Russell's interpretation of Leibniz on relations was forged in a profoundly neo-Hegelian crucible, where substance was equated with the subject of a proposition, whose predicates were indiscriminately supposed to be attributes, qualities or monadic states.[3] States were interpreted as properties internal to substance, and Leibniz's doctrine of the causal independence of substances was taken to mean that relations are somehow internalized in the substances. This was natural, according to the doctrine popularized by Hermann Lotze and Francis Bradley, where relations are interpreted as "internal", meaning that they are reducible to properties of the relata. Russell had taken the latter doctrine on board in the process of constructing his own theory of space in 1897–98, and when he set about studying Leibniz in earnest, in December 1898 he discovered in the *Discourse on Metaphysics* and letters to Arnauld what looked to him like a precise statement of it. This led him to identify what he thought to be the key to understanding all Leibniz's philosophy: that the predicate is contained in the subject is a consequence of Leibniz's commitment to the subject–predicate logic, leading immediately to his denial of relations, and the consequent causal independence of substances. Russell therefore followed Hegel in regarding the doctrine of pre-established harmony as a futile attempt to paper over a doctrine that is bound to collapse into a necessitarian monism, notwithstanding Leibniz's claims to the contrary. For Russell's own philosophy, moreover, it immediately indicated that the way forward in the theory of space was to reject the doctrine of internal relations, and develop a logic of relations where the latter are conceived as irreducible and purely external to their relata.

The situation is of course complicated by the fact that at the time Russell took on the interpretation of Leibniz, he and G. E. Moore were in the process of trying to throw off the yoke of both Hegel and Kant. The doctrines of these two were often combined in the doctrines of the neo-Hegelians, and particularly in Lotze. A second complicating factor is that, owing to the very success of Russell's own contributions to logic and philosophy, we now have a conception of relations that is so indebted to his that we cannot help presupposing this conception when we try to understand Leibniz.

Both these difficulties can be illustrated by reference to a passage that Russell took to present an explicit statement by Leibniz of his denial of relations, the much-quoted passage from his Fifth Letter to Clarke in their controversy:

> The ratio or proportion of two lines *L* and *M* may be conceived in three ways: as a ratio of the greater *L* to the lesser *M*, as a ratio of the lesser *M* to the greater *L*, and finally as something abstracted from both, that is, as the ratio between *L* and *M*, without considering which is the antecedent and which the consequent, which the subject and which the object. . . . In the first way of considering them, *L* the greater is the subject, in the second, *M* the lesser is the subject, of that accident that the philosophers call relation or respect. But which of them will be the subject in the third sense? It cannot be said that both of them, *L* and *M* together, are the subject of such an accident, for in that case we should have an accident in two subjects, with one leg in one, and the other leg in the other, which is contrary to the notion of accidents. Thus we must say that this relation, in this third way of considering it, is indeed outside the subjects; but that, being neither a substance nor an accident, it must be a purely ideal thing, the consideration of which is nonetheless useful. (GP VII 401/L 704)

Given Leibniz's account of truth, according to which in every true proposition the predicate is contained in the subject, Russell took Leibniz to be saying that relational propositions could not express truths, since they are not in subject–predicate form, and should therefore

[3] I have given an extended treatment of Russell's doctrine in two long papers, "The Hegelian Roots of Russell's Critique of Leibniz" (Arthur 2018b), and an as-yet-unpublished paper I finished in early 2019, "Russell's Rejection of the Relational Theory: Leibniz, Lotze, and Internal Relations" (Arthur, unpublished).

be regarded as ideal. *L* may be regarded as having the accident 'greater than *M*', and *M* may have the accident 'greater than *L*', but there is no such thing as the relation abstracted from both. Russell assimilated this to Francis Bradley's neo-Hegelian doctrine that relations are never wholly external, and always inadequately represent the whole universe,[4] and to Hermann Lotze's doctrine that "every relation must be analysable into adjectives of the related terms" (Russell 1900, 46).[5]

When seen against the backdrop of Scholastic disputes about relations, however, Leibniz's text appears in a different light. The injunction against a relation being an accident of both its relata at once is reported by Thomas Aquinas, and (in the words of Sydney Penner) "can be found in a wide array of scholastics diverse in philosophical temperament," including not only Avicenna and Aquinas, but "Bonaventure, Scotus, Ockham, Suárez, and many others" (Penner 2016, 58).[6] That is, as Mugnai explains, it was upheld by both nominalists and realists of every stripe, all of whom agreed that there is nothing outside the mind corresponding to such polyadic relations. What was in dispute between them was not whether an accident such as 'greater than *M*' can truly be predicated of *L*, or 'is the father of Solomon' to David, but whether there is anything giving a foundation *in re* for *L*'s being greater than *M* or for David's being the father of Solomon. Thus according to the influential realist philosopher, Walter Burley (*c.*1275–1344), the foundation in the former case would be the *quantity* possessed by each of *L* and *M*, and in the latter case it would be the *generative power* in David by virtue of which he fathered Solomon. In opposition to the nominalist strictures of William of Ockham (*c.*1287–1347), who eschewed universals and made no reference to the foundations of relations, Burley held that a relation is a real form existing outside the mind and independently of the individual of which it is predicated. It is an accidental form inhering in its subject and entailing a reference to the other relatum (the "term"). So although for him a relation is predicated truly of an individual by virtue of its foundation, it inheres in only one individual, its subject.[7] In sum, if the "denial of relations" simply followed from upholding a subject–predicate logic, then the issue should have been moot throughout the history of logic: because an accident cannot be in more than one subject, (polyadic) relations cannot exist in reality, end of story. Instead, as is evident,

[4] "Nothing in the whole and in the end can be external, and everything less than the Universe is an abstraction from the whole, an abstraction more or less empty, and the more empty the less self-dependent. Relations and qualities are abstractions, and depend for their being always on a whole, a whole which they inadequately express, and which remains always less or more in the background." (Bradley 1893, 371)

[5] In his (1900, 13) Russell recognized that Leibniz did not neglect relational propositions, but concludes his discussion on a Lotzean note: "Thus relations and aggregates have only a mental truth; the true proposition is one ascribing a predicate to God and to all others who perceive the relation" (14). This tallies with what he wrote in January 1899 immediately after 'discovering' these doctrines in Leibniz: "Wherever there is really a relation, the attempt to find a corresponding predicate of each term will, at best, force us to admit a corresponding relation between the two predicates. To hold, therefore, as Lotze does, that a relation is not properly *between* terms, but consists in a pair of predicates of the terms, or, as he expresses it, in internal states of the related things, is to overlook entirely the nature of relations. For it will be necessary that these internal states are in turn somehow related to each other, and this necessity forces Lotze, as it forced Leibniz, to a *deus ex machina*, thinly veiled under a mathematical symbol, whose adjective, in some mysterious way, the correlation is supposed to be." (Russell 1989, 144-45).

[6] Both Mugnai (2012, 176) and Penner (2016, 57) quote Thomas Aquinas: "some people said, as Avicenna notes, that numerically the same relation is in both extremes. But that cannot be, since one accident is not in two subjects" (*In quatuor libros Sententiarum*, I, d. 27, q. 1, art. 1, ad 2).

[7] "Unlike modern logicians, Burley denies that a relation is a two-place predicate, and views it instead as a monadic function, arguing that like the other accidental forms, relation inheres in a single substrate and makes reference to another thing without inhering in it" (Conti 2016, §7).

the issue of whether there exist relations in reality was a live topic among logicians of various schools, all of whom subscribed to a subject–predicate logic, and none of whom denied the truth of relational statements.

Moreover, it is not the case that Leibniz was so beholden to syllogistic logic that he did not recognize its limitations; indeed, he made valiant attempts to construct an algebraic logic to form the basis not only of syllogistic logic, but asyllogistic inferences too. Thus Leibniz claimed that even the validity of what logicians call an "inversion of relation" could be demonstrated in this way by a change of its terms that would render the inference asyllogistic, citing as an example "David is the father of Solomon, therefore Solomon is the son of David"—a proposition that involves the very form of judgement that Russell said he "is unable to admit as valid" (1900, 13). Moreover, as Russell himself observed, even "judgments of subject and predicate are themselves relational," a fact that Leibniz knew very well, as it is embodied in his logic of containment (entailment), *de continente et contento*.

We may certainly grant Russell his point about the prejudice against polyadic relations stemming from a certain collusion between the Indo-European grammar of subjects and predicates and an ontology that recognizes only substances and accidents. But it must be recognized that not all subjects of propositions are substances; certainly the line *L* discussed above is not a substance, and where traditional philosophy would count, say, rocks and chairs as substances, for Leibniz these are mere *substantiata*, subjects of propositions, but not substances. And while predicates for him are contained in concepts, and all the predicates applicable to a substance are contained in its complete individual concept, he does not equate such predicates (which are items of thought) with the accidents of a substance or with the states that are its modifications *in re* at any of the instants of its existence.[8] So even scholars who agree that Leibniz intended to dispense with relations in his fundamental ontology reject any attempts like Russell's and Couturat's to derive his metaphysics from his logic[9]—attempts to "wring metaphysical blood from a logico-linguistic turnip," as Cover and O'Leary-Hawthorne put it (1999, 64).

A2.2 The Relational Properties Interpretation

Not all commentators agree that Leibniz intended to eliminate relations from his fundamental ontology, however. That claim has been vigorously contested above all by Hidé Ishiguro. In the first edition of her *Leibniz's Philosophy of Logic and Language* (1972a), she maintained that Leibniz's thesis of the ideality of relations is to be understood as applying to relations conceived in abstraction from their relata, as abstract entities, but as in no way undermining the genuine relatedness of things. Indeed, as she contends, Leibniz repeatedly stresses the connectedness of all substances in the universe, and all phenomena too. And on her reading, any such relation of connection is not for Leibniz a merely mental phenomenon (as it would be on the reading of Mates and others), but founded in the modifications of substances themselves. Thus although Leibniz denies

[8] Cf. Mates: "the accidents of a substance are not to be confused with the concepts (properties, attributes) under which it falls... The accidents of a monad are in the domain of reality, not in the region of ideas; they are 'ways of being' of substances" (Mates 1986, 197).

[9] Cf. Rescher ([1981] 2003, 79): "Leibniz is concerned to establish not the logical eliminability of relations but their metaphysical dispensability at the level of individual substances. It needs to be stressed that what is at issue is not a strictly logical doctrine, but a metaphysical one."

reality to relations as entities in themselves, according to her he does not thereby deny the reality of *relational facts*, such as the fact of two states of a substance occurring one after the other, or *relational properties*, such as the property of being the son of Adam. She cites such passages as the following, from Leibniz's letter to de Volder:

> ... time is no more nor less an ideal thing [*ens rationis*] than space is. To co-exist or to pre-exist is something real; though these should not be, I confess, as real as substances or matter are, as is widely assumed. (to De Volder, June 23, 1699: GP II 183/ L 519)

As Ishiguro interprets such texts, Leibniz's assertions about the ideality of relations apply only to space and time considered as entities, and do not apply to coexistence or pre-existence as relational facts, thus casting his theories of space and time in an altogether more realistic light. Moreover, Ishiguro observes, the texts cited by its proponents do not provide unambiguous support for the reducibility thesis. For it cannot be uncritically assumed that Leibniz understands the term 'relation' is the same way as we do. Leibniz's concept of 'predicate,' Ishiguro urges, is more grammatical than metaphysical (1990, 124), and cannot simply be assumed to exclude what we would now call "open sentences with one free variable"—i.e. propositional functions like xRa or $(\exists y)xRy$ (Ishiguro 1972a, 99). Rescher agrees with Ishiguro: "a property like $(\forall x)(\exists y)xRy$—having R to something—is inherently relational in its meaning content. Thus in resorting to it one has not eliminated relationality in its *logical* aspect" (Rescher [1981] 2003, 88).

In my (1985) I had openly embraced Ishiguro's interpretation of the ideality of relations, and argued further for the compatibility of this interpretation with Leibniz's understanding of the ideality of continuous quantities. In doing so, however, I also accepted Ishiguro's thesis that Leibniz recognized not only relational facts, but also *relational properties*. But the idea that Leibniz countenanced relational properties has been subjected to some withering criticisms by Mates, as well as by Cover and Hawthorne (1999). In between, Mugnai, in his thorough study of Leibniz's philosophy of relations in historical context (1992), found the relational properties interpretation inconsistent with the scholastic tradition in which Leibniz's work was embedded. But although Mugnai expresses substantial agreement with Mates's interpretation, he does not accept that Leibniz tried to eliminate relations, only to explain their mental character.[10]

Let us begin with Mates's criticisms. He objects that Leibniz, and many of his commentators too, have frequently ignored the distinctions "between a sentence and the proposition it is supposed to express and between a predicate and the corresponding property" (Mates 1986, 213). Thus "various formal features of the linguistic entities are uncritically projected onto the propositions and properties they are supposed to express" (213), with the result that commentators happily discuss whether given propositions are of "subject–predicate form" or "relational," and distinguish relational from non-relational properties, without giving any workable criterion for distinguishing them. As Mates makes clear, one of the main targets of these criticisms is the concept of a "relational property," relied on by Ishiguro, Hintikka, and Rescher—even though all these authors had also warned against reading off metaphysical conclusions from the linguistic surface.

Nevertheless, Mates assures us, despite Leibniz's own "ever-present carelessness about use and mention" (210), there is one thing we can say with confidence about his views in

[10] "Rather than seeking to *eliminate* relations by having recourse to logico-grammatical analyses using reduplicative operators (*eatenus, quatenus,* et *eo ipso*, etc.), Leibniz intends to illustrate the nature of relations as mental objects resolving relations of connection into implications..." (Mugnai 1992, 123).

this connection: he could not have held the view that every predicate (i.e. every open sentence with one free variable) expresses a property, since this would have rendered some of his best known principles and doctrines, such as the Identity of Indiscernibles, utterly trivial (214). For "if the individuals *A* and *B* are different, and if 'different from *B*' expresses a property, then that will be a property that *A* has and *B* lacks" (214), from which it follows trivially that no two different individuals *A* and *B* can have all their properties in common; and, therefore, that all identicals are indiscernible (135, 214, 250).[11] This is a problem for Ishiguro's interpretation, because there does not appear to be any good reason why the predicate 'different from *B*' should not express a relational property. It is a predicate of the form *xRb* (attributing the property *Rb* to some subject *x*), and Ishiguro explicitly says that such predicates ascribe relational properties to the subject *x*.[12]

There is a similar difficulty for the relational properties interpretation with another of Leibniz's principal doctrines, his denial of purely extrinsic denominations. For, as Mates points out, 'different from *B*' is a prime example of what Leibniz (following Scholastic usage) called an "extrinsic denomination," that is, something which is true of an individual by virtue of something extrinsic to that individual. Leibniz himself characteristically expresses the distinction between intrinsic and extrinsic denominations in terms of changes in the subject, extrinsic denominations being those "which arise and perish without any change in the subject itself, but only because a change comes about in something else.... This is how my similitude to someone else originates, it arises even without any change in me, solely by a change in the other person" (A VI 4a, 308/Mates 1986, 225). While acknowledging such extrinsic denominations, however, Leibniz denies that any of them are in fact purely extrinsic. To cite his favourite example, "no one becomes a widower in India by the death of his wife in Europe unless a real change occurs in him" (GP VII 321/L 365; Mates 1986, 225). But again, Mates observes, according to the relational properties interpretation, "the point could have been established simply by observing that after her demise the man has the property expressed by 'is a man whose wife has died,' which he lacked beforehand, and that therefore he has changed" (Mates 1986, 214). Similarly, in commenting on Socrates' astonishment (in the *Theaetetus*, 155c) at having become smaller than Theaetetus without having changed, Leibniz ought to have noted that it is quite obvious that Socrates now has the relational property 'smaller than Theaetetus' that he formerly lacked (215).[13] Yet Leibniz appears to mean a less trivial type of change than this, for he says that "place and position, quantity (number, proportion) are only relations, resulting from other things which *per se* constitute or terminate change,... [and] which need a foundation from the predicament of quality or an intrinsic accidental denomination" (C 9/Mates 1986, 223).

[11] This is the same objection that Russell had already made to Leibniz: "A differs from B, in the sense that they are different substances; but to be thus different is to have a relation to B. This relation must have a corresponding predicate of A. But since B does not differ from itself, B cannot have the same predicate. Hence A and B will differ as to predicates, contrary to the hypothesis" (Russell 1900, 58). Russell, however, took the same logic to entail also that there cannot be more than one substance at all.

[12] "Every open sentence with one free variable, which could express a true proposition when a name of an individual is put in the variable place, corresponds to a predicate of an individual. Some of the predicates which have logical forms like *xRa* or (∃*y*)*xRy* are grammatically monadic predicates but ascribe relational properties to the subject" (Ishiguro 1972a, 99). Cf. (Ishiguro 1990, 132).

[13] Plaisted (2002) seems quite unfazed by this criticism, apparently perfectly happy to have the Identity of Indiscernibles follow as an immediate consequence of the doctrine that there are no purely extrinsic denominations, interpreted as making every extrinsic denomination a relational property of the individual subject. See esp. pp. 11–12.

This criticism is supported by Mugnai, who shows that Leibniz's language here is consistent with the medieval tradition. That relations need to be based on a foundation—an accident or quality intrinsic to that of which they are predicated and which is not itself a relational accident—was insisted on by realists like Walter Burley, as we have seen.[14] The fact that a relational accident such as 'is the father of Solomon' alludes to another individual was not in question.

A2.3 Extrinsic Denominations: Mates's Interpretation

Now, as Mates observes, all Ishiguro's relational properties—'is the father of Solomon', 'is taller than Socrates', etc.—will be *extrinsic denominations* according to the standard meaning of this scholastic term, since in ascribing something to a given individual they always involve reference to at least one other subject extrinsic to the first. Accordingly, Mates proposes to interpret Leibniz's doctrine that "there are no purely extrinsic denominations" as the claim that all extrinsic denominations—that is, everything that Ishiguro regards as a relational property—"is reducible, in some sense of 'reducible,' to non-relational properties of that and other individuals" (219); or more precisely, that any proposition "I is a D," ascribing an extrinsic denomination D to an individual I, must follow from a set of propositions "I_i is a D_i," where each of the D_i is intrinsic, i.e. makes no "reference, via a name or quantified variable, to individuals other than the [I_i]" (218), and "the same accidents are the ground for applying the denominations D_i to the I_i, and the denomination D to I" (219).[15] Approving Mates's account, Cover and O'Leary-Hawthorne render it as "extrinsic denominations are relational properties or concepts, intrinsic denominations are monadic properties or concepts, and no individual falls under an irreducibly extrinsic denomination" (1999, 70–1).

Like Rescher before him and Mugnai after, Mates derives support for this analysis from his reading of Leibniz's attempts to reform language (Mates 1986, 174–83, 213–26). In the *Nachlass* there are a good many "curious entries" in which Leibniz offers what he terms "resolutions," "analyses", or "reductions" of various Latin sentences (180). Notably, these reductions are firmly in the tradition of Ockham's nominalism-conceptualism, where statements relating two individuals are parsed as combinations (not always conjunctions) of statements in subject–predicate form: for example, "'Paris is Helen's lover' is reduced to 'Paris loves and by that very fact (*eo ipso*) Helen is loved.' There are therefore two propositions collected into one compendious one" (C 286/Mates 1986, 179); and "'Peter is similar to Paul'... is reduced to the propositions 'Peter is *A* now' and 'Paul is *A* now'" (C 244/Mates 1986, 180). The correspondence with Ockham is close. Thus, as the Ockham scholar Paul Spade reports, the Venerable Inceptor argues in his *Summa of Logic,* II.11

[14] In his (2012, 175) Mugnai gives the following quotation from Burley: "a relation inheres in a substance through some more perfect accident only. And this accident through which the relation inheres in the subject is called the foundation of the relation" (Burley, *Expositio super librum Praed.*, fol. 29vb).

[15] Elaborating, Mates says that in effect he is "interpreting denominations in general as properties, and extrinsic denominations, perhaps, as what some recent commentators are calling 'relational properties'" (219). But Mates's definition of a denomination of an individual *I* as "*purely extrinsic* if and only if it is not reducible to intrinsic denominations of I_i" (219) seems to need emending, since what he says elsewhere indicates that it should be reducible to intrinsic denominations "of that and other individuals."

(as well as elsewhere) "that we can account for the truth of "Socrates is similar to Plato" without having to appeal to a relational entity or accidental form called "similarity":

> For example, for the truth of "Socrates is similar to Plato," it is required that Socrates have some quality and that Plato have a quality of the same species. Thus, from the very fact that Socrates is white and Plato is white, Socrates is similar to Plato and conversely. Likewise, if both are black, or hot, [then] they are similar without anything else added.[16]

Mates quotes a number of similar 'nominalistic rephrasings' that Leibniz gave, among them "*A is red* and *B is red* and therefore *A* is similar (in this respect) to *B*" (A VI 4, 944/Mates 1986, 181), and Leibniz's definition of similarity in the *Initia rerum*:

> *Similars* are things having the same quality. Hence, if two things are different, they can be distinguished only when they are co-present. (GM VII 19/L 667)

The example of similitude is particularly relevant since, as we saw, similitude is mentioned by Leibniz as an extrinsic denomination: Socrates could cease to be similar to Plato in respect of whiteness (for instance, if Plato got sunburned), without any apparent change in his own properties. Mates takes this to be a paradigm case of Leibnizian reduction of relations, stressing that the term "reduction" in such cases does not require the result to be logically equivalent to the original (as was required on the accounts of Broad (1975, 39) and Rescher (1967, 71 ff.)), but only for it to be entailed by it. Thus, with respect to the above example of Socrates and Theaetetus, the real intent of Leibniz's doctrine that there are no purely extrinsic denominations would be captured as follows: supposing that Theaetetus' being taller than Socrates is true by virtue of Theaetetus' being six feet tall and Socrates' being five feet tall, then the relational proposition "Theaetetus is taller than Socrates" is reducible to the two non-relational propositions "Theaetetus is six feet tall" and "Socrates is five feet tall," since the truth of the first is entailed by the truth of the second and third taken together, and is grounded in the same accidents (Mates 1986, 217–18).

Implicit in Mates's analysis is the idea that we can explicate Leibniz's doctrine of the ideality of relations without falling foul of any use-mention confusion. The restatement of Leibniz's views in terms of the reducibility of *propositions* allows us to avoid the whole question of whether there are any such things as relational properties. Thus, given Mates's reduction of the relational proposition "Theaetetus is taller than Socrates", the question of whether 'is taller than Socrates' expresses a relational property becomes otiose. Consequently, however, although Mates's analysis allows us to avoid the pitfalls of use-mention confusion, it does not so much resolve as sidestep the question of whether Leibniz thinks relational predicates are included in the complete concept of an individual, or whether there are real relational properties corresponding to such relational predicates. This leaves Mates with some salient difficulties. On the one hand he purports to have shown that Leibniz could not have regarded every relational predicate as ascribing a relational property to an individual. But in order to support his thesis that Leibniz held all relations to be unreal, Mates needs the much stronger claim that no relational predicate really ascribes a property to an individual. Yet, on the other hand, Mates himself reminds us that Leibniz's doctrine is not that there are no extrinsic denominations,[17] but that there are no *purely*

[16] This passage is quoted in translation from the passage in the SEP article on Ockham (Spade and Panaccio 2019, §4).

[17] —at least, not in his usual formulations; an exception is GP VII 321 (L 365, Mates 225), "because there are no extrinsic denominations," but this involves his usual example of the man becoming a widower in India on the death of his wife in Europe, and so probably does not indicate any difference in doctrine.

extrinsic denominations (219, n.34). Thus, accepting his claim that relational predicates such as 'being taller than Socrates' are extrinsic denominations, it would follow that Leibniz does not deny such relations, he only denies that there can be such a relation without some underlying accidents by virtue of which it is true. To use Leibniz's own words, he does not insist that no substance really has a relation to any other, but rather that "Every substantial thing, be it soul or body, has its own characteristic relation to every other, and the one must differ from the other by intrinsic denominations" (A VI 6, 110/Mates 226).

A2.4 Relational Accidents as Tropes

These difficulties are amplified by the fact that Mates has another, independent objection to the relational properties interpretation which pushes in quite a different direction from his reductionist reading. This is that in supposing that relational properties are something real possessed by the subjects, that interpretation confuses *properties*, which (for Mates) are only concepts that individuals fall under, with the underlying *accidents* of the individuals by virtue of which they have these properties, and on which relational truths about the individuals are founded. This criticism depends upon Mates's agreement with Kenneth Clatterbaugh's attribution to Leibniz of a doctrine of individual accidents, according to which "Leibniz considers the accidents of any individual substance to be themselves individual, that is, to belong to that substance only" (Clatterbaugh 1973, 65). Thus there is a crucial difference between properties—such as being strong, or being taller than Socrates, which are concepts that various individuals may fall under—and the particular conditions that individuals have at particular times, their accidents, by virtue of which they have these properties. Thus a relational property, such as 'is taller than Socrates', is a concept that Theaetetus and a great many other individuals fall under. But his falling under this concept follows immediately from the individual accidents possessed by him and by Socrates—whatever they might be—that make it true. Thus we do not have to posit the relational property 'is taller than Socrates' as a real entity in addition to these accidents.[18] Leibniz, on this reading, is not committed to the reality of relational properties.[19]

Consequently, no two individuals can share all their accidents, trivially, since no two individuals share any accidents. (These are what we now call 'tropes'.) But the principle of the Identity of Indiscernibles concerns not accidents but properties, and it is in principle possible for two individuals to share all these. That they cannot is a consequence of the fact that there are no purely extrinsic denominations. If *a* is different from *b*, or to the left of *b*, etc., it must be so because of a difference in the underlying accidents: e.g. in the case of place, their individual situations. The justification is in the connection of all things. Leibniz argues that because there are no changes at all that do not effect changes in every other substance in the universe, the change in Theaetetus' accidents will be reflected in a change of Socrates' individual accidents. Mates appeals to this doctrine of individual accidents to interpret a famous passage of Leibniz on relations:

[18] Mates notes, however, that Leibniz is not always consistent about the status of accidents, sometimes calling them "abstractions" and insisting that "only substances are concrete" (GP II 458/ Mates 1986, 171, n.6; and 196 ff.).

[19] As we shall see below, Dennis Plaisted manages to derive the opposite conclusion from the premise that relational accidents are individual, namely, that they are contained in their subjects as properties.

> You will not, I believe, admit an accident which is in two subjects at the same time. Thus concerning relations, I hold that paternity in David is one thing, and filiation in Solomon another, but the relation common to both is a merely mental thing, of which the modifications of the individuals are the foundation.
>
> (to Des Bosses, April 1714; GP II 486/LDB 327, Mates 211)

As Mates points out, in passages like this Leibniz refers to relations in two different ways. In one sense, a relation is supposed to be a kind of accident that is in two substances at once, "and since in reality there is no such thing as that kind of accident, relations in this sense are declared to be 'merely mental' or 'merely ideal'" (211). Binary relations are, as Leibniz says, "that according to which two things are thought of at the same time," "a relation is the *concogitatabilitas* of two things."[20] But "in a second sense, a relation is a legitimate kind of accident,... indicated by such terms as 'David's paternity' and 'Solomon's filiation,' or 'paternity-in-David' and 'filiation-in-Solomon.' When the word is used in this sense," Mates declares, "Leibniz has no quarrel with relations."[21]

This admission of Mates's, however, seems to me both perfectly correct and impossible to reconcile with his stated position that Leibniz thought all relations dispensable. It also seems to contradict his interpretation of Leibniz's doctrine that that there are no purely extrinsic denominations. For, according to his interpretation, David's paternity of Solomon, being an extrinsic denomination or "relational property" of David, must be reducible to (in the sense of being entailed by) non-relational properties grounded in the individual accidents of David. Yet he acknowledges that David's paternity is an individual accident of David (1986, 213), even though it is a relational one. Thus if Leibniz regarded some relations as individual accidents that he "has no quarrel with" (and therefore presumably do not need to be reduced), then he could not have held, as Mates claims he did, that all relations should be eliminated by recourse to statements ascribing only intrinsic (in the sense of non-relational) accidents to substances. Yet the case for interpreting Leibniz as holding that certain relations are individual accidents is strong.

One type of such relations that figure prominently in chapter 2 are the "relations of situation" he refers to in his exchange with Clarke, which, as he makes clear there, are individual accidents, not to be confused with the *places* and *spaces* abstracted from them. The latter are conceived by the mind as being purely *extrinsic*, with the result that we think that we can distinguish things by such extrinsic denominations. This is an illusion: "To be in a place is not a bare extrinsic denomination; indeed, there is no denomination so extrinsic that it does not have an intrinsic denomination for a foundation" (to De Volder, April 1702; GP II 240/ LDV 239). Places distinct from things are *abstracta*, and as such are perfectly similar to one another. Not so the situations, which are accidental denominations specific to the individuals in question.[22]

[20] The quotations are from A VI 4a, 28 and A VI 2, N. 58 (Mates 1986, 225).

[21] (Mates 1986, 211). Again on p. 213, Mates says that such relations as the friendship of Orestes towards Pylades, or David's paternity of Solomon, "are really individual accidents of the related individuals" (213); "David's paternity and Solomon's filiation (as distinguished from paternity and filiation in general) are the 'modifications of singulars' (i.e. of David and Solomon) on which the 'merely mental' relation is founded." (210, n.4). Mugnai agrees that Leibniz admits relational accidents; but on his reading, they "inhere in a given subject, but only as 'modes of being' of that subject; in the last analysis, what properly inheres is the accident on which the relational accident is 'grounded'" (1992, 117).

[22] I made this point earlier in my (1994b); it has since also been made at greater length by Dennis Plaisted in his (2002), esp. 36–9 and 70–4.

This, I propose, is entirely in keeping with what Leibniz himself said about nominalism and the reduction of abstract terms. In an unpublished draft written probably in 1688, but which may be taken as representative, Leibniz says:

> I see no other way of avoiding these difficulties than by considering *abstracta* not as real things (*res*) but as abbreviated ways of talking (*compendia loquendi*)—so that when I use the name *heat* it is not required that I should be making mention of some vague subject, but rather that I should be saying that something is hot—and to that extent I am a nominalist, at least provisionally.... Geometricians, too, do not use definitions of *abstracta* but reduce them to *concreta*: thus Euclid does not use his own definition of *ratio* but rather that in which he states when two quantities are said to have the same, greater, or lesser, ratio.
>
> (*De realitate accidentium*; A VI 4a, 996/Mates 1986, 171)

Notice how this agrees with what he says in the correspondence with Clarke almost thirty years later. We do not have to assume *abstracta* like ratio or place to be things: it is enough to define when two quantities have the same ratio or place, and this can be done without circularity in terms of other notions: in the case of place, I propose, in terms of situations.

But are situations themselves the intrinsic denominations that Leibniz claims are necessary? This has been argued at length by Dennis Plaisted in his book, where he claims that what are usually considered extrinsic denominations are in actual fact intrinsic.[23] The nub of his argument is that extrinsic denominations must be included in the complete concept of the individual in order to make sense of Leibniz's various claims.[24] We can readily grant this concerning extrinsic denominations as predicates, as Mugnai has argued in detail.[25] But the containment of a predicate in a subject-concept is not the same as the containment of a property (in the sense of trope) in the individual in question. Thus, as we have seen, Leibniz reduces the statement "David is the father of Solomon" to "David is a father and *eo ipso* Solomon is a son," where the relational accident of 'fatherhood' is attributed to David, and that of 'sonship' to Solomon. Plaisted asserts that this means that "for Leibniz, <paternity> and <sonship> are the very properties of the related individuals that provide the foundation for the common relation in this case" (2002, 11). But this does not fit Leibniz's way of talking about foundations, which, as we have seen, are "modifications" of individual substances from which relational accidents "result." Plaisted seems to be conflating relational accidents as denominations that may inhere in an individual's concept with accidents as tropes that are modifications of the individual *in re*.

We can illustrate this by reference to situations: the situation of a body can only be determined by reference to those bodies with respect to which it is situated: it is therefore a paradigm of an extrinsic denomination, since it alludes to such bodies. But if two

[23] Thus Plaisted interprets Leibniz's doctrine that there are no purely extrinsic denominations (NPE) as "the claim that there are no extrinsic denominations (where 'extrinsic denomination' is here understood as a relational accident) that are not in what they denominate. And so, because denominations that are in what they denominate are rightly counted as intrinsic denominations, NPE is basically the assertion that what are commonly considered extrinsic denominations are actually intrinsic denominations" (Plaisted 2002, 69).

[24] "There are no extrinsic denominations (which truly denominate that individual) that are not included in that individual's complete concept" (Plaisted 2002, 11).

[25] Cf. Mugnai's discussion in his (1992), where he argues that "the complete concept of each individual contains all the denominations, both intrinsic and extrinsic (i.e. relational)," and that in this sense relational properties may be said to *inhere in* the complete concept of a given individual (133).

bodies share "the same situation" they do not share individually the same situation (situation-as-trope), since this depends on the individual accidents of each body. Here Leibniz uses the language of the realists: the situation of each (as trope) must result from the different qualities intrinsic to each, that is, from a foundation consisting in "the modifications of the individuals." They "demand a foundation from the category of quality," as he explains in "On the Principle of Indiscernibles," and this foundation consists in the degree of expressions in the substances in the bodies in question:

> To be in a place seems, abstractly at any rate, to imply nothing but position. But in actuality, that which has a place ought to express place in itself; so that distance and the degree of distance involves also a degree of expressing in itself the remote thing, either a degree of affecting it or of receiving an affection from it. So, in fact, situation really involves a degree of expressions. (C 9/Leibniz 1995b, 133)

Moreover, Plaisted's reading, like the relational properties interpretation, does not sit well with how Leibniz describes changes of denominations. In this case he has a distinctive position that differs from both the realists and the nominalists. As we saw already, Leibniz claimed that no extrinsic denomination may change without there being a change in its intrinsic denomination, which is something not even realists about relations would have asserted. No change would occur, for example, in the man in India whose wife had died without him knowing; a realist like Burley would say that there was a real change in the wife brought about by her death (a definite change in her intrinsic qualities!), but that there would be no corresponding foundation in the man for a change of relation in him, despite the change of extrinsic denomination by which the relational predicate 'is a widower' would become applicable to him. Similarly, in a passage quoted by Mugnai, Leibniz claims that "in virtue of the universal connection of things, the Emperor of China as known by me differs in intrinsic qualities from how he was when not yet known by me."[26] This presents a grave difficulty for Plaisted's interpretation, since he interprets each relational accident as being itself the foundation for its being truly attributed to its subject. It makes little sense to claim that "being known by me" should be identified with any intrinsic qualities of the Emperor of China, nor that this would become a property of his because of the "interconnection of things."

Leibniz appealed to the interconnection of things in the course of sketching his theory of time, and I give an explicit rendering of it in my exposition of it in §1.1. As applied to this case, it would go as follows: all simultaneous states reflect one another, and my earlier state, reflecting the Emperor of China as yet unknown by me, in containing the reason for my later state when I know him, also contains the reason for his later state when I know him. So his later state differs from his earlier one in that the earlier one reflects a state containing the reason for the later while the later one does not. Such a connection with the theory of time is consistent with Leibniz's further remark: "Further, there is no doubt that each thing undergoes a change at the very same time, and time is needed in order for the Emperor, once not known by me, to become known by me."[27] On this analysis, *the intrinsic quality or modification that changes is the state of each substance*, together with the reasons it contains. Thus the Emperor's change of state is the foundation for the relation 'is known by me' to

[26] In his (1992, 53) Mugnai quotes these passages from the *Vorausedition*, Ve 1086. He gives the original Latin in Mugnai (2012, 203).

[27] In drawing the distinction between relations of comparison and those of connection in one text from 1688–90, Leibniz writes: "Relations between two things seem to be either only rational or real, that is, relations of Essence or of Existence. Real relations are those either of Position (namely of time and situation), or of Influx; since when some change is made through one thing, it is even impeded in something" (A VI 4a, 944).

become truly applicable to him at the later time, and thus to be included (at least implicitly) in his complete individual concept. Nevertheless, the reason for this change is expressed much more distinctly in Leibniz's state, in which the grounds for the change in his knowledge are to be found.

The same account is applicable to Leibniz's account of motion. As discussed in §3.3, motion for Leibniz is essentially relative, and is therefore an extrinsic denomination, since it necessarily makes reference to other bodies relative to which it is moving. But it is not a purely extrinsic denomination. There must be a subject of motion in order for a true motion to be distinguished from a merely apparent one, and this will be that body in which the cause of the change of situation is most intelligibly located. As he wrote to Clarke, "when the immediate cause of the change is in the body, that body is truly in motion, and then the situation of other bodies with respect to it will be changed consequently, though the cause of that change be not in them" (Fifth Paper, §53; GP VII 404/LC 74).

A2.5 Leibniz's Linguistic Rephrasings

The foregoing interpretation of relational accidents is in keeping with the realists' demand that they should have a foundation *in re*. But it is in tension with the Ockhamist tenor of Leibniz's "reductions" of relational statements noted above. These generally involve an analysis of a relational statement into a connected pair of statements each assigning a monadic predicate to the subject in question. In keeping with this, Mates, Cover and O'Leary-Hawthorne see Leibniz as intending to reduce all relational accidents to monadic properties of substances. In his (1992) Mugnai also gives an answer similar to Mates's, based on Leibniz's "nominalistic rephrasings": sentences containing *symmetrical* relations are reduced to pairs of sentences each of which ascribes a predicate which does not denote a relational accident; sentences containing *asymmetrical relations*, to pairs of sentences each of which ascribes a predicate denoting a relational accident, and these "in their turn, are 'modes of being' generated by the simultaneous existence of the subjects to which they refer, with their fundamental properties" (1992, 103). Thus, when Leibniz says in his discussion in "On the Principle of Indiscernibles" (from which I already quoted above),

> And in the universe, place and position, quantity (number and proportion) are only relations, resulting from other things which per se constitute or terminate change.... Considering the matter more accurately, I saw that they are only mere results, which themselves do not constitute any intrinsic denomination, and thus are only relations which need a fundament from the predicament of quality or an intrinsic accidental denomination. (C 9/Mates 1986, 223)

Mates interprets this as a call for the reduction of all relations—not just abstract ones such as place or quantity, but relational accidents too—to 'non-relational' qualities. As we have seen, this is assimilated by him to a notion of reduction according to which Leibniz would analyse "Theaetetus is taller than Socrates" into, for instance, "Theaetetus is 6 feet tall" and "Socrates is 5 feet tall." But the trouble with this example is not just that it attempts to solve an ontological issue by linguistic means alone: it is that the kind of reduction envisaged by Mates and others on the model of Leibniz's linguistic parsings does not match the kind of nominalism he professes. For *lengths and heights are not absolutely intrinsic qualities of things, any more than are places and ratios.*

This is a fundamental criticism of Mates's whole position, and of the relation-reduction interpretation generally. For although Mates proposes that Leibniz's states be regarded as composites of simple properties or their negations,[28] he gives no examples of what such simple properties would be. It is widely supposed that these monadic or "non-relational" properties would be the primitives of the "alphabet of human thought" he dreamed of in his youth. But those were conceived by Leibniz as the divine attributes, assuming they could be identified.[29] Yet in all the examples of linguistic rephrasings he offers, the reduced properties are such as having a specific colour, height or length—and none of these is primitive, according to Leibniz. Thus all *quantities* such as height and length presuppose another object as a standard, so that for him they all involve relations of comparison, and are thereby *extrinsic*; nor is colour an intrinsic quality, but a secondary quality (which, by making implicit reference to an observer, is thereby necessarily extrinsic). On the other hand, according to Leibniz's "provisional" profession of nominalism, one would expect that a statement ascribing a certain height to an individual, as if it were an absolute property, would be treated syncategorematically, to be circumvented by locutions in which it is stated when two concrete individuals will have the same, greater, or lesser, height. Thus Mates's reductions appear to go in precisely the opposite direction to those extolled by Leibniz. This conclusion, note, applies whether we are considering a symmetrical relation, such as 'is the same height as', or an asymmetrical one, such as 'is taller than'. It would therefore appear that Leibniz's linguistic parsings of relational statements do not map onto his nominalistic commitments in the way that many of his modern commentators have supposed.

A comparison with Locke's position will help solidify this point. Unlike the majority of Leibniz's contemporaries, Locke held a nominalist position, according to which relations are mere "marks" which "lead the thoughts beyond the subject... to something distinct from it." In support he offered the (traditional) argument that relations can change without any corresponding change in the subject: "Titius, whom I consider today as a father, ceases to be so tomorrow, only by the death of his son, without any alteration made in himself" (A VI 6, 227/RB 227).[30] Thus for Locke the fact that relative terms qualify the subject means only that they "necessarily lead it to other ideas than are supposed to really exist in that thing," but they have no reality beyond the subject. Absolute terms, like 'black', do not so lead to ideas other than those "supposed to really exist in the thing." Leibniz will have none of this. The term 'black' appears to be absolute only if we stay within the limits of everyday knowledge. But, he insists, in fact "there is no term which is so absolute or so detached that it does not involve relations and is not such that a complete analysis of it would lead to other things and indeed to all other things" (A VI 6, 228) "Consequently," he adds, "we can say that 'relative terms' *explicitly* indicate the relationship they contain"—leaving unspoken the corollary, that absolute terms do so *implicitly*.[31]

Moreover, Leibniz himself never to the best of my knowledge claimed that the examples of reductions he gives of relational statements are also supposed to be exemplars of how to reduce or eliminate extrinsic denominations, nor do we need to assume that they are. This

[28] "We shall suppose that a possible state of a possible individual is a composite 'complete' property, in which each simple property P or its complement $\bar{P}$, but not both, is a factor" (pp. 8–81). (One wonders here what has become of Mates's careful distinction between property and predicate.)

[29] By the time of his *Introduction to a Secret Encyclopedia* of1679 (A VI 4, 528), Leibniz had ceded that the project of identifying such primitives was beyond human ken, and that even if we found them, we would not be able to tell; so he contented himself with taking what were primitive relative to any given inference. See Arthur (2014, ch. 2) for discussion.

[30] Locke's text has 'Caius' for 'Titius'. (Locke 1690, 291/ 1997, 290). As Mugnai reports, this is a stock example in the tradition; see his (2012) for examples.

[31] Leibniz makes that corollary explicit elsewhere: "I find among notions no predicates wholly absolute, or not involving connection with others" (GP II 249/Russell 1900, 212).

is put into particularly sharp relief in the case of Leibniz's analysis of "*A* is simultaneous with *B*" as "*A* exists today" and "*B* exists today." This fits Mugnai's characterization of the way Leibniz deals with symmetric relations in his linguistic parsings. But "existing today" is hardly an "absolute property" or "intrinsic foundation" of the individuals in question. It rather seems to contain implicit reference to a particular time ("today"), and therefore to be an extrinsic denomination. Moreover, Leibniz's way of treating times in expressions such as "is at the same time as" is—consistently with his commitment to a provisional nominalism—to reduce "*A* is simultaneous with *B*" to relations of connection between states of the substances *A* and *B*, with each state representing or involving the other. Similarly, he does not reduce "*a* is earlier than *b*" by proposing that it be replaced by "*a* is earlier and *eo ipso b* is later," with "earlier" and "later" denoting relational accidents of substances, to be expounded in terms of some mysterious monadic modifications of these substances. Rather *a* and *b* are themselves modifications of substances, namely their states.[32] Thus, as Mugnai acknowledges, it is Leibniz's ontology of substances with states that are representational which requires him to say with the realists that there is no change in an extrinsic denomination without a change in the foundation, since the change of relations is necessarily reflected or represented in the state, however obscurely. This, I believe, is consistent with what Leibniz did explicitly say about the realist notion of foundations in the passage in *De realitate accidentium* from which we quoted already, where he expresses his provisional nominalism:

> But if we should deny any reality in accidents, as if they are nothing but relations, we will again become stuck. For a relation, since it results from a state of things, will never arise or perish except by some change made in its foundation.... I may say therefore that a substance changes, or has different attributes at different times, for this does not put into doubt whether some reality arises or perishes in the change; and there is no need to raise the issue whether there are various realities in a substance that are the foundations of its various predicates (though, indeed, if it *is* raised, it is difficult to adjudicate). It suffices to posit only substances as real things [*res*] and to assert truths about these. (A VI 4, 996/Mates 1986, 171)

Thus, I propose, Leibniz's reluctance to commit himself about what properties might be the fundaments of relational accidents is a symptom of the fact that it is an issue he does not need to resolve. And he does not need to resolve it is because he has provided a way of sidestepping the issue, by introducing his own idea of what is intrinsic to a substance: namely a state, or perception, which inherently reflects or expresses all other simultaneous existents, and contains the reasons for those succeeding it. This conclusion accords with Mugnai's in his (2012), where he grants that the states of substances, being representational, can themselves be regarded as the modifications of substances:

> Thus, the picture that Leibniz offers of the world is that of an aggregate of individual substances, each with its intrinsic modifications, which do not have any physical contact

[32] As I argue in chapter 1, §1.1, Leibniz's own treatment of time also tells against another interpretation of what he meant by intrinsic modifications offered by Mugnai in his (1992). For there he adopts Rescher's view that there is an ontological difference for Leibniz between relations denominated of one individual substance (*intramonadic* relations) and those relating different substances (*intermonadic* relations). On this basis Mugnai suggests that 'the love of Paris for Helen' can be considered "an absolute modification of Paris, insofar as it does not go beyond the limits of Paris's soul" (124–5). But this distinction between intramonadic and intermonadic relations does not seem well founded in Leibniz's explicit treatment of time, where relations between states of different substances are not treated any differently than those of the same substance. See also my (1985) and (1994b).

> or influence on one another. These modifications mainly consist of representations and in a continuous passage from representation to representation. Relations simply result from the existence of the individual substances with their modifications...
>
> (Mugnai 2012, 205)

A2.6 Real Relations and the Divine Understanding

The crucial question here, of course is how relations are supposed to "result" from substances and their modifications. Leibniz's talk of *concogitabilitas*, being thought together at the same time, seems to indicate that a relation (in the sense of what is "neither a substance nor an accident") is merely mental. This is how Russell interpreted him, assimilating his view to Kant's and Lotze's, so that space and time would be things of the mind only. On such a view, a binary relation would be produced in the mind on the occasion of someone's comparing the two items supposed to enter into the relation. Leibniz seems to have allowed this for what he termed "relations of comparison," such as similarity. He wanted more than this, however, for "relations of connection," those which do indeed serve to connect the individuals concerned. It is all very well for a relation of situation, for example, to occur to someone perceiving or considering things existing at the same time. But Leibniz understood a situation as that mode by which anyone would necessarily understand things existing at the same time: it is not something derived from anyone's perceptions, but a condition for their possibility (it is thus a transcendental condition, as De Risi has observed.) As such it has to have a foundation that does not depend on any individual perceiver. Leibniz was very clear about this. In a passage from his mature years quoted by both Mates and Mugnai he wrote:

> Besides substances, or ultimate subjects, there are the modifications of the substances, which can be produced and destroyed in their own right. And finally there are also relations, which are not produced in their own right but result when other things are produced, and have reality without regard to our understanding—they are truly there when no one is thinking them. For their reality comes from the divine understanding, without which nothing would be true. Thus there are two things that are realized only through the divine understanding: all the eternal truths and, of the contingent ones, those that are respective. ("On Temmik"; Mates 1986, 224; Mugnai 1992, 154–64)

This idea of a foundation for relations in the divine understanding is not just a view Leibniz happened upon in trying to tackle the problem of relations, but was his consistent view from about 1676 onwards, at least in the case of space and time. I have charted this in detail in chapters 1 and 2, so I shall restrict myself here to supplying some supplementary commentary. Thus in a fragment from mid-1685 Leibniz writes

> Now the root of time is in the first cause, potentially containing in itself the successions of things, which makes everything either simultaneous, earlier or later. It is the same with place, for the first cause makes everything have some distance. Therefore whatever is real in space and time consists in God comprising everything. (A VI 4, 629/LLC 275)

and in a marginal note in the *Specimen Inventorum* (*c*.1686–89):

> In order to understand the nature of time it is essential for us to consider change, that is, the contradictory predicates of the same thing in a different respect, which respect is nothing but the consideration of time. Space and time are not things, but real relations.
>
> (A VI 4, 1621/LLC 313)

This doctrine of real relations is not to be confused with the scholastic notion of "real accidents," which Leibniz firmly rejected. Where adherents of that doctrine accorded that status to extension, so that extension (and other qualities depending on it) could exist without a substance to inhere in (as in the Thomist understanding of the miracle of transubstantiation in the Eucharist), for Leibniz extension and duration are accidents requiring a substance in which to exist. Space and time, on the other hand, are not existents; but even so, they have reality, which they derive from the divine understanding. They are "orders of existing," ordering possible existents as well as actual existents, without existing themselves.[33] Although they are neither substances nor attributes, they should not be regarded as a third class of existing entities, à la Gassendi.

Leibniz gives one of his fullest accounts of the doctrine of real relations in a fragment written on his journey to Italy in about 1689:

> Time and place, or, duration and space, are real relations, i.e. orders of existing. Their foundation in reality is divine magnitude, to wit, eternity and immensity. For if to space or magnitude is added appetite, or, what comes to the same thing, endeavour, and consequently action too, already something substantial is introduced, which is in nothing else than in God, or the primary unity. That is to say, real space in itself is something that is one, indivisible, immutable; and it contains not only existences but also possibilities, since in itself, with appetite removed, it is indifferent to different ways of being dissected. But if appetite is added to space, it makes existing substances, and thus matter, i.e. the aggregate of infinite unities. (A VI 4, 1641/LLC 335)

Thus time and space are ideal entities, encompassing all possibilities of what might exist as well as what does. But they are also real relations or orders of existing, whose reality derives from divine eternity and immensity, respectively. This does not make them actual existents, however; that requires the introduction of endeavour or appetition, which differentiates primary matter into existing substances and their successions of states. Thus the relations we perceive, although constructed in our perception, are founded in the divine understanding.

In the case of time, writing to the Electress Sophie, Leibniz distinguishes duration from time by comparing duration with matter. Each is an aggregate of presupposed entities: as "matter is an aggregate of innumerable simple substances," so "the duration of created things, like actual motion, consists in a cluster (*amas*) of momentaneous states" (to Sophie; GP VII 562; October 31, 1705). Time, on the other hand, being a continuous quantity, consists in a "uniformly regulated continuity": "even though it is nothing but a supposition and an abstraction, it forms the basis of eternal truths and the sciences of the necessary [*des sciences necessaires*]" (GP VII 564). In fact,

> It is only a principle of relations [*rapports*], a foundation of the order of things insofar as one conceives their successive existence, or, without which they would exist simultaneously [*ensemble*]. It must be the same with space. That is the foundation of the relation [*rapport*] of the order of things insofar as they exist together [*ensemble*]. Each of these foundations is true [*veritable*], even though it is ideal. (GP VII 564)

[33] Leibniz records Wilkins' definition of order in his notes under "Transcendental Mixed Relations Pertaining to Discrete Quantity": "An order is a relation according to earlier and later" (A VI 4a, 33). There is also a beautiful short note from perhaps around 1695, transcribed for me by Osvaldo Ottaviani: 'L'*ordre* est le rapport d'une variete de rapports, qui naissent d'une multitude de terms ou ingrediens. De sorte que par ce moyen on peut distinguer chacun de ces ingrediens de tous les autres' (LH 4 3, 5e, Bl. 20); which might be translated: '*Order* is the relation of a variety of relations which originates from a multitude of terms or ingredients; so that by this means each of these ingredients can be distinguished from all the others.'

In the case of space, place is an equivalence class (*avant la lettre*) of situations: something is in the same place as something else if the second thing would have all the same relations of situation to other bodies as does the first. However, in talking about the "same situation" we are indulging in some abstraction from particulars, since situations are individual accidents or tropes: each body has its own particular situation, and such a situation is a *relational accident* of that body.

Now propositions ascribing such relational accidents to a body or embodied substance do not need to be reduced to ones involving only non-relational predicates, since they are already in subject–predicate form. If a relational predicate truly applies to its subject, it does so because it is entailed by the concept of the subject. On the other hand, the relational predicates are not intrinsic denominations, as Plaisted asserts they are, since they refer to other bodies. A body's relation of situation is a paradigm example of an *extrinsic denomination*: it only has this accident by virtue of the coexisting bodies to which it is situated. But *there are no purely extrinsic denominations*: something in the body must account for, or be the foundation of, this denomination. What is in a body are its *constitutive unities*, and these are *simple substances*. And what are in these simple substances are *states* or *representations, perceptions*. In short, the foundations of a body's relation of situation lie in the states of the simple substances it contains, in that the bodies to which it is situated are so represented in these states. The state is the modification by virtue of which the situation is truly denominated of the individual embodied substance. Although the situation is an extrinsic denomination, it is included in the complete concept of the individual substance because the complete concept encodes or includes all that would happen to the substance (i.e. all its accidents) at every stage of its existence; this corresponds to the fact that the relation of situation of an embodied substance to coexisting substances derives from the mutual representation of these substances in each other's states. This is how relations such as situation *result* from the positing of simple substances, and do not need to be separately posited.

The doctrine of real relations, finally, explains how Leibniz is able to combine a nominalist-conceptualist position according to which all relations are mental, that is, supplied by the mind in perception, with an underlying realism about the spatial and temporal orders based on the divine understanding: in our veridical perceptions the reality of these relations are realizations of truths in the divine understanding.

> ...time is no more nor less an ideal thing [*ens rationis*] than space is. To co-exist or to pre-exist is something real; though these should not be, I confess, as real as substances or matter are, as is widely assumed. (to de Volder, June 23, 1699: GP II 183/ L 519)

When Leibniz says here that relations of succession are 'something real', he is not abandoning his thesis of the ideality of relations, nor is he abandoning his nominalism, according to which the only things that God has to create are the *res permanentes* and their modifications, from which all else will result. But the order of succession that results from the ordering of states by reason-inclusion, together with the appetition taking one into the next, gives an order corresponding to that existing in the divine understanding, which therefore accounts for the reality of the order of succession.

In sum, there are several senses in which Leibniz understands "relation," and these have to be kept distinct. A relation in distinction from its relata is not an existing thing, but merely denotes what can be thought together, a *concogibilitas*. It is an ideal thing, or *ens rationis*. There is also relation in the sense of a relational accident, which is true of an individual substance by virtue of that substance's modifications. These modifications are states or perceptions, within which the relation may be represented. Insofar as these states encode the relation, they may be said to constitute an order or real relation. This corresponds to what is in the divine understanding, and accounts for the reality or independence of the relation from human thought. The spatial relations of embodied substances exist objectively (in our sense of that word) because they are a necessary, implicit consequence of the relatedness of monads encoded in their states. The same goes for the causal order and the temporal order, which are founded in reason-inclusion. All these orders are therefore *real relations*.

APPENDIX 3

Excerpts from Leibniz's Writings on *analysis situs*

Since so little of Leibniz's work on *analysis situs* exists in English translation, I give here some representative samples: the last four (philosophically important) paragraphs of the 1679 *Characteristica geometrica*, six subsequent manuscripts in full, and finally some short extracts from three late manuscripts.

A3.1 *Geometrical Characteristic*—Final Paragraphs[1] 10 August, 1679

(104) Moreover, all the common definitions of the straight line are not yet sufficiently perfect, in that, since there is still always a need for a demonstration, it can be doubted that such a straight line is possible. It seems that this, however, should be supposed to be among the simplest things, and so there is a need for a definition that makes it immediately evident that a straight line is possible. If you define it as the shortest [*minimum*], it still remains in doubt whether there is a shortest [path] from one point to another. If you define it as that whose points cannot be further separated, you are presupposing distance ôr a minimum path. For to be separated is to acquire greater distance. Equally, it is one thing to exhibit a material straight line, ôr to draw it, and another to grasp it in thought.[2]

(105) When two points are perceived simultaneously [*simul*], by that very fact their situation to one another is perceived. But any two situations between two points are similar, and so they can be distinguished solely by co-perception ôr magnitude. When two points are perceived simultaneously, therefore, what distinguishes situation is the magnitude of a certain extensum, by the very fact that some extensum is perceived.

(106) A **straight line** is an extensum which is perceived by the very fact that two points are perceived simultaneously.

(107) When one point is perceived, and then another is perceived separately, no variety can be observed.

(108) If *A.B* are perceived simultaneously, and in turn *C.D* are perceived simultaneously, there is a distinction of situation. But what is perceived when *A.B* are perceived and what is perceived when *C.D* are perceived, are similar: for it is clear that nothing can be observed in them individually which will not be observed in both. Thus if what is perceived when *A.B*

[1] *Characteristica Geometrica*. From LH XXXV, I, 11, Bl. 1–16. These paragraphs, numbered (by Leibniz) §§104–8 and enclosed in brackets, were not included by Gerhardt in his edition (GM V 141–68). As Echeverría has noted (CG 226, n.77), in omitting them Gerhardt was probably following to the letter Leibniz's own instruction in the margin: "*inclusa omittantur vel aliorsum referantur* [the enclosed are to be omitted, or assigned to another context]." He did not cancel them however, and there is a clear continuity of content with the preceding paragraphs, to which they add philosophical clarification.

[2] In what follows I am translating *recta* as 'straight line', and *linea* as 'line', whether straight or curved.

are perceived and what is perceived when *C.D* are perceived can be distinguished from each other, they are distinguished solely by magnitude. Thus when *A.B* are simultaneously perceived they are simultaneously perceived as having some magnitude. When two things are perceived to exist simultaneously in space, by that very fact a path is perceived from one to the other. And since they are congruent, by that very fact is conceived the path of one into the place of the other.[3] Now, two points are congruent to one another. So what is perceived when two points are simultaneously perceived is thus a Line, that is, the path of a point. It is also clear that this path is such that whether it goes from *A* to *B* or from *B* to *A*, the place in which *A* will be, will be congruent to that of *B*, and will be congruently disposed; that is, the place of the point coming from *A* will have the same relation to *A* as will the place of the point coming from *B* to *B*; what is more, the place of the point coming from *A* will have the same relation to *B* as will the place of the corresponding point coming from *B* to *A*. When we conceive two points as simultaneously existing, and ask why we would say they are simultaneously existing, we will think that they are simultaneously perceived, or at least that they can be simultaneously perceived. When we perceive something as existing, by that very fact we perceive it to be in space, that is, that there can exist indefinitely many other things absolutely indiscernible from it. Or, what is the same thing, we perceive that it can move, ôr that it can be first in one place and then in another, and because it cannot be in several places at the same time nor move in an instant, we thereby perceive this place as a continuum. But since it is still indefinite what direction it moves in, for it can be moved in many ways that are mutually indiscernible from one another, it follows that the soul is determined to some certain motion if we suppose another thing beyond it, congruent to the first, for by that very fact one is thought to be able to come into the place of the other. And since there are many ways this can happen, only one is determined, for the consideration of which nothing needs to be assumed except the positing of these two things. That is, when two congruent things are posited in space, a path is posited from one to the other, connecting them. The simplest position, however, is that of a point. For even if you posit something else, nevertheless, since from among the diverse motions from one to the other, only one is determined by which the corresponding points move determinately towards one another, it is clear that the soul will finally fall into the consideration of two points, it being self-evident that these are congruent to each other.[4]

An **extensum** is a continuum. In an extensum parts can be made. In an extensum parts can be made which exist simultaneously. In an extensum parts can be made in infinitely many ways. A part of an extensum is an extensum. In one extensum there can exist many extensa. In one extensum there can exist infinitely many extensa. In one extensum there can exist infinitely many similar extensa. In one extensum there can exist infinitely many congruent extensa. If something exists in an extensum and is congruent with it, it coincides with it. If something exists in an extensum and is not congruent with it, an infinity of things can exist in the same extensum which do not coincide with the first, but are congruent with it.[5] Two moveable things which are assumed in an extensum can be congruent with each other, that is, at different times they can be so located that the earlier state can not be

[3] Here I have followed the change made to Echeverría's transcription by De Risi (2007, 411, fn.87).

[4] Here two passages are cancelled: "Thus I think first: a point *A* can exist per se. A point exists in space. A point coexists with other points. A point can move. A point can coexist with other points. A point..."; then "A point is the simplest thing in extension [*extensum*]. One point cannot be discerned from another only if the two points cannot be discerned from one another taken in turn. Therefore another point can also exist by itself. Another point exists. Any *Extensum* exists."

[5] Here Leibniz has crossed out the following: "If something exists in a bounded extensum, it is successively congruent to it, that is to say, it exists in a bounded extensum to which something

discerned from the later. The locus of an extensum is itself extended. The locus of an extensum is congruent with an extensum. Locus is immovable.

A3.2 *A New Characteristic for Expressing Situation and Motion*[6] September 1679

In order to perfect the new Characteristic for expressing Situation and Motion that I have devised, certain things should be more distinctly expressed, and some should also be demonstrated.

Every point is congruent to each other point. That is, $A \cong B$, ôr $Y \cong Z$.[7]

Therefore every point is congruent with one and the same point. That is, $Y \cong A$. Because one can substitute for *Z*, that is, any point whatever, the proposed point *A*.

Any two points whatever have a **situation** to one another, that is to say, there is a path from one to the other. The situation of the first point *A* to the second point *B*, I denote *A.B*. This is the same as saying that the two proposed points can be understood as the extrema of a certain extensum connecting them.

Hence it is clear that all points can be understood to be in one and the same extensum, which is the unbounded ôr absolute extensum, namely space; in which space, that is, there is nothing but extension, and therefore other things can be understood as having an assignable situation in it, but it itself cannot be understood as situated anywhere. A point, on the contrary, is that in which nothing can be understood to be situated, ôr a point is the simplest thing that can be understood in the extensum.

Since therefore all the points in the universe are congruent with one another, and all possible points are in unbounded space, it follows that unbounded space will be the locus of all points congruent to a given point, that is, the locus of all points taken absolutely. To express this symbolically [*speciose*], let there be a congruence $Y \cong A$, and the locus of all the *Y* will be unbounded space.

We have said that when two points are proposed there is some situation between them. For I define **situation** in such a way that it is a relation of two things in respect of extension, determined by their co-existence. But the relation which is determined solely from the nature of the two things in respect of their magnitude, that is, their co-essence, is their **ratio**.

And it cannot be doubted that something is determined with respect to them, for among all those things assumed determinately, for example, when two extensa are assumed, and according to something determined in each, namely magnitude, or coexistence ôr position, or something else when they are considered in a determinate mode, for example, in the simplest mode of all, it is necessary that some determinate consideration emerges, whose object is that which I have called relation.

Or, more briefly and clearly: since two things exist now, they also co-exist with one another, and since they exist in a determinate mode, they also co-exist with one another in a determinate mode; and since this mode is absolutely determinate, it will also be determinate

congruent can be assumed in another movable extensum to which another thing congruent to it in the movable extensum cannot co-exist in the same time. If however many things [exist] in the same immobile extensum…"

[6] LH XXXV, vol. 1, No. 11, f. 48–9; Echeverría's transcription (CG 246–55). I give a slight variant of his suggested title: *Caractéristique nouvelle pour exprimer la position et le movement.*

[7] As I noted in §2.3, Leibniz uses the symbol 'γ' in his 1679 manuscripts for 'is congruent to'. '$\cong$' is my approximation to the symbol for congruence he uses later.

with respect to an extension determinately assumed. But this mode is situation, therefore the situation is determinate.

Hence, when nothing can be understood in points but their determinate mode of existing with respect to extension, ôr **position** (for **situation** is as it were com-position of many things), it is clear that in two co-existing points nothing determinate of either their co-essence or their co-existence can be understood but the situation between them. For they do not have any other point of comparison since all [points] are congruent to one another. Thus by writing *A.B*, we express the relation of the point *A* to the point *B*, which cannot be usefully understood except as their situation. Therefore *A.B* will be the mutual situation of the points *A* and *B*.

When two points are proposed in one and the same unbounded space, they will also be in some one bounded space. For since their coexistence is determinate, it also suffices for the space in which they exist to be bounded, and it is evident that it can be understood as their coexistence, even if certain things existing in that unbounded space are not considered. Whence it can be understood that any two points whatever are connected by some extensum, namely, one through which it can continuously pass from one to the other—that is to say, there is a path from one to the other.

The situation of different points to one another can differ, ôr *A.B* can differ from *C.D*. That is, given two points having the situation *A.B* to each other, there can be two others having a different situation to each other, *Y.Z*. For otherwise it would be necessary for all points to coincide with each other, that is to say, all points would be one and the same. For the fact that one point *A* does not coincide with another *C* cannot be demonstrated in any other way than that there is some other point *B* with respect to which they have a different situation, so that *A.B* non ≅ *C.B*.

The situation of one point to another can change. This is clear from the preceding. For the situation of a second point can be different from the first, therefore also the situation of the first can be different than it now is, since it does not differ in fact from the second, and so what is possible for the second, is also possible for it.

The place [*locus*] of a thing is that in which it is situated. Here I understand one thing to be in another if each of its extrema is congruent to the extremum of a part of the other. But every extremum of a point, a line, a surface, is itself a point, a line, a surface.

The points of a determinate extensum have a determinate situation to one another.

Therefore two points connected to a determinate extensum have a determinate situation to one another.

There can be two points having that situation to one another that two other points have to one another, such as *A.B* ≅ *C.D*; if this were not the case, it could be demonstrated that they coincide. But this being admitted, I ask whether it can be demonstrated from this that *A* ∞ *C* and *B* ∞ *D* ('*A* ∞ *C*' signifies that *A* and *C* coincide[8]), or whether *A* ∞ *D* and *B* ∞ *C*. For no reason can be given why one rather than the other; thus either coincidence will not follow from this, or it will follow that all four coincide. But in fact from one congruence the congruences of four things cannot be inferred. This assertion of congruence signifies

[8] Again, I have replaced the symbol Leibniz uses here for coincidence (a 'γ' with a vertical line through it) by the symbol ∞ which he uses later. I have also silently corrected a typo in the Latin that is also silently corrected by Parmentier in his French translation.

nothing other than that some extensum can be moved, ôr that the extensum from the place whose bounds are *A* and *B* can be transferred into the place whose bounds are *C* and *D*. And this can also be shown from the fact that unbounded space is indifferent with respect to the proposed extensum. In the same way it is proved that there can be a thousand points having a situation to one another which another thousand have to one another. And so this can be written thus:

$$A.B.C.D.\text{ etc.} \cong (A).(B).(C).(D).\text{ (etc.)}$$
$$\text{or } A.B.C.D.\text{ etc.} \cong {}_yA.\ {}_yB.\ {}_yC.\ {}_yD.\text{ etc.}$$

There can be given a point *A* which has the same given situation to two given points: *BC* Likewise there can be given a point *C* which has the same (given) situation to a given point *A* that a given point *B* has to the point *A*.

That is, if $A.B \cong C.D$ (which is possible without coincidence, by the preceding reasoning), and $A \infty D$ (ôr $A.A \cong AD$), it does not follow from this that $B \infty D$, that is, that *B* and *D* coincide.

Otherwise, it would follow only from the fact that *A.B.C.D.* etc. ≅ *L.M.N.O.* etc. and $A \infty L$, that $B \infty M$, $C \infty N$, and $D \infty O$, etc.; for there would be equal reason for all of them, and we would have *A.B.C.D.* etc. ∞ *L.M.N.O.* etc.

This can also be demonstrated by means of motion. For let there be two congruent but not coincident bodies, *ABCD* and *LMNO*, and let these be moved until the points *L* and *A* coincide (where the points *L* and *A* are homologous or corresponding points, as is clear from the arrangement of the letters), that is, so that $A \infty L$. It is clear that this can happen if the bodies are only touching at *A* and *L*, and are not coincident. The same thing is clear from contact alone, without any appeal to motion, if we suppose that two congruent bodies having no coincident parts touch one another at some point, and that the two points of contact correspond. Also one can understand one body to be touched by another much bigger body, so that when the superfluous part is removed from the bigger one something congruent to the smaller one is carved out and positioned congruently to the point of contact. But one can obtain an analytic demonstration, and one that is the most general of all these possibilities, if, when the analysis has been prosecuted sufficiently far, it is apparent that the contrary cannot be demonstrated.

The path of a point is a line. A path is a continuous successive locus. From this it is clear that the intersection of two lines is a point, since two points in motion collide with one another. It can happen, however that two lines have a common part, and then they will to this extent constitute a single line. But we will discuss this more carefully in its place.

A straight line is a line determined by two points.

The centre of a sphere must be shown to be unique. This is shown from the fact that the centre of four points is unique; that is to say, the intersection of four spheres is a unique point. We may show this from the fact that the intersection of two straight lines is a unique point. From which it is clear also that from four given points a sphere is determined. That is to say, if four points are given any three of which are not situated in a straight line, one can find a sphere whose surface passes thought them.

It may also be demonstrated from the definition of the straight line *AB*, namely that which is determined by the two points *A* and *B*. Supposing *A* to have the same relation to some three points *C*, *D*, and *E*, and likewise *B* to have the same relation to them, then any point whatever has the same relation to *C*, *D*, and *E*. For whatever is determined from these two points also has the same relation to these three.

The demonstration is similar for the plane. For since it is determined by three points *A*, *B*, and *C*, and since it is possible for any one of these three to have the same relation to two given points (not to three, unless *A*, *B*, and *C* fall on one straight line), it is also what is determined from them.

A3.3 *First Elements of Geometry*[9] January 1680

There are, in sum, two things to be considered in geometry, extension and situation. On the one hand, there is the **extensum**, a continuum in which parts that are simultaneous can be assigned; and on the other, there is the **continuum**, in which the parts are indefinite, ôr in which parts are merely designated by the mind. But when parts are designated in an extensum there already arise more extensa, possessing boundaries and situation. For, on the one hand, **situation** is nothing other than that state of a thing by which it happens that it is understood to exist in a certain way [*modo*] together with [*simul*] other extensa, that is, a mode of coexisting; whereas a **bound** is that in an extensum which has the same situation as another thing has in another extensum; for instance, the convex surface of an inner orb and the concave one of an outer orb revolving on the inner one, for it is evident that even though these two surfaces are different, since when one is moved, the other remains at rest, they do not however differ in situation, but are congruent.

Space is that in which, regarded in itself, nothing else can be considered except that which is extended. A **point** is that in which, regarded in itself, nothing else can be considered except that which has situation. Thus space is unbounded, for otherwise something else could be observed apart from it by which it would be bounded; that is to say, apart from extension there would be boundaries observable in it. A point, on the other hand, is indivisible, otherwise parts could be considered in it, and therefore, apart from situation, also extension. And so space is that which is simplest in extension, a point that which is simplest in situation.

Moreover, when one point is assumed, nothing further is determined, but when two points are assumed, or even more, by that very fact something further is determined, which is unique with respect to the given points. Now that which is determined when two points are assumed is the simplest extensum which passes through them, and this we call a **straight line.** Evidently, when the mind designates two points in space, by that very fact it conceives an unbounded straight line passing through them. For it is one thing to consider the individual points separately, and another to consider both as existing together (*simul*) and therefore to consider the situation they have to one another; and so when two things are regarded as existing at the same time, it is necessary that something further results from them. When we consider one of these two points exactly as if we ourselves were in it, and the other situated with respect to it, whatever must be considered by the mind to be determined thereby is called the **direction**, namely, that straight line produced indefinitely from one towards the other. Moreover the straight line is uniform, ôr everywhere similar to itself; otherwise apart from the two points by the consideration of which it is determined, something else would have to be assumed by which the species of dissimilarity could be explained. Hence a part of a part and a part of a whole are similar in the straight line. And since straight lines are equal if their extrema match up, they are congruent (otherwise there would be two straight lines between the two points, and so the straight line would not be determined by two points, contrary to the definition). Hence it follows that two equal straight lines are similar. Now when one straight line is equal to another, or greater or

[9] *De primis Geometriae Elementis*, LH XXXV, XII, I, Bl. 324. From Echeverría's transcription (CG 2 xiv, 276–84).

smaller than it, a part could be assumed at least in one of them that is equal to a part of the other, indeed, to the whole of the other, which, since they are similar (because equal), and a part of a straight line is similar to the whole, and what are similar to a third thing are similar to each other, any two straight lines will be similar to each other. I call **similar** those things which cannot be distinguished when they are perceived separately, but only provided both are perceived together (*simul*), that is, if their situation to one another is perceived. The straight line is the shortest line from one point to another since if it were shorter something simpler than it would be determined, which could not happen, since it is determined as the simplest line between two points alone.

A **plane** is that which is determined by three points which are not themselves mutually determined, ôr one of which is not already determined by the other two, ôr which are not in the same straight line. For the taking of two points by that very fact determines a straight line passing through them, and therefore also all of its points and sets them off from all the other points of the whole space; but if therefore nothing new is assumed, nothing new will be determined in space either. And so it is clear that in order to obtain a plane one must assume a new point outside this straight line. One can also define a **plane** as the path of a straight line touching two other straight lines, although the consideration of points alone is simpler. From these things, moreover, it is also clear that a plane is everywhere similar to itself; that is, the parts of a plane whose bounds are similar are themselves similar to one another, which is not the case for a gibbous [*gibba*] surface. It is clear also that there is only one unbounded plane passing through three points. A solid ôr body, or *Stereon*, is what is determined by four points that are assumed not to exist in the same plane. But when this is unbounded it is nothing other than the whole of space. Because with four points assumed, by that very fact are determined all the straight lines that can be drawn from them in turn; and all the straight lines that can be drawn from the points of the straight lines in turn; and other points beyond its surface, and so points that are not bounds—that is to say, points to which straight lines that are drawn from another point outside this figure must pass through the surface of this solid figure. Therefore there is no point in the whole of space outside this solid which is not already determined, for it is already determined in a straight line, namely in the straight line already determined by passing through two points (ôr existing in this determinate solid). Hence it follows that there is no other or fourth dimension, that is, there is no unbounded extensum that could be determined by five points that is not already determined by four points, since it is not possible to find five points which are not already in the same solid, that is, in the same figure of three dimensions, and so a figure of four dimensions is impossible.

Moreover, those things that can be understood as equal to, greater than or smaller than one another, that is, those for which part of one is at least equal to the other, we call **things of the same dimension**. And whatever is of the same dimension as a straight line we call a **line**. And a line different from a straight line is called bent [*flexa*], whether it consists of several straight lines, or is a line no part of which is straight. Whatever is of the same dimension as a plane is called a **surface**, and a surface other than a plane is called gibbous [*gibba*], whether it is composed of planes or is everywhere curved. Now in a solid it is not possible—as it is in other dimensions—for two species to be determined, one uniform and the other non-uniform [*disquiformis*]. For instance, in plane surfaces, if the bounds are similar then the surfaces themselves are similar, but in gibbous surfaces this does not hold. On the other hand, in any solids whatever, if the bounds are similar then the solids are also similar. The reason for this is that a surface not only has bounds itself, but also the whole surface can be the bound of something of a higher dimension, and bounds can be dissimilar. Therefore, it can be dissimilar to another not only through its own boundaries but also through itself, insofar as the bound can be that of the other thing. But a solid cannot be the bound of something of higher dimension.

A3.4 From *First Principles of Geometry* (II)[10] [1682–85?][11]

[Part 1]

A **point** is that which is the simplest of all things having a situation. Thus every point is congruent to every other point, and multiple things having a situation to each other cannot be assumed in one point, but coincide with one another. From which we infer these expressions:

$A \sim B$, that is, A is similar to B, or conversely.
$A = B$, that is, A is equal to B, or conversely.
$A \cong B$, that is, A is congruent to B, or conversely. If B is in A, then we will have $A \infty B$.

An **extensum** is a continuum whose parts exist at the same time.

A **continuum** is a whole two co-integrant parts of which have something in common; which is called the bound of each of them. Now I call those parts **co-integrant** which taken together equal the whole, and have no part in common. Thus a bound is not a part, nor is it homogeneous with the things it bounds, otherwise it would be a part.

[Part 2]

A is understood to **be in** *B* if *B* can be made by increasing *A*, or conversely, if *A* can be made by decreasing *B*. Thus a part is in the whole; but also a point, which is not a part, is in a line; they have in common this thing that they are both in, and what can be understood to be generated by the increasing or decreasing of the same thing, can be called homogeneous. Or they could also be said to be of the same matter, for the point of a body and a body have the same matter, as do a point of space and space, a moment and an hour, a motion and a tendency to move.[12]

An **indivisible** is that which has no part and by the increasing of which something can be made that has parts, so that by increasing a point a line can be generated, by increasing a moment, an hour. Conversely, by decreasing an hour, just before it disappears into nothing, it ends in a moment.

A **continuum** is that whose parts are indefinite.

A **part** is what is in another thing with which it is homogeneous.

To **be in** something or to be a **constituent** of a thing is another way of saying that it is an immediate requisite of it. Thus a point of a body is its immediate requisite, for at once and without any reasoning it is understood that this point is required in order to constitute the body; a part of a thing is also an immediate requisite of it.

[10] LH 35 I 5 Bl. 5–6 (MT N.30 (39884)).

[11] On the dating, Siegmund Probst relates that "This is written on a paper produced in 1678; the use of the paper by Leibniz is recorded often in 1682–1685."

[12] "tendency to move" translates *nisus*. Leibniz had originally written *conatus*, "endeavour."

A **minimum sign** ôr **indivisible** is said to be that which can be the immediate requisite of another thing which has no other immediate requisite. Thus a point in an extensum, a moment in time.

Those things are **of the same kind ôr matter** whose simplest immediate requisites ôr minima are indiscernible from each other.

A **transformation** is a change of a thing into another with the minima remaining the same and unchanged.

Those things are **homogeneous** which are similar or can be rendered similar by a transformation, according to some mode of consideration.

Those things are **equal** which are congruent or can be rendered congruent by a transformation.

A **part** is a homogeneous immediate requisite of another thing which is called the **whole.**

Point and body are of the same kind ôr matter, yet they are not homogeneous, since a body can be similar to another body, but a point cannot.

That is similar whose [*breaks off*]

[Part 3]

Every surface, when produced infinitely, perhaps bending back on itself, is divided into two parts, so that a straight line cannot be drawn from a point of one part to a point of the other part without cutting the surface. Every line, when produced infinitely, perhaps bending back on itself, divides the surface on which it is drawn into into two parts, so that a line cannot be drawn from a point of one part to a point of the other part that does not cut the former line. If the line or surface is bending back on itself, the part towards which the concavity is turned is called *inside*, the other part, *outside*. Indeed, it can happen that the convexity that was once enclosed is inverted, when it has a new contrary bending. Thus in a case of bending on itself, the smaller part is said to be inside, the greater outside. Or it is asked, when the circumference of a smaller circle divides a spherical surface into 2 parts, which should be said to be outside, and which inside?

A3.5 *Notions Used in Explaining Geometry*[13] [April 1687][14]

Similar things are those which cannot be distinguished taken one at a time. **Congruent** things are those which cannot be distinguished unless a third thing is assumed. Those things are **congenious** with one another that are produced by a single operation, or are similar to these things, like a point and a line. And indeed the former may be called intrinsic *congenia*, or even simply *congenia*, and in geometry they are those things which are said to **be in** [*inesse*] something.

One thing is said to come out of another ôr to be **transformed** when nothing can be understood to be in the one thing in which there is not something that is also in the other. So if an equilateral triangle is made from a square, no part can be assumed in the square in which one cannot find a part that is also in the triangle. If on the other hand a straight line is

[13] LH 35 I 5 Bl. 49 (MT N. 46 (39942)).

[14] On the dating, Siegmund Probst relates that "This is written on the same sort of paper as a letter from Leibniz to Bourgeauville, who replied to it (A I, 4 N. 527) on 12/22 April 1687."

changed into a curved one, even if there is no part common to both of them, still points have remained in both of them, and therefore no arc can be taken in which there is not some point which is in the straight line that is made from the arc.

In explaining geometry we assume the following notions: Similar, congruent, being in (that is, as in a place), and being determined. Things that are congruent are also similar. Hence there arises **transformation**, that is, one thing is said to come out of another by a transformation when nothing can be taken in the one in which there is not something that is also in the other.

Homogeneous things are those which are similar or can be rendered similar by a transformation.

Equal things are those which are congruent or can be rendered congruent by a transformation.

If something is in another homogeneous thing, then what is in it is called a **part**; what it is in is called a **whole**.

A **point** is what is in another thing while nothing can be in it. Hence all points (with respect to the consideration of place, ôr being in) are similar and congruent.

Those things are said to have a **situation to one another** which are in the same thing, and this is held to be determined from the given thing that they are in, and not from its being unchanged.

A3.6 *On Homogeneity and Homogony*[15] [1693?][16]

Homogeneous things are those which can be made similar without affecting magnitude [*salva magnitudine*]. Symbolic or **metasymbolic** things are those one of which disappears into another.[17] **Homogonous**[18] things are those one of which can disappear into the other through a continuous change;[19] such are not only **homogeneous** things, that is, those that are similar or assimilable, such as straight line and curved line, triangle and square, sphere and cylinder; but also **heterogeneous** things, such as point and line, and surface and space, indeed also, moment and time, hour and day.

Continuous change is a **flux**.[20] As a line is made by the flux of a point, so homogeneous things [*sic*] are changed into each other by a flux, such as a point into a line, a circle into a square, even into a cone. A **transition** is different from a change or a flux, and is even made

[15] *De homogeneis et de homogonis*, LH 35 I 14 Bl. 67 (MT N. 88 (40938)).

[16] This piece is dated 1686 in the Ritterkatalog in Hannover, but there is no record of any justification for this dating. The paper has watermarks, but according to Siegmund Probst these are almost unreadable; although there is a small piece of watermark that can be seen, and this *might* match with a watermark on the paper that Leibniz used in 1693, or even with one on the paper used in 1697. De Risi argues (private communication) that a 1697 date coheres with the fact that here Leibniz seems to be settling on the term *homogona* (which he still uses in the *Initia Rerum*), whereas in the *Specimen Geometriae luciferae* and other papers probably from the early to mid-90s he uses *συγγενής* or *congeneum*.

[17] Marginal note by Leibniz: '(*Metasymbola, anasymbola*)'.

[18] Above the '*gona*' of '*homogona*', Leibniz has written '*bola*', suggesting the alternate term '*homobola*'; see note 21 below. He also notes in the margin: "*Homometra* [are] comparable, *Heterometra* [are] incomparable."

[19] This is identical to the definition Leibniz gives of homogonous things in the first essay of the *Initia Rerum* of 1715 (GM VI 20/L 668).

[20] Marginal note by Leibniz: "What Euclid called *homologa* are more properly called *homoptera*." See Euclid, *Elements*, V, def. 12.

in essences, such as number into number. But such is not continuous, and cannot be made mentally by us, but by *God*; in fact, however, it also occurs in essences, ôr from one species to another, such as in ellipses. It must be seen whether there is a transition from essences to essences without homogony,[21] such as, perhaps, from one plant into another; indeed you tend from a plant through an animal to a human being—although perhaps in all these cases there is a homogony unknown to us. Just as there is continuity of quantities, so there is also continuity of forms, the seat of which, ôr space, will be the science of *God*, seeing as the space of places is nothing other than the immensity of divine operation. And of course local space is the continuous order of coexisting; and the mode by which things are distinguished in this order, since we distinguish things through situation. If there are two homogonous coexistents, and while the one *A* remains as it is, the other *B*, remaining the kind of thing it was per se, changes continuously with respect to others until it is happens that some given thing in *B* coincides with a certain given thing in *A*, with a necessarily determined mode of change for this; the situation of *B* to *A* will be determined. In a change of situation there is a continuous change of those things with which something coincides.

Space is the order of coexisting.

Situation is a relation [*respectus*] of one thing to another in the order of coexisting.

Every change is continuous. Situation can hold for future things as well as for those coexisting now, for future things are determined from present ones, and conversely.

A3.7 *Scheda on Situation and Extension*[22] [1695?][23]

(1) **An extensum** is a continuum containing those things that coexist.

(2) Those things **have situation** which are in the same extensum (just as those things have position which are in the same continuum).

(3) A **point** is what has situation to whatever thing is posited in an extensum, but does not have extension. See (10) below.

(4) Any two points whatsoever are in the same extensum. For two points have situation to each other by (3). Therefore they are in the same extensum by (2).

(5) Two extensa that have a common point constitute one extensum. For by having a common point they are continuous. And so by (1) they are one extensum.

(6) Any points whatsoever are in the same extensum. Let there be points *A*, *B*, *C*, *D*, etc. *A* and *B* are in the same extensum by (4), which would be *AB*. Similarly *B* and *C* are in the same extensum, which would be BC. Therefore *AB* and *BC* have the common point *B* and so constitute one extensum *ABC* by (5). Similarly, there is an extensum *CD* by (4), which has a common point with the extensum *ABC*, therefore they constitute the common extensum *ABCD*. And so on.

(7) *Def.* Universal Space is the locus of all points.

[21] Above '*homogona*' Leibniz wrote '*homobola*'.

[22] LH XXXV, 1, 14, Bl. 90; from De Risi's transcription (2007, 588–9). On this manuscript Leibniz has written: "This scheda is good." I have followed De Risi and others in leaving the Latinized Greek term *scheda* untranslated; it means, roughly, a small tract.

[23] De Risi suggests a date of 1695 mainly on the grounds of maturity of expression, and because it uses the term *congeneum*, equivalent to the Greek συγγενής that he uses in the *Specimen Geometriae luciferae*, for which a date of 1695 has been proposed.

(8) There is a Universal Space. For there is a locus of all points by (6); therefore there is a space by (7).

(9) If two things have situation in the same thing *A*, which contains whatever coexists, they will be in some extensum which is in that same thing *A*. For generally if two things are in the same thing *A*, they will be in some continuum which is in *A*; but it is an extensum in the same place, since whatever is in *A* coexists, by hypothesis, and so also is something in a continuum. Therefore by (1) this continuum is an extensum.

(10) Whatever has situation in a point is itself a point. Let there be a point *A*, such that a thing *B* having a situation is assumed in it. I say that *B* coincides with *A*. For we have two things having situation in the same thing *A*, namely *A* and *B*. Therefore they will be in a certain extensum which is in *A* by (9), which is absurd, by (3). A point was held to be so defined that anything having situation that could be assumed in it, would be it itself.

(11) In any extensum whatsoever another extensum can be assumed. Because in any continuum whatsoever another continuum can be assumed homogeneous to it.

(12) Let there be an extensum having something in common with another extensum; this is transformed in such a way that what is in common is continually diminished until nothing more can be subtracted from it, and when it reaches the point where it is necessary that if anything further is subtracted from it nothing in common will be left, then the common extremity will be a point. Thus it is demonstrated also that when an extensum is given, so is a point.

(13) A point is congeneous[24] with an extensum. For a point can be made from an extensum by continuous change, as demonstrated in prop. [12].[25]

([14]) Whatever is congeneous with an extensum is either an extensum or a point. This is evident from the demonstration in prop. [12].

A3.8 Extracts from Late Manuscripts

1. From *Magnitudinis nomine*...[26] January 1712...

You may define **distance** as the magnitude of the shortest path from one point to another.

We need to find out whether there is such a distance, and whether such a path is unique.

First of all it can be asserted that the distance cannot be greater than a straight line, otherwise it would not be the shortest path. It can also be added that if the shortest path is unique, it will be a straight line. For the unique path from *A* to *B* is determined solely by *A* and *B*, therefore any point on the path has a unique situation to *A* and *B*, for any other point of space would not fall on this line, and another point on the same path would differ from it, since it would be before or after it. But any point situated uniquely falls on the straight line through *A* and *B*, and so the shortest path falls on it, but it falls between the endpoints, otherwise it would not be the shortest path.

...

[24] In the 1680s (MT #46 and MT#106) Leibniz had used the term *congenia* for congeneous things.

[25] Leibniz's numbering goes a little awry here. He twice alludes to a demonstration in 'prop. 11', when it should be proposition (12), and numbers the last proposition (15) instead of (14).

[26] From ***Magnitudinis nomine***..., XXXV, 1, 8, 2 verso. From De Risi's transcription (2007, 600).

2. From *Spatium absolutum*...[27] August–September 1714

Absolute space contains all places. A point contains only itself.

A **part** is what is in another (the **whole**) and can be removed even if nothing else is removed.

Equals are those whose quantity is the same, ôr, those one of which can be substituted for the other without altering the quantity [*salva quantitate*].

The **smaller** is what is equal to a part of another (the **greater**).

Two magnitudes are **homogeneous** if, when one repeatedly subtracts everywhere from one something equal to the other, there finally remains something smaller than the other.

A **continuum** is that any part of which has something in common with some other part of the same thing.

An **extensum** is what has continuity and coexistence of parts.

An extensum whose quantity is finite is usually called a **magnitude**.

A **finite quantity** is that which admits of being bounded—even though, however, a finite quantity also admits a certain unbounded extensum, such as the Torricellian solid.

An **unbounded** extensum is that in which two points can be assumed between which the shortest path drawn in the same extensum becomes greater than any given line however great.

A **path** is the locus of all the places of a moving thing and nothing besides.

An **extremum** of a magnitude is that which it can have in common with another magnitude in such a way that they have no part in common.

Whenever an extremum is in fact common to two things, it will be called a **section**. More strictly, however, it is understood when a whole is not cut into a part by dint of its own construction (as, for instance, when two globes touch one another in a point, or two cylinders in a line, so that they are said to compose a whole), but as a result of some new data.

Absolute space is **uniform**, that is, everywhere similar (congruent) to itself. Thus when one cuts a solid out of it that is enclosed by a surface, it will be uniform inside.

Inside, that is, if the surface or extremity of the body is not reached.

Hence a plane is also uniform inside, since what is inside it is also inside a cut body, and nothing else pertains to the construction of a plane that its being related the same way on both sides. By the same token, a straight line which is a section of a plane having the same relationship on both sides is uniform inside.

Those things are uniformly congruent (or are similar) inside whose extremities are congruent (similar)....

Analysis Situs[28]

A **point** is that which is simplest in situation, and therefore a point is a place [*locus*] in which there is no other place.

Unbounded **absolute space** is that which is amplest in situation. Therefore every place is in the same absolute space. Absolute space is a continuum, otherwise some place could be interposed which would not be in it. Absolute space is uniform, since variety would arise from determinations, which is not consistent with its being the amplest.

[27] From ***Spatium absolutum***..., XXXV, 1, 12, Bl. 6 recto, 12 verso. From De Risi's transcription (2007, 608–10).

[28] LH XXXV, 1, 12, 19 recto (De Risi 2007, 609–10).

There are arbitrarily many places in a common bounded place, namely those which can be assumed in absolute space.

A **line** is the path of a point, ôr a continuous successive place [*locus*]. From any point to any other is a path. From any point to any other point through any other point is a line. Different lines can be drawn from one point to another: let *AB* be a line from a point *A* to a point *B*, and now let there be a point *C* outside *AB*, then there will be a new path *ACB*.

Distance is the shortest line from one place to another: *A* and *B* are more distant from one another than *A* and *C* if there is a path between the latter pair that is less than any path between the former pair. For the path from one point to another can be increased to infinity, but it cannot be diminished. Those things are less distant which can be included in a smaller common place.

A **straight line** is the distance between points. A straight line from one point *A* to another *B* proceeds through points whose situation to *A* and *B* is unique, for in this way it proceeds by a maximally determined line. The situation of one point to another remains if rigid connecting extensa are supposed; and straight lines cannot then be moved.

If a point has a situation to two points that is not unique, it will be movable in such a way that it will conserve its situation to these unmoved points, and will describe a line which conserves the same distance from any point on the line passing through those two points. This will afterwards be shown to be the circumference of a circle....

3. From *Rectam definio*...[29] 1715

A.B **unic.** signifies: If $A.B \cong A.C$, then $B \infty C$....

A **point** is the simplest place [*locus*], ôr a place in which there is no other place. So if *B* is in *A*, and for this reason $B \infty A$, this is the same as *A*'s being a **point**.

Space is the amplest place, ôr a place in which there is every other place. If *B* is a place, and for this reason *B* is in *A*, *A* will be **space**. If $Y \cong A$, then $\bar{Y}$ will be **space**.

If $Y.A \cong Y.B$, then (if it is not the case that $A \infty B$) then $\bar{Y}$ will be a **plane**. Hence the intersection of two equal spherical surfaces falls in a plane.

Likewise, if $Y.A \cong Y.B \cong Y.C$, then $\bar{Y}$ will be a **straight line**. Hence the intersection of three equal spherical surfaces, ôr that of two equal circles, falls in a **straight line**.

Hence a straight line is the intersection of two planes. $Y.A \cong Y.B$, since *Y* is in a plane along *A.B*, and $Y.A \cong Y.M$, since *Y* is in a plane along *A.M*. Whence, dividing in a different way, $Y.A \cong Y.B$ and $Y.M \cong Y.A$. Therefore $Y.A \cong Y.B \cong Y.C$. Therefore $\bar{Y}$ will be a straight line.

If $Y.A \cong Y.B \cong Y.C \cong Y.D$, $\bar{Y}$ will be a **point**, or *Y* will be unique. That is to say, the intersection of four equal spherical surfaces, or of three equal circles, falls in a point.

But the preceding are definitions of the straight line and the plane, but this is not the definition of a point, for which we have another definition: and so it should be demonstrated. Now the demonstration follows from the fact that, supposing $Y.A \cong Y.B \cong Y.C \cong Y.D$, $\bar{Y}$ is the intersection of two straight lines, namely *Y* is in the straight line along *A*, *B*, *C*, and in the straight line along *B*, *C*, *D*. Therefore it is in the intersections of the straight lines, which is unique.

It must still be demonstrated that the section of two unequal spherical surfaces is in a plane, and that the section of three unequal spherical surfaces is in a straight line; and that the section of four unequal spherical surfaces is in a point. This goes as follows: the spherical

[29] From ***Rectam definio***..., LH XXXV, 1, 12, 17^r, & 17^v. From De Risi's transcription (2007, 613, 614).

surfaces around *A* with radius *AY*, and around *B*, with $BY \cong AY$, intersect in a surface in a plane; but if in the same straight line *A.B* you take another point *C*, and from it you describe a spherical surface with radius *CY*, it will be the same straight line as before (this should be shown). Therefore any intersection of any two spherical surfaces is in a plane. And it goes the same way in the other cases.

That is, whether the straight line or axis is *AB* or *AC*, it is the same axis of rotation, since all the same points remain at rest. . . .

APPENDIX 4

Leibniz's Rome Manuscripts on Copernicanism

Leibniz made these notes on the Copernican system during his stay in Rome in 1689, as established both by contents and by watermark. The second, longer, draft of *De praestantia* was first published by Gerhardt as a footnote to the published version of the *Tentamen* together with a letter 'Ad H. P. B.' (GM VI 144–7). "H. P. B." has been convincingly identified by Domenico Bertoloni Meli as Antonio Baldigiani S.J., and so *De praestantia II* would appear to be the supplement to that letter, in which Leibniz asks the addressee to give his opinion. In a letter to Melchisedech Thévenot of 1691 Leibniz later mentions that he had met Vincenzo Viviani in Florence (and this must have been in the first week of so of December 1689, after Leibniz's arrival there and before Viviani left for Rome), where they "discussed at great length together the question of whether there might be a means of making the Roman Curia listen to reason on the matter of Copernicus."[1] He also says in a letter to Huygens in 1694 that he had "communicated a little paper' to Viviani about the relativity of motion in this connection (GM II 199/L 419). So it is assumed that this paper for Viviani was the supplement Leibniz sent in his letter to Baldigiani.

A4.1 On Motion Taken in Mathematical Rigour Summer 1689

I have demonstrated that when motion is taken in mathematical rigour, there is no principle for determining which of two bodies is absolutely at rest or is moving with such an absolute motion.[2]

Thus it can happen that a body can exert some percussion or other force of moving and yet not be moving absolutely in mathematical rigour, but remain at rest,—for instance, when a ball thrown by hand in a ship runs in a horizontal straight line from the bow towards the stern, and meanwhile the ship is moving with equal speed in the opposite direction, from the stern towards the bow; there will be no absolute motion of the ball, even if the ball also received an impulse from the hand and made an impact on another ball at rest on the ship.

[1] This was in a letter to Melchisedech Thévenot of September 3, 1691 (A I 7, 352–3), quoted by Meli (1988, 33). As Meli notes, Viviani was Galileo's last disciple, and made strenuous efforts to have the latter's *Dialogo* removed from the index. Baldigiani was a Jesuit professor of mathematics in Rome, and also "consultor of the Congregation of the Index and Qualificator of the Holy Office, that is, a censor" (1988, 34–5); this explains the reverential tone Leibniz diplomatically adopts in his references to the censors in these papers.

[2] *De motu in rigore mathematico accepto* (A VI 4, 2065–7).

And, specifically, I have demonstrated this to be a wonderful law of nature, that all phenomena, not only of bodies moving freely, but also those colliding with one another, take place in the same way, wherever rest or absolute determinate motion finally exists.

Hence it follows that whenever one system is preferred to another this can only be accepted in the sense that one is more intelligible than the other and is better accommodated to the things it is proposed to explain; and therefore when one hypothesis is said to be true and the other false, this is to be understood of it in its own particular respect. Thus the Spherical doctrine of Ptolemy is just as true as the Theoretical planetary system of Copernicus, because here the true way of explaining, which is the right and appropriate and best one, is the same for both; and it would be just as mistaken on either side to stick the motion of the Earth into the Spherical doctrine or to use the standing still of the Earth in the Theoretical system. Thus whoever is in a ship when nothing happens with respect to the bank wrongly denies that the above-mentioned ball is moving; on the other hand, anyone standing on the bank who sees the ball does not rightly deny that it is at rest, since it is not a matter of change within the ship, while the change can be defined as absolutely nothing when it comes to the new question concerning the bank itself.

Thus the observational astronomers [*historici*] and the writers of the sacred scriptures themselves not only would be able to, but should be able to, treat the Earth as stationary. To say this is more worthy of God, the author of the book containing the thesaurus of knowledge, than to assert that he spoke less accurately or in a popular idiom, or that here he diverged from a literal meaning.

And because it is permitted, without violating the Censure [*salva Censura*], to employ the Copernican system rather than others as the most apt hypothesis for understanding the planetary motions—while to defend its truth, as we have explained, is really nothing other than to defend its utmost intelligibility—so in this sense, anyone can be in the highest degree Copernican *salva Censura*.

Indeed it is of no little concern not only to the Church, but also to the sciences and to the honour of Italy, for this finally to be acknowledged at some point; nor in this way will there be a fear that the Censors would be forced into a retraction at some time by new discoveries, or into vainly opposing the torrent of the age and the public voice of the learned; and it would oppose the false allegations of those who say that the truth is oppressed; and therefore distinguished scholars, in Italy no less than among other peoples, could profit from the light of the age, and come upon great discoveries, which now are being snatched away from them. Nor do I believe that the very fine men to whom the right of censure belongs had anything else in mind.

And indeed so superior is the Copernican system for explaining the phenomena of the planets that it must be admitted that an astronomer who did not understand it would be groping around in the greatest darkness. And every day so many new discoveries confirm its harmony and simplicity that not to wish to utilize it would be to obscure the glory of God himself by taking away the occasion for recognizing his admirable wisdom.

And we ourselves, having taken the universal laws of planetary motion discovered by Kepler in his principles, namely, the harmonic circulation of celestial matter as deferent orbits of planetary fluid around the Sun, and the rectilinear approach of the planet as something heavy and towards the Sun as if that were a magnet, and having geometrically resolved them, have, if I am not mistaken, contributed something towards obtaining the best explanation of the phenomena, that is, the truest hypothesis.

A4.2 On the Superiority of the Copernican System: First Draft Summer 1689

Cicero famously said "Time destroys the fabrications of opinions; those grounded on nature it confirms with justice."[3,4] We experience this ourselves concerning the system of the world. For with the new discoveries made every day the Pythagorean Hypothesis approaches to the utmost verisimilitude. So that now there is hardly anyone among the most intelligent mathematicians who would not admit to preferring it, if only he were permitted to by the decree of his superiors.

To me it seems that the censure on Galileo ought not to be taken as condemning anyone at all who believes the Copernican hypothesis to be truer, but only those who fight for it rashly, and audaciously judge the Holy Scripture to have spoken less properly.

Today, moreover, every intelligent person admits that the Copernicans are supported by so many newly discovered arguments that the rashness can no longer be imputed to them, and the more prudent among them admit and defend the idea that the Sacred Scripture had spoken properly and appropriately, and, indeed, should not and could not have spoken in any other way, even if these writings were imagined to have been produced by the Copernican school.

Thus it seems to be in the interest both of the literary republic and of the Church itself, that in this argument there should no longer be thought to be a barrier to scholars, nor an opportunity for false accusations by opponents, as if the truth in Italy were oppressed by some intolerable servitude.

On closer reflection, this gives me the opportunity to present certain thoughts which seem to us to reveal in good measure the secrets of the system and physical causes of the celestial motions, which I now touch on briefly.

For I saw that Kepler, drawing on Tycho's observations and his own meditations, had discovered certain general and very beautiful laws of Celestial Motions, soundly satisfying the phenomena, from which it easily appeared that there should be certain physical reasons as a basis; I myself had long brooded over the investigation of causes, and finally by drawing from an unexpected light, almost beyond hope, I seem to have been granted a large part of my wish.

These are the planetary laws: 1) A (primary) planet is carried in an ellipse with the Sun in its focus; 2) the times of the same planet are proportional to the sectors of the ellipse marked out by the Sun; 3) the squares of the periodic times of different planets are as the cubes of the average distances from the Sun.

From the second law I have demonstrated that a planet and therefore its aether are being carried harmonically in a circle, that is, such that with its distances from the Sun existing in an arithmetic progression, the velocities of circulation are in a harmonic progression; and vice versa. For I have discovered a general theorem that if a moving body is moved with a motion composed of a harmonic circulation around some centre, together with a certain paracentric motion with respect to the same centre, then it describes areas proportional to the times; and vice versa. And conversely, if it describes these areas in this way, then its motion can be resolved into the stated circulation.

Next I have investigated what the law of paracentric motion must be, that is, that of a motion of attraction to the Sun.

[3] *De praestantia systematis Copernicani, I* (A VI 4c, 2068–9).

[4] Cicero, *De Natura Deorum*, II. 2, 5.

A4.3 On the Superiority of the Copernican System: Second Draft

De praestantia systematis Copernicani, II: (A VI 4c, 2070–75)

Cicero famously said "Time destroys the fabrications of opinions; those grounded on nature it confirms with justice."[5] We experience this ourselves concerning the best way of explaining the system of the world, which by new discoveries every day brings it about that scarcely anyone who is a distinguished mathematician would not admit to a preference, if it were permitted by the decree and censure of his superiors. Whence at one time already Christopher Clavius, celebrated mathematician of the Society of Jesus, when in his old age he understood the new celestial discoveries made with the telescope, and especially the detection of the moons of Jupiter, declared himself delivered from the received astronomy. For he saw the very great force of the analogy, which afterwards the new discovery of the ring and satellites of Saturn rendered so manifest, that it could hardly be resisted by him any longer. And Claude Deschales, a man of the same Society of Jesus who was very well versed in these studies, frankly confessed that hardly any other hypothesis could be hoped for which would satisfy the phenomena so beautifully and fully.[6] Indeed, almost everyone concedes today that the Copernican hypothesis in philosophy is not absurd, as was once believed. Riccioli himself rejects all the common arguments against it that have been brought forward, except the one that is derived from the motion of heavy bodies or projectiles.[7] But that this also has no force has been shown by Gassendi, Fr. Stefano Degli Angeli and Giovanni Alfonso Borelli.[8]

As for what pertains to the theological argument taken on the authority of the sacred scriptures, both Fr. Marin Mersenne of the Order of Minims and Fr. Honoré Fabry of the Society of Jesus have admitted that nothing prevents the Church, having once acknowledged the efficacy and weight of natural reasons, from declaring that the words of the sacred scriptures can be accepted in the same way that the words of all mathematicians are accepted, who, even though they are followers of the new system, nevertheless always say, and should say, that the Sun moves, sets and rises, whenever they are not treating the theorem of the planets professionally.

Meanwhile the subject of the censure is deservedly the rashness of those who seem to feel less reverently about the sacred scripture, as if, that is, it had not spoken accurately enough, since its aim is not to teach philosophy but the way to salvation. For it is more complimentary and truer to acknowledge that all of the sciences lie in the sacred scriptures as hidden treasures, and that astronomical matters are spoken about just as correctly as all others. And this can be asserted without damaging the new system. For the sacred authors would not otherwise have been able to express what was on their minds without absurdity, even if the new system is supposed true a thousand times. And it would be ridiculous for an observational astronomer [*Historicus*], whatever opinion he finally followed in mathematics, to declare that it is not the Sun but the Earth that rises and sets.

But it will be worthwhile to mention at this point something that has not yet been sufficiently noticed before now, which will plainly be capable of destroying all difficulty in this business, and vindicating the Copernican system of all censure. This is that everyone

[5] Cicero, *De Natura Deorum*, II. 2.

[6] C. F. M. De Chales, *Cursus seu mundus mathematicus*, Leiden, 1674.

[7] Giambattista Riccioli, *Almagestum novum*, Bologna 1651; *Astronomia Reformata*, Bologna 1665.

[8] Pierre Gassendi, *Epistolae tres. De motu impresso a motore translato* (Lyons 1658); Borelli (1667); Stefano Degli Angeli, *De Infinitarum Cochlearum Mensuris ac Centris Gravitatis*, Venice: Ioannem La Noù, 1661.

concedes that the Copernican system can be taught as a hypothesis, understanding also that they admit that of all hypotheses none is simpler or more intelligible than the Copernican. But it should be recognized as true (although this will appear paradoxical to many people) that there is no other reality in motions than a hypothetical one, nor is there any other sign of a true system in the nature of things than that everything is explained most aptly by it. When, therefore, the Copernican system is conceded to be the simplest, by that very fact it is conceded that it is true. To understand this more precisely, it must be known that motion is necessarily taken in such a way that it involves something respective, and no phenomena can be given by which it would be determined absolutely to be in motion or at rest. For motion consists in change of situation ôr place; and place itself again involves something relative, even according to the opinion of Aristotle, who defined it as the surface of the surrounding body. Hence in all rigour every system can be defended, so that not even an angel could determine anything absolutely with metaphysical certainty. For it is itself a condition for the general laws of motion that everything happens in the same way among phenomena, and it is impossible to judge whether and to what extent a given body is at rest or moves, unless one has a reason for its greater explicability. And this is so true that not even a force of acting is a certain indication of absolute motion.

For instance, if a ball thrown by hand in a ship runs in a horizontal straight line from the bow towards the stern, and meanwhile the ship is moving with equal speed in the contrary direction, that is, from the stern towards the bow; there will be no motion of the ball absolutely speaking, for absolutely the ball remains in the same place in space as it appears to a spectator on the bank, who supposes it motionless. And still the ball has a respective motion, namely with respect to those who are in the ship; and granting (by hypothesis) that it is at rest absolutely and in all rigour, still something on the ship that opposes it could get broken.

And so, just as often happens with the ship and the things that are in it, such as occurs with the impression of the hand or the breaking of some glass jar on the ship which had presented a strong obstruction to the ball, it is considered that we all say truly and correctly that the ball is moving; but with respect to the bank we say, again truly and correctly, that the ball is at rest, since it always keeps the same situation to all the immobile points on the bank. And those on the bank who attribute motion to the ball would be just as mistaken as those on the ship who ascribe rest to it. In the same way we say that it is true in its own way and with respect to the spherical doctrine of Ptolemy no less than to the theoretical doctrine of Copernicus, and so it is just as inapt to insert the motion of the Earth into the explanation of the primum mobile of the spherical doctrine, as to want to teach the theory of the planets by means of innumerable epicycles and eccentrics which the new hypothesis, or rather the revived hypothesis of antiquity, shows to the intellect with admirable simplicity. From which it is clear that those who say the Copernican Hypothesis is true, ought to feel or be understood in this way, that it is the best, that is, the most apt for explaining the phenomena, and that no other truth has a place in a matter which by its nature involves something respective. When these things are correctly understood, anyone can be a Copernican without violating the censure. And in this way we also save what the censors said, that the Copernican hypothesis is absurd in philosophy, if, namely, it is perversely applied in the explaining of those things to which one's intelligence is rather disturbed at its being applied.

Indeed, it is of no little concern for both piety and the sciences for these things finally to be acknowledged at some time. In order to shut the mouth of those people who by bringing together faith and reason do injury to one or both, the most recent Lateran Council decreed that they are deservedly to be opposed. It suffices, then, to condemn the rashness of those who are not afraid to say that the sacred scripture spoke less accurately concerning astronomical matters. But the liberty of declaring about the truer hypothesis can be left to other people apart from the rest; from which the fact of the matter shows that the new system as it is explained today is neither absurd, nor should one be afraid to defend it, but is advanced with very strong arguments, and outside the Catholic faith many people of great learning and piety incline to its defence. And so with this explanation of the original censure the scruple which many people wrongly had would be destroyed, and [these considerations] would prevent the Censors from needing to revoke the decree at some point, or oppose in vain the torrent of a proficient age and the public voice of learned people; and also to prevent the occurrence of the false allegations which charge that the truth is oppressed among the Catholics, and turn away upright minds from communion with the church. Likewise it is important to the honour of Italy, for in this way outstanding talents could flourish in the light of the age no less there than among other peoples, and fall upon the outstanding discoveries that now are often snatched away by others. Nor do I believe that the very fine men to whom the right of censure belongs had anything else in mind.

And indeed so superior is the Copernican system for explaining the phenomena of the planets that it must be admitted that an astronomer who did not understand it would be groping around in mere darkness. And every day so many new discoveries confirm its harmony and simplicity that it is to be feared that anyone who did not wish to utilize it would be obscuring the glory of God himself by taking away the occasion for recognizing his admirable wisdom in so many of his works. Indeed it seems it can be said especially now (if ever) from what a certain most recent light has revealed concerning the physical causes of planetary motion, whose universal laws are explained with marvellous fortune by the harmonic circulation around the Sun of the aetherial matter carrying the planet, and by a certain solicitation of the planet towards the Sun as if by a magnet, as of a heavy body towards the centre; which are not only assumed by supposition, but also demonstrated by the geometrical regression of the phenomena. So that concerning the best system it already seems that it there is hardly any place left for doubting.

Glossary of Technical Terms

accident (*accidens*): a quality that is not permanent or intrinsic to an individual but which is applicable at some times and not at others; opposed to **attribute** (q.v.).

arcwise connectedness: a technical term from modern topology; a space is arcwise- or path-connected (these being equivalent for the space concerned here) if a continuous path can be drawn from any one point in the space to another; a notion which Echeverría and De Risi have seen to be anticipated by Leibniz in sections §§60–8 of the *Characteristica geometrica*.

assignable (*assignabilis*): an **assignable quantity** is one to which a finite measure can be assigned; an **assignable instant** or **point** is one that is at an assignable interval from another such instant or point.

attribute (*attributus*): a quality that is intrinsic to an individual, and permanent, not accidental (see **accident**).

categorematic: a designation from Scholastic philosophy; if a term is understood **categorematically**, it refers to a particular entity. As such it is opposed to a **syncategorematic** term (q.v.). For example, a categorematically infinite number is a number greater than all finite numbers, while 'infinitely many' understood in the syncategorematic sense would mean that however many are taken, there are more than that. See **infinity**.

co-integrant parts (*partes cointegrantes*):parts which taken together, equal the whole, and which are such that if they have something in common, this is not itself a part.

connectedness: a technical term from modern topology; a topological space is said to be connected if it is not the union of two disjoint nonempty open sets of points. This definition is not strictly applicable to Leibniz, in that he could accept neither open nor closed infinite sets of points, since he denied infinite collections.

contiguous: contiguous parts, segments, or intervals, are those whose boundaries are touching, but which do not share a boundary in common.

contiguum: a whole all of whose parts are successively contiguous.

continuous: even in modern mathematics this is not a univocal notion. A space may be said to be continuous if it is **Dedekind-complete** (q.v.), if it is **connected** (q.v.), if it is **arcwise connected** or path-connected (q.v.), to name just some properties that might be intended; see De Risi (2019). See **continuum** for Leibniz's understanding of it.

continuum

mathematical continuum: one whose parts are merely potential, and from which it is not composed; the boundaries (e.g. points or instants) in it designate sections (points or instants) at which it could be divided. The possible parts in it must satisfy the following two conditions: (1) any two parts that are **co-integrant** (q.v.) must have something in common that is not itself a part; and (2) there must be *partes extra partes*, i.e. parts that are not co-integrant and have nothing in common, not even a boundary.

physical continuum: that in which there are no two actually assignable points or instants between which another cannot be actually located.

Dedekind-completeness: a technical term from modern mathematics for the continuity (in the sense of completeness, or absence of gaps) of the real line; it is based on the idea of a Dedekind cut, and on a one–one correspondence between points in a geometrical line and the real numbers. A Dedekind cut is the unique point that divides all the points on a straight line into two classes such that every point in the first class lies to the left of every point in the second. One such cut corresponds, for example, to the real number which divides all the rational numbers into those which are strictly greater than $\sqrt{2}$, and all those strictly less than it. Leibniz gives definitions that come quite close to the idea of a Dedekind cut in the points on a line, without, however, showing any inclination to establish a correspondence with the real numbers.

density: in modern mathematics an interval is **dense** if between any two elements there is a further element; the rational numbers, for instance, are dense.

extensum (*extensum*, [*une*] *étendue*): an extended thing in general; this is a technical term analogous to **continuum** for a continuous thing in general.

fibration: a technical term from modern topology; a fibration is a generalization of the notion of one topological space being 'parameterized' by another. Leibniz's claim that we may construct a (biunivocal and continuous) correspondence between a family of ellipses and the points of a segment can be seen as an anticipation of this notion (see De Risi 2019, 132). From this point of view an **extensum** (q.v.) may be regarded as 'constituted' by an infinity of points (or more generally, objects of one dimension less) 'taken in order'.

homeomorphism: in modern topology a homeomorphism is a continuous function between topological spaces that has a continuous inverse function—i.e. it is a bicontinuous transformation, such as that by which a doughnut can be continuously transformed into a coffee cup. But a continuous deformation such as a line into a point is not a homeomorphism; for Leibniz such deformations would be transformations between **homogonous** objects (q.v.), but not **homogeneous** ones (q.v.).

homogeneous: two things are homogeneous to one another if a third thing may be found, which is equal to the first and similar to the second: for instance, any two parts of an **extensum** (q.v.) are homogeneous.

homogonous: homogonous things are those one of which can be reduced to the other 'by a process of continuous change', that is, by a bi-continuous transformation, whether these things are **homogeneous** (q.v.) (for instance, a circle and a square), or **heterogeneous** (q.v.) (for instance, a line and a point, a surface and a line). Whereas homogeneous things can be transformed into one another by a simple *continuous transformation*, however, heterogeneous things cannot: a point and a line are of different dimensions. The closest modern analogue to a transformation between homogonous things, then, is a one-parameter family of continuous transformations (see De Risi 2007, 162–3, n.34).

homomorphism: a structure-preserving mapping between two structures of the same type; applied to Leibniz's idea of expression, where the same universe is expressed (with different degrees of expression or of distinct perception) from the different points of view of each monad.

hypercategorematic: a term apparently of Leibniz's own coining: see **infinity**.

immensum (*immensum*): the divine attribute of being extended per se, which is maximal and indivisible; it is the basis of space, but cannot itself be divided into parts; rather finite **extensa** (q.v.) result from it by the introduction into it of limits in the form of spatial boundaries.

infinitesimal (*infinitesimalis*): an infinitely small quantity. Leibniz understands this **syncategorematically** (q.v.), that is, as meaning that no matter how small a quantity is taken, it is smaller still. In this sense a finite line may be regarded as fictionally composed of infinitely many infinitesimals, on the understanding that if sufficiently many sufficiently small finite quantities are taken, the error may be made as small as desired, and thus zero.

infinity (*infinitum*): the **actual infinite**, as applied to the created world and to possibles, is understood **syncategorematically** (q.v.). Thus to say that there are infinitely many parts of matter is to say that no matter how many parts may be taken, there are in fact more; but there is no infinite number of them taken collectively. It is the same in mathematics; there are infinitely many odd numbers, but there is no infinite number of odd numbers, no **categorematic infinite** (q.v.);

likewise, there are no **actually infinitely large** quantities or **actually infinitely small** ones, since their concepts entail a contradiction with the part-whole axiom. But one may nevertheless use terms apparently denoting them, as **fictions**. On such a syncategorematic understanding, **infinitesimal** quantities (q.v.), for instance, 'are not at all fixed and determined, but can be taken as small as one wishes, and so have the effect of the rigorously infinitely small' (to Pierre Varignon, 1702; GM IV 92). See **infinitesimal**.

instant (*instans*): the endpoint of a temporal interval having no temporal extent, analogous to a point of space; there are never more than two such **geometric instants** immediately next to each other in the same time;

an **actual instant** is one at which change occurs; and no two such actual instants can be immediately next to one another, since there are always further changes occurring in the states separating them; they are separated by **unassignable** intervals (q.v.).

isomorphism: a structure-preserving mapping between two structures of the same type that can be reversed by an inverse mapping. This has been suggested as capturing Leibniz's idea of expression or representation; although **homomorphism** (q.v.) seems a closer match. An ideal isomorphism would be a representation of every facet of the universe, without residue.

moment (*momentum*): an arbitrarily small temporal interval. Leibniz does not always carefully distinguish a moment from an **instant** (q.v.) in the sense of the endpoint of a temporal interval.

momentaneous (*momentaneus*): enduring for only a **moment** (q.v.).

partes extra partes: this is a traditional formula for one property of a continuous quantity, that 'there are parts outside one another', or as Leibniz rephrases it, 'that two of its parts can be assumed (but not ones that together equal the whole) which have nothing in common, not even a minimum' (GM V 184).

requisite (*requisitum*): a necessary condition for something that is prior by nature to it;

mediate requisite: a requisite whose priority is one of *existence*, and requires intervening changes; mediate requisites are efficient causes that produce their effects;

immediate requisite: a requisite whose priority is one of *essence*, and which involves no intermediate changes; examples are parts, extrema, and generally things which *are in* a thing.

situation (*situs*): situation is a mode of coexisting; it is 'a certain relation of coexisting among many things, and it is known through other things, the intermediaries, that is, those things having a simpler relation of coexistence to the former things' (GM VII 25).

syncategorematic: a designation from Scholastic philosophy; if a term is understood **syncategorematically**, this connotes that it derives its meaning from the context, but does not refer to an actual entity. As such it is opposed to a **categorematic** term (q.v.).

unassignable (*inassignabilis*): an unassignable quantity is one smaller than any assignable; on Leibniz's **syncategorematic** (q.v.) interpretation, an unassignable interval is one that may be made so small that the resultant error is smaller than any that is pre-assigned.

Bibliography

Primary Sources

Aristotle. 1984. *The Complete Works of Aristotle*, 2 vols., Revised Oxford Translation, ed. Jonathan Barnes. Princeton: Princeton University Press; abbr. **BA**.

Barrow, Isaac. 1670. *Lectiones Geometricae*. London: John Dunmore.

Barrow, Isaac. 1683. *Lectiones Mathematicae XXIII*. London: Wells.

Barrow, Isaac. 1860. (Includes *Lectiones Mathematicae, Lectiones Opticae* and *Lectiones Geometricae*). *The Mathematical Works of Isaac Barrow*. Ed. Cambridge: Cambridge University Press.

Barrow, Isaac. 1916. *The Geometrical Lectures of Isaac Barrow*. Transl., with notes and proofs, by J. M. Child. Chicago: Open Court.

Baumgarten, Alexander. 2013. *Metaphysics. A Critical Translation with Kant's Elucidations, Selected Notes, and Related Materials*. Ed. and transl. Courtney D. Fugate and John Hymers. London. New York: Bloomsbury Publishing.

Borelli, Giovanni. 1667. *De vi percussionis*. Bologna: Ex typographia Iacobi Montii.

Bradley, Francis H. 1893. *Appearance and Reality*. Revised ed. Oxford: Oxford University Press, 1969.

Clauberg, Johannes. 1691. *Opera omnia philosophica*, ed. Johannes Theodor Schalbruch. 2 vols. Amsterdam; reprint Hildesheim: Georg Olms, 1968.

Clavius, Christophorus François de Foix. 1589. *Euclidis elementorum libri XV: Accessit XVI*. Cologne: Peter Cholinus.

Clavius, Christophorus François de Foix. 1612. *Opera mathematica*. 5 vols. Mainz: Reinhard Eltz.

Descartes, René. 1644. *Principia Philosophiae*. Amsterdam: Elzevir. Transl. in (Descartes 1985).

Descartes, René. 1964–76. *Oeuvres de Descartes*, 12 vols., Nouvelle présentation, ed. Charles Adam and Paul Tannery. Paris: J. Vrin; abbr. **AT**.

Descartes, René. 1985. *The Philosophical Writings of Descartes*, Vol. 1. Ed. and transl. John Cottingham, Robert Stoothof, and Dugald Murdoch. Cambridge: Cambridge University Press.

Diogenes Laertius. 1925. *Lives of the Eminent Philosophers*. Book VIII. Transl. Robert Drew Hicks. http://en.wikisource.org/wiki/Livesofthe_EminentPhilosophers/BookVIII#Archytas (accessed June 10, 2020).

Euclid. 1933. *The Elements of Euclid*, transl. Isaac Toddhunter, London: J. M. Dent.

Fine, Oronce. 1536. *In sex priores libros Geometricorum elementorum Euclidis Megarensis demonstrationes*. Paris: Simon de Colines (republished 1544, 1551).

Gerhardt, C. I. 1875. Zum zweihundertjährigen Jubiläum der Entdeckung des Algorithmus der höheren Analysis durch Leibniz, *Monatsberichte der Königlich Preußischen Akademie der Wissenschaften* 1875, 595–608.

Hegel, Georg W. 1896. *Lectures on the History of Philosophy*, transl. E. S. Haldane and Frances M. Simson, 3 vols.. Reprinted Routledge and Kegan Paul, 1955.

Hobbes, Thomas. 1839a. *The English Works of Thomas Hobbes of Malmesbury*, Vol. 1, *Concerning Body*, collected and edited by Sir William Molesworth. London: John Bohn.

Hobbes, Thomas. 1839b. *Opera Philosophica*. Ed. Sir William Molesworth. London: John Bohn; reprint. Darmstadt: Scientia Verlag Aalen, 1966.

Huygens, Christiaan. 1673. *Horologium Oscillatorium sive de motu pendulorum ad horologia adapta demonstrationes geometricae*. Paris: F. Muguet.

Huygens, Christiaan. [1654] 1929. *De motu corporum ex percussione. Appendice; Oeuvres Complètes de Christiaan Huygens*, XVI. The Hague: Martinus Nijhoff.

Huygens, Christiaan. 1986. *The Pendulum Clock, or Geometrical Demonstrations Concerning the Motion of Pendula as Applied to Clocks*. Transl. of *Horologium Oscillatorium* (1673) with notes by Richard J. Blackwell. Ames: Iowa State University Press.

Leibniz, Gottfried Wilhelm. 1689. *Tentamen de Motuum Coelestium causis*, *Acta Eruditorum*, February 1689, 82–96. Reprinted at GM VI 144–61.

Leibniz, Gottfried Wilhelm. 1695."*Specimen Dynamicum*", *Acta Eruditorum*, April 1695. Reprinted at GM VI 234–46.

Leibniz, Gottfried Wilhelm. 1710. *Essais de Théodicée sur la bonté de Dieu, la liberté de l'homme et l'origine du mal.*

Leibniz, Gottfried Wilhelm. 1840. *God. Guil. Leibnitii quae extant Latina Gallica Germanica Omnia*. Ed. Joannes Eduardus Erdmann. Berlin: Olms, 1840; abbr. **E**.

Leibniz, Gottfried Wilhelm. 1849–63. *Leibnizens Mathematische Schriften*. Ed. C. I. Gerhardt. Berlin and Halle: Asher and Schmidt; reprint Hildesheim: Olms, 1971, 7 vols; abbr. **GM**.

Leibniz, Gottfried Wilhelm. 1875–90. *Der Philosophische Schriften von Gottfried Wilhelm Leibniz*. Ed. C. I. Gerhardt. Berlin: Weidmann; reprint Hildesheim: Olms, 1960, 7 vols; abbr. **GP**.

Leibniz, Gottfried Wilhelm. 1903. *Opuscules et fragments inédits de Leibniz*. Ed. Louis Couturat. Paris: Alcan; reprint Hildesheim: Olms, 1966; abbr. **C**.

Leibniz, Gottfried Wilhelm. 1923–. *Sämtliche Schriften und Briefe*. Ed. Akademie der Wissenschaften der DDR. Darmstadt and Berlin: Akademie-Verlag; abbr. **A**, e.g. (A VI 2, 229).

Leibniz, Gottfried Wilhelm. 1969. *Philosophical Papers and Letters*, 2nd. ed. Transl. and ed. Leroy Loemker. Dordrecht: D. Reidel; abbr. **L**.

Leibniz, Gottfried Wilhelm. 1981. *New Essays on Human Understanding*. Transl. and ed. Peter Remnant and Jonathan Bennett. Cambridge: Cambridge University Press; abbr. **RB**.

Leibniz, Gottfried Wilhelm. 1985. *Theodicy: Essays on the Goodness of God, the Freedom of Man, and the Origin of Evil*. Transl. E. M. Huggard of 1710. La Salle, Ill: Open Court; abbr. **H**.

Leibniz, Gottfried Wilhelm. 1989. *Philosophical Essays*. Ed. and transl. Roger Ariew and Daniel Garber. Indianapolis and Cambridge: Hackett; abbr. **AG**.

Leibniz, Gottfried Wilhelm. 1991. *Phoranomus; seu de Potentia et Legibus Naturae* (July 1689), édité par André Robinet, *Physis*, v. 28, n. 3, 1991, 429–541, & v. 28, n. 23, 1991, 797–885.

Leibniz, Gottfried Wilhelm. 1992. *De Summa Rerum: Metaphysical Papers 1675–1676*. Translated and with an introduction by G. H. R. Parkinson. New Haven: Yale University Press; abbr. **DSR**.

Leibniz, Gottfried Wilhelm. 1995a. *La caractéristique géométrique*. Ed. Javier Echeverría, transl. Marc Parmentier, with annotations, introduction, and postface. Paris: J. Vrin; abbr. **CG**.

Leibniz, Gottfried Wilhelm. 1995b. *Philosophical Writings*. Ed. G. H. R. Parkinson, transl. Mary Morris and G. H. R. Parkinson. London: J. M. Dent; Rutland, Vermont: Charles Tuttle, 1995; abbr. **MP**.

Leibniz, Gottfried Wilhelm. 1998. *Philosophical Texts*. Transl and ed. R. S. Woolhouse and Richard Francks. Oxford/New York: Oxford University Press; abbr. **WFT**.

Leibniz, Gottfried Wilhelm. 2001. *The Labyrinth of the Continuum: Writings on the Continuum Problem, 1672–1686*. Ed., sel., and transl. R. T. W. Arthur. New Haven: Yale University Press; abbr. **LLC**.

Leibniz, Gottfried Wilhelm. 2007. *The Leibniz-Des Bosses Correspondence*. Selected, ed. and transl., with an introductory essay by Brandon Look and Don Rutherford. New Haven: Yale University Press; abbr. **LDB**.

Leibniz, Gottfried Wilhelm. 2008. *Ricerche generali sull'analisi delle nozioni e della verità*. a cura di Massimo Mugnai. Pisa: Edizioni della Normale.

Leibniz, Gottfried Wilhelm. 2013. *The Leibniz-De Volder Correspondence*. Translated, edited and with an Introduction, by Paul Lodge. New Haven: Yale University Press; abbr. **LDV**.

Leibniz, Gottfried Wilhelm. 2016a. *The Leibniz-Arnauld Correspondence, with selections from the correspondence with Ernst, Landgrave of Hessen-Rheinfels*. Text established and translated and with an introduction by Stephen Voss. New Haven: Yale University Press; abbr. **LAV**.

Leibniz, Gottfried Wilhelm. 2016b. *The Leibniz-Stahl Controversy*. Transl., ed., and with an introduction by François Duchesneau and Justin E. H. Smith. New Haven: Yale University Press; abbr. **LSC**.

Leibniz, Gottfried Wilhelm. 2021. *Mathesis Texts*. Leibniz texts prepared from manuscript sources by the ANR MATHESIS project in collaboration with the Leibniz Research Centre in Hannover (Leibniz-Archiv); appearing under the Creative Commons licence **CC-by-NC 4.0**; http://www.gwlb.de/Leibniz/Leibnizarchiv/Veroeffentlichungen/Mathesis.pdf (accessed March 21, 2021); abbr. **MT** and number of text.

Leibniz, Gottfried Wilhelm and Samuel Clarke. 1956. *The Leibniz-Clarke correspondence; together with extracts from Newton's* Principia *and* Opticks. Ed. H. G. Alexander. Manchester: Manchester University Press; abbr. **LC**.

Leibniz, Gottfried Wilhelm and Burchard De Volder. 2016. *Leibniz-De Volder Correspondance*. Traduite, annotée et précédée d'une introduction par Anne-Lise Rey. Paris: Vrin.

Locke, John. 1690. *An Essay Concerning Humane Understanding*. 2nd ed. Accessed through Project Gutenberg, produced by Steve Harris and David Widger, 2004. London: Elizabeth Holt for Thomas Basset.

Mach, Ernst. 1919. *The Science of Mechanics: A Critical and Historical Account of its Development*, 4th ed. Transl. Thomas J. McCormack. Chicago/London: Open Court.

Mariotte, Edme. 1673. *Traitté de la Percussion ou Chocq des Corps*. Paris: Estienne Michallet.

Newton, Isaac. 1687. *Philosophiae Naturalis Principia Mathematica*. 1st ed. London: Joseph Streater. Produced by Jonathan Ingram, Keith Edkins and the Online Distributed Proofreading Team at http://www.pgdp.net (accessed June 15, 2021).

Newton, Isaac. 1713. *Philosophiae Naturalis Principia Mathematica* (2nd ed., further increased and emended). Cambridge: Cambridge University Press.

Newton, Isaac. 1962. *Unpublished Scientific Papers of Isaac Newton*. Ed. Rupert and Marie Boas Hall. New York: Cambridge University Press.

Newton, Isaac. 1964. *The Mathematical Works of Isaac Newton*, I. Ed. Derek T. Whiteside. New York: Johnson.

Newton, Isaac. 1999. *The Principia: Mathematical Principles of Natural Philosophy*. Transl. and ed. I. B. Cohen and Anne Whitman. Berkeley: The University of California Press.

Newton, Isaac. 2004. *Newton: Philosophical Writings*. Ed. Andrew Janiak. Cambridge: Cambridge University Press.

Newton, Isaac. 2008. *The Mathematical Papers of Isaac Newton*, VII, 1691–1695. Ed. Derek T. Whiteside. Cambridge: Cambridge University Press.

Ockham, William. 1984. *Quæstiones in Libros Physicorum Aristotelis, Opera Philosophica et Theologica*, Vol. 6, ed. Stephan Brown. St. Bonaventure, Franciscan Institute.

Pardies, Ignace-Gaston. 1670. *Discours du movement local.* Paris: Edme Martin.

Peletier du Mans, Jacques. 1557. *In Euclidis elementa geometrica demonstrationum, Libri 6.* Lyon: Jean de Tournes.

Proclus of Lycia. 1533. *Commentarium Procli editio prima quae Simonis Grynaei opera addita est Euclidis elementis graece editis.* Basle: J. Hervagius.

Proclus of Lycia. 1873. *Procli Diadochi In Primum Euclidis Elementorum Librum Commentarii.* Ed. Gottfried Friedlein. Leipzig: B. G. Teubner.

Proclus of Lycia. 1970. *A Commentary on the First book of Euclid's Elements.* Transl. Glenn R. Morrow. Princeton: Princeton University Press (reissued in 1992, with a new foreword by I. Mueller).

Secondary Sources

Adams, Robert Merrihew. 1994. *Leibniz: Determinist, Theist, Idealist.* Oxford: Oxford University Press.

Ademollo, Francesco, Vincenzo De Risi, and Fabrizio Amellini (eds.). 2021. *Thinking and Calculating. Essays on Logic, its History and its Applications*, Dordrecht: Springer.

Adomaitis, Laurynas. 2019. "Equivalence of hypotheses and Galilean censure in Leibniz: A conspiracy or a way to moderate censure?", *Revue d'Histoire des Sciences*, **1**, 63–86.

Adomaitis, Laurynas. 2020. *Leibniz's Logical Foundations of Physics: The Path to a Reform.* Doctoral Thesis, Scuola Normale Superiore, Pisa.

Aiton, E. J. 1972. *The Vortex Theory of Planetary Motions.* London: Macdonald.

Anfray, Jean-Pascal. 2007. "La théorie du temps de Leibniz", pp. 91–114 in *Le temps*, ed. Alexandre Schnell (Paris: Vrin, 2007).

Antognazza, Maria Rosa (ed.). 2018. *The Oxford Handbook of Leibniz*, Oxford: Oxford University Press.

Arthur, R. T. W. 1985. "Leibniz's Theory of Time", pp. 263–313 in *The Natural Philosophy of Leibniz*, ed. Kathleen Okruhlik and James R. Brown, D. Reidel: Dordrecht.

Arthur, R. T. W. 1987. "Space as an Order of Situations", *Abstracts, LMPS '87*, Vol. 2, Moscow: Institute of the Academy of Sciences of the USSR, 21–3.

Arthur, R. T. W. 1994a. "Space and Relativity in Newton and Leibniz", *British Journal for the Philosophy of Science*, **45**, 219–40.

Arthur, R. T. W. 1994b. "Relations of Time and Space", pp. 9–16 in *Leibniz und Europa*, ed. Herbert Breger. Hanover: Gottfried-Willhelm-Leibniz-Gesellschaft.

Arthur, R. T. W. 1995. "Newton's Fluxions and Equably flowing Time", *Studies in the History and Philosophy of Science*, **26**, 323–51.

Arthur, R. T. W. 2008. "Leery Bedfellows: Newton and Leibniz on the Status of Infinitesimals", pp. 7–30 in (Goldenbaum and Jesseph, ed. 2008).

Arthur, R. T. W. 2009 "Actual Infinitesimals in Leibniz's Early Thought", pp. 11–28 in *The Philosophy of the Young Leibniz, Studia Leibnitiana Sonderhefte* **35**, ed. Mark Kulstad, Mogens Laerke, and David Snyder. Stuttgart: Franz Steiner.

Arthur, R. T. W. 2010. "Minkowski's Proper Time and the Status of the Clock Hypothesis", pp. 159–79 in *Space, Time, and Spacetime*, ed. Vesselin Petkov. Berlin/Heidelberg: Springer-Verlag.

Arthur, R. T. W. 2013a. "Leibniz's Syncategorematic Infinitesimals, Smooth Infinitesimal Analysis, and Second Order Differentials", *Archive for History of Exact Sciences*, **67**, 553–93.

Arthur, R. T. W. 2013b. "Leibniz's Theory of Space", *Foundations of Science*, **18**, 3, 499–528, 2013. Published online in 2012 as DOI: 10.1007/s10699-011-9281-4 (accessed June 15, 2021).

Arthur, R. T. W. 2013c. "'Leibniz's *Mechanical Principles*' (c. 1676): Commentary and Translation", *The Leibniz Review*, **23**, 101–16.

Arthur, R. T. W. 2013d. "Can Thought Experiments Be Resolved by Experiment? The Case of 'Aristotle's Wheel", 107–122 in *Thought Experiments in Philosophy, Science and the Arts*, ed. Melanie Frappier, Letitia Meynell, and James Robert Brown. New York and Abingdon: Routledge.

Arthur, R. T. W. 2014. *Leibniz*. (Classic Thinkers Series). Cambridge: Polity Press.

Arthur, R. T. W. 2015a. "The Relativity of Motion as a Motivation for Leibnizian Substantial Forms", pp. 143–60 (chapter 10) in *Leibniz's Metaphysics and Adoption of Substantial Forms*, ed. Adrian Niță. Springer Verlag.

Arthur, R. T. W. 2015b. "Newton and Leibniz on the Relativity of Motion". Online version of the *Oxford Handbook of Newton*, ed. Chris Smeenk and Eric Schliesser. Oxford: Oxford University Press.

Arthur, R. T. W. 2016. "Leibniz's Causal Theory of Time Revisited", *The Leibniz Review*, **26**, 151–78.

Arthur, R. T. W. 2018a. *Monads, Composition and Force: Ariadnean Threads through Leibniz's Labyrinth*. Oxford: Oxford University Press.

Arthur, R. T. W. 2018b. "The Hegelian Roots of Russell's Critique of Leibniz", *The Leibniz Review*, **28**, 9–42.

Arthur, R. T. W. 2018c. "Leibniz's Syncategorematic Actual Infinite", pp. 155–79 in Nachtomy and Winegar, ed. 2018.

Arthur, R. T. W. 2019a. "Leibniz in Cantor's Paradise", ch. 3 in *Leibniz and the Structure of Science: Modern Perspectives on the History of Logic, Mathematics, Epistemology*, ed. Vincenzo De Risi. Dordrecht: Springer.

Arthur, R. T. W. 2019b. "Mario Bunge on causality: some key insights and their Leibnizian precedents", pp. 185–204 in *Mario Bunge: A Centenary Festschrift*, ed. Michael R. Matthews. Springer Verlag.

Arthur, R. T. W. 2019c. *The Reality of Time Flow: Local Becoming in Modern Physics*. Frontiers Collection. Cham, Switzerland: Springer.

Arthur, R. T. W. 2021. "Leibniz as a precursor to Chaitin's Algorithmic Information Theory", pp. 153–76, chapter 9 in *Information and the History of Philosophy*, ed. Chris Meyns. London: Routledge.

Arthur, R. T. W. (unpublished). "Russell's Rejection of the Relational Theory: Leibniz, Lotze, and Internal Relations", chapter in a planned book with Nicholas Griffin, *Russell on Leibniz*.

Axworthy, Angela. 2013. "The ontological status of geometrical objects in the commentary on the *Elements of Euclid* of Jacques Peletier du Mans (1517–1582)", *Fondation Maison des sciences de l'homme*, FMSH-WP-2013-**41**, août.

Axworthy, Angela. 2014. "Sixteenth-century Interpretations of the Nature and Use of Motion in Geometry", presentation for a workshop on *The Mechanization of Geometry. From Antiquity to the Modern Age*, MPIWG, June 23, 2014.

Axworthy, Angela. 2017. "La notion géométrique de flux du point à la Renaissance et dans le commentaire des Éléments de Jacques Peletier du Mans", pp. 453–64 in *Miroir de l'amitié. Mélanges offerts à Joël Biard à l'occasion de ses 65 ans*, ed. Christophe Grellard. Paris: Vrin.

Axworthy, Angela. 2018. "The debate between Peletier and Clavius on superposition", *Historia Mathematica*, **45**, 1–38.

Axworthy, Angela. 2021. *Motion and Genetic Definitions in Sixteenth-century Commentaries on Euclid's Elements*. Forthcoming with Birkhäuser (in the series Frontiers in the History of Science).

Barbour, Julian. 1974. "Relative-distance Machian Theories", *Nature*, **249**, 38–29.

Barbour, Julian. 1999. *The End of Time: The Next Revolution in Physics*. Oxford: Oxford University Press.

Barbour, Julian and Bertotti, B. 1977. "Gravity and Inertia in a Machian Framework", *Nuovo Cimento* **38B**, 1–27.

Barbour, Julian and Bertotti, B. 1982. "Mach's Principle and the Structure of Dynamical Theories", *Proceedings of the Royal Society*, **382**, 295–306.

Belkind, Ori. 2013. "Leibniz and Newton on Space", *Foundations of Science*, **18**, 467–97. Published online in April 2012 at DOI: 10.1007/s10699-011-9280-5 (accessed June 15, 2021).

Bertoloni Meli, Domenico. 1988. "Leibniz on the Censorship of the Copernican System", *Studia Leibnitiana*, **20**, 1, 19–42.

Bertoloni Meli, Domenico. 1993. *Equivalence and Priority: Newton versus Leibniz*. Oxford: Clarendon Press.

Bohm, David. 1957. *Causality and Chance in Modern Physics*. Philadelphia: University of Pennsylvania Press.

Bouquiaux, Laurence. 2017. "Equivalence des hypothèses et relativité du mouvement dans la *Dynamica*", *Studia Leibnitiana*, **49**, 1, 54–74.

Brancato, Mattia. 2015. *Erhard Weigel and his Influence on Leibniz's Philosophy of Mathematics*. Doctoral Thesis under Gianfranco Mormino, Università degli Studi di Milano.

Broad, C. D. 1975. *Leibniz: An Introduction*. Ed. C. Lewy. Cambridge: Cambridge University Press.

Brower, Jeffrey. 2005. "Medieval Theories of Relations", *The Stanford Encyclopedia of Philosophy* (Winter 2018 Edition), Edward N. Zalta (ed.), URL = <https://plato.stanford.edu/archives/win2018/entries/relations-medieval/>.

Bussotti, Paolo. 2015. *The Complex Itinerary of Leibniz's Planetary Theory Physical Convictions, Metaphysical Principles and Keplerian Inspiration*. Cham, Switzerland: Birkhäuser/Springer.

Butts, Robert E. 1985. "Leibniz on the Side of the Angels", pp. 207–26 in *The Natural Philosophy of Leibniz*, ed. Kathleen Okruhlik and James R. Brown. Dordrecht: D. Reidel.

Čapek, Milič (ed.). 1976. *Concepts of Space and Time*. Boston: D. Reidel.

Cassirer, Ernst. 1902. *Leibniz' System in seinen wissenschaftlichen Grundlagen*. Marburg: N. G. Elwert'sche Verlagsbuchhandlung.

Collier, John D. 1999. "Causation is the Transfer of Information", pp. 279–331 in *Causation and Laws of Nature*, ed. Howard Sankey. Dordrecht: Kluwer; paginated according to a downloaded version, 1–32.

Coopersmith, Jennifer. 2017. *The Lazy Universe: An Introduction to the Principle of Least Action*. Oxford: Oxford University Press.

Couturat, Louis. 1902. "Sur la métaphysique de Leibniz", *Revue de métaphysique et de morale* **10**, janvier, 1–25.

Cover, Jan. 1997. "Non-Basic Time and Reductive Strategies", *Studies in History and Philosophy of Science*, **28**, 2, 289–318.

Cover, Jan. 1989. "Relations and Reduction in Leibniz", *Pacific Philosophical Quarterly*, **70**, 3, 185–211.

Cover, J. A. and John O'Leary-Hawthorne. 1999. *Substance and Individuation in Leibniz*. Cambridge: Cambridge University Press.

Debuiche, Valèrie. 2009. "La notion d'expression et ses origines mathématiques",*Studia Leibnitiana*, Bd. 41, H. 1, 88–117.

Debuiche, Valèrie. 2013. "L'expression leibnizienne et ses modèles mathématiques", *Journal of the History of Philosophy*, **51**, 3, 409–39.

Debuiche, Valèrie and David Rabouin 2019. "On the Plurality of Spaces in Leibniz", chapter 5 in (De Risi, ed. 2019), 171–201.

De Risi, Vincenzo. 2005. "Leibniz on Geometry: Two Unpublished Texts with Translation and Commentary", *The Leibniz Review*, **15**, 127–51.

De Risi, Vincenzo. 2007. *Geometry and Monadology: Leibniz's* Analysis Situs *and Philosophy of Space*. Basel/Boston/Berlin: Birkhäuser.

De Risi, Vincenzo. 2012. "Leibniz on Relativity: The Debate between Hans Reichenbach and Dietrich Mahnke on Leibniz's Theory of Motion and Time", pp. 143–85 in *New Essays in Leibniz Reception: Science and Philosophy of Science 1800–2000*, ed. Ralf Krömer and Yannick Chin-Drian. Basel: Springer.

De Risi, Vincenzo. 2014. "Francesco Patrizi e la nuova geometria dello spazio", pp. 269–327 in *Locus-Spatium*, ed. Delfina Giovannozzi and Marco Veneziani. Firenze: Olschki.

De Risi, Vincenzo. 2015. *Leibniz on the Parallel Postulate and the Foundations of Geometry*. Basel/Boston: Birkhäuser.

De Risi, Vincenzo (ed.). 2015. *Mathematizing Space*. Cham, Switzerland: Springer.

De Risi, Vincenzo. 2016. "Francesco Patrizi and the New Geometry of Space", pp. 55–106 in *Boundaries, Extents and Circulations* (*Studies in History and Philosophy of Science*, Vol. 41, ed. K. Vermeir and J. Regier. Cham: Springer (English version of De Risi 2014).

De Risi, Vincenzo. 2018. "Analysis Situs, the Foundations of Mathematics and a Geometry of Space", pp. 247–58 in (Antognazza, ed. 2018).

De Risi, Vincenzo. 2019. "Leibniz on the Continuity of Space", chapter 4, pp. 111–69 in (De Risi, ed. 2019).

De Risi, Vincenzo (ed.). 2019. *Leibniz and the Structure of Sciences: Modern Perspectives on the History of Logic, Mathematics, Epistemology*. Boston Studies in the Philosophy of Science, Vol. 337. Cham, Switzerland: Springer.

Di Bella, Stefano. 2005. *The Science of the Individual: Leibniz's Ontology of Individual Substance*. Dordrecht: Springer.

Di Bella, Stefano. 2018. "The Complete Concept of an Individual Substance", 119–36 in (Antognazza, ed. 2018).

Earman, John. 1989. *World Enough and Space-Time: Time: Absolute versus Relational Theories of Space and Time*. Cambridge, MA: MIT Press.

Earman, John, Clark Glymour, and John Stachel (eds.). 1977. *Foundations of Space-Time Theories*. Minneapolis: University of Minnesota Press.

Ehlers, Jürgen. 1973. "The Nature and Structure of Space-time", 71–91 in *The Physicist's Conception of Nature*, ed. J. Mehra. Dordrecht: D. Reidel.

Feingold, Mordechai. 1993. "Newton, Leibniz, and Barrow Too: An Attempt at a Reinterpretation", *Isis*, **84**, 310–38.

Fichant, Michel. 1994. *G. W. Leibniz: La réforme de la dynamique*. Ed., présentation, traductions et commentaires par Michel Fichant. Paris: J. Vrin.

Fichant, Michel. 1995. "De la puissance à l'action, la singularité stylistique de la Dynamique", *Revue de Métaphysique et de Morale*, **100**, 49–81.

Frankfurt, Harry (ed.). 1972. *Leibniz: A Collection of Critical Essays*. Garden City, NY: Doubleday Anchor.

van Fraassen, Bas. [1970] 1985. *An Introduction to the Philosophy of Time and Space*. Morningside ed. New York: Columbia University Press.

Friedman, Michael. 1983. *Foundations of Space-Time Theories*. Princeton: Princeton University Press.

Friedman, Michael. 2010. "A Post-Kuhnian Approach to the History and Philosophy of Science", *The Monist*, **93**, 4, 497–517.

Futch, Michael J. 2008. *Leibniz's Metaphysics of Time and Space*. Dordrecht: Springer.

Futch, Michael J. 2011. Review of *La Métaphysique de Temps chez Leibniz et Kant*, by Adrian Nita, *The Leibniz Review*, **21**, 171–4.

Garber, Daniel. 1995. "Leibniz: Physics and Philosophy", pp. 270–352 in *The Cambridge Companion to Leibniz*, ed. N. Jolley. New York: Cambridge University Press.

Garber, Daniel. 2009. *Leibniz: Body, Substance, Monad*. Oxford: Oxford University Press.

Garber, Daniel. 2012. "Leibniz, Newton and Force", pp. 33–47 in *Interpreting Newton*, ed. E. Schliesser and A. Janiak. Cambridge: Cambridge University Press.

Giovanelli, Marco. 2011. "Leibniz Equivalence: On Leibniz's (Bad) Influence on the Logical Empiricist Interpretation of General Relativity", preprint submiitted to *Studies in History and Philosophy of Science*, URL: http://philsci-archive.pitt.edu/id/eprint/9676 (accessed 2021-07-06)

Goodman, Nelson and Willard v. O. Quine. 1947. "Steps Toward a Constructive Nominalism," *Journal of Symbolic Logic* **12**, 4, 105–122.

Goldenbaum, Ursula and Douglas Jesseph (eds.). 2008. *Infinitesimal Differences: Controversies between Leibniz and his Contemporaries*. Berlin and New York: De Gruyter.

Gorham, Geoffrey. 2007. "Descartes on Time and Duration", *Early Science and Medicine*, **12**, 28–54.

Gorham, Geoffrey. 2008. "Cartesian Temporal Atomism: A New Defence, a New Refutation", *British Journal for the History of Philosophy*, **16**, 3, 625–37.

Grünbaum, Adolf. 1963. *Philosophical Problems of Space and Time*. New York: Knopf.

Hartz, Glenn A. 1992. "Leibniz's Phenomenalisms", *The Philosophical Review*, **101**, 3, 511–49.

Hartz, Glenn A. and J. A. Cover. 1988. "Space and Time in the Leibnizian Metaphysic", *Noûs*, **22**, 4, 493–519.

Hellman, Geoffrey and Stewart Shapiro. 2018. *Varieties of Continua: From Regions to Points and Back*. Oxford: Oxford University Press.

Hintikka, Jaakko. 1972. "Leibniz on Plenitude, Relations and the 'Reign of Law'", pp. 155–90 in (Frankfurt, ed. 1972).

Huggett, Nick (ed.). 1999. *Space from Zeno to Einstein*. Cambridge, MA/London: Bradford Books.

Huggett, Nick. 2006. "Motion in Leibniz's Physics and Metaphysics", Chapter 3: Leibniz, of unpublished ms. available on the internet at: https://philpapers.org/archive/HUGCL.pdf (accessed June 15, 2021).

Huggett, Nick and Carl Hoefer. 2019. "Absolute and Relational Theories of Space and Motion", *The Stanford Encyclopedia of Philosophy* (Spring 2019 Ed.), Edward N. Zalta (ed.), http://plato.stanford.edu/archives/spr2019/entries/spacetime-theories/ (accessed June 15, 2021).

Ishiguro, Hidé. 1972a. *Leibniz's Philosophy of Logic and Language*. Ithaca, NY: Cornell University Press.

Ishiguro, Hidé. 1972b. "Leibniz's Theory of the Ideality of Relations", pp. 191–214 in Frankfurt, ed. (1972).

Ishiguro, Hidé. 1990. *Leibniz's Philosophy of Logic and Language*. 2nd ed. Cambridge: Cambridge University Press.

Jalabert, Jacques. 1985. *La théorie leibnizienne de la substance*. Paris: Presses Universitaires de France, 1947; reprinted by Garland Publishing of New York and London.

Jalobeanu, Dana. 2011. "The Cartesians of the Royal Society: The Debate Over Collisions and the Nature of Body (1668-1670)", pp. 103–29 in (Jalobeanu and Anstey, ed. 2011).

Jalobeanu, Dana and Peter R. Anstey (eds.). 2011. *Vanishing Matter and the Laws of Motion: Descartes and Beyond*. NY/Abingdon: Routledge.

Janssen, Michel. 2012. "The Twins and the Bucket: How Einstein Made Gravity Rather Than Motion Relative In General Relativity", *Studies in History and Philosophy of Modern Physics*, **43**, 159–75.

Jardine, Nicholas. 1984. *The Birth of the History and Philosophy of Science*. Cambridge: Cambridge University Press.

Jauernig, Anja. 2008. "Leibniz on Motion and the Equivalence of Hypotheses", *The Leibniz Review*, **18**, 1–40.

Jauernig, Anja. 2009. "Leibniz on Motion: A Reply to Edward Slowik", *The Leibniz Review*, **19**, 139–47.

Jesseph, Douglas. 1999. *Squaring the Circle: The War between Hobbes and Wallis*. Chicago: University of Chicago Press.

Kneale, Martha. 1972. "Leibniz and Spinoza on Activity", pp. 215–37 in (Frankfurt, ed. 1972).

Knobloch, Eberhard. 2002. "Leibniz's Rigorous Foundation of Infinitesimal Geometry by means of Riemannian Sums", *Synthese*, **133**, 59–73.

Koyré, Alexandre. 1965. *Newtonian studies*. London: Chapman & Hall.

Kuhn, Thomas S. 1957. *The Copernican Revolution*. Cambridge, MA: Harvard University Press.

Kulstad, Mark. 1980. "A Closer Look at Leibniz's Alleged Reduction of Relations", *Southern Journal of Philosophy*, **8**, 4, 417–32.

Kulstad, Mark, Mogens Lærke, and David Snyder (eds.). 2009. *The Philosophy of the Young Leibniz*. Stuttgart: Franz Steiner.

Lacey, Hugh. 1968. "The Causal Theory of Time: A Critique of Grünbaum's Version", *Philosophy of Science* **35**, 332–54.

Lærke, Mogens. 2009. "*De Origine Rerum ex Formis* (April 1676): A Quasi-Spinozistic Parallelism in *De Summa Rerum*", pp. 203–19 in (Kulstad, Lærke, and Snyder, eds. 2009).

Lærke, Mogens. 2018. "All the Forms of Matter: Leibniz, Regis and the World's Infinity", pp. 115–29 in Nachtomy and Winegar, ed. 2018.

Layzer, David. 1990. *Cosmogenesis: The Growth of Order in the Universe*. Oxford: Oxford University Press.

Le Poindevin, Robin. 1990. "Relationalism and Temporal Topology: Physics or Metaphysics?", *The Philosophical Quarterly*, **40**, 419–32.

Levanon, Tamar. 2010. "A Reply to Anja Jauernig's [2008]", *The Leibniz Review*, **20**, 139–50.

Levey, Samuel S. 2002. "Leibniz and the Sorites", *The Leibniz Review*, **12**, 25–49.

Levey, Samuel S. 2003. "The Interval of Motion in Leibniz's *Pacidius Philalethi*", *Noûs* 37: 3 371–416.

Levey, Samuel S. 2005. "Leibniz on Precise Shapes and the Corporeal World", in Donald Rutherford and J. A. Cover (eds.), *Leibniz: Nature and Freedom* (New York: Oxford University Press, 2005), 69–94.

Levey, Samuel S. 2008. "Archimedes, Infinitesimals and the Law of Continuity: On Leibniz's Fictionalism", pp. 107–33 in (Goldenbaum and Jesseph, ed. 2008).

Levey, Samuel S. 2010. "*Dans les corps il n'y a point de figure parfaite*: Leibniz on Time, Change and Corporeal Substance", *Oxford Studies in Early Modern Philosophy*, 5, 146–70.

Levey, Samuel S. 2012. "On Time and the Dichotomy in Leibniz", *Studia Leibnitiana*, XLIV, 1, 33–59.

Lodge, Paul. 2003. "Leibniz on Relativity and the Motion of Bodies", *Philosophical Topics*, **31**, 1 & 2, 277–308.

Luna Alcoba, Manuel. 1996. "G. W. Leibniz: *Geschichte des Kontinuumproblems*", *Studia Leibnitiana*, **28**, 183–98.

Machamer, Peter and R. G. Turnbull (eds.). 1976. *Motion and Time, Space and Matter*. Columbus: Ohio State University Press.

Mates, Benson. 1986. *The Philosophy of Leibniz: Metaphysics and Language*. New York/ Oxford: Oxford University Press.

McCullough, L. B. 1978. "Leibniz and Traditional Philosophy", *Studia Leibnitiana*, **10**, 2, 254–70.

McDonough, Jeffrey. 2008. "Leibniz's Two Realms Revisited", *Noûs*, **42**, 4, 673–96.

McDonough, Jeffrey and Zeynep Soysal. 2019. "Leibniz's Formal Theory of Contingency", pp. 17–43 in Katherine Dunlop and Samuel Levey (eds.), *From Leibniz to Kant*. Mentis.

McGuire, J. E. 1976. "'*Labyrinthus Continui*': Leibniz on Substance, Activity and Matter", pp. 290–326 in (Machamer and Turnbull, ed. 1976).

McRae, Robert. 1976. *Leibniz: Perception, Apperception, and Thought*. Toronto: University of Toronto Press.

Marshall, Douglas Bertrand. 2011. "Leibniz: Geometry, Physics, and Idealism", *The Leibniz Review*, **21**, 9–32.

Mehlberg, Henryk. 1935/1937. "Essai sur la théorie causale de temps", *Studia Philosophica*, **I**, 119–260; II, 111–231.

Mercer, Christia. 2001. *Leibniz's Metaphysics: its Origins and Development*. Cambridge/ New York: Cambridge University Press.

Mondadori, Fabrizio. 1975. "Leibniz and the Doctrine of Inter-World Identity", *Studia Leibnitiana*, 7, 21–57.

Mormino, Gianfranco. 1993. *Penetralia Motus: La Fondazione relativistica della meccanica in Christiaan Huygens, con l'edizione del Codex Hugeniorum 7A*. Florence: La Nuova Italia Editrice.

Mormino, Gianfranco. 2011. *Leibniz entre Huygens et Newton: force centrifuge et relativité du mouvement dans les lettres de 1694*, pp. 697–705 in *Natur und Subjekt. IX. Internationaler Leibniz-Kongreß: Vorträge*, ed. Herbert Breger, Jürgen Herbst, and Sven Erdner. Hannover: Druckerei Hartmann.

Mugnai, Massimo. 1973. "Der Begriff der Harmonie als metaphysische Grundlage der Logik und Kombinatorik bei Johann Heinrich Bisterfeld und Leibniz", *Studia Leibnitiana*, **5**, 43–73.

Mugnai, Massimo. 1992. *Leibniz' Theory of Relations*. Studia Leibnitiana Supplementa XXVIII, Stuttgart: Franz Steiner Verlag.

Mugnai, Massimo. 2000. "Two Leibniz Texts with Translations", *The Leibniz Review*, **10**, 135–7.

Mugnai, Massimo. 2001. *Introduzione all filosofia di Leibniz*. Torino: Einaudi.

Mugnai, Massimo. 2012. "Leibniz's Ontology of Relations: A Last Word?", pp. 171–208 in *Oxford Studies in Early Modern Philosophy*, Vol. 6, ed. Daniel Garber and Donald Rutherford (Oxford: Oxford University Press, 2012).

Mugnai, Massimo. 2018. "Theory of Relations and Universal Harmony", pp. 27–44 in *The Oxford Handbook of Leibniz*, ed. Maria Rosa Antognazza. Oxford: Oxford University Press.

Nachtomy, Ohad and Reed Winegar (eds.). 2018. *Infinity in Early Modern Philosophy*. Cham: Springer Verlag.

Newton-Smith, William. 1980. *The Structure of Time*. London: Routledge & Kegan Paul.

Nikulin, Dmitri. 2008. "Imagination and Mathematics in Proclus", *Ancient Philosophy*, **28**, 153–72.

Ottaviani, Osvaldo. 2016, "Divine Wisdom and Possible Worlds: Leibniz's Notes to the Spinoza-Oldenburg Correspondence and the Development of Leibniz's Metaphysics". *Studia Leibnitiana*. **48**, 1, 15–41.

Ottaviani, Osvaldo. 2021. "Leibniz's Imaginary Bridge: The Analogy between Pure Possibles and Imaginary Numbers in the Paris Writings", chapter 5, pp. 133–67 in *Oxford Studies in Early Modern Philosophy*, ed. Donald Rutherford. Oxford: Oxford University Press.

Palmerino, Carla Rita. 2016. "*Geschichte des Kontinuumproblems* or Notes on Fromondus's *Labyrinthus*? On the True Nature of LH XXXVII, IV, 57 r°–$58v^{o}$", *The Leibniz Review*, **26**, 63–98.

Parkinson, G. H. R. 1969. *Logic and Reality in Leibniz's Metaphysics*. Oxford: Clarendon Press.

Pasini, Enrico. 1988. "Die private Kontroverse des GW Leibniz mit sich selbst." Handschriften über die Infinitesimalrechnung im Jahre 1702. In Leibniz. *Tradition und Aktualität*. Hannover: Leibniz-Gesellschaft, 695–709.

Pasnau, Robert. 2011. *Metaphysical Themes: 1274–1671*. Oxford: Oxford University Press.

Penner, Sydney. 2016. "Why do Medieval Philosophers reject Polyadic Accidents?", pp. 55–79 in *The Metaphysics of Relations*, ed. Anna Marmodoro and David Yates. Oxford: Oxford University Press.

Rabouin, David. 2015. "Leibniz's Rigorous Foundations of the Method of Indivisibles, Or How to Reason with Impossible Notions", pp. 347–64 in *Seventeenth-Century Indivisibles Revisited*, ed. Vincent Jullien (Science Networks. Historical Studies, Vol. 49). Cham, Switzerland: Birkhäuser.

Rabouin, David. 2018. "Logic of Imagination. Echoes of Cartesian Epistemology in Contemporary Philosophy of Mathematics and Beyond", *Synthese*, **195**, 4751–83.

Rabouin, David. 2020. "Mathematics and Imagination in Early Modern Times: Descartes and Leibniz' *Mathesis Universalis* in the Light of Proclus' Commentary of Euclid's *Elements*, preprint from *Knowledge and Imagination in Early Modern Philosophy*, ed. Koen Vermeir. Dordrecht: Springer (forthcoming).

Rabouin, David and Richard T. W. Arthur. 2020. "Leibniz's syncategorematic infinitesimals II: their existence, their use and their role in the justification of Differential Calculus", *Archive for History of Exact Sciences*, 1–43. DOI: 10.1007/s00407-020-00249-w (accessed June 15, 2021).

Reichenbach, Hans. 1924. *Die Bewegungslehre bei Newton, Leibniz und Huyghens*, *Kant-Studien*, **29**, 416–38.

Reichenbach, Hans. 1958. *The Philosophy of Space and Time*. Transl. Maria Reichenbach and John Freud. New York: Dover Publications.

Rescher, Nicholas. 1967. *The Philosophy of Leibniz*. Englewood Cliffs, NJ: Prentice-Hall.

Rescher, Nicholas. 1977. "Leibniz and the Plurality of Space-Time Frameworks", *Rice University Studies*, **63**, 97–106; reprinted in (Rescher 2003, 92–105).

Rescher, Nicholas. 1979. *Leibniz: An Introduction to his Philosophy*. Totowa, NJ: Rowman and Littlefield.

Rescher, Nicholas. 1996. "Leibniz and Possible Worlds", *Studia Leibnitiana*, **28**, 129–62; reprinted in (Rescher 2003, 1–44).

Rescher, Nicholas. 2003. *On Leibniz*. Pittsburgh: University of Pittsburgh Press.

Robb, Alfred A. 1921. *The Absolute Relations of Time and Space*. Cambridge: The University Press.

Roberts, John T. 2003. "Leibniz on Force and Absolute Motion", *Philosophy of Science*, **70**, 553–73.

Russell, Bertrand. 1900. *A Critical Exposition of the Philosophy of Leibniz* Cambridge: Cambridge University Press; 2nd ed. 1937; reprint London: Routledge, 1992.

Russell, Bertrand. 1914. *Our Knowledge of the External World* London: Allen & Unwin, 1952 ed.

Russell, Bertrand. 1915. "On the Experience of Time", *The Monist*, **25**, 212–33.

Russell, Bertrand. 1936. "On Order in Time", *Proceedings of the Cambridge Philosophical Society*, **32**, May, 216–28.

Russell, Bertrand. 1989. *The Collected Papers of Bertrand Russell*, Vol. 2, *Philosophical Papers, 1896–99*. Ed. Nicholas Griffin and Albert C. Lewis. London: Routledge.

Rutherford, Donald. 1995. *Leibniz and the Rational Order of Nature*. Cambridge: Cambridge University Press.

Rutherford, Donald. 2005. "Leibniz on Spontaneity", pp. 156–80 in *Leibniz: Nature and Freedom*, ed. Donald Rutherford and J. A. Cover. Oxford University Press.

Schepers, Heinrich. 2018. "Space and time", pp. 410–22 in (Antognazza, ed. 2018).

Silva, Camilo. 2016. "Y a-t-il vraiment une théorie causale de temps chez Leibniz?", pp. 323–34 in (Li 2016, IV).

Sleigh, Robert, Jr. 1990. *Leibniz & Arnauld: A Commentary on Their Correspondence*. New Haven: Yale University Press.

Slowik, Edward. 2006. "The 'Dynamics' of Leibnizian Relationism: Reference Frames and Force in Leibniz's Plenum", *Studies in the History of Modern Physics*, **37**, 617–34.

Slowik, Edward. 2009. "Another Go-Around on Leibniz and Rotation", *The Leibniz Review*, **19**, 131–7.

Slowik, Edward. 2013. "Leibniz and the Metaphysics of Motion", *Journal of Early Modern Studies*, **2**, 56–77.

Slowik, Edward. 2016. *The Deep Metaphysics of Space: An Alternative History and Ontology Beyond Substantivalism and Relationism*. Cham, Switzerland: Springer.

Slowik, Edward. 2019. "Cartesian Holenmerism and its Discontents: Or, on the 'Dislocated' Relationship of Descartes's God to the Material World", *Journal of the History of Philosophy*, **57**, 2, 235–54.

Smolin, Lee. 1997. *The Life of the Cosmos*. New York/Oxford: Oxford University Press.

Spade, Paul Vincent and Panaccio, Claude. 2019. "William of Ockham", *The Stanford Encyclopedia of Philosophy* (Spring 2019 Ed.), ed. Edward N. Zalta, https://plato.stanford.edu/archives/spr2019/entries/ockham/ (accessed June 15, 2021).

Stein, Howard. 1967. "Newtonian Space-time," *Texas Quarterly*, 10, Autumn, 174–200.

Stein, Howard. 1977. "Some Philosophical Prehistory of General Relativity", pp. 3–49 in (Earman, Glymour, and Stachel, ed. 1977).

Stein, Howard. 2002. "Newton's Metaphysics", pp. 256–307 in *Cambridge Companion to Newton*, ed. I. Bernard Cohen and George E. Smith. Cambridge: Cambridge University Press.

Strong, E. W. 1970. "Barrow and Newton", *Journal of the History of Philosophy*, **8**, 2, 155–72.

Swoyer, Chris. 1995. "Leibnizian expression", *Journal of the History of Philosophy*, **33**, 1, 65–99.

Torretti, Roberto. 1983. *Relativity and Geometry*. (Dover ed., 1996). Mineola, NY: Dover.

Westfall, Richard S. 1977. *The Construction of Modern Physics*. Cambridge: Cambridge University Press.

Weyl, Hermann. 1949. *Philosophy of Mathematics and Natural Science*. Princeton: Princeton University Press.

Whipple, John. 2010. "The Structure of Leibnizian Simple Substances", *British Journal for the History of Philosophy*, **18**, 3, 379–410.

Whipple, John. 2011. "Continual Creation and Finite Substance in Leibniz's Metaphysics", *Journal of the History of Philosophical Research*, **36**, 1–30.

Whitrow, G. J. 1980. *The Natural Philosophy of Time*, 2nd ed. Oxford: Clarendon Press.

Williams, M.E. W. 1979. "Flamsteed's Alleged Measurement of Annual Parallax for the Pole Star", *Journal for the History of Astronomy*, **10**, 2, June, 102–16.

Winnie, John. 1977. "The Causal Theory of Space-time", pp. 134–205 in (Earman, Glymour, and Stachel, ed. 1977).

Winterbourne, A. T. 1982. "On the Metaphysics of Leibnizian Space and Time", *Studies in the History and Philosophy of Science*, **13**, 3, 201–14.

Name Index

Subject Index

The manufacturer's authorised representative in the EU for product safety is Oxford University Press España S.A. of el Parque Empresarial San Fernando de Henares, Avenida de Castilla, 2 – 28830 Madrid (www.oup.es/en or product.safety@oup.com). OUP España S.A. also acts as importer into Spain of products made by the manufacturer.

www.ingramcontent.com/pod-product-compliance
Lightning Source LLC
Chambersburg PA
CBHW071407180926
51690CB00037B/22
9780192849076